AF773754

IET ENERGY ENGINEERING 156

Lithium-ion Batteries Enabled by Silicon Anodes

Other volumes in this series:

Volume 1 **Power Circuit Breaker Theory and Design** C.H. Flurscheim (Editor)
Volume 4 **Industrial Microwave Heating** A.C. Metaxas and R.J. Meredith
Volume 7 **Insulators for High Voltages** J.S.T. Looms
Volume 8 **Variable Frequency AC Motor Drive Systems** D. Finney
Volume 10 **SF_6 Switchgear** H.M. Ryan and G.R. Jones
Volume 11 **Conduction and Induction Heating** E.J. Davies
Volume 13 **Statistical Techniques for High Voltage Engineering** W. Hauschild and W. Mosch
Volume 14 **Uninterruptible Power Supplies** J. Platts and J.D. St Aubyn (Editors)
Volume 15 **Digital Protection for Power Systems** A.T. Johns and S.K. Salman
Volume 16 **Electricity Economics and Planning** T.W. Berrie
Volume 18 **Vacuum Switchgear** A. Greenwood
Volume 19 **Electrical Safety: a guide to causes and prevention of hazards** J. Maxwell Adams
Volume 21 **Electricity Distribution Network Design, 2nd Edition** E. Lakervi and E.J. Holmes
Volume 22 **Artificial Intelligence Techniques in Power Systems** K. Warwick, A.O. Ekwue and R. Aggarwal (Editors)
Volume 24 **Power System Commissioning and Maintenance Practice** K. Harker
Volume 25 **Engineers' Handbook of Industrial Microwave Heating** R.J. Meredith
Volume 26 **Small Electric Motors** H. Moczala *et al.*
Volume 27 **AC-DC Power System Analysis** J. Arrillaga and B.C. Smith
Volume 29 **High Voltage Direct Current Transmission, 2nd Edition** J. Arrillaga
Volume 30 **Flexible AC Transmission Systems (FACTS)** Y-H. Song (Editor)
Volume 31 **Embedded generation** N. Jenkins *et al.*
Volume 32 **High Voltage Engineering and Testing, 2nd Edition** H.M. Ryan (Editor)
Volume 33 **Overvoltage Protection of Low-Voltage Systems, Revised Edition** P. Hasse
Volume 36 **Voltage Quality in Electrical Power Systems** J. Schlabbach *et al.*
Volume 37 **Electrical Steels for Rotating Machines** P. Beckley
Volume 38 **The Electric Car: Development and future of battery, hybrid and fuel-cell cars** M. Westbrook
Volume 39 **Power Systems Electromagnetic Transients Simulation** J. Arrillaga and N. Watson
Volume 40 **Advances in High Voltage Engineering** M. Haddad and D. Warne
Volume 41 **Electrical Operation of Electrostatic Precipitators** K. Parker
Volume 43 **Thermal Power Plant Simulation and Control** D. Flynn
Volume 44 **Economic Evaluation of Projects in the Electricity Supply Industry** H. Khatib
Volume 45 **Propulsion Systems for Hybrid Vehicles** J. Miller
Volume 46 **Distribution Switchgear** S. Stewart
Volume 47 **Protection of Electricity Distribution Networks, 2nd Edition** J. Gers and E. Holmes
Volume 48 **Wood Pole Overhead Lines** B. Wareing
Volume 49 **Electric Fuses, 3rd Edition** A. Wright and G. Newbery
Volume 50 **Wind Power Integration: Connection and system operational aspects** B. Fox *et al.*
Volume 51 **Short Circuit Currents** J. Schlabbach
Volume 52 **Nuclear Power** J. Wood
Volume 53 **Condition Assessment of High Voltage Insulation in Power System Equipment** R.E. James and Q. Su
Volume 55 **Local Energy: Distributed generation of heat and power** J. Wood
Volume 56 **Condition Monitoring of Rotating Electrical Machines** P. Tavner, L. Ran, J. Penman and H. Sedding
Volume 57 **The Control Techniques Drives and Controls Handbook, 2nd Edition** B. Drury
Volume 58 **Lightning Protection** V. Cooray (Editor)
Volume 59 **Ultracapacitor Applications** J.M. Miller
Volume 62 **Lightning Electromagnetics** V. Cooray
Volume 63 **Energy Storage for Power Systems, 2nd Edition** A. Ter-Gazarian
Volume 65 **Protection of Electricity Distribution Networks, 3rd Edition** J. Gers
Volume 66 **High Voltage Engineering Testing, 3rd Edition** H. Ryan (Editor)
Volume 67 **Multicore Simulation of Power System Transients** F.M. Uriate
Volume 68 **Distribution System Analysis and Automation** J. Gers
Volume 69 **The Lightening Flash, 2nd Edition** V. Cooray (Editor)
Volume 70 **Economic Evaluation of Projects in the Electricity Supply Industry, 3rd Edition** H. Khatib
Volume 72 **Control Circuits in Power Electronics: Practical issues in design and implementation** M. Castilla (Editor)
Volume 73 **Wide Area Monitoring, Protection and Control Systems: The enabler for Smarter Grids** A. Vaccaro and A. Zobaa (Editors)
Volume 74 **Power Electronic Converters and Systems: Frontiers and applications** A. M. Trzynadlowski (Editor)
Volume 75 **Power Distribution Automation** B. Das (Editor)
Volume 76 **Power System Stability: Modelling, analysis and control** B. Om P. Malik
Volume 78 **Numerical Analysis of Power System Transients and Dynamics** A. Ametani (Editor)
Volume 79 **Vehicle-to-Grid: Linking electric vehicles to the smart grid** J. Lu and J. Hossain (Editors)
Volume 81 **Cyber-Physical-Social Systems and Constructs in Electric Power Engineering** S. Suryanarayanan, R. Roche and T.M. Hansen (Editors)
Volume 82 **Periodic Control of Power Electronic Converters** F. Blaabjerg, K.Zhou, D. Wang and Y. Yang

Volume 86 **Advances in Power System Modelling, Control and Stability Analysis** F. Milano (Editor)
Volume 87 **Cogeneration: Technologies, Optimisation and Implentation** C. A. Frangopoulos (Editor)
Volume 88 **Smarter Energy: from Smart Metering to the Smart Grid** H. Sun, N. Hatziargyriou, H. V. Poor, L. Carpanini and M. A. Sánchez Fornié (Editors)
Volume 89 **Hydrogen Production, Separation and Purification for Energy** A. Basile, F. Dalena, J. Tong and T. N. Veziroğlu (Editors)
Volume 90 **Clean Energy Microgrids** S. Obara and J. Morel (Editors)
Volume 91 **Fuzzy Logic Control in Energy Systems with Design Applications in Matlab/Simulink®** T. H. Altaş
Volume 92 **Power Quality in Future Electrical Power Systems** A. F. Zobaa and S. H. E. A. Aleem (Editors)
Volume 93 **Cogeneration and District Energy Systems: Modelling, Analysis and Optimization** M. A. Rosen and S. Koohi-Fayegh
Volume 94 **Introduction to the Smart Grid: Concepts, technologies and evolution** S. K. Salman
Volume 95 **Communication, Control and Security Challenges for the Smart Grid** S. M. Muyeen and S. Rahman (Editors)
Volume 96 **Industrial Power Systems with Distributed and Embedded Generation** R Belu
Volume 97 **Synchronized Phasor Measurements for Smart Grids** M. J. B. Reddy and D. K. Mohanta (Editors)
Volume 98 **Large Scale Grid Integration of Renewable Energy Sources** A. Moreno-Munoz (Editor)
Volume 100 **Modeling and Dynamic Behaviour of Hydropower Plants** N. Kishor and J. Fraile-Ardanuy (Editors)
Volume 101 **Methane and Hydrogen for Energy Storage** R. Carriveau and D. S-K. Ting
Volume 104 **Power Transformer Condition Monitoring and Diagnosis** A. Abu-Siada (Editor)
Volume 106 **Surface Passivation of Industrial Crystalline Silicon Solar Cells** J. John (Editor)
Volume 107 **Bifacial Photovoltaics: Technology, applications and economics** J. Libal and R. Kopecek (Editors)
Volume 108 **Fault Diagnosis of Induction Motors** J. Faiz, V. Ghorbanian and G. Joksimović
Volume 110 **High Voltage Power Network Construction** K. Harker
Volume 111 **Energy Storage at Different Voltage Levels: Technology, integration, and market aspects** A. F. Zobaa, P. F. Ribeiro, S. H. A. Aleem and S. N. Afifi (Editors)
Volume 112 **Wireless Power Transfer: Theory, Technology and Application** N. Shinohara
Volume 114 **Lightning-Induced Effects in Electrical and Telecommunication Systems** Y. Baba and V. A. Rakov
Volume 115 **DC Distribution Systems and Microgrids** T. Dragičević, F. Blaabjerg and P. Wheeler
Volume 116 **Modelling and Simulation of HVDC Transmission** M. Han (Editor)
Volume 117 **Structural Control and Fault Detection of Wind Turbine Systems** H. R. Karimi
Volume 119 **Thermal Power Plant Control and Instrumentation: The control of boilers and HRSGs, 2nd Edition** D. Lindsley, J. Grist and D. Parker
Volume 120 **Fault Diagnosis for Robust Inverter Power Drives** A. Ginart (Editor)
Volume 121 Monitoring and Control using Synchrophasors in Power Systems with Renewables I. Kamwa and C. Lu (Editors)
Volume 123 **Power Systems Electromagnetic Transients Simulation, 2nd Edition** N. Watson and J. Arrillaga
Volume 124 **Power Market Transformation** B. Murray
Volume 125 **Wind Energy Modeling and Simulation Volume 1: Atmosphere and plant** P. Veers (Editor)
Volume 126 **Diagnosis and Fault Tolerance of Electrical Machines, Power Electronics and Drives** A. J. M. Cardoso
Volume 128 **Characterization of Wide Bandgap Power Semiconductor Devices** F. Wang, Z. Zhang and E. A. Jones
Volume 129 **Renewable Energy from the Oceans: From wave, tidal and gradient systems to offshore wind and solar** D. Coiro and T. Sant (Editors)
Volume 130 **Wind and Solar Based Energy Systems for Communities** R. Carriveau and D. S-K. Ting (Editors)
Volume 131 **Metaheuristic Optimization in Power Engineering** J. Radosavljević
Volume 132 **Power Line Communication Systems for Smart Grids** I. R. S Casella and A. Anpalagan
Volume 139 **Variability, Scalability and Stability of Microgrids** S. M. Muyeen, S. M. Islam and F. Blaabjerg (Editors)
Volume 145 **Condition Monitoring of Rotating Electrical Machines** P. Tavner, L. Ran, C. Crabtree
Volume 146 **Energy Storage for Power Systems, 3rd Edition** A. G. Ter-Gazarian
Volume 147 **Distribution Systems Analysis and Automation 2nd Edition** J. Gers
Volume 152 **Power Electronic Devices: Applications, failure mechanisms and reliability** F Iannuzzo (Editor)
Volume 153 **Signal Processing for Fault Detection and Diagnosis in Electric Machines and Systems** Mohamed Benbouzid (Editor)
Volume 155 **Energy Generation and Efficiency Technologies for Green Residential Buildings** D. Ting and R. Carriveau (Editors)
Volume 157 **Electrical Steels, 2 Volumes** A. Moses, K. Jenkins, Philip Anderson and H. Stanbury
Volume 158 **Advanced Dielectric Materials for Electrostatic Capacitors** Q Li (Editor)
Volume 159 **Transforming the Grid Towards Fully Renewable Energy** O. Probst, S. Castellanos and R. Palacios (Editors)
Volume 160 **Microgrids for Rural Areas: Research and case studies** R. K. Chauhan, K. Chauhan and S. N. Singh (Editors)
Volume 166 **Advanced Characterization of Thin Film Solar Cells** N Haegel and M Al-Jassim (Editors)
Volume 167 **Power Grids with Renewable Energy** Storage, integration and digitalization A. S. Sallam and O. P. Malik
Volume 171 **Utility-scale Wind Turbines and Wind Farms** A. Vasel-Be-Hagh and D. S.-K. Ting
Volume 172 **Lighting interaction with Power Systems, 2 volumes** A. Piantini (Editor)
Volume 193 **Overhead Electric Power Lines: Theory and practice** S. Chattopadhyay and A. Das
Volume 905 **Power system protection, 4 volumes**

Lithium-ion Batteries Enabled by Silicon Anodes

Edited by
Chunmei Ban and Kang Xu

The Institution of Engineering and Technology

Published by The Institution of Engineering and Technology, London, United Kingdom

The Institution of Engineering and Technology is registered as a Charity in England & Wales (no. 211014) and Scotland (no. SC038698).

© The Institution of Engineering and Technology 2021

First published 2021

This publication is copyright under the Berne Convention and the Universal Copyright Convention. All rights reserved. Apart from any fair dealing for the purposes of research or private study, or criticism or review, as permitted under the Copyright, Designs and Patents Act 1988, this publication may be reproduced, stored or transmitted, in any form or by any means, only with the prior permission in writing of the publishers, or in the case of reprographic reproduction in accordance with the terms of licences issued by the Copyright Licensing Agency. Enquiries concerning reproduction outside those terms should be sent to the publisher at the undermentioned address:

The Institution of Engineering and Technology
Michael Faraday House
Six Hills Way, Stevenage
Herts, SG1 2AY, United Kingdom

www.theiet.org

While the author and publisher believe that the information and guidance given in this work are correct, all parties must rely upon their own skill and judgement when making use of them. Neither the author nor publisher assumes any liability to anyone for any loss or damage caused by any error or omission in the work, whether such an error or omission is the result of negligence or any other cause. Any and all such liability is disclaimed.

The moral rights of the author to be identified as author of this work have been asserted by him in accordance with the Copyright, Designs and Patents Act 1988.

British Library Cataloguing in Publication Data
A catalogue record for this product is available from the British Library

ISBN 978-1-78561-955-7 (Hardback)
ISBN 978-1-78561-956-4 (PDF)

Typeset in India by Exeter Premedia Services Private Limited
Printed in the UK by CPI Group (UK) Ltd, Croydon

Contents

List of figures

PEFM increase the polarity and thus improve the swelling of the polymer. Therefore the PEFM has a lower contact angle with water. (C) The swelling tests of PEFM and PFM polymer film in the 1 M $LiPF_6$ EC/DEC (1:1) electrolyte. The electrolyte uptake in PEFM is three times higher than that in PFM and is at the same level as that for conventional nonconductive PVDF binder. (D and E) The scratch tests of PEFM/Si (D) and PFM/Si (E) composite electrodes. The SEM image of the scratch morphology of PEFM shows a very smooth scratch path with little crack. (F) The PEFM/Si electrode displays higher friction coefficiency than that of the PFM/Si, indicating the higher binding strength of PEFM. Both have the same level of caustic emission. (G) Nanoindentation force of the PEFM is lower than the PFM at a given indentation depth. PEFM with the ether side chains shows lower modulus, thus higher ductility, than that of the PFM. 261

Figure 8.10 Electrochemical performance of alginate-based nano-Si electrodes (electrode density = 0.50 g/cm^3,weight ratio of Si:C = 3:1). (A) The molecular structure of alginate acid. Partially sodium ion replacement of the proton on the carboxylic acids group forms the Na alginate binder. The chemical composition of the Na alginate binder is very similar to the CMC binders. (B) Reversible Li-extraction capacity and CE of the nano-Si electrodes versus cycle number for Li insertion level fixed to 1 200 mA·hour/g Si (Ah, ampere hour). (C) Galvanostatic discharge profiles of the nano-Si electrode at different current densities between 0 and 1 V. (D) Differential capacity curves of the nano-Si electrode in the potential window of 0 to 1 V versus Li/Li^+collected at the rate of 0.025mV/s after the first galvanostatic charge-discharge cycle. Q/m, specific capacity;V, voltage. All electrochemical measurements were performed at room temperature in two-electrode coin-type half-cells. In (A), the capacity is reported for the Si contribution only. In (D), the capacity is normalized by the total weight of Si and C additives. 263

Figure 8.11 Design and structure of the self-healing electrode. (A) Scheme 1: schematic illustration of the design and behavior of a conventional silicon electrode that shows failure of the electrode because of cracking in particles and polymer binder, which results in loss of electrical contact. Scheme 2: schematic illustration of the design and behavior of our stretchable self-healing electrode that shows the maintaining of electrical contact between the broken particles and no cracks in the

List of tables

About the editors

Chunmei Ban is an associate professor at University of Colorado Boulder, USA. Prior to her appointment at CU Boulder, she was a Senior Scientist at National Renewable Energy Laboratory, Golden, CO. She has pioneered in developing surface modification science and engineering for intermetallic alloy anodes and multivalent metal anodes, and has published extensively in this field. She is a member of the IET Energy Series Board.

Kang Xu is an ARL fellow and team leader of the Extreme Electrochemistry Team at USA Army Research Lab. Kang is one of the world leaders in developing electrolyte materials and has published more than 250 papers in this field, including two comprehensive reviews. He serves as a co-founder of Research on Extreme Batteries (CREB), advisory board of ACS Applied Materials, associate editor for Energy & Environmental Materials and Electrochemistry, and guest-editor of ECS Interface.

Foreword

Replacing intercalation graphite anodes with silicon and silicon-based anode materials

"Lithium-ion batteries have revolutionised our lives since they first entered the market in 1991. They have laid the foundation of a wireless, fossil fuel-free society, and are of the greatest benefit to humankind." So stated the Nobel announcement on October 9th, 2019. I and my colleagues, John Goodenough and Akira Yoshino were honored to receive this award on behalf of all the researchers and companies that made this possible. Our work is not yet finished, as consumers demand increased energy density with lower cost and higher safety.

Lithium-ion batteries are based on intercalation chemistry, in which the lithium ions are held in host materials originally sulfides and now oxides as the cathode and graphitic carbon as the anode. In intercalation reactions, there is essentially no change in the structures of the electrode materials apart from a slight lattice expansion. The excellent cycling stability of the transition metal dichalcogenides established the first prototype rechargeable lithium batteries, in which titanium disulfide, with metallic conductivity, was used as the cathode and lithium metal as the anode. However, the battery's lifetime was limited by the formation of dendritic lithium, and so the lithium metal was replaced by aluminum giving the first Li-ion battery. Even so, the lifetime was limited by the degradation of the anode. The first commercially successful Li-ion batteries used non-graphitic petroleum-coke as the anode, which was immediately replaced by graphitic carbon in only a few years. Here again, the reaction is intercalation and the graphite expands by less than 10% allowing extended reversible reaction. It delivers a capacity of 372 mAh/g and 756 Ah/L, and forms an effective stable SEI layer. However, it takes over 70 gm of carbon to store 7 gm of lithium and the carbon takes up 50% of the volume of the cell, so must be replaced, as must the overly reactive electrolyte; neither of these have changed significantly over the last 30 years.

A number of elements readily react with lithium. Beyond the aluminum used by Exxon in the original lithium cells, tin and silicon have become favorites because both react with nearly four lithium atoms ensuring much higher capacities than graphitic carbon. Of these two, silicon has become the favorite of researchers because of its lighter weight and greater availability, which should result in a lower cost. The volumetric capacity of Si is slightly higher than that of Li-metal itself with a molar volume of 81 moles Si(?)/L vs 77 moles Li/L in lithium metal. This should allow for anode volumetric capacities more than double that of graphitic carbon. However, the reaction of Si with Li occurs by a conversion reaction in which the structure is destroyed, and must be rebuilt on charge. This reaction occurs with around a 300% expansion, which is expected to destroy the SEI so that it must be reformed on each cycle. Thus, many efforts are focused on how to handle this reactivity. One approach that uses nanostructured architecture, e.g. including 1D nanowires, 2D thin film and 3D nanoparticles, increases the reversible cycling capacity and cycling life. However, with knowledge about surface and interphasial chemistry absent at the time, these early efforts of only using nanostructured Si materials have not succeeded in practical applications. Furthermore, in a traditional Si-based anode, a polymer binder binds Si and carbon black particles together to enable an electrical path throughout the anode. During electrochemical cycling, stress generated by the volumetric changes causes cracking in Si electrodes, resulting in loss of electrical contact and subsequent loss of capacity. The quest to revealing the full potential of Si anodes is lengthy and onerous, with much Edisonian trial-and-error efforts.

A number of books have been devoted to the LIB science and technologies in the last three decades, covering the fundamentals of electrochemistry, development of electrolytes and intercalation electrode materials. However, there is no book giving a comprehensive overview of silicon and silicon-based anode materials for their contribution to develop next-generation high-energy low-cost LIBs. With the high demand for developing high-capacity and low-cost anode materials, this books not only summarizes the knowledge achieved over the last 20 years for silicon and silicon-based alloy derivates, but also provides insight for developing silicon-based anode materials for next-generation LIBs. Twelve chapters cover five key areas: chemistry and electrochemistry, synthesis and nano-engineering, surface and interface science, mechanical properties and commercialization. The book provides a professional reference for researchers and technologists in the field of electrochemical energy storage, and will also be of use for the training new scientists and engineers.

I have been active in lithium battery research for 50 years, and have seen lithium batteries come from a dream to dominating electrochemical energy storage. We still have a long way to go in optimizing their output. They still deliver much less than 50% of their theoretical energy density. We, as researchers, should be able to double that leading to cells achieving 500 Wh/kg and 1kWh/L over more than a thousand cycles and at a cost under $100/kWh.

M. Stanley Whittingham
Vestal, NY
August 10, 2020

Chapter 1

Overview of the development of silicon anodes for lithium-ion batteries

M.N. Obrovac[1]

1.1 Introduction

1.1.1 Silicon-based Li-anode materials

Si-based anode materials are currently by far the most studied negative electrode (anode) materials for the purpose of increasing Li-ion battery energy density. Si has nearly three times the volumetric capacity as graphite, the anode material conventionally used in Li-ion batteries today. Such an increase in anode capacity could theoretically increase the energy density of Li-ion batteries by as much as 34 percent [1]. Si-based anodes are also potentially less expensive than graphite: in 2017 the cost of Si in the US was about 2.6 USD/kg [2], whereas the cost of battery-grade graphite was about 9–17 USD/kg, depending on whether spherical natural graphite or high-performance synthetic battery graphite is used [3]. This price advantage becomes more interesting when considered in more meaningful units of capacity normalized cost: 7 USD/kAh for Si, 24–46 USD/kAh for graphite. Based on these numbers, it costs 3.5–6.5 less to store lithium in Si compared to graphite. Moreover, batteries using a Si-based anode would theoretically have higher energy density and therefore could be presumably be sold at a higher cost per kWh than a conventional graphite cell, increasing the cost advantage for Si further. Of course, processing steps would be needed to convert raw Si into a battery-grade material, increasing its costs from the 2.6 USD/kg value of raw Si. However, considering this vast price difference it seems highly likely that Si-based anodes could offer higher energy density at reduced cost. As an example, battery-grade SiO_x has been estimated to cost ~100 USD/kg [4]. This estimate is likely high. However, even this high figure corresponds to about 40 USD/kAh in units of capacity normalized cost, which is still competitive with graphite, while additionally providing an energy density advantage. The possibility of enabling higher energy density with similar costs or even at a cost reduction compared to cells with conventional anodes is a strong driver for research and development work in this field.

[1]Departments of Chemistry and Physics, Dalhousie University, Canada

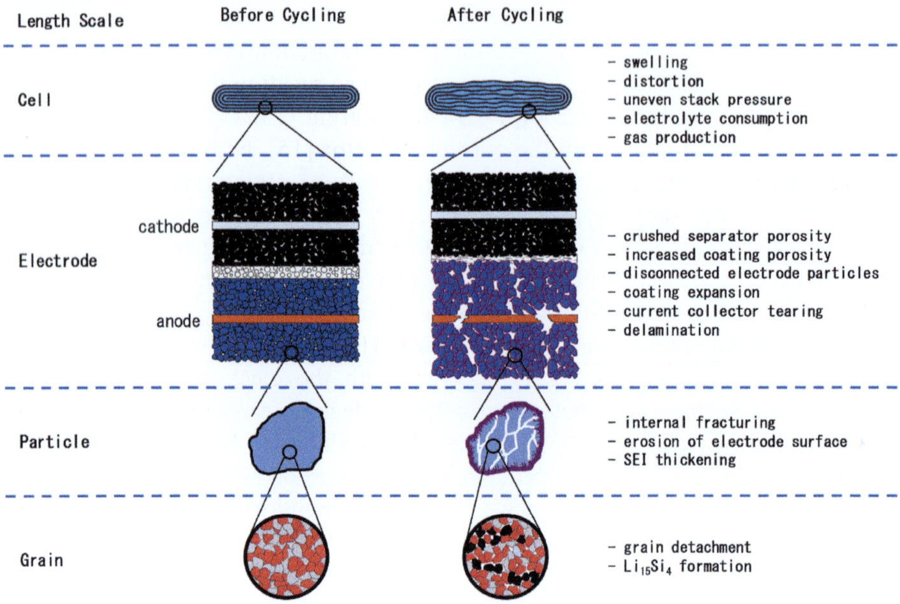

Figure 1.1 *An illustration showing the effects at all length scales of the volume expansion of silicon during cell cycling*

Despite the perceived advantages of Si anodes based on their high lithium storage capacity, their implementation in commercial cells is difficult. This is because of the large volume changes of Si during cycling: 280 percent expansion during lithiation and a corresponding 280 percent contraction during delithiation [5]. As illustrated in Figure 1.1, the volume expansion of Si anodes has negative consequences at all length scales, ranging from the battery scale down to the atomic scale.

Cell-scale: Si expansion can cause cell swelling and buckling of the jellyroll. The use of additives for improving Si cycling, such as fluoroethylene carbonate (FEC), can exacerbate these effects by gas production [6, 7]. Uneven stack pressure caused by swelling and/or the formation of gas can lead to uneven current distribution, which is known to be a cause of lithium plating [8, 9], that can further result in cell fade and thermal stability/safety issues. Electrolyte consumption by continual solid electrolyte interphase(SEI) growth can also lead to sudden and rapid capacity fade [7, 10].

Electrode-scale: Expansion of electrodes with a high loading of Si-anode materials can result in the tearing of the anode current collector and crushing of the cell separator, resulting in immediate cell failure. Volume changes during Si-anode cycling can cause crack formation in the electrode, leading to increased electrode porosity [7, 11]. Both processes of continual electrode porosity increase and continual SEI growth can cause continuous irreversible volume expansion of the anode during cycling [7, 12], resulting in a reduction in energy density as cycling continues. This process can also result in the electrical disconnection of alloy particles from the coating [13, 14] or delamination of the coating from the current collector [15], both leading to cell fade.

Particle-scale: Si-alloy particles are typically hard and brittle. Electrode calendering can result in Si-alloy particle fracture and cell fade [16]. Expansion and contraction of Si-alloy particles during cycling can disrupt the SEI layer, leading to continuous reactions with the electrolyte and continuous growth of the SEI layer [11]. Erosion of the Si-anode particle surface by continual reactivity with the electrolyte causes increased surface area that can accelerate electrolyte reactivity [14]. Increased surface area of active materials has also been linked to thermal stability/safety issues. The build-up of SEI products at the particle surface can increase impedance [11, 15]. The build-up of electrolyte decomposition products at the negative electrode has also been linked to thermal stability/safety issues. The continual expansion and contraction of the Si-anode particles during cycling can further cause internal fracturing within the Si-anode particles, leading to internal electrical disconnections and fade [15].

Grain/atomic-scale: The full lithiation of Si grains can cause $Li_{15}Si_4$ formation, which is thought to induce particle fracture via stresses along 2-phase boundaries, resulting in cell fade [1, 7]. Within the grains of multiphase Si-anode materials (e.g. SiOx, Si alloy, Si/C), the decoupling of the active Si phase from other phases within the particle during cycling leads to electrical disconnection of the active Si phase and cell fade. The decoupling of the active Si phase can also lead to $Li_{15}Si_4$ formation, which, as just mentioned, is thought to lead to cell fade.

The above consequences of Si-anode material expansion lead to reduced coulombic efficiency (CE), cell fade, and possible thermal stability/safety issues. In overcoming these issues, it has been discovered that solutions encompassing the entire battery are required, including advanced Si alloys, binders, slurries, electrode formulations, electrolytes, the use of compatible cathode materials, and overall cell design. Major advances have been made in the past 15 years in all of these areas; however, many issues still remain. Nevertheless, the technology has reached a point where Si-based electrode materials are currently in use in top-performing long-life and high-energy-density electric vehicle (EV) batteries (including Tesla car batteries) and in consumer electronic cells (e.g. Panasonic and Hitachi Maxel-Kopin) [17]. From this it can be concluded that the use of Si-anode materials now presents practical advantages that result in longer cycle life, higher energy density, and possibly lower cost commercial cells than if graphite is used alone. These small beginnings are the culmination of years of work by many.

After introducing some basic concepts about graphite and Si, this chapter presents a roughly chronological overview of how Si-anode technology has progressed to its current state, citing some key articles. However, the mention of an early work does not necessarily indicate that an earlier work might not exist. While some articles are from the academic literature, many are from the patent literature. In this particular field especially, a significant amount of important discoveries key to its development have taken place behind-the-scenes by brilliant researchers and teams of industrial researchers in battery and battery materials companies. In many cases, "firsts" reported in the academic literature have been discovered or invented years before in industry. Many of these industrial researchers have not had the opportunity to publish their discoveries other than in patents and therefore their names might be unfamiliar. Some companies at which pioneering breakthroughs in fundamental Si electrochemistry, Si materials, binders, electrolytes, and cell design include (alphabetically) 3M Company, Fujifilm Celltec Co., LG Chemical, Panasonic, Samsung SDI, Sanyo, and others. If they have not done so already, interested readers are encouraged to read the fascinating patent literature in this field, which continues to be excellent today.

1.1.2 Conventional graphite anodes

Before embarking on a review of Si-anode technology, it is necessary to review the properties of graphite anodes, which are conventionally used in Li-ion cells. The electrochemical lithiation of graphite proceeds via intercalation between graphene layers, forming a series of staged compounds until the fully lithiated LiC_6 composition is reached. This intercalation process is highly reversible and is associated with only small volume changes (~10 percent volume expansion). As a result, graphite anodes do not have many of the issues associated with Si anodes. Typical characteristics of graphite and graphite electrodes are listed in Table 1.1.

High-loading (~4 mAh/cm^2), low-porosity (<20 percent) graphite electrodes that have long cycle life are relatively easy to manufacture using low levels of conductive diluents. Inexpensive binders may be used, such as polyvinylidene difluoride (PVDF) or carboxymethyl cellulose/styrene-butadiene rubber (CMC/SBR). The role of these binders is primarily to control slurry rheology and to maintain good coating adhesion and mechanical properties. The low expansion of graphite has allowed today's Li-ion cells to operate with very thin current collectors (~6 μm)

Table 1.1 Characteristics of graphite and graphite electrodes

Theoretical properties of graphite					
Density	$V_{average}$	RC (mAh/g)	RC (Ah/L)	ξ (%)	
2.26 g/ml	125 mV	372	764	10	
Typical graphite electrode characteristics (energy cell)					
Coating thickness (single side)	Active loading (single side)	Current collector thickness	Porosity	Active volume	ICE
~80 μm	~4 mAh/cm^2	~6 μm	~25 %	>70%	>90%

and separators (~15 μm), increasing cell energy density. The use of low surface area (~1 m^2/g) graphites has improved thermal stability, contributing to cell safety [18, 19]. Utilizing low surface area particles and carbon coatings results in low reactivity with electrolytes, resulting in high initial coulombic efficiency (ICE) (>90 percent) and low fade [18]. Many thousands of 100 percent depth of discharge (DOD) cycles have been achieved for full Li-ion cells using graphite electrodes [20, 21]. To achieve this level of performance requires CEs on the order of 99.99 percent to be maintained throughout cycling [21]. In addition to all these properties, graphite is relatively inexpensive (9–17 USD/kg) [3]. Advances in the development of battery-grade natural graphites will likely reduce costs further.

The goal of Si-anode technology is to improve on the energy density of conventional Li-ion cells, while maintaining cell performance and hopefully reducing costs or at providing a cost value that is commensurate with the energy density increase. Considering the already excellent characteristics of graphite, this is a difficult proposition. Nevertheless, there are few other paths for increasing Li-ion cell energy density, and Si-anode technology remains one of the most promising ways of achieving this goal.

1.1.3 Why Si?

Out of many possible host materials, Si anodes have emerged as the leading negative electrode technology for Li-ion batteries. There are many contributing factors that make Si attractive. Foremost is its potential to increase cell energy density. The energy of a Li-ion cell is the product of the amount of Li stored in the host materials and the average potential difference between the cathode and the anode. Therefore, to be useful as a negative electrode host, anode materials need to store a substantial amount of lithium at a low potential. The volumetric and specific capacities of some alloying elements are shown in Figure 1.2. The specific capacity of Si is the highest. However, volumetric capacity is more important than specific capacity for most applications and, according to Figure 1.2(a), a number of elements have

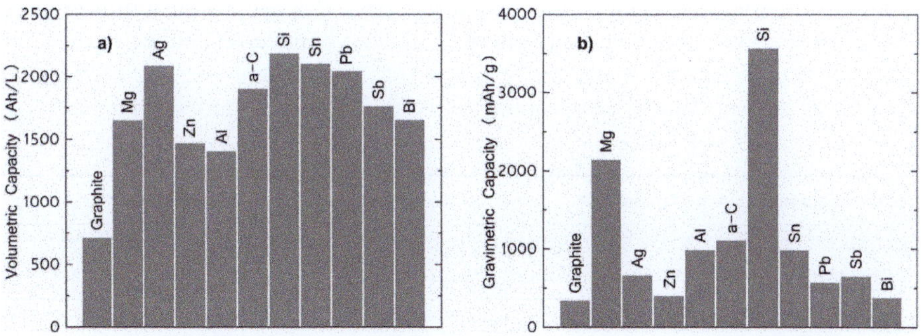

Figure 1.2 (a) Volumetric capacities calculated at the state of full lithiation and (b) gravimetric capacities of selected elements. Reprinted with permission from [1]. Copyright 2014 American Chemical Society.

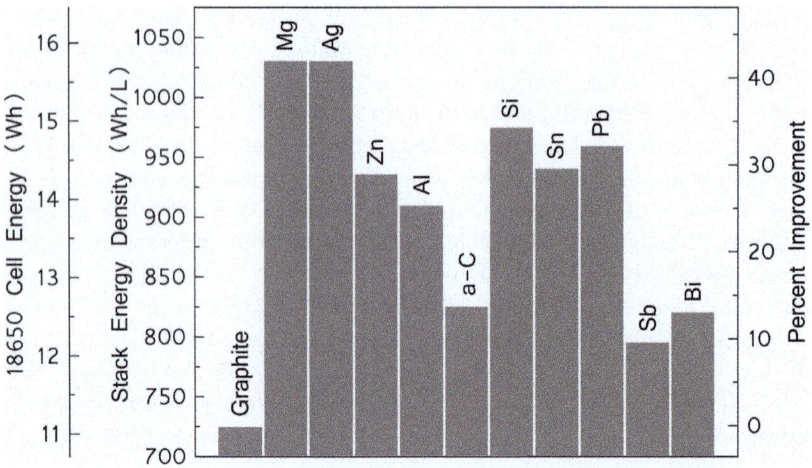

Figure 1.3 *Calculated stack energies for different negative electrode active materials. The right-hand side axis indicates the percent improvement in energy over a graphite electrode. A scale was also provided with 18650 cell energies based on the model. Reprinted with permission from [1]. Copyright 2014 American Chemical Society.*

volumetric capacities similar to Si. Combining the volumetric capacity with average potential, one can estimate the maximum cell energy that would result if graphite were replaced with various host elements [1]. This is shown in Figure 1.3. All of the elements shown would result in a substantial energy increase compared to graphite, some elements resulting in an even greater potential increase in energy density compared to Si. Therefore, why is Si so attractive? This can be arrived at by a process of elimination.

Energy improvement: The energy improvement in Figure 1.3 is arrived at by assuming that the negative electrode is a nonporous coating of the pure element. Such electrodes are likely not practical because of the high stresses induced by volume expansion during lithiation. Practical electrodes will probably be in the form of traditional porous composite coatings of powders including graphite as a calendering lubricant. This will reduce the potential energy improvement substantially, to ~20 percent in the case of Si anodes. In the case of Al, Sb, and Bi, the energy improvement will be less than this, making the development of these electrodes less attractive.

Li-plating potential: Li-ion battery anodes should operate above ~50–80 mV, since operation at lower potentials increases the risk of lithium plating. The lithiation of Mg occurs near 0 V, making it an unattractive host material.

Potential range: The potential change of an anode during lithiation should be small. While a gradual potential change of the anode during lithiation is desirable, to allow easy state of charge determination, large potential swings are not desirable, since they require the use of expensive circuit elements to keep constant potentials supplied to electronics. For this reason, Li-ion battery negative electrodes are typically restricted to cycling in approximately the 0.08–1 V potential range (80 mV lower limit to avoid Li plating). This restriction makes amorphous carbon and Pb less attractive.
Toxicity: Low toxicity is desirable for obvious reasons. Sb and Pb are less attractive for these reasons.
Electrolyte reactivity: Zn, Sn, and Pb have been found to catalyze electrolyte decomposition reactions, making them less attractive as host materials [1]. However, the reactivity of Sn can be overcome when it is alloyed with other elements.
Lithiation kinetics: Mg and Zn have poor lithiation kinetics, making them impractical for use as negative electrodes.
Cost: The high cost of Ag makes it less attractive as a host element. Si and Mg have the lowest capacity normalized cost (since Mg has an infinite gravimetric capacity for Li, the theoretical capacity normalized cost of Mg is zero).

With the above considerations, the most attractive host elements are Si and Sn. Anodes based on both these elements have received much research attention. Sn-based anodes have been used commercially in Sony's Nexelion battery [22], but to the author's knowledge, only Si-based alloys are used in commercial Li-ion cells today. This is likely due to cost. At a price of ~20 USD/kg [23] and having a capacity of 993 mAh/g, the capacity normalized cost of Sn is about 20 USD/kAh, almost three times greater than that of Si (~7 USD/kAh). In addition, Si can be processed much more easily compared to Sn. For instance, Si can be easily ground, amorphized, and alloyed by ball milling methods, whereas the ductility of Sn makes this more difficult. Bulk amorphous Si (a-Si) is stable at room temperature, whereas bulk amorphous Sn is not. Si may also be heated to high temperatures and is chemically resistant, which makes anode materials, such as Si/C composites, relatively easy to make. In comparison, the low melting point of Sn makes the synthesis of such composites much more difficult. These processing difficulties for Sn would increase the cost of using Sn-based materials further, making Si-based materials all the more attractive.

1.2 Electrochemical insertion of Li in Si

Research in alloy negative electrode materials predates Li-ion technology by decades. Early secondary Li battery technology (1960s–90s) was primarily based on ambient room temperature nonaqueous electrolyte cells or "medium temperature"

(~400 °C) molten salt cells. Both technologies used a lithium metal negative electrode (solid Li foil for ambient temperature cells and molten lithium for medium temperature cells) vs an intercalation host positive electrode (e.g. FeS_2, MoS_2) [24, 25]. However, the use of lithium metal as a negative electrode was found to be problematic. In medium temperature cells, liquid lithium migrated through ceramic separators creating shorts [24]. In ambient temperature cells, repeated cycling was found to result in the formation of high surface area dendritic lithium, which also caused internal shorting and, additionally, thermal stability-related safety issues [26, 27]. This led to the investigation of the use of elements or compounds that can act as hosts for lithium as negative electrode materials [24, 28]. Si was intensively studied for this purpose because of its high capacity for Li.

Impressive cycling of Si was achieved in medium temperature cells with molten salt electrolytes, such as the LiCl-KCl eutectic composition. In these electrolytes, no solid decomposition products are formed during cycling [29]. Therefore, electrode materials maintain a pristine surface, regardless of any volume changes they may undergo. This allows the maintenance of electrical connectivity even to relatively large powdered electrodes with no binders or conductive additives. Under such conditions, impressive cycling performance could be achieved for Si powder electrodes.

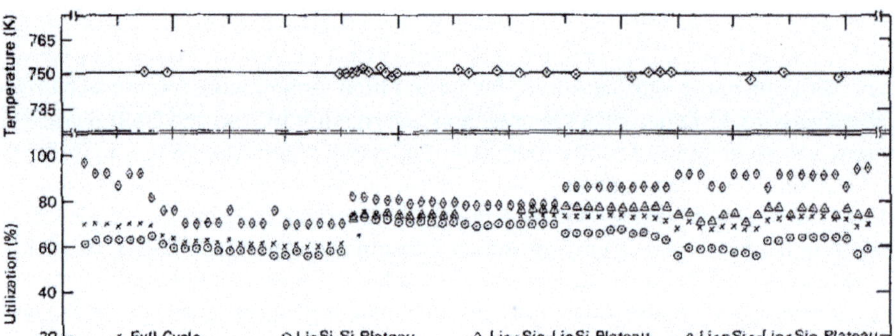

Figure 1.4 *Cycling performance of a powdered Si electrode cycling in a LiF-LiCl-KCl eutectic melt at different current densities. Note the mention of the $Li_{15}Si_4$ phase in the figure, assigned based only on stoichiometry at the time. Reproduced from [30]. © IOP Publishing. Reproduced with permission. All rights reserved.*

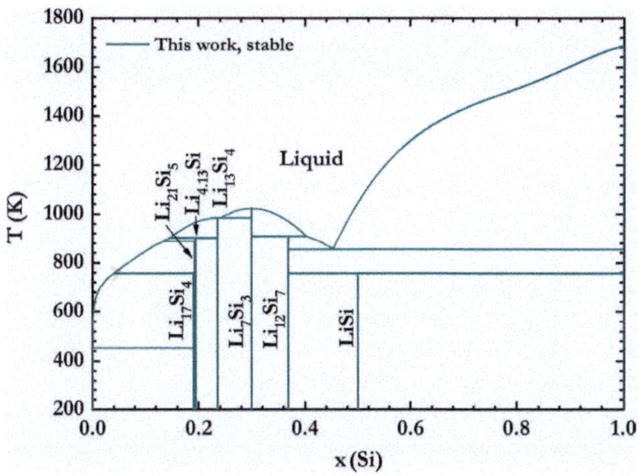

Figure 1.5 *The Li-Si binary phase diagram. Reprinted from [31] with permission from Elsevier.*

For instance, Figure 1.4 shows results from a 1977 study that achieved 140 cycles with no fade at 100 percent DOD at 477 °C for loose Si powder electrodes [30]. In such cells, Li-Si chemistry was found to have very different behavior than conventional intercalation electrode materials, into which Li can be inserted and removed with little disruption to the host structure. At high temperature, Li inserts electrochemically into Si via an alloying reaction that roughly follows the Li-Si phase diagram, shown in Figure 1.5. According to equilibrium thermodynamics, the lithiation of Si should occur along the 2-phase regions in the Li-Si diagram between different Li-Si intermetallic compounds. During the equilibrium lithiation of Si, this would result in the formation of each intermetallic compound in turn, resulting in a series of potential steps, each corresponding to a 2-phase region in the diagram [32]. This process requires massive structural rearrangement, the breaking of Si–Si bonds, and the formation of Li–Si bonds.

Figure 1.6 shows the medium temperature (415 °C) potential vs capacity curve for the lithiation of Si in a LiCl-KCl eutectic molten salt [33]. Most of the plateaus correspond in composition to the 2-phase regions in the equilibrium phase diagram in Figure 1.5. However, some differences are apparent. Most notably, no features are present for the LiSi phase in the medium temperature cell potential profile. This is due to the slow formation kinetics of this phase [31]. As the temperature is reduced, the formation of other phases become kinetically hindered, until at room temperature the potential profile of Si bears no resemblance to what is expected from the equilibrium phase diagram, as will be discussed in more detail further in this section.

An interesting aspect of the medium temperature cycling of Si is the good cycling that can be obtained. During cycling, the Si particles remain pristine. Such pristine particles evidently can remain in good electrical contact as they expand and shrink during lithiation and delithiation. They also exhibit good cycling despite the presence of 2-phase regions in their potential profiles. In such regions, particles

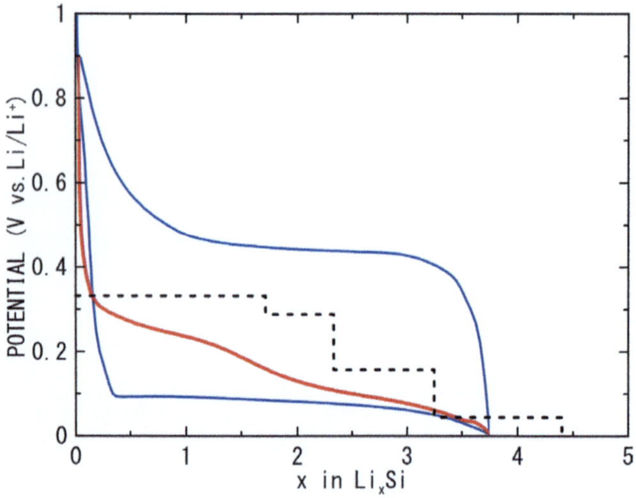

Figure 1.6 *The potential profiles of Si: at 415 °C (dashed line) and at 30 °C (blue first cycle, red second lithiation). The potential profile of Si at 415 °C is reprinted from [33] with permission from Elsevier.*

will consist of two phases with vastly different molar volumes. As such, there are enormous stresses produced along the 2-phase boundaries as they pass through the particle during lithiation. Stresses along 2-phase boundaries are known to cause particle fracture in alloy anodes, resulting in electrical disconnection of the active material and cell fade [1]. However, as shown in Figure 1.4, excellent capacity retention can be achieved with Si powder electrodes regardless of the many 2-phase regions encountered in its potential profile. This is likely due to the inertness of the molten salt electrolyte used, allowing any new surfaces formed by fracturing to remain pristine. In this way, any Si fragments could remain in electrical contact regardless of how much the Si particles fracture.

Good reversibility is much more difficult to obtain for Si alloys in ambient temperature cells with organic electrolyte solvents and Li-Si electrochemistry at ambient temperatures is much more complex. Figure 1.7 shows a state-of-the art potential profile as measured in 1995 of an Si electrode with a PVDF binder cycled in a Li half-cell with a 1 M $LiN(CF_3SO_2)_2$ in ethylene carbonate/propylene carbonate (EC/PC) electrolyte [34]. Graphite is highly reversible when cycled under such conditions. In contrast, although a large lithiation capacity can be obtained for the Si electrode, its reversible capacity (RC) is tiny in comparison. This behavior is typical for bulk Si electrodes when no special binders or electrolyte additives are used. During lithiation, the Si particles expand and push against each other, maintaining electrical contact and allowing a large first lithiation capacity to be achieved. During delithiation the particles contract and separate from each other, resulting in the loss of electrical contact [35]. In addition, the large volume changes disrupt the SEI layer, continually exposing fresh surfaces to the electrolyte and causing the SEI to continually grow

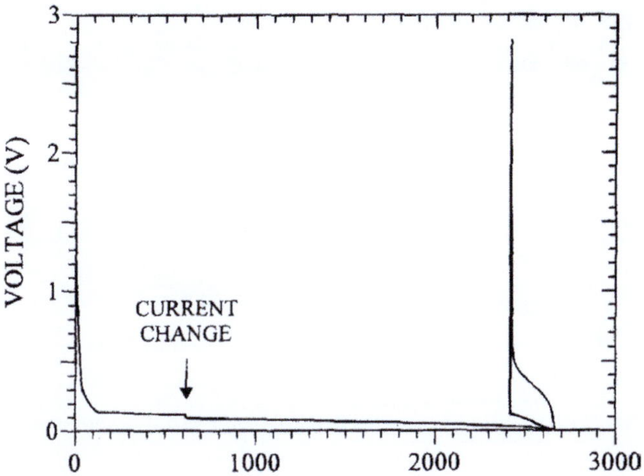

Figure 1.7 *A potential profile of Si that represents the state-of-the-art in 1995. The electrode utilized PVDF binder and was cycled in a cell with a 1 M LiN(CF_3SO_2)$_2$ in EC/PC electrolyte. Reproduced from [34]. © IOP Publishing. Reproduced with permission. All rights reserved.*

[36]. This causes irreversible lithium loss, and the thick SEI electrically insulates the particles, causing a rapid rise in cell polarization. The difficulties in accomplishing even a single charge/discharge cycle with conventional binders and electrolytes is the primary reason that little was known about the electrochemistry of Si near room temperatures until the 2000s. In addition, relatively little work was done in the field, since the practical utilization of Si anodes seemed unlikely.

To overcome the limitations posed by the electrical disconnection of Si particles from electrodes, early studies used electrodes containing large amounts of conductive additives (e.g. carbon black). Such electrodes are not practical for commercial use, but they at least enabled Si cycling to an extent that allowed its basic electrochemistry to be observed, starting with the discovery that amorphous lithiated silicon (a-Li_xSi) was formed during the lithiation of crystalline Si (c-Si) [37]. Upon further lithiation, a-Li_xSi was found to spontaneously crystallize below 50 mV to form $Li_{15}Si_4$ [5], apparently the highest lithiated phase of Si obtainable by room temperature electrochemical lithiation. During the delithiation of $Li_{15}Si_4$ in ambient temperature cells, a-Si was found to form [5]. $Li_{15}Si_4$ was a phase not previously known although it was speculated to form in medium temperature cells, based on reaction stoichiometry alone (e.g. see Figure 1.4). The $Li_{15}Si_4$ structure is shown in Figure 1.8. It is isostructural with $Cu_{15}Si_4$ and $Li_{15}Ge_4$. This structure has been described simply as the closest packing of two types of atoms of different sizes [38]. It is also the Li-Si phase with the lowest Li content in which all Si atoms are isolated. In the $Li_{15}Si_4$ structure, all of the Si are in equivalent crystallographic sites and are coordinated by 12 Li atoms at distances ranging from 2.61 to 2.87 Å [39]. The insertion of Li in Si therefore results in the complete destruction of the Si lattice

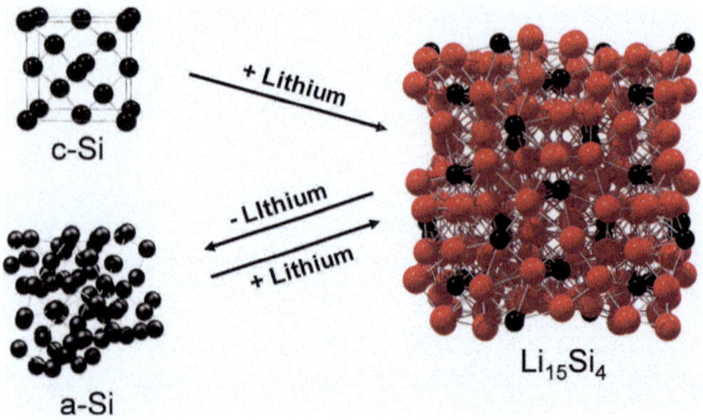

Figure 1.8 *An illustration of the structural changes that occur as c-Si is fully lithiated to form $Li_{15}Si_4$ and then subsequently delithiated to form a-Si*

with all of the Si–Si bonds broken and a massive increase in the crystal volume per Si atom, from 20 Å/Si for c-Si to 76 Å/Si for $Li_{15}Si_4$, corresponding to a volume expansion of 280 percent. Being the highest lithiated phase of Si attainable at room temperature, $Li_{15}Si_4$ sets the maximum capacity of Si at about 3 578 mAh/g or 2 194 Ah/L. Volumetrically, this capacity is nearly three times that of graphite. Although $Li_{15}Si_4$ was initially assumed to be a metastable phase, according to recent calculations, it is the most stable Li-Si phase at this composition at room temperature [31], but its stability range is small. However, at higher temperatures, past studies suggest that $Li_{15}Si_4$ can form metastability. A phase corresponding to a composition of $Li_{15}Si_4$ was often suspected as early as 1976 in medium temperature cells [30, 40, 41]. In 1990, H. Okamoto remarked: "Because of persistent observations of 'Li_4Si' or '$Li_{15}Si_4$' ... a phase with one of these formulas may exist in a limited temperature range or may be stabilized by impurity" [41]. From consideration of these early medium temperature studies, it is possible that $Li_{15}Si_4$ may be formed electrochemically at temperatures as high as 400 °C.

The electrochemistry of Si electrodes at room temperatures was much more complex than expected at the time: none of the phases present in the equilibrium phase diagram shown in Figure 1.5 were observed during the lithiation or delithiation of Si; a new crystalline phase formed during lithiation; and the a-Si formed after one charge/discharge cycle was completely different than the c-Si starting material. Although the suppression of equilibrium phases during the lithiation of metals by slow kinetics at room temperature had been anticipated (see Figure 1.9) and observed previously [32], the spontaneous crystallization of intermetallic phases, as in the case of $Li_{15}Si_4$ formation, was something new and its consequences were later found to be profound.

The advent of more advanced binders and electrolytes allowed the potential profile of Si to be measured in detail [42], as shown in Figure 1.10a, with corresponding differential capacity curves shown in Figure 1.10b. This potential profile

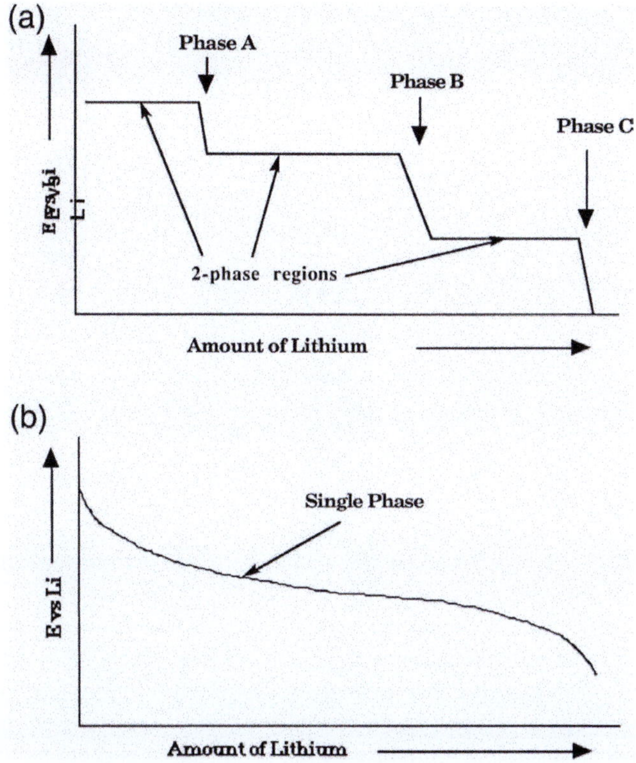

Figure 1.9 *Illustration of potential profiles during the lithiation of a metal when (a) the equilibrium phase diagram is followed and (b) when phase formation is suppressed by slow kinetics. Reprinted from [32] with permission from Elsevier.*

is very strange indeed, showing significant potential history-dependent hysteresis. Eventually, a comprehensive picture was developed to explain this behavior. The corresponding phase behavior is shown in Figure 1.10c. The lithiation of c-Si (Region I) was found to commence at about 170 mV [42], with the formation of a-Li_xSi ($x \approx 3.5$) in a 2-phase region [43]. During lithiation, a-Li_xSi forms on the surface of an Si particle. The reaction front then proceeds from the surface to the core of the particle, accompanied by massive volume expansion. Figure 1.11 shows this behavior nicely. Li diffusion is preferred in the <110> direction, resulting in anisotropic volume expansion of single crystal Si [45], as shown in Figure 1.12. After the Si is fully converted to a-Li_xSi (Region II), this phase lithiates further until a composition of $Li_{3.75}Si$ and a potential of about 50 mV is reached. At this point, the lithiated Si spontaneously crystallizes to form $Li_{15}Si_4$. This phase was found to have a small solid solution range (Region III) [43, 46]. Delithiation of $Li_{15}Si_4$ proceeds in a $Li_{15}Si_4$/a-Li_zSi 2-phase region, where $z \approx 2.0$ (Region IV) [43], resulting in a characteristic peak in the delithiation differential capacity curve at about 0.45

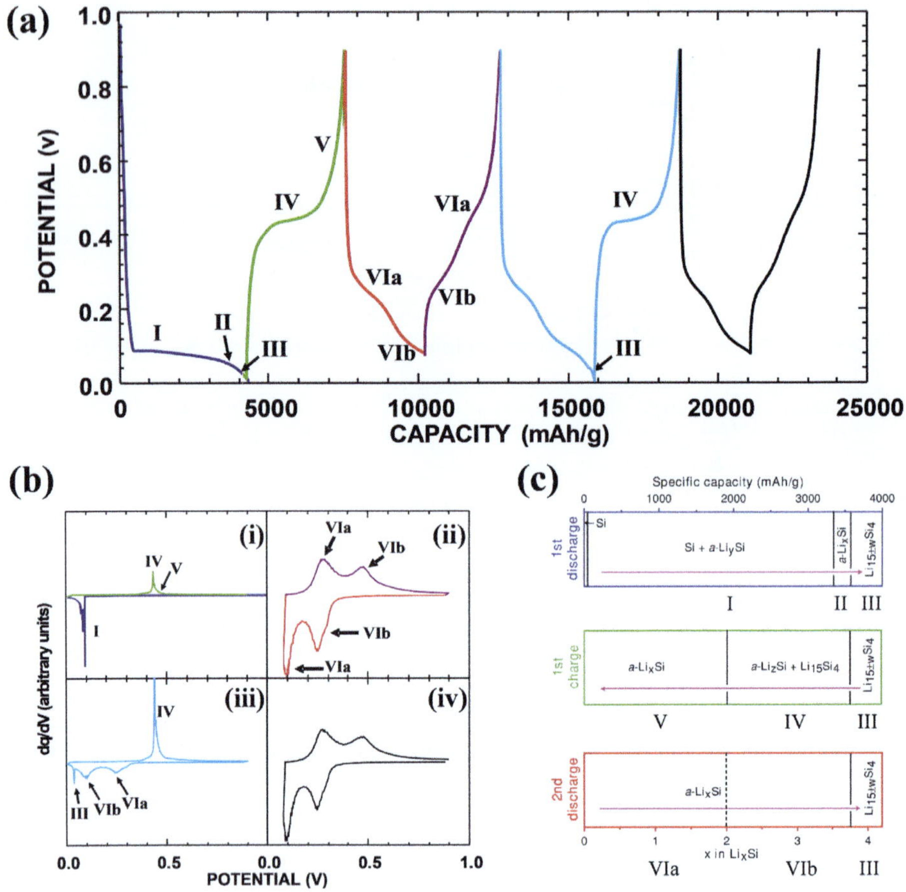

Figure 1.10 *(a) Potential profile of Si. (b) Differential capacity curves, corresponding to the potential profile shown in (a). (c) The phase behavior of Si during lithiation/delithiation shown in phase diagram format. (a) and (b) reproduced from [42]. (c) Adapted from [43].* *© IOP Publishing. Reproduced with permission. All rights reserved.*

V. Further delithiation is in a single-phase region (Region V) and results in the formation of a-Si. During the subsequent cycle, a-Si is lithiated in two single-phase regions (regions VIa and VIb), forming a-Li_xSi [47, 48]. These regions correspond to first inserting Li into sites surrounded predominantly by Si neighbors and then, at lower potentials, inserting Li into sites surrounded predominately by Li neighbors. As shown in Figure 1.10b,ii, if the potential during lithiation is not permitted to reach 50 mV, then lithium may be inserted and removed from a-Li_xSi strictly in a single-phase region. If, however, the potential is allowed to drop below 50 mV, as shown in Figure 1.10b,iii, then $Li_{15}Si_4$ phase will form, whose presence is indicated

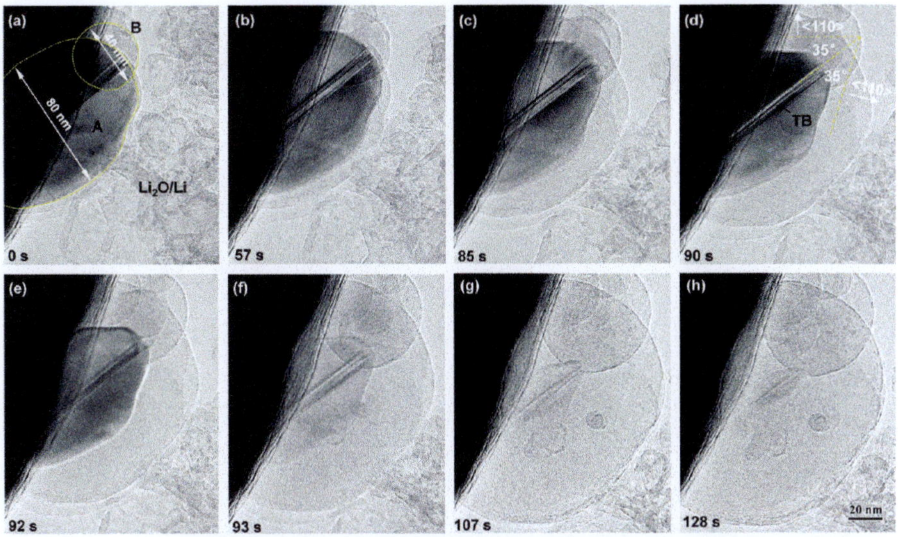

Figure 1.11 *A sequence of TEM images showing the electrochemical lithiation of two small Si nanoparticles. The core-shell structure of the c-Si/a-Li$_x$Si 2-phase region that forms during lithiation can be clearly seen. Reprinted with permission from [44]. Copyright 2012 American Chemical Society.*

by its characteristic peak at 0.45 V during the subsequent delithiation. This process can be repeated, as shown by the differential capacity curve of the next cycle in Figure 1.10b,iv.

The elucidation of the basic electrochemistry of Si was a major advance. Its implications govern the design of all practical Si-based materials, as discussed in Section 1.3.

1.3 Si-based materials design

Already before the electrochemistry of Si was fully understood, the problems associated with using alloy negative materials were well known. It was known that all alloys expand enormously during lithiation, which can cause particle disconnection, particle fracture, and the disruption of the SEI layer—all leading to cell fade. Silicon is no different than any of the other elements that alloy with lithium in this respect. All metals were found to expand at the same rate during lithiation: ~9 ml expansion per mole of Li inserted, corresponding to the molar volume occupied by Li in metal alloys [49]. This is illustrated in Figure 1.13. Therefore, problems with respect to volume expansion are the same for all alloys and require similar solutions to mitigate its effects. Furthermore, it was also known that alloy lithiation and delithiation in 2-phase regions can be detrimental to cycling because of the enormous stresses

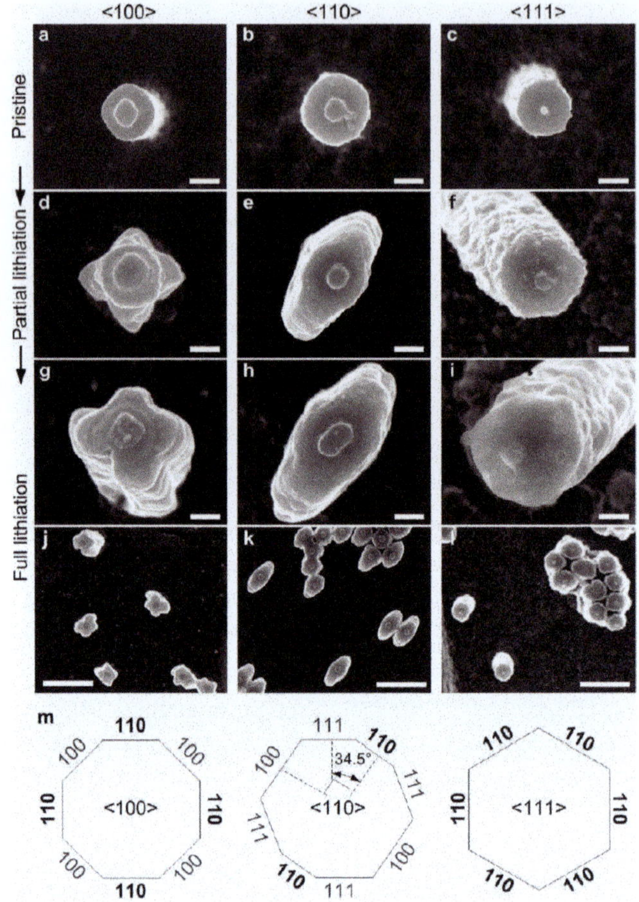

Figure 1.12 *Sequences of images showing the lithiation of Si pillars with different crystallographic orientations, illustrating the preference for lithiation (and hence volume expansion) in the <111> direction. The crystallographic orientation along the axial pillar direction is indicated at the top of each column of images. Scale bars for (a–i) are 200 nm and for (j–l) are 2 μm. Panel (m) shows the crystallographic orientation of the pillar facets as viewed from above. Reprinted with permission from [45]. Copyright 2012 American Chemical Society.*

that develop along phase boundaries, which can cause particle fracture [50, 51]. Such 2-phase regions could be avoided by making the active phase nano-sized. In the nano-regime, the alloy grains would not form equilibrium phases during lithiation, resulting in avoidance of 2-phase regions.

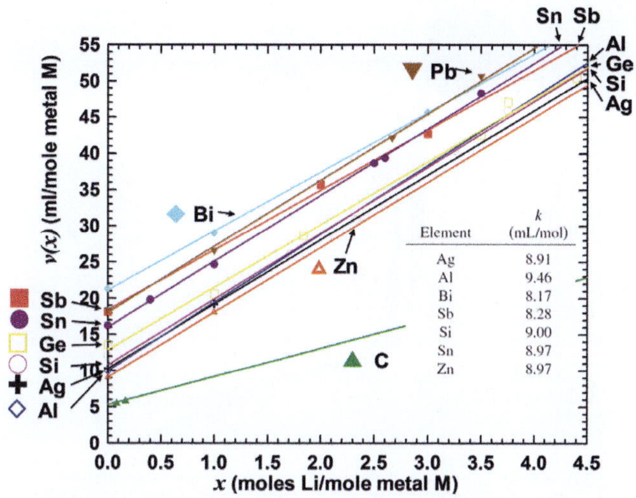

Figure 1.13 *The molar volume of lithium alloys vs lithium content. The slopes of the lines correspond to the volume occupied by lithium in the alloys and are listed in the inset table. Adapted from [49]. © IOP Publishing. Reproduced with permission. All rights reserved.*

The lithiation mechanism discovered for Si, involving electrochemical amorphization and spontaneous crystallization of $Li_{15}Si_4$ was unique among the elements. In one sense, the use of Si-based anodes was advantageous compared to the other elements: a-Si is not difficult to make, lithiates in a single-phase region, and much of its potential profile occurs in single-phase regions. However, if the electrode potential becomes below 50 mV, $Li_{15}Si_4$ will form and a 2-phase $Li_{15}Si_4$-Li_xSi region will be encountered during subsequent delithiation, as shown in Figure 1.10. As with the 2-phase regions encountered for other alloys, this often results in cell fade [1, 7, 52–54]. In addition, the $Li_{15}Si_4$ phase was found to be particularly unstable and reactive toward electrolyte [54].

Therefore Si-alloy design typically involves strategies to mitigate the effects of volume expansion and also to suppress $Li_{15}Si_4$ formation. In this way, a knowledge of the electrochemistry of Si plays a key role in materials design.

1.3.1 $Li_{15}Si_4$ suppression

In most studies, the formation of $Li_{15}Si_4$ is associated with cell fade, due to particle fracture in the $Li_{15}Si_4$-Li_xSi 2-phase region and the electrolyte reactivity of the $Li_{15}Si_4$ phase itself. However, in some studies excellent cycling has been achieved despite complete conversion to the $Li_{15}Si_4$ phase every cycle [55]. This occurs, for example, in the case of half-cells with nano-Si (n-Si) electrodes that contain large amounts of high surface area carbon, so that good electrical contact can be maintained despite volume expansion or any particle fracture that might take place. This situation is reminiscent of the cycling shown for Si powder with no binder in

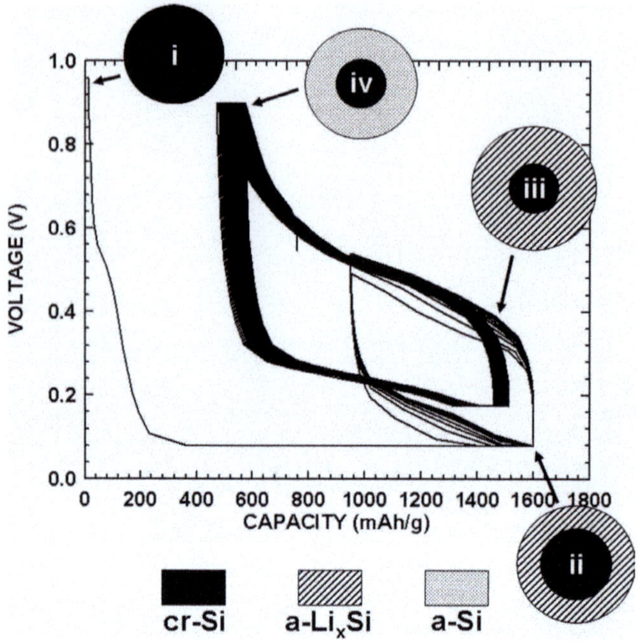

Figure 1.14 *An illustration showing how bulk c-Si particles can be converted to a-Si/c-Si core/shell particles by careful navigation of the Si potential profile. Reproduced from [42]. © IOP Publishing. Reproduced with permission. All rights reserved.*

medium temperature cells, in which particle fracture must be occurring due to the multiple 2-phase regions encountered during cycling, but good capacity retention is maintained regardless, since good electrical contact is maintained among the particles. Therefore, one method of mitigating the effects of $Li_{15}Si_4$ formation is simply to maintain good electrical connection to the Si phase, regardless of volume expansion or any particle fracturing that may take place. This can be done by combining high surface area n-Si with large amounts high surface area conductive carbons. However, high surface area electrodes typically consume substantial amounts of electrolyte that would result in fade in full cells. The use of high surface area anode materials is also associated with poor thermal stability, which may lead to safety issues.

Another method of $Li_{15}Si_4$ suppression not related to materials design is to avoid it altogether by restricting the negative electrode to potentials above 50 mV. However, this potential is very close to the lower cutoff potential used for graphite. As discussed in Section 1.6, blending Si with active graphite is typically required to obtain high energy density electrodes. In addition, Si is currently added to graphite in small amounts. In such anodes most of the capacity comes from graphite and therefore in most applications the cycling of Si anodes needs to be compatible with typical graphite cycling protocols. Nevertheless, there is no doubt that cycling Si

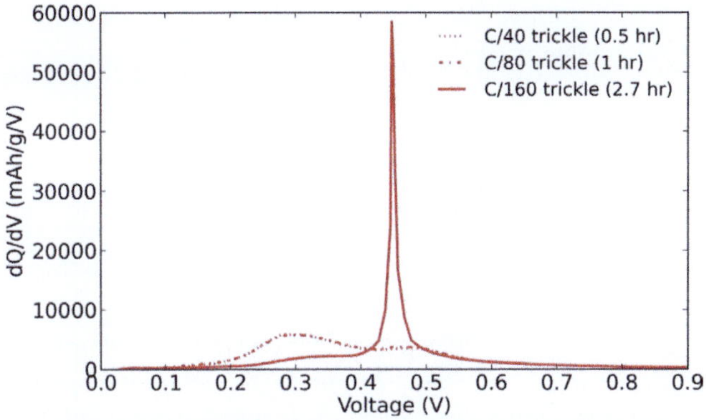

Figure 1.15 *Differential capacity of the delithiation of 50–70 nm Si particles at C/4. The "trickle" currents represent the minimum current reached during a 5 mV potential hold prior to delithiation. Higher currents do not result in the formation of* $Li_{15}Si_4$ *phase due to polarization effects. Reproduced from [7]. © IOP Publishing. Reproduced with permission. All rights reserved.*

anodes in a manner that avoids $Li_{15}Si_4$ formation can lead to impressive cycling. For instance, by careful navigation of the Si potential profile, bulk c-Si can be converted to c-Si/a-Si core/shell particles and cycled many times, while completely avoiding $Li_{15}Si_4$ formation [42]. This is shown in Figure 1.14. In this method, c-Si is lithiated to the desired capacity to match a cathode. This converts some of the c-Si to a-Si. If the material is used in this state, a high irreversible capacity (IC) would result, since the a-Si has substantial capacity below 50 mV. Therefore, the electrode is put through a series of formation cycles at high rate, so as to force the electrode below 170 mV and produce more a-Si during each cycle. This results in more high potential capacity and less IC. After this formation, good cycling can be achieved by limiting the cell capacity so that only the amorphous shell on the resulting c-Si/a-Si particles is cycled. In a properly balanced full cell, this cycling protocol is not difficult. The main issue with this method of cycling is that the anode must be cycled above 170 mV, which is above much of graphite's capacity. This makes the method incompatible for use with Si/graphite blended electrodes.

$Li_{15}Si_4$ suppression has often been claimed to occur if Si is made nano-sized [56, 57]. However, a large number of reports show that $Li_{15}Si_4$ forms regardless of size [7, 44, 55, 58]. These conflicting reports are likely due to the sensitivity of $Li_{15}Si_4$ formation toward cell polarization. Since $Li_{15}Si_4$ forms at only about 50 mV, little polarization is needed to suppress its formation. Furthermore, as a consequence of its high surface area, n-Si can react with considerable amounts of electrolyte, forming relatively thick SEI layers [7]. Therefore, n-Si can be especially prone toward increased polarization. Figure 1.15 shows an example of how cell polarization can suppress $Li_{15}Si_4$ formation in n-Si. $Li_{15}Si_4$ formation only becomes apparent after holding the electrode at 5 mV for longer than one hour. This does not mean that

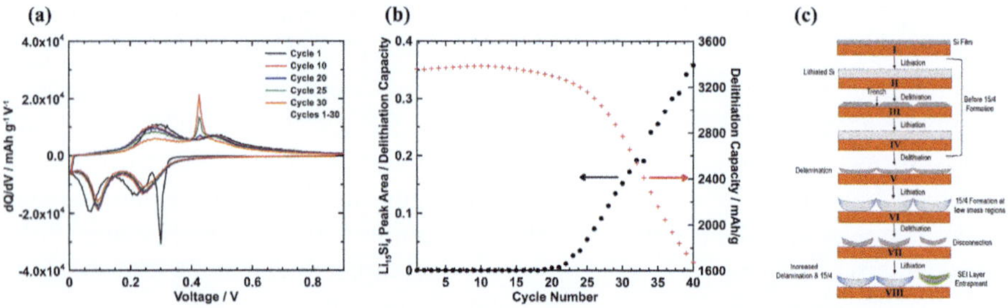

Figure 1.16 *(a) The differential capacity curve of a 275-nm thick Si thin film on a Cu-substrate. (b) The* $Li_{15}Si_4$ *delithiation peak area, and delithiation capacity vs cycle number of the cell in (a). (c) An illustration showing the microstructural evolution of Si thin film during cycling. Reproduced from [52]. © IOP Publishing. Reproduced with permission. All rights reserved.*

$Li_{15}Si_4$ cannot be suppressed by making Si nano-sized. The threshold potential for $Li_{15}Si_4$ formation should indeed be affected by size. For instance, it should be impossible for clusters of 16 Si atoms or less (<1 nm) to form $Li_{15}Si_4$, since this is less than the number of Si atoms in the $Li_{15}Si_4$ unit cell. However, a measurement of the relation between the $Li_{15}Si_4$ formation potential and size would be extremely difficult, owing to cell polarization and the presence of passivating SEI and oxide layers (which could affect the Si potential via stress-potential coupling). From a practical standpoint, regardless of whether making Si nano-sized can suppress $Li_{15}Si_4$ or not, high surface area n-Si can form excessive SEI, resulting in high IC if it is allowed to come in contact with electrolyte [7].

Considering that n-Si particles as small as 30 nm were found to form $Li_{15}Si_4$ during cycling [55], a report that <2-μm thick films of Si sputtered onto Cu foil did not form $Li_{15}Si_4$ during cycling [59] was quite surprising, since these thin film thicknesses are two orders of magnitude larger than the n-Si particles that form $Li_{15}Si_4$. Therefore, the reason why the films did not form $Li_{15}Si_4$ was not due to finite size effects. Indeed, it was later shown that with its Cu substrate, $Li_{15}Si_4$ formation is completely suppressed in a Si thin film, but $Li_{15}Si_4$ forms if the Cu substrate was removed and the Si film was cycled on its own [52]. Therefore, the $Li_{15}Si_4$ suppression in thin films was due to the presence of the substrate. As an Si thin film expands during lithiation, there are large compression forces opposing the volume expansion (up to 1.5 GPa) because of the presence of a fixed substrate [60]. This causes the potential during lithiation to shift toward lower values. This effect is caused by stress-potential coupling and can cause the potential to be depressed by as much as 150 mV [60]. With such shifts in potential, the 50 mV $Li_{15}Si_4$ formation potential becomes shifted below 0 V, resulting in the complete suppression of $Li_{15}Si_4$ formation [52]. Long-term cycling of Si thin films on Cu was found to eventually form $Li_{15}Si_4$, due to disconnection of the film from the substrate, thus relieving the stress between the Si film and the substrate [52].

$Li_{15}Si_4$ formation and its consequences for Si thin film cycling are shown in Figure 1.16 [52]. As a Si thin film attached to a Cu substrate is cycled, $Li_{15}Si_4$ formation is initially completely suppressed, as shown in Figure 1.16(a), from the complete absence of a $Li_{15}Si_4$ delithiation peak for cycles 1–10. The $Li_{15}Si_4$ suppression is due to stress-potential coupling with the Cu substrate that shifts the differential capacity curve toward lower potentials during lithiation. During this time, there is no capacity fade (Figure 1.16b) and scanning electron microscope (SEM) images show that the film forms cracks, but is completely bonded to the substrate, as illustrated in Figure 1.16cI–IV. At cycle 20, a small $Li_{15}Si_4$ peak appears in the differential capacity curve. Simultaneously, the cell begins to fade. With continued cycling the cell fades more severely and the $Li_{15}Si_4$ delithiation peak area (also shown in Figure 1.16b) increases simultaneously. SEM images show that $Li_{15}Si_4$ formation is coincident with the Si film detaching from the substrate, which leads to electrical disconnection, as illustrated in Figure 1.16cV–VIII. During cycling, the differential capacity lithiation peaks have shifted toward higher potential, as the compressive stresses are relieved between the Si film and copper substrate because of film detachment. There is no shift in the differential capacity peaks during delithiation during cycling, since any tensile stresses that form can be relieved by fracture, as shown in Figure 1.16cIII, even when the film is still connected to the substrate.

The behavior illustrated in Figure 1.16 implies that the lack of $Li_{15}Si_4$ formation at low potentials indicates that the active Si is constrained by a substrate, while the presence of $Li_{15}Si_4$, as indicated by a ~0.45 V delithiation peak in differential capacity curves, indicates the presence of unconstrained active Si. This has implications that extend to all Si-based alloys. For instance, suppression of $Li_{15}Si_4$ formation by stress-potential coupling has been observed in Si-M (M = transition metal) alloys due to the stress between an active Si phase and inactive metal silicide phases in the alloys [53]. The same mechanism is likely responsible for $Li_{15}Si_4$ suppression in SiO_x, where a Si active phase is constrained by a Li_4SiO_4 inactive matrix. In general:

- If a Si-based alloy has a differential capacity curve that contains no $Li_{15}Si_4$ delithiation peak, then the Si in the alloy is likely connected to a fixed substrate that suppresses $Li_{15}Si_4$ formation via stress-potential coupling. If $Li_{15}Si_4$ remains suppressed during cycling, then good contact to the active Si phase is being maintained and good cycling will likely result.
- If an Si alloy has a large $Li_{15}Si_4$ delithiation peak in its differential capacity curve, this implies the alloy contains Si that is free to expand (i.e. is not connected to a fixed substrate). Particle fracture associated with the $Li_{15}Si_4$/a-Li_xSi 2-phase region may cause capacity fade.
- If an alloy initially has no $Li_{15}Si_4$ delithiation peak in its differential capacity, but then a $Li_{15}Si_4$ delithiation peak forms and grows during subsequent cycling, it implies that active Si regions that were initially well connected to a substrate become detached as cycling progresses. This could result in the loss of electrical contact with active Si and capacity fade.

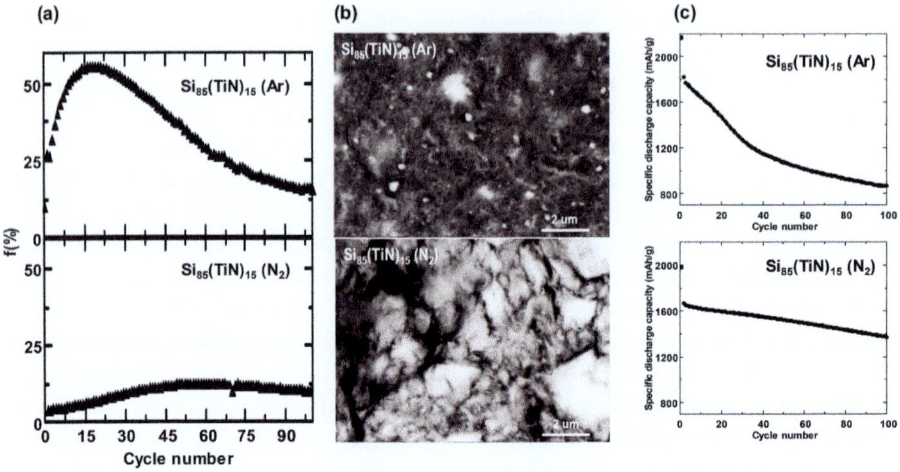

Figure 1.17 *Cycling results of two Si-Ti-N alloys with identical composition, but different microstructures. The $Si_{85}(TiN)_{15}(Ar)$ alloy has a larger Si grain size than the $Si_{85}(TiN)_{15}(N_2)$ alloy. (a) The fraction of active Si in the alloy that forms $Li_{15}Si_4$ as a function of cycle number. (b) Cross-sectional SEM images of electrodes showing the relative amount of alloy (white in these images) that remains intact after 50 cycles. (c) Cycling performance of the two alloys. Reprinted from [61] with permission from Elsevier.*

The latter point is illustrated in Figure 1.17. Figure 1.17a shows the percent active Si that forms $Li_{15}Si_4$ during cycling of two $Si_{85}(TiN)_{15}$ samples. $Si_{85}(TiN)_{15}(Ar)$ is made by ball milling Si and TiN directly together and has a large active Si grain size, whereas $Si_{85}(TiN)_{15}(N_2)$ is made by reactive gas milling of Si and Ti in $N_2(g)$ and has a smaller active Si grain size [61]. The $Si_{85}(TiN)_{15}(Ar)$ sample shows substantial $Li_{15}Si_4$ formation that increases drastically during cycling (the decrease after ~17 cycles is due to fade). This indicates the presence unconstrained active Si. The SEM image of this sample after cycling (Figure 1.17b) shows that it has completely disintegrated, presumably due to strain from the $Li_{15}Si_4$/a-Li_xSi 2-phase region. Severe capacity fade results (Figure 1.17c). In contrast, the $Si_{85}(TiN)_{15}(N_2)$ sample has smaller active Si grains, allowing greater Si surface area in contact with the inactive phase. This results in almost complete $Li_{15}Si_4$ suppression during cycling (Figure 1.17a). This alloy remains completely intact during cycling and capacity retention is much improved (Figure 1.17b and c).

A method of $Li_{15}Si_4$ suppression that has not been fully explored is doping Si with elements that disrupt the crystallization of this phase. To the author's knowledge, only Sn and Zn are known to have this effect [62]; however, other $Li_{15}Si_4$ suppressing dopants may exist.

1.3.2 Fracture suppression

It had been shown very early that nano-sized alloy particles can cycle without fracture formation [63]. This behavior was very nicely demonstrated for Si by theory and by directly observing crack formation during the lithiation of Si nanoparticles, nanowires, and nanopillars [44, 64–66]. a-Si nanoparticles with sizes less than 870 nm were found not to fracture during lithiation [66]. It was found that c-Si must be much smaller in size (300 nm for wires or pillars [64, 65] or 150 nm for Si nanoparticles [44]) in order to suppress fracture during lithiation. This is due to internal stresses associated with the c-Si/a-Li_xSi 2-phase region that occurs during the first lithiation of c-Si, as shown in Figure 1.11. This 2-phase region does not occur for a-Si, allowing its particle size to be larger while still avoiding fracture [66]. $Li_{15}Si_4$ formation was observed to occur during the lithiation of the smallest of these particles [44]. Unfortunately, to the author's knowledge, no similar study has shown the critical size for fracture formation during the delithiation of $Li_{15}Si_4$ by direct observation. Nevertheless, these studies are highly important, since they suggest fracture formation can be avoided regardless of 2-phase formation (in the case of c-Si, at least) as long as the size of the active Si phase is made small.

The major issues leading to the fade of Si-alloy cells stem from volume expansion during lithiation. One of the earliest approaches to mitigate volume expansion effects of alloys in general is to dilute the active phase with an inactive metal [67, 68], an inactive oxide matrix (e.g. SiO [69–71]) or in a carbon or graphite matrix [34, 72]. This reduces particle fracture by reducing the overall volume expansion during lithiation. These different types of alloys will be discussed separately in the following sections.

1.3.2.1 Si-transition metal (Si-M) alloys

Silicon-rich transition metal alloys are generally in the form of active Si/inactive metal silicide multiphase mixtures. Si-M alloys can also comprise carbon, oxygen, nitrogen or other elements. Si-M alloys made from the solidification of melts have been studied for use as negative electrode materials since the 1970s [73]. However, alloys formed by this method have a rather coarse grain structure, which leads to fracture along grain boundaries and capacity fade. Therefore, nanostructured active/inactive alloys are preferred [68]. Using this strategy, the active Si regions can be small (to avoid fracture) and connected to a fixed inactive matrix (to avoid $Li_{15}Si_4$ formation), while the alloy as a whole can be micron-sized, so as to present a small surface area to the electrolyte.

Nanostructured Si-M alloys can easily be made by scalable methods; most typically ball milling [74–76] or less often by melt spinning [77]. Early academic studies explored Si-M alloys by combinatorial sputtering [78]. All these methods often produce metastable alloy mixtures when noncongruently melting phases occur in the Si-M system, most notably in the case of Fe and Mn [79–81]. As is typical with nonequilibrium synthesis methods, phase composition and microstructure depend heavily on synthesis conditions. Therefore, it was found that combinatorial sputtering methods are not ideal for predicting the behavior of materials made by

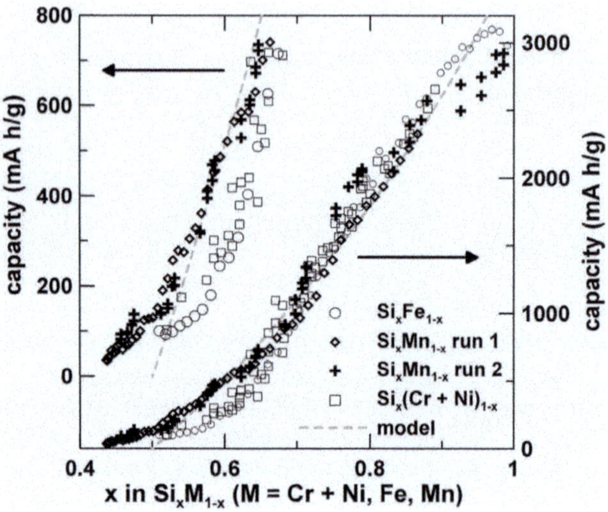

Figure 1.18 *Capacity vs x in sputtered* Si_xM_{1-x} *(M = Mn, Fe, Cr + Ni) thin films. The dashed line was calculated assuming a sample composition* Si_xM_{1-x} *is made up of a mole fraction of (2x-1) active Si and (2-2x) inactive SiM for 0.5 < x < 1. Reproduced from [78]. © IOP Publishing. Reproduced with permission. All rights reserved.*

bulk methods, especially since sputtered thin films additionally can have significant stress-potential coupling effects from the substrate [53, 82].

In many early studies it was believed that Si-M alloys lithiate via a disproportionation or intercalation mechanism that follows the equilibrium phase diagram [76, 83]. However, this is not the case. Most metal silicide phases are inactive towards lithiation at room temperature. Therefore, most Si-M alloys can be accurately modeled as a simple mixture of an active Si phase and inactive silicide phases [53, 78, 79, 81, 82, 84]. This is illustrated in Figure 1.18 for a number of alloys made by combinatorial sputtering. Such models also work well for ball milled alloys. Only the silicides of the noble transition metals Cu and Ni have been found to be electrochemically active and this only occurs when they are nano-sized [85–88]. This activity must be taken into account when predicting the capacity of Si alloys with these metals.

When implementing Si alloys in full cells, there is a trade-off in energy density for reducing volume expansion (and increasing cycle life) by introducing an inactive phase [49]. For a set target volumetric energy density, cells with alloy/graphite negative electrodes will require a higher fraction of alloy content as the inactive content in the Si alloy increases, which also leads to an overall higher electrode volume expansion [89]. Therefore, disruption of the SEI layer will still occur (albeit to a much lesser extent than pure Si), causing lithium consumption and erosion of the alloy surface [14], as shown in Figure 1.19A. This effect can be suppressed to a great degree with the use of proper electrolyte additives, as will be discussed in

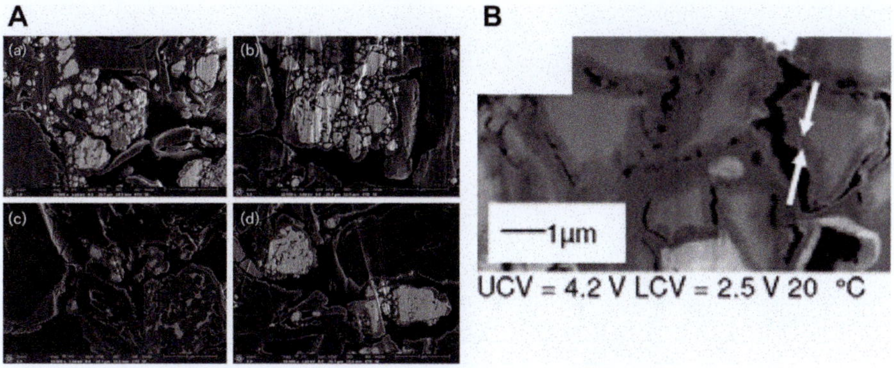

Figure 1.19 *(A) Field emission scanning electron microscope cross-sectional images of Si-alloy/graphite negative electrodes taken from pouch cells, showing surface erosion during cycling. (a) Without FEC after 30 cycles, (b) with FEC after 30 cycles, (c) without FEC after 75 cycles, and (d) with FEC after 75 cycles. (B) Annular dark field image of cross sections of C-SiO particles in the discharge state showing surface erosion after 500 cycles. (A) Reproduced with permission from [14]. (B) Reproduced from [90]. © IOP Publishing. Reproduced with permission. All rights reserved.*

Section 1.4, so that hundreds of cycles can be obtained in full cells with little fade. Good cycle life, ease of manufacture, low IC, and high ICE are attributes of Si-M alloys that make them attractive for use in commercial cells [7].

1.3.2.2 SiO

Silicon monoxide, SiO (or SiO_x, $0 \leq x \leq 1$), has been known since the nineteenth century [91], although the synthesis of phase pure samples was probably not accomplished until the 1940s [92]. SiO is commonly made by the thermal evaporation of Si and SiO_2 under vacuum at about 1 300–1 400 °C and rapidly quenching the resulting vapors [92]. The phase purity and oxygen content are dependent on the temperature and quenching rate. The structure of SiO has been under debate for as long as it has been known. This is because SiO contains regions of Si and SiO_2 that are extremely small and difficult to characterize. Recently, angstrom-beam electron diffraction measurements have allowed a detailed picture of the local structures in SiO, almost at the atomic scale [93]. According to the model developed from these studies, shown in Figure 1.20, SiO consists of small regions of Si surrounded SiO_2, where the scale of the inhomogeneity is ~1–2 nm. Accordingly, the silicon in this structure may be bonded to only other Si atoms (in the bulk of the Si regions), to 1–3 oxygen atoms (at the interface between the Si regions), and to 4 oxygen atoms (in the bulk of the SiO_2 regions) [93]. As a signature of SiO, x-ray photoelectron spectra indicate the presence of Si in the full range of oxidation states from 0 to 4+ [94].

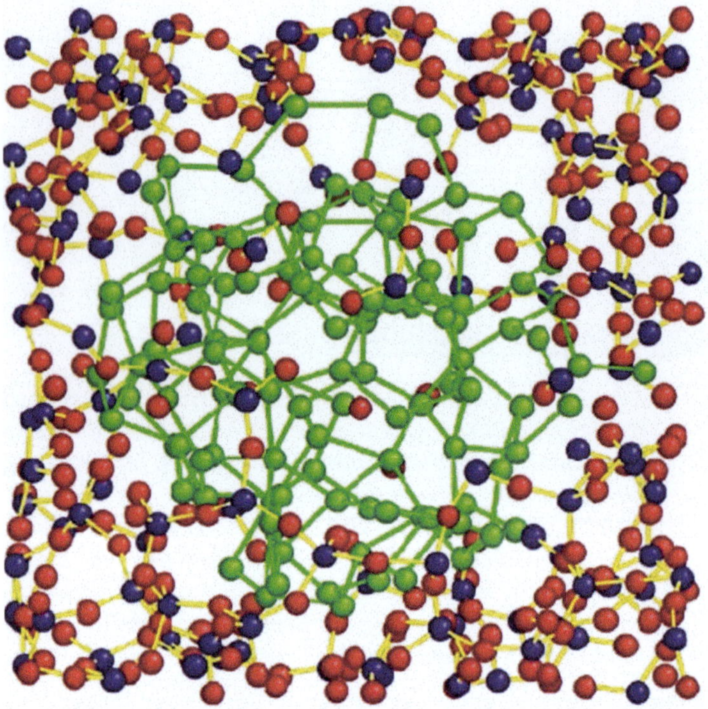

Figure 1.20 *Reconstructed heterostructure model of amorphous SiO. The inner part corresponds to an a-Si cluster and the outer part is amorphous SiO_2 matrix. The blue, red, and green circles denote Si and O in amorphous SiO_2 and Si in the Si cluster, respectively. Reproduced from [93].*

Heating SiO at temperatures higher than 800 °C results in the disproportionation of SiO, eliminating all silicon suboxides, so that only Si and SiO_2 are present [94].

The use of SiO as an anode material in Li-ion batteries was demonstrated as far back as 1994 [69–71], but did not appear in the academic literature until the 2000s [95]. The reaction of SiO with lithium can be summarized as follows [96–98]:

First lithiation:

$$\left(15 + \tfrac{x}{4}\right) \mathrm{Li} + 4\mathrm{SiO}_x \rightarrow \left(1 - \tfrac{x}{4}\right) \mathrm{Li}_{15}\mathrm{Si}_4 + x\mathrm{Li}_4\mathrm{SiO}_4 \quad (1.1)$$

Subsequent cycles (below 0.9 V):

$$\left(1 - \tfrac{x}{4}\right) \mathrm{Li}_{15}\mathrm{Si}_4 + x\mathrm{Li}_4\mathrm{SiO}_4 \leftrightarrow (4 - x)\,\mathrm{Si} + x\mathrm{Li}_4\mathrm{SiO}_4 \quad (1.2)$$

where the fully lithiated state of Si is assumed to be $Li_{15}Si_4$ and the oxygen content of SiO is explicitly stated as SiO_x here. Above 0.9 V (at potentials up to 2 V), all the lithium can be extracted from Li_4SiO_4 [97]; however, this is not the normal cycling

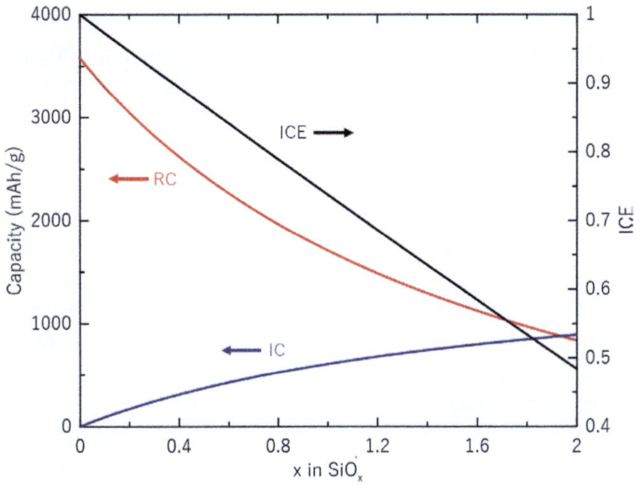

Figure 1.21 *The theoretical RC, first cycle IC, and ICE of SiO_x as a function of the oxygen content, according to the model described in [97]*

regime of an anode in a full Li-ion cell [7]. According to (1.1) and (1.2), the following relations can be derived, which are adapted from [97]:

$$\text{Reversible capacity} = \frac{\frac{15}{16}(4-x)F}{(28.1\ \text{g/mol})+x(16.0\ \text{g/mol})} \tag{1.3}$$

$$\text{Irreversible capacity} = \frac{xF}{(28.1\ \text{g/mol})+x(16.0\ \text{g/mol})} \tag{1.4}$$

$$\text{ICE} = \frac{15(4-x)}{60+x} \tag{1.5}$$

where F is Faraday's constant in units of mAh/mol (26 802 mAh/mol) and ICE is the initial coulombic efficiency. These relations are plotted in Figure 1.21 and have been found to model the behavior of SiO accurately [97]. However, in the author's experience, some commercial SiO samples have less IC and higher ICE than is predicted from this model. This may be achieved by prelithiation, for instance, which has been used to reduce the IC of SiO [99]. An attractive feature of SiO is its thermal stability. Indeed, disproportionation of SiO by heating at 1 000 °C has been found to result in improved cycle life [94]. This property also enables SiO to be carbon coated or incorporated into carbon composites by high-temperature processing methods [99, 100].

A drawback of SiO is its higher cost compared to graphite, due to the vacuum processing method in which it is commonly made [4]. Therefore, significant research effort has been made in less expensive synthesis routes, such as solution phase methods [101–103] and reactive-gas milling [104]. Another drawback of SiO is its low ICE due to the irreversible formation of Li_4SiO_4. For instance, according to Figure 1.21, a typical SiO_x with $x = 1$ has an ICE of only 73 percent (compared to ~85 percent for Si-metal alloys or >90 percent for graphite). As just mentioned, this can be somewhat addressed by prelithiation; however, the products formed by this process (e.g. lithium silicates) may not be water stable. This limits the extent of prelithiation and/or inhibits the use of water-based binders. Although SiO has less volume expansion than pure Si, the formation of Li_4SiO_4 during the initial lithiation step is associated with volume expansion, resulting in SiO having a relatively large volume expansion per active Si than other Si alloys [105]. Nevertheless, SiO is thought to be one of the best Si-alloy materials in terms of cycling performance [99]. Analysis of post-cycled electrodes show that SiO has similar surface erosion issues that Si alloys also exhibit during cycling [90, 106], as shown in Figure 1.19B. However, because SiO particles are typically much larger than ball milled alloy particles (~10 vs. 0.5 μm), their surface area is less and therefore their surface reaction rate with electrolyte is also likely less. To this author's knowledge, a direct comparison of the cycling performance and surface erosion rates between Si alloys and SiO has not been reported.

1.3.2.3 Si-C

Si-C and especially Si-graphite composites are conceptually highly attractive. In such materials, Si regions can be protected from the electrolyte by a carbon matrix phase, which also serves to reduce volume expansion. Moreover, the carbon matrix is electrically conducting, has low reactivity with the electrolyte, is deformable during electrode calendering, provides additional RC, and can have low IC and high CE. As such, numerous studies of Si-C materials have been made [34, 107–111]. However, the tendency for Si and C to react by ball milling or at temperatures above ~800 °C to form inactive Si-C poses a difficult problem [112]. Although Si-C composites can be made at lower temperatures, low temperature carbons tend to be disordered and contain hydrogen, leading to low capacity and hysteresis [113]; while heating carbons at high enough temperatures to remove hydrogen or induce graphitization causes Si and C to react to form inactive Si-C. Nevertheless, impressive Si-C composites have been made. In particular, such materials can be designed so that Si grains reside in voids in the carbon matrix that are large enough to accommodate the volume expansion of Si during lithiation, resulting in zero volume expansion particles. An example of such a material is shown in Figure 1.22. In this material, Si

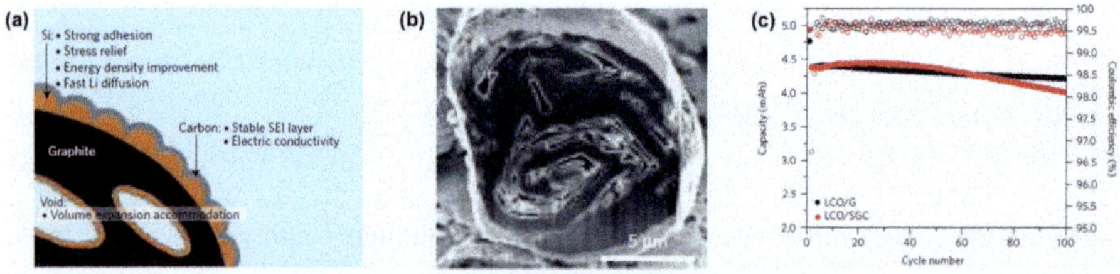

Figure 1.22 *Cross-sectional (a) schematic view and (b) SEM image of a Si-nanolayer-embedded graphite/carbon hybrid (SGC) particle. (c) Cycling performance of $LiCoO_2$ vs SGC or graphite full cells. Reprinted with permission from [114], Copyright 2016.*

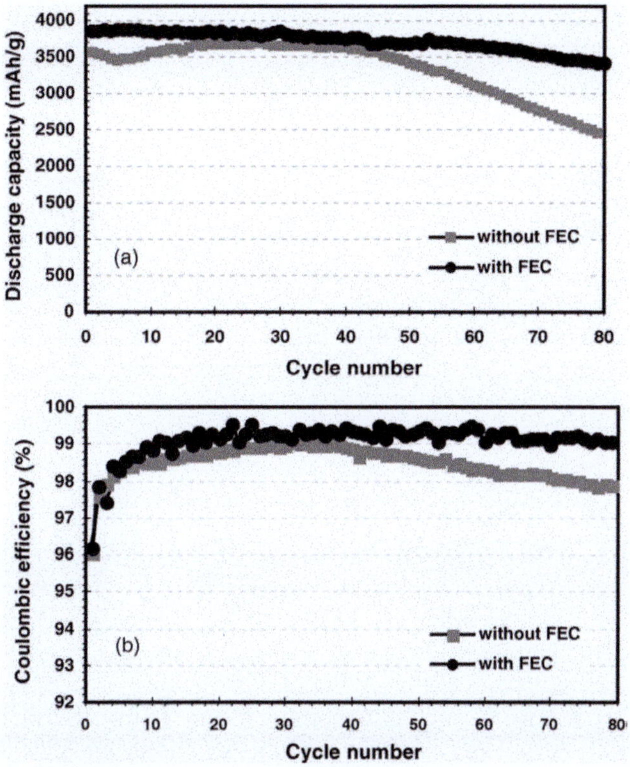

Figure 1.23 *(a) Discharge capacity retention and (b) CE of Si thin-film electrode|electrolyte|Li half-cells vs. cycle number. Reprinted from [120] with permission from Elsevier.*

is deposited by CVD processing directly into the internal porosity of natural graphite particles, thus achieving a Si-graphite composite with low volume expansion and excellent cycling characteristics [114]. Much research continues in this promising area, especially to make Si-C materials with low cost that can operate in conventional Li-ion cells with no special binders or electrolyte additives.

1.4 Electrolyte additives

All materials react with electrolyte if they are brought to potentials outside the electrolyte's electrochemical stability window. This includes both active and inactive materials [115]. In conventional anode materials, such as graphite, this results in the decomposition of the electrolyte at the active material surface. These decomposition reactions consume lithium, resulting in cell fade, while the electrolyte decomposition products form a SEI layer on active material surfaces. Fortunately, with the right selection of electrolyte components, the SEI formed can be nonporous, electrically

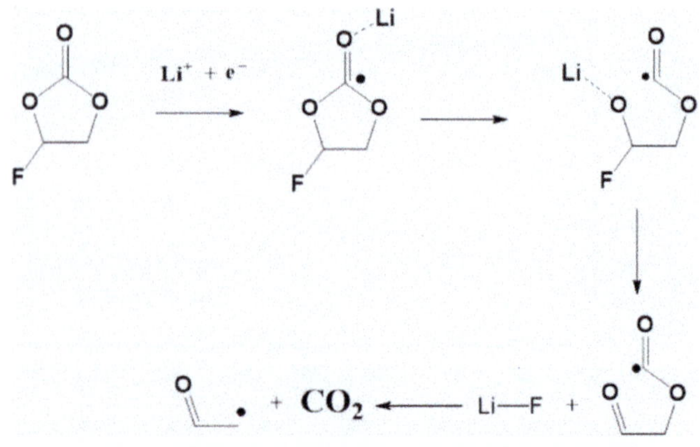

Figure 1.24 *Reduction mechanism of FEC at the surface of Si alloy and graphite particles. Reproduced from [10]. © IOP Publishing. Reproduced with permission. All rights reserved.*

insulating, and ionically conducting. These properties slow further decomposition of electrolyte (reducing the fade rate), while enabling continued cell operation [116]. Under such conditions, the SEI layer grows parabolically with time and the cell fade accordingly slows at a rate that is proportional to the square root of time [116–118].

Si alloys can have a much different fade behavior than conventional materials. The significant volume expansion/contraction during lithiation/delithiation of most Si alloys causes repeated disruption of the SEI layer during cycling. This can cause electrolyte to come in contact with exposed alloy surfaces, causing electrolyte decomposition and lithium consumption. In this case the SEI growth is both time dependent (i.e. with the same SEI parabolic growth rate as conventional materials) and cycle number dependent (proportional to the number of times the SEI is disrupted) [10]. The magnitude of the time-dependent fade of Si alloys has been shown to be similar to that of graphite [10]. Therefore, strategies to address the Si-alloy fade associated with electrolyte decomposition are mainly focused on reducing the cycle-dependent component of capacity fade caused by SEI layer disruption. In addition, the $Li_{15}Si_4$ phase has been found to be particularly reactive towards electrolyte [46, 54]. One method of reducing fade in Si alloys is to design materials in which the Si is not exposed to the electrolyte and that have no or little volume expansion (e.g. the Si-C composite materials discussed in Section 1.3.2.3). Another method is to use special electrolyte additives, which is the topic of this section.

A major advance in the development of electrolytes for use with Si alloys was the introduction of FEC as an electrolyte additive [119, 120]. Additions of FEC to carbonate electrolytes seems to improve the capacity retention generally of all Li-metal alloy electrodes. An early example of FEC's ability to significantly improve alloy capacity retention is shown in Figure 1.23. The mechanism by which FEC can improve capacity retention has been under debate since its discovery. It is generally agreed that LiF is a FEC reduction product that becomes incorporated in the

SEI when FEC is used as an additive [10, 121]. Such a reaction scheme is shown in Figure 1.24. Another reduction product of FEC is poly(vinylene carbonate) or poly(VC), which is known to form passivating and flexible films on anode surfaces from the reduction of VC [122–124]. In some studies, VC additions to the electrolyte have been found to result in superior Si-alloy electrode capacity retention and CE than when FEC is used [122, 125]. Recently it was proposed that VC actually forms more robust poly(VC) surface films on Si alloys than FEC, but these films are highly resistive [122]. This results in excellent cycling performance at low rates, but leads to fade resulting from increased cell polarization at higher cycling rates. In contrast, while the surface films formed by FEC are less passivating than for VC, the incorporation of LiF with poly(VC) may improve Li^+ diffusion, resulting in less impedance and improved cycling at moderate to high cycling rates. In addition to the above proposed mechanisms that only involve electrolyte species, it is well to keep in mind that surface erosion of Si alloys seems to occur for all alloys to varying degrees in different electrolytes [14, 90], for example as shown in Figure 1.19. Therefore, the Si alloy also becomes incorporated as a component of the SEI [14, 90].

In addition to FEC and VC, many other electrolyte salts, solvents and additives have been suggested that also can significantly improve the cycling retention of Si alloys, including methylene EC [126], $LiNO_3$ [127], and CO_2 [128]. Nevertheless, none of these additives or FEC, completely passivate Si-alloy surfaces from reactions with electrolyte. Such reactions are only slowed. As a result, such additives are continually consumed during cycling, resulting in the build-up of SEI components, alloy surface erosion, and continual electrode thickness dilation [12, 129, 130], as shown, for example, in Figure 1.25. Eventually, this can cause complete depletion of additives, salts or solvents from the electrolyte, resulting in sudden cell failure [10]. Therefore, electrolyte additives and electrolyte design in general are critical fields for the practical utilization of Si-alloy materials.

1.5 Binders

The large volume changes that accompany Si-alloy lithiation and delithiation poses particularly difficult problems for maintaining electrode integrity and electrical contact to the alloy particles. These problems are a combination of the volume changes themselves and the SEI coating on the particles that result. For instance, as shown in Figure 1.4, loose Si powder confined in a metal mesh can be lithiated and delithiated many times with little capacity loss in a molten salt. In this case, the Si surfaces remain pristine and the loose contact between particles is sufficient to maintain electrical connectivity. Figure 1.7 shows the result when the same experiment is conducted at ambient room temperature with a conventional PVDF binder and electrolyte. Instead of evenly coating particles, PVDF binder is known to forms a porous net of fine ~30 nm polymer fibers throughout the electrode coating [131]. During lithiation the Si particles expand, causing continual SEI growth on the bare particle surfaces. However, the particle expansion also places large compressive forces on

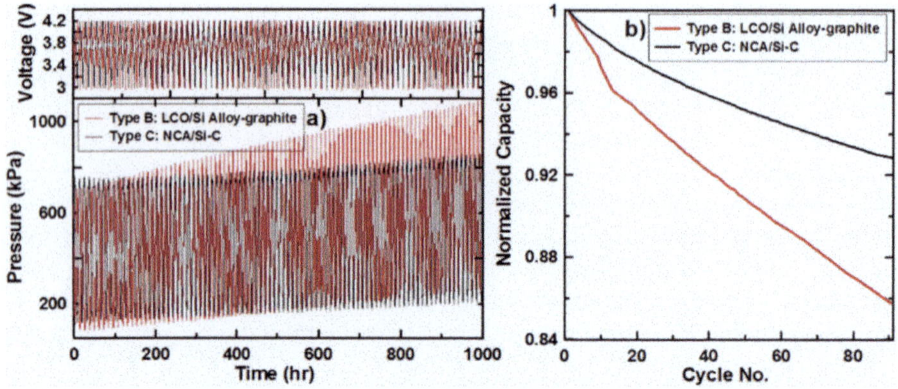

Figure 1.25 *Long-term in situ pressure measurements of Type B LCO/Si-alloy graphite (shown in red) and Type C NCA/Si-C cells (shown in black). (a) Pressure vs. time (bottom panel) with the corresponding voltage vs time profiles (top panel). (b) The normalized capacity of Type B and C cells corresponding to the in situ pressure measurements shown in (a). The cells were cycled at 40 °C with currents corresponding to C/6 between 3 and 4.2 V. Reproduced from [12].* *© IOP Publishing. Reproduced with permission. All rights reserved.*

the particles, maintaining electrode electrical connectivity. During delithiation, the Si particles contract. The compressive forces that kept the particles together despite SEI formation is relieved. As a result, the Si particles become electrically isolated from each other by the ever thickening SEI layer. Therefore, in order to maintain good cycling characteristics, binders for Si-alloy materials must maintain electrode mechanical integrity and electrical connectivity despite the problematic combination of alloy volume changes and the continual growth of insulating SEI passivating layers. Binders for Si-alloy electrodes that have been found to work can be roughly classified into two groups: those that are electrochemically active, and those that are electrochemically inactive. These will be discussed separately in the two following sections.

1.5.1 Electrochemically active binders

Perhaps the first type of binders found to be effective for alloy electrodes in general were conducting polymers, such as poly(p-phenylene) and poly(acetylene), whose use was described in the 1980s [132, 133]. These polymers can undergo electrochemical reduction to become both ion and electrically conducting. As a result, they can form an "artificial SEI" around the alloy particles and were even observed to enable cycling in electrolytes that otherwise decompose on the alloy surface [133]. Thus, the binder serves to maintain structural integrity, impede SEI growth, and to provide ion and electronic conduction to the alloy particles. This work was well

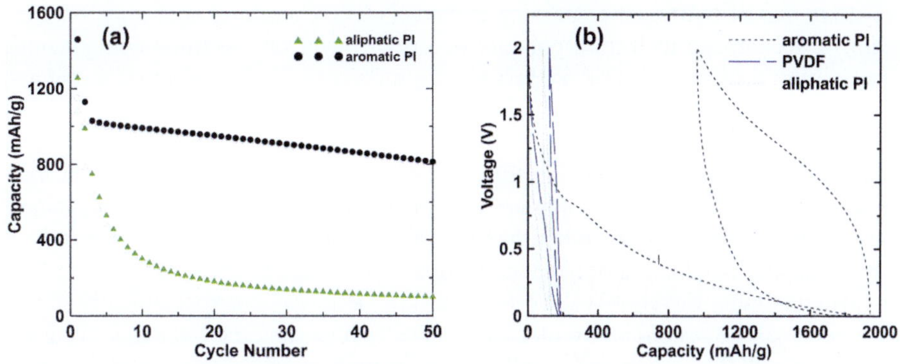

Figure 1.26 *(a) Cycling performance of Si-alloy electrodes with aliphatic and aromatic PI binders. (b) The potential profile of aromatic PI, aliphatic PI, and PVDF binders vs Li. Reproduced with permission from [141]. © IOP Publishing. Reproduced with permission. All rights reserved.*

ahead of its time, as such concepts are still being proposed today [46, 134, 135]. Although this work was done in the late 1980s, reversible lithium insertion in Pb with only 5 percent loss in 320 cycles was demonstrated in electrodes using conductive binders [133]. This is impressive even by today's standards. Since then conducting polymers have also shown effectiveness as binders for Si-alloy electrodes [136, 137].

Polyimides are other important electrochemically active binders for Si-alloy negative electrodes. As is often the case in this field, the use of polyimide binders for Si-alloy electrodes was known and used widely in industry (since at least 2001, e.g. see [138, 139]) long before being reported in academic journals (~2008 [140]). Although polyimides include both aromatic and aliphatic varieties, only the aromatic polyimides were found be useful as binders in Si-alloy electrodes, while Si-alloy electrodes with aliphatic polyimides had similarly poor cycling performance as PVDF binders [141], as shown in Figure 1.26a. Like the conducting polymers, aromatic polyimides can undergo electrochemical reduction. For instance, in the form of thin films, typical aromatic polyimides are known to undergo a one or two electron reduction [142]. Surprisingly, when used as a binder in composite electrodes the electrochemistry of polyimide is very different than that of thin polyimide films. Figure 1.26b compares the potential profiles of aromatic polyimide with aliphatic polyimide and PVDF. When used as a binder in a composite electrode, aliphatic polymers were found not to be electrochemically active, while aromatic polyimide was found to undergo a ~35 electron reduction (corresponding to ~1 900 mAh/g) [141] and have significant RC (~900 mAh/g). This magnitude of reduction capacity could only be possible if polyimide was becoming fully reduced to its constituent elements, which was indeed found to be true by ex-situ IR spectroscopy [143]. In addition, the reversible potential profile of reduced polyimide was found to closely resemble that of hydrogen containing carbons [143]. Therefore, polyimide binders seem to decompose to form a conductive carbon layer on electrode components. Considering the good cycling properties this imparts to Si-alloy electrodes, the as-formed carbon layer presumably provides good electronic conduction and inhibits SEI growth on Si alloys. Subsequently other aromatic polymers were found that function via the same mechanism, for instance phenolic resin [144]. It was further noticed that the conductive polymers also have much higher reduction capacities than is impossible to explain other than their full decomposition [145]. Therefore, it is possible that conducting polymers also function via this mechanism.

Although the irreversible electrochemical reduction of electrochemically active polymers may be beneficial for electrode capacity retention, the large first cycle IC associated with this reduction (~1 000 mAh/g) is problematic. One method of reducing this IC is by carbonizing polymer binders thermally instead of electrochemically. For instance, Si-alloy electrodes with carbonized PVDF, PI, and PAN binders have all shown to have good cycling performance [146–148]. This method has its own drawbacks, since it requires the entire electrode to be heated to high temperatures (~600–900 °C), which may not me compatible with many Si-alloy materials. In addition, polymers can lose 80 percent or more of their mass during carbonization as CO_2, CO, hydrocarbons and other gases; requiring excessive amounts of binder to be used to form a sufficient amount of carbon for this method to be effective [148].

1.5.2 Electrochemically inactive binders

Electrochemically inactive binders useful for Si-alloy electrodes are mainly comprised of aliphatic polymers with functional groups that can bond with metal surfaces. One of the first types of these binders was CMC. CMC has been used since the 1990s as a thickener for Li-ion battery negative electrode aqueous slurries with SBR latex binder [149]. In the early 2000s, Si-alloy negative electrodes with CMC/SBR binder were found to have much superior cycling performance compared to those with PVDF binder [150]. However, in these early studies SBR was considered to be the binder, while CMC was only included as a slurry thickener/rheology modifier. Later, it was found that electrodes using CMC alone had superior cycling performance than CMC/SBR electrodes [151]. At about the same time CMC was being studied as a Si-alloy binder, polyacrylic acid (PAA) and its salts were discovered to be excellent binders for Si alloys [152] and were reported in the academic literature a few years later [153]. This was an important discovery, since PAA had comparable performance to the electrochemically active binders like PI, but was electrochemically inactive and did not contribute to cell IC. Moreover, PAA was inexpensive and compatible with water-based slurries.

Both CMC and PAA binders contain carboxyl functional groups. It was found that as the carboxyl group concentration increased, the binder performance also increased [154], as is shown in Figure 1.27. Such groups have the ability to form ester bonds with the hydroxyl groups that form on the surfaces of metals [155], as shown in Figure 1.28. This prompted the investigation of binders based on polymers with high concentrations of carboxyl groups, such as alginate and poly-γ-glutamate

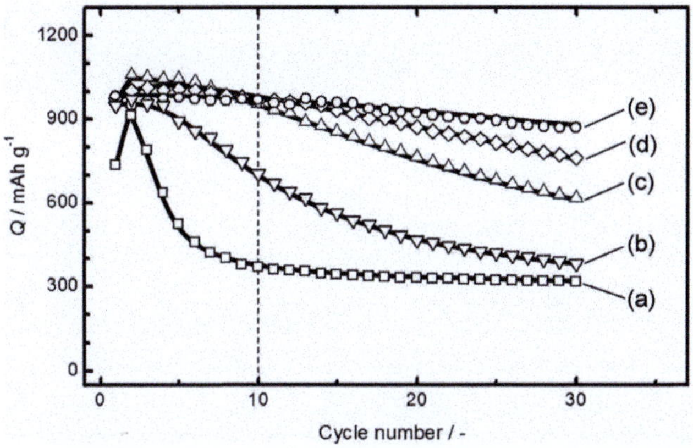

Figure 1.27 *Variation in reversible oxidation capacities of silicon graphite composite electrodes with (a) 10 wt% PVDF and (b) 10 wt% CMCNa, and (c) 10, (d) 20, and (e) 30 wt% PAANa binder. Reprinted with permission from [154]. Copyright 2012 American Chemical Society.*

[156, 157]. Even citric acid was found to be an excellent binder for Si alloys, demonstrating that the carboxyl group concentration alone, and not the length of the polymer chain, is responsible for good binder properties [158]. The mechanism by which these carboxyl functionalized binders work was found to be quite complex. Because of their bond forming ability with metal surfaces, they were found to form a continuous coating around alloy particles in cast electrodes. This coating is thought to form an artificial SEI layer, that can help passivate Si-alloy surfaces from reaction with the electrolyte [159]. Moreover, hydrogen bonding in carboxyl functionalized binders was found to give them the ability to self-heal, to maintain a connected binder network during Si-alloy expansion and contraction [160]. This may explain how these binders can continue to function even when severe Si-alloy surface erosion occurs during cycling.

A drawback of heavily carboxylated binders, such as PAA, is their hydrophilicity. For example, NaPAA is a superabsorbent polymer with the ability to absorb up to several thousand times its weight in water and expand many hundreds of percent in size [161]. This property causes cast electrode coatings to shrink during curing. This can be a serious issue for electrode coating machines that coat one side of the electrode at a time, since PAA shrinkage can cause one-sided electrodes to scroll during curing. Another issue is that fully cured electrodes need to be handled in especially moisture-free conditions. As a result of these issues, CMC is often chosen as a preferred binder, even though its carboxyl concentration is lower than PAA.

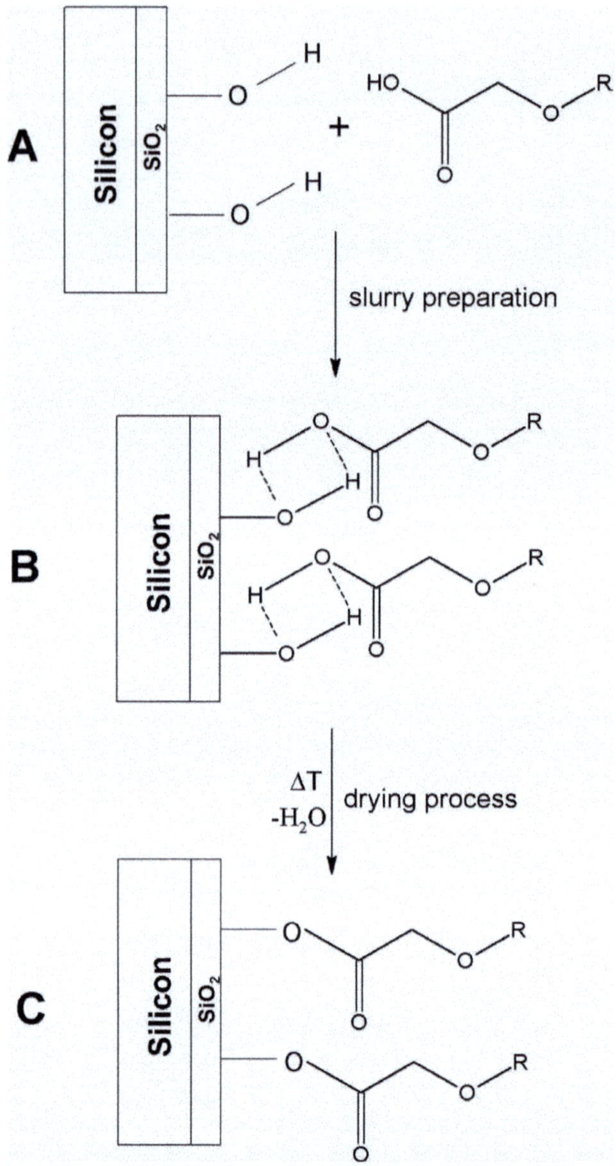

Figure 1.28 *Reaction mechanism proposed in [155] describing how binders with carboxylic acid groups can bind with the partially hydrolyzed native oxide layer on silicon alloy surfaces. Reproduced from [155].* *© IOP Publishing. Reproduced with permission. All rights reserved.*

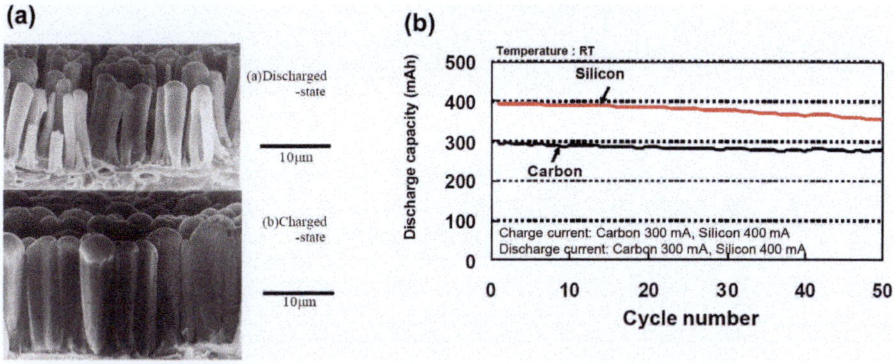

Figure 1.29 *(a) Cross-sectional SEM images of a colulmnar a-Si thin film electrode in its discharged and charged state. (b) Cycle characteristics of a cell using a $LiCoO_2$ cathode and the columnar a-Si thin film anode. Reproduced from [162]. © IOP Publishing. Reproduced with permission. All rights reserved.*

1.6 Implementation of Si alloys in Li-ion cells

In early Si alloy and alloy anode material research in general, it was assumed that these high capacity materials would be used as a replacement for graphite in Li-ion cells [63]. This is not an altogether bad assumption in the case of thin film alloy anodes. Thin Si-alloy films have low surface areas, reducing electrolyte reactivity, and being nonporous, they can have high volumetric capacities. An issue with thin films, however, is that they can introduce huge stresses on foil substrates during volume expansion. This can even lead to tearing of the current collector when thin film active masses suitable for commercial energy cells are used. However, such problems can be overcome by careful design of the film itself. For instance, Sanyo demonstrated a 400 mAh Li-ion battery having thin film Si anode in 2003 [162]. As shown in Figure 1.29 the thin film comprises columnar Si in which the spaces between the columns can accommodate the volume expansion of Si. Nevertheless, a commercialized version of a Li-ion battery with a thin film alloy negative electrode has not materialized. One can only speculate that the cost in making such thin films might be a reason for this.

In the case of Si-alloy powders, unless alloy materials are used that have no volume expansion (e.g. the Si-C alloys discussed in Section 1.3.2.3), the assumption that such alloy powders can entirely replace graphite is probably a poor one. This is because, unlike graphite, Si alloys tend to be hard, brittle and irregularly shaped. Therefore, as cast, neat alloy coatings have high porosity (~50–60 percent) [16]. This amount of empty space in the electrode will reduce its volumetric capacity substantially [1]; possibly to a volumetric capacity that is even less than a conventional calendered graphite electrode (~20 percent porous). Calendering can be applied to reduce Si-alloy electrode porosity, but unlike graphite which is a calendering

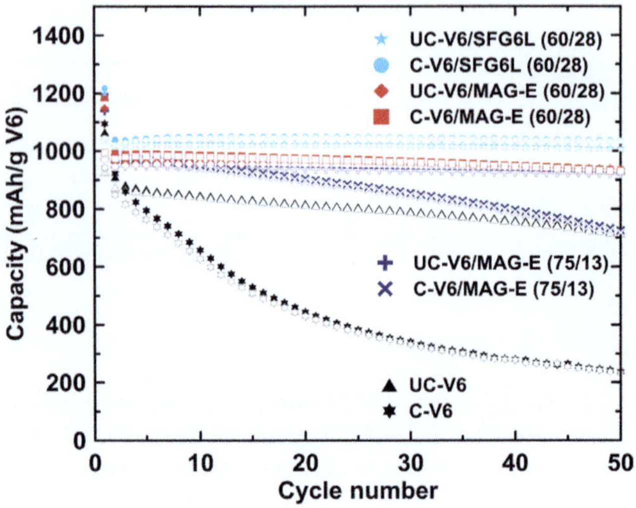

Figure 1.30 *Cycling characteristics of electrodes containing Si alloy (V6) blended with different types of graphite in the indicated volume ratios. C = calendered electrodes; UC = uncalendered electrodes. Reproduced from [16]. © IOP Publishing. Reproduced with permission. All rights reserved.*

lubricant and can easily deform to reduce porosity, Si-alloy electrodes tend to reduce porosity by particle fracture. This creates fresh alloy surfaces that are not covered by binder, leading to cell fade [16], as shown in Figure 1.30. Therefore, Si-alloy powders are probably not likely to replace graphite completely, but can be blended

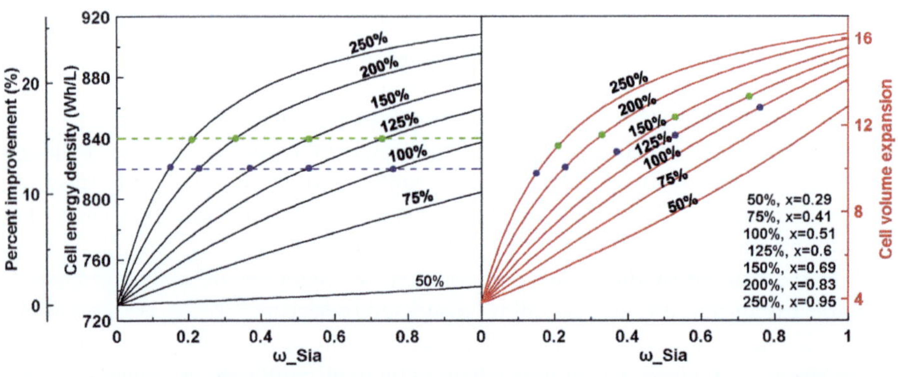

Figure 1.31 *Cell energy density and volume expansion vs the Si-alloy/graphite weight fraction (ω_Sia) for alloys containing different amounts of active Si. Percentages on the curves correspond to the volume expansion ratio of Si alloys. Reprinted from [89] with permission from Elsevier.*

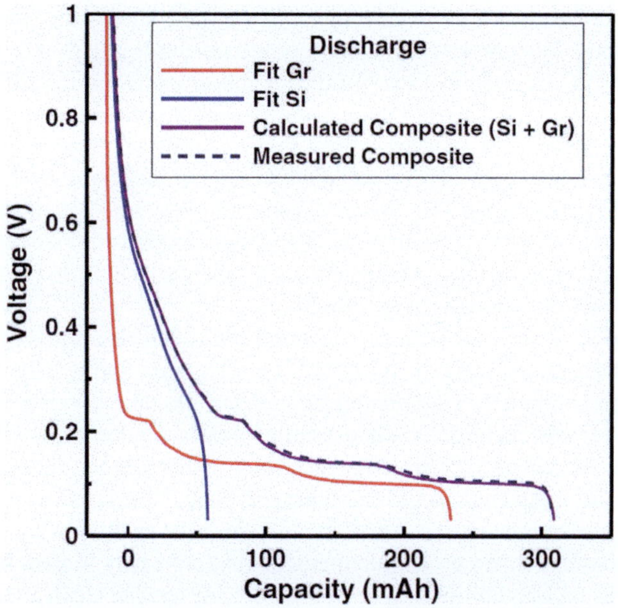

Figure 1.32 *Voltage vs capacity results of dV/dQ fits for charge (a) and discharge (b) of a graphite-SiO composite negative electrode. The measured voltage vs capacity of the composite electrode is shown in purple dotted line. The fit graphite and SiO components, shown in red and blue, add to make the calculated composite negative electrode shown in solid magenta. Reproduced from [12]. © IOP Publishing. Reproduced with permission. All rights reserved.*

with graphite in an amount that does not affect calendering properties. As shown in Figure 1.30 electrodes with as high as a 60/28 v/v alloy/graphite ratio can be calendered and still retain good cycling properties. The maximum amount of Si alloy that can be blended with graphite depends on particle size and how well the alloy and graphite are dispersed [16]. Depending on the volumetric capacity of the alloy, Si-alloy/graphite blends can result in cell energy density improvements of up 20 percent or more [89]. At a given energy density improvement, the volume expansion of such blends becomes smaller as the Si-alloy capacity (or alloy volume expansion) becomes larger. This effect is shown in Figure 1.31.

Despite these promising energy density improvements, the level of Si alloys being blended with graphite in commercial cells is only about 5 wt% [17]. According to Figure 1.31, this level of Si-alloy content could only improve cell energy density by less than 5 percent. If energy density improvement is the goal, this small level of Si alloy addition hardly seems worth the effort. It is likely that the Si alloy serves another purpose in these cells than energy density improvement. For instance, cells for electric vehicle applications are typically not cycled to 100 percent DOD to extend cycle life. If an 80 percent DOD is assumed, then 20 percent of the negative electrode is not accessed during cycling. Figure 1.32 shows the potential profile of a commercial SiO/graphite composite electrode during discharge (delithiation) from [12], where the contributions from SiO and graphite are individually shown. According to this figure, the SiO does not appreciably delithiate until the DOD has reached about 80 percent. In other words, this electrode seems to be designed such that the SiO component is almost completely inactive during reversible cycling to 80 percent DOD. This could, for instance, provide a very stable emergency capacity reserve in the cell (e.g. to prevent over-discharge), which is not susceptible to fade from volume changes during cycling (which in this instance could be even less than the volume changes experienced by graphite).

The previous discussion is pure conjecture; however, regardless of the intended purpose by the manufacturer, it can clearly be seen from Figure 1.32 that Si alloys added in small amounts to graphite negative electrodes take very little part in DOD restricted reversible cycling. To significantly increase cell energy will require Si-alloy contents beyond the 5 percent level, which will bring Si alloys into a regime where they must contribute to the RC during cycling. In this author's opinion, this can only be implemented in practical commercial cells with Si-alloy materials that have a low cost, are made by simple methods that are easily scaled, have a similar volume expansion as graphite, and do not require any special electrolyte additives or binders. Considering the impressive progress that continues to be made in Si-alloy materials, I have little doubt that such materials are possible. I even suspect that they are being made today.

References

[1] Obrovac M.N., Chevrier V.L. 'Alloy negative electrodes for Li-ion batteries'. *Chemical Reviews*. 2014;**114**(23):11444–502.

[2] *Silicon price by type U.S [online]*. 2018. Available from www.statista.com/statistics/301564/us-silicon-price-by-type [Accessed June 2019].

[3] Deign J. *China plans graphite megafactories to meet booming demand for battery storage [online]*. Available from www.greentechmedia.com/articles/read/china-builds-graphite-megafactories-for-battery-storage [Accessed June 2019].

[4] Choi J.W., Aurbach D. 'Promise and reality of post-lithium-ion batteries with high energy densities'. *Nature Reviews Materials*. 2016;**1**(4):1–16.

[5] Obrovac M.N., Christensen L. 'Structural changes in silicon anodes during lithium insertion/extraction'. *Electrochemical and Solid-State Letters*. 2004;**7**(5):A93–6.

[6] Berckmans G., De Sutter L., Kersys A., *et al*. 'Electrical characterization and micro x-ray computed tomography analysis of next-generation silicon alloy lithium-ion cells'. *World Electric Vehicle Journal*. 2018;**9**(3):43–56.

[7] Chevrier V.L., Liu L., Le D.B., *et al*. 'Evaluating Si-based materials for Li-ion batteries in commercially relevant negative electrodes'. *Journal of the Electrochemical Society*. 2014;**161**(5):A783–91.

[8] Downie L.E., Krause L.J., Burns J.C., Jensen L.D., Chevrier V.L., Dahn J.R. 'In situ detection of lithium plating on graphite electrodes by electrochemical calorimetry'. *Journal of the Electrochemical Society*. 2013;**160**(4):A588–94.

[9] Petzl M., Kasper M., Danzer M.A. 'Lithium plating in a commercial lithium-ion battery—a low-temperature aging study'. *Journal of Power Sources*. 2015;**275**:799–807.

[10] Petibon R., Chevrier V.L., Aiken C.P., *et al.* 'Studies of the capacity fade mechanisms of $LiCoO_2$/Si-alloy graphite cells'. *Journal of the Electrochemical Society*. 2016;**163**(7):A1146–56.

[11] Radvanyi E., Porcher W., De Vito E., Montani A., Franger S., Jouanneau Si Larbi S. 'Failure mechanisms of nano-silicon anodes upon cycling: an electrode porosity evolution model'. *Physical Chemistry Chemical Physics*. 2014;**16**(32):17142–53.

[12] Louli A.J., Li J., Trussler S., Fell C.R., Dahn J.R. 'Volume, pressure and thickness evolution of Li-ion pouch cells with silicon-composite negative electrodes'. *Journal of the Electrochemical Society*. 2017;**164**(12):A2689–96.

[13] Trask S.E., Pupek K.Z., Gilbert J.A., *et al.* 'Performance of full cells containing carbonate-based LiFSI electrolytes and silicon-graphite negative electrodes'. *Journal of the Electrochemical Society*. 2016;**163**(3):A345–50.

[14] Krause L.J., Brandt T., Chevrier V.L., Jensen L.D. 'Surface area increase of silicon alloys in Li-Ion full cells measured by isothermal heat flow calorimetry'. *Journal of the Electrochemical Society*. 2017;**164**(9):A2277–82.

[15] Su X., Wu Q., Li J., *et al.* 'Silicon-based nanomaterials for lithium-ion batteries: a review'. *Advanced Energy Materials*. 2014;**4**(1):1300882.

[16] Du Z., Dunlap R.A., Obrovac M.N. 'High energy density calendered Si alloy/graphite anodes'. *Journal of the Electrochemical Society*. 2014;**161**(10):A1698–705.

[17] Schmuch R., Wagner R., Hörpel G., Placke T., Winter M. 'Performance and cost of materials for lithium-based rechargeable automotive batteries'. *Nature Energy*. 2018;**3**(4):267–78.

[18] Joho F., Novák P., Spahr M.E. 'Safety aspects of graphite negative electrode materials for Lithium-ion batteries'. *Journal of the Electrochemical Society*. 2002;**149**(8):A1020–4.

[19] Richard M.N., Dahn J.R. 'Accelerating rate calorimetry study on the thermal stability of lithium intercalated graphite in electrolyte. I. experimental'. *Journal of the Electrochemical Society*. 1999;**146**(6):2068–77.

[20] Ma X., Harlow J.E., Li J., *et al.* 'Hindering rollover failure of Li[Ni0.5Mn0.3Co0.$_2$]O2/graphite pouch cells during long-term cycling'. *Journal of the Electrochemical Society*. 2019;**166**(4):A711–24.

[21] Harlow J.E., Ma X., Li J., *et al.* 'A wide range of testing results on an excellent Lithium-ion cell chemistry to be used as benchmarks for new battery technologies'. *Journal of the Electrochemical Society*. 2019;**166**(13):A3031–44.

[22] *Sony's new nexelion hybrid, lithium ion batteries [online]*. Available from www.sony.net/SonyInfo/News/Press/200502/05-006E [Accessed August 2019].

[23] Statista. *Price for refined tin from 2014 to* 2030 [online]. Available from www.statista.com/statistics/469855/tin-price-forecast [Accessed August 2019].

[24] Vissers D.R., Anderson K.E. 'The characterization of porous Li-Al alloy electrodes'. *Proceedings of the Symposium and Workshop on Advanced Battery Research and Design: ANL-76-8*; 1976. pp. 176–88.

[25] Besenhard J.O., Eichinger G. 'High energy density lithium cells'. *Journal of Electroanalytical Chemistry and Interfacial Electrochemistry*. 1976;**68**(1):1–18.

[26] von Sacken U., Nodwell E., Sundher A., Dahn J.R. 'Comparative thermal stability of carbon intercalation anodes and lithium metal anodes for rechargeable lithium batteries'. *Journal of Power Sources*. 1995;**54**(2):240–5.

[27] Xu W., Wang J., Ding F., *et al.* 'Lithium metal anodes for rechargeable batteries'. *Energy & Environmental Science*. 2014;**7**(2):513–37.

[28] Rao B.M.L., Francis R.W., Christopher H.A. 'Lithium-aluminum electrode'. *Journal of the Electrochemical Society*. 1977;**124**(10):1490–2.

[29] Seefurth R.N., Sharma R.A. 'Investigation of the reaction between liquid lithium and lithium chloride-potassium chloride melts'. *Journal of the Electrochemical Society*. 1975;**122**(8):1049–53.

[30] Seefurth R.N., Sharma R.A. 'Investigation of lithium utilization from a lithium-silicon electrode'. *Journal of the Electrochemical Society*. 1977;**124**(8):1207–14.

[31] Braga M.H., Dębski A., Gąsior W. 'Li–Si phase diagram: enthalpy of mixing, thermodynamic stability, and coherent assessment'. *Journal of Alloys and Compounds*. 2014;**616**(8):581–93.

[32] Huggins R.A. 'Lithium alloy negative electrodes'. *Journal of Power Sources*. 1999;**81-82**:13–19.

[33] Wen C.J., Huggins R.A. 'Chemical diffusion in intermediate phases in the lithium-silicon system'. *Journal of Solid State Chemistry*. 1981;**37**(3):271–8.

[34] Wilson A.M., Dahn J.R. 'Lithium insertion in carbons containing nanodispersed silicon'. *Journal of the Electrochemical Society*. 1995;**142**(2):326–32.

[35] Ryu J.H., Kim J.W., Sung Y.-E., Oh S.M. 'Failure modes of silicon powder negative electrode in lithium secondary batteries'. *Electrochemical and Solid-State Letters*. 2004;**7**(10):A306–9.

[36] Oumellal Y., Delpuech N., Mazouzi D., *et al.* 'The failure mechanism of nano-sized Si-based negative electrodes for lithium ion batteries'. *Journal of Materials Chemistry*. 2011;**21**(17):6201–8.

[37] Limthongkul P., Jang Y.-I., Dudney N.J., Chiang Y.-M. 'Electrochemically-driven solid-state amorphization in lithium–metal anodes'. *Journal of Power Sources*. 2003;**119-121**:604–9.

[38] Kripyakevich P.I., Gladyshevskii E.I. 'Structures of the Cu15Si4 type'. *Zhurnal Strukturnoi Khimii*. 1961;**2**:573–7.

[39] Zeilinger M., Baran V., Van Wüllen L., Häussermann U., Fässler T.F. 'Stabilizing the phase $Li_{15}Si_4$ through lithium–aluminum substitution in

$Li_{15-x}Al_xSi_4$ (0.4 < x < 0.8)—single crystal X-ray structure determination of $Li_{15}Si_4$ and $Li_{14.37}Al_{0.63}Si_4$'. *Chemistry of Materials*. 2013;**25**(20):4113–21.

[40] Sharma R.A., Seefurth R.N. 'Thermodynamic properties of the lithium-silicon system'. *Journal of the Electrochemical Society*. 1976;**123**(12):1763–8.

[41] Okamoto H. 'The Li-Si (lithium-silicon) system'. *Bulletin of Alloy Phase Diagrams*. 1990;**11**(3):306–12.

[42] Obrovac M.N., Krause L.J. 'Reversible cycling of crystalline silicon powder'. *Journal of the Electrochemical Society*. 2007;**154**(2):A103–8.

[43] Li J., Dahn J.R. 'An in situ X-ray diffraction study of the reaction of Li with crystalline Si'. *Journal of the Electrochemical Society*. 2007;**154**(3):A156–61.

[44] Liu X.H., Zhong L., Huang S., Mao S.X., Zhu T., Huang J.Y. 'Size-dependent fracture of silicon nanoparticles during lithiation'. *ACS Nano*. 2012;**6**(2):1522–31.

[45] Lee S.W., McDowell M.T., Choi J.W., Cui Y. 'Anomalous shape changes of silicon nanopillars by electrochemical lithiation'. *Nano Letters*. 2011;**11**(7):3034–9.

[46] Key B., Bhattacharyya R., Morcrette M., Seznéc V., Tarascon J.-M., Grey C.P. 'Real-time NMR investigations of structural changes in silicon electrodes for Lithium-ion batteries'. *Journal of the American Chemical Society*. 2009;**131**(26):9239–49.

[47] Li J., Smith A., Sanderson R.J., Hatchard T.D., Dunlap R.A., Dahn J.R. 'In situ ^{119}Sn Mössbauer effect study of the reaction of lithium with Si using a Sn probe'. *Journal of the Electrochemical Society*. 2009;**156**(4):A283–8.

[48] Chevrier V.L., Dahn J.R. 'First principles studies of disordered Lithiated silicon'. *Journal of The Electrochemical Society*. 2010;**157**(4):A392–8.

[49] Obrovac M.N., Christensen L., Le D.B., Dahn J.R. 'Alloy design for lithium-ion battery anodes'. *Journal of the Electrochemical Society*. 2007;**154**(9):A849–55.

[50] Courtney I.A., Dahn J.R. 'Electrochemical and in situ X-ray diffraction studies of the reaction of lithium with tin oxide composites'. *Journal of the Electrochemical Society*. 1997;**144**(6):2045–52.

[51] Courtney I.A., Dahn J.R. 'KeyfactorscontrollingthereversibilityofthereactionoflithiumwithSnO_2andSn_2BPO6glass'. *Journal of the Electrochemical Society*. 1997;**144**(9):2943–8.

[52] Iaboni D.S.M., Obrovac M.N. 'Li15Si4 formation in silicon thin film negative electrodes'. *Journal of the Electrochemical Society*. 2016;**163**(2):A255–61.

[53] Du Z., Hatchard T.D., Dunlap R.A., Obrovac M.N. 'Combinatorial investigations of Ni-Si negative electrode materials for Li-ion batteries'. *Journal of the Electrochemical Society*. 2015;**162**(9):A1858–63.

[54] Gao H., Xiao L., Plümel I., *et al*. 'Parasitic reactions in nanosized silicon anodes for lithium-ion batteries'. *Nano Letters*. 2017;**17**(3):1512–9.

[55] Schott T., Robert R., Ulmann P.A., *et al*. 'Cycling behavior of silicon-containing graphite electrodes, part A: effect of the lithiation protocol'. *The Journal of Physical Chemistry C*. 2017;**121**(34):18423–9.

[56] Gauthier M., Mazouzi D., Reyter D., *et al.* 'A low-cost and high performance ball-milled Si-based negative electrode for high-energy Li-ion batteries'. *Energy & Environmental Science*. 2013;**6**(7):2145–55.

[57] Wang F., Wu L., Key B., *et al.* 'Electrochemical reaction of lithium with nanostructured silicon anodes: A study by in-situ synchrotron x-ray diffraction and electron energy-loss spectroscopy'. *Advanced Energy Materials*. 2013;**3**(10):1324–31.

[58] McDowell M.T., Ryu I., Lee S.W., Wang C., Nix W.D., Cui Y. 'Studying the kinetics of crystalline silicon nanoparticle lithiation with in situ transmission electron microscopy'. *Advanced Materials*. 2012;**24**(45):6034–41.

[59] Hatchard T.D., Dahn J.R. 'In situ XRD and electrochemical study of the reaction of lithium with amorphous silicon'. *Journal of the Electrochemical Society*. 2004;**151**(6):A838–42.

[60] Sethuraman V.A., Srinivasan V., Bower A.F., Guduru P.R. 'In situ measurements of stress-potential coupling in lithiated silicon'. *Journal of the Electrochemical Society*. 2010;**157**(11):A1253–61.

[61] Cao S., Bennett J.C., Wang Y., Gracious S., Zhu M., Obrovac M.N. 'Si–TiN alloy Li-ion battery anode materials prepared by reactive N2 gas milling'. *Journal of Power Sources*. 2019;**438**:227003.

[62] Hatchard T.D., Obrovac M.N., Dahn J.R. 'A comparison of the reactions of the SiSn, SiAg, and SiZn binary systems with L3i'. *Journal of the Electrochemical Society*. 2006;**153**(2):A282–7.

[63] Besenhard J.O., Yang J., Winter M. 'Will advanced lithium-alloy anodes have a chance in lithium-ion batteries?' *Journal of Power Sources*. 1997;**68**(1):87–90.

[64] Ryu I., Choi J.W., Cui Y., Nix W.D. 'Size-dependent fracture of Si nanowire battery anodes'. *Journal of the Mechanics and Physics of Solids*. 2011;**59**(9):1717–30.

[65] Lee S.W., McDowell M.T., Berla L.A., Nix W.D., Cui Y. 'Fracture of crystalline silicon nanopillars during electrochemical lithium insertion'. *Proceedings of the National Academy of Sciences*. 2012;**109**(11):4080–5.

[66] McDowell M.T., Lee S.W., Harris J.T., *et al.* 'In situ TEM of two-phase lithiation of amorphous silicon nanospheres'. *Nano Letters*. 2013;**13**(2):758–64.

[67] Anani A., Huggins R.A. 'Multinary alloy electrodes for solid state batteries I. A phase diagram approach for the selection and storage properties determination of candidate electrode materials'. *Journal of Power Sources*. 1992;**38**(3):351–62.

[68] Turner R.L., McClure D.J., Krause L.J. 'Electrode for a lithium battery'. US Patent No. 6,203,944 B1, Filed Mar. 26; 1998.

[69] Iwasaki F., Tahara K., Ishikawa H. 'Nonaqueous electrolyte secondary battery and method of producing the same'. U.S. Patent No.5,506,075, Filed Mar. 3; 2000.

[70] Tahara K., Iwasaki F., Yahagi S. 'Nonaqueous electrolyte secondary battery'. European Patent Application No. EP1494302A1, Filed June 24; 1996.

[71] Inoue N., Inoue H., Sugiyama K. 'Nonaqueous secondary battery'. U.S. Patent No. 6,090,506, Filed Aug.1; 1997.

[72] Wang C.S., Wu G.T., Zhang X.B., Qi Z.F., Li W.Z. 'Lithium insertion in carbon-silicon composite materials produced by mechanical milling'. *Journal of the Electrochemical Society*. 1998;**145**(8):2751–8.

[73] Eichinger G. 'Electrochemical formation of lithium iron(copper)silicide alloys'. *Journal of Applied Electrochemistry*. 1979;**9**(3):315–20.

[74] Hatchard T.D., Genkin A., Obrovac M.N. 'Rapid mechanochemical synthesis of amorphous alloys'. *AIP Advances*. 2017;**7**(4):045201.

[75] Kim I.-seok, Kim I., Kumta P.N., Blomgren G.E. 'Si/TiN nanocomposites novel anode materials for Li-ion batteries'. *Electrochemical and Solid-State Letters*. 1999;**3**(11):493–6.

[76] Wang G.X., Sun L., Bradhurst D.H., Zhong S., Dou S.X., Liu H.K. 'Innovative nanosize lithium storage alloys with silica as active centre'. *Journal of Power Sources*. 2000;**88**(2):278–81.

[77] Obrovac M.N., Christensen L. 'Electrode compositions based on an amorphous alloy having a high silicon content'. *U.S. Patent No. 7,732,095B2, Filed Nov. 21*. 2006'Electrode compositions based on an amorphous alloy having a high silicon content'U.S. Patent No. 7,732,095B2, Filed Nov. 21.

[78] Fleischauer M.D., Topple J.M., Dahn J.R. 'Combinatorial investigations of Si-M (M = Cr + Ni, Fe, Mn) thin film negative electrode materials'. *Electrochemical and Solid-State Letters*. 2005;**8**(2):A137–40.

[79] Fleischauer M.D., Mar R., Dahn J.R. 'Method to predict phase formation and specific capacity for lithium in codeposited silicon-transition metal thin films'. *Journal of the Electrochemical Society*. 2007;**154**(3):A151–5.

[80] Cao Y., Bennett J.C., Dunlap R.A., Obrovac M.N. 'Li insertion in ball milled Si-Mn alloys'. *Journal of the Electrochemical Society*. 2018;**165**(9):A1734–40.

[81] Du Z., Dunlap R.A., Obrovac M.N. 'Structural and electrochemical investigation of FexSi1-x thin films in Li cells'. *Journal of the Electrochemical Society*. 2016;**163**(9):A2011–6.

[82] Du Z., Ellis S.N., Dunlap R.A., Obrovac M.N. 'Ni_xSi 1-x alloys prepared by mechanical milling as negative electrode materials for lithium ion batteries'. *Journal of the Electrochemical Society*. 2016;**163**(2):A13–18.

[83] Weydanz W., Wohlfahrt-Mehrens M., Huggins R. 'A room temperature study of the binary lithium–silicon and the ternary lithium–chromium–silicon system for use in rechargeable lithium batteries'. *Journal of Power Sources*. 1999;**81–82**:237–42.

[84] Wang Y., Cao S., Kalinina M., *et al*. 'Lithium insertion in nanostructured $_{Si1-}$xTi_x alloys'. *Journal of the Electrochemical Society*. 2017;**164**(13):A3006–10.

[85] Du Z., Liu H., Ellis S.N., *et al*. 'Electrochemistry of CuxSi1 x alloys in Li cells'. *Journal of the Electrochemical Society*. 2016;**163**(7):A1275–9.

[86] Du Z., Hatchard T.D., Bissonnette P., Dunlap R.A., Obrovac M.N. 'Electrochemical activity of nano-$NiSi_2$ in Li cells'. *Journal of The Electrochemical Society*. 2016;**163**(10):A2456–60.

[87] Wang G.X., Sun L., Bradhurst D.H., Zhong S., Dou S.X., Liu H.K. 'Nanocrystalline NiSi alloy as an anode material for lithium-ion batteries'. *Journal of Alloys and Compounds*. 2000;**306**(1–2):249–52.

[88] Zhou Y.-N., Li W.-J., Chen H.-J., Liu C., Zhang L., Fu Z. 'Nanostructured NiSi thin films as a new anode material for lithium ion batteries'. *Electrochemistry Communications*. 2011;**13**(6):546–9.

[89] Du Z., Li J., Daniel C., Wood D.L. 'Si alloy/graphite coating design as anode for Li-ion batteries with high volumetric energy density'. *Electrochimica Acta*. 2017;**254**:123–9.

[90] Kajita T., Yuge R., Nakahara K., *et al.* 'Improvement in cycle performance and clarification of deterioration mechanism of lithium-ion full cells using SiO Anodes'. *Journal of the Electrochemical Society*. 2013;**160**(10):A1806–10.

[91] Mabery C.F. 'On the composition of certain products from the Cowles electrical furnace'. *American Chemical Journal*. 1887;**9**:11–15.

[92] DeWet Erasmus H., Persson J.A. 'Preparation and properties of silicon monoxide'. *Journal of the Electrochemical Society*. 1949;**95**(6):316–8.

[93] Hirata A., Kohara S., Asada T., *et al.* 'Atomic-scale disproportionation in amorphous silicon monoxide'. *Nature Communications*. 2016;**7**(1):11591.

[94] Park C.-M., Choi W., Hwa Y., Kim J.-H., Jeong G., Sohn H.-J. 'Characterizations and electrochemical behaviors of disproportionated SiO and its composite for rechargeable Li-ion batteries'. *Journal of Materials Chemistry*. 2010;**20**(23):4854–60.

[95] Yang J*et al.* 'SiOx-based anodes for secondary lithium batteries'. *Solid State Ionics*. 2002;**152-153**:125–9.

[96] Nagai R., Kita F., Yamada M., Katayama H. 'Development of highly reliable high-capacity batteries for mobile devices and small- to medium-sized batteries for industrial applications'. *Hitachi Review*. 2011;**60**:28–32.

[97] Al-Maghrabi M.A., Suzuki J., Sanderson R.J., *et al.* 'Combinatorial studies of $_{Si1-xOx}$ as a potential negative electrode material for Li-ion battery applications'. *Journal of the Electrochemical Society*. 2013;**160**(9):A1587–93.

[98] Yamamura H., Nobuhara K., Nakanishi S., Iba H., Okada S. 'Investigation of the irreversible reaction mechanism and the reactive trigger on SiO anode material for lithium-ion battery'. *Journal of the Ceramic Society of Japan*. 2011;**119**(1395):855–60.

[99] Chen T., Wu J., Zhang Q., Su X. 'Recent advancement of SiOx based anodes for lithium-ion batteries'. *Journal of Power Sources*. 2017;**363**:126–44.

[100] Morita T., Takami N. 'Nano Si Cluster-SiO_xC composite material as high-capacity anode material for rechargeable lithium batteries'. *Journal of the Electrochemical Society*. 2006;**153**(2):A425–30.

[101] Park E., Yoo H., Lee J., Park M.-S., Kim Y.-J., Kim H. 'Dual-size silicon nanocrystal-embedded SiO(x) nanocomposite as a high-capacity lithium storage material'. *ACS Nano*. 2015;**9**(7):7690–6.

[102] Lv P., Zhao H., Gao C., Du Z., Wang J., Liu X. 'SiO–C dual-phase glass for lithium ion battery anode with high capacity and stable cycling performance'. *Journal of Power Sources*. 2015;**274**:542–50.

[103] Park M.-S., Park E., Lee J., *et al.* 'Hydrogen silsequioxane-derived Si/SiO(x) nanospheres for high-capacity lithium storage materials'. *ACS Applied Materials & Interfaces.* 2014;**6**(12):9608–13.

[104] Cao Y., Bennett J.C., Dunlap R.A., Obrovac M.N. 'A simple synthesis route for high-capacity SiO_x anode materials with tunable oxygen content for lithium-ion batteries'. *Chemistry of Materials.* 2018;**30**(21):7418–22.

[105] Miyachi M., Yamamoto H., Kawai H., Ohta T., Shirakata M. 'Analysis of SiO anodes for lithium-ion batteries'. *Journal of the Electrochemical Society.* 2005;**152**(10):A2089–91.

[106] Choi I., Lee M.J., Oh S.M., Kim J.J. 'Fading mechanisms of carbon-coated and disproportionated Si/SiOx negative electrode (Si/SiOx/C) in Li-ion secondary batteries: Dynamics and component analysis by TEM'. *Electrochimica Acta.* 2012;**85**(9):369–76.

[107] Yang J., Wang B.F., Wang K., Liu Y., Xie J.Y., Wen Z.S. 'Si/C composites for high capacity lithium storage materials'. *Electrochemical and Solid-State Letters.* 2003;**6**(8):A154–6.

[108] Kim I.-seok., Blomgren G.E., Kumta P.N. 'Si–SiC nanocomposite anodes synthesized using high-energy mechanical milling'. *Journal of Power Sources.* 2004;**130**(1–2):275–80.

[109] Wang M.-S., Fan L.-Z., Huang M., Li J., Qu X. 'Conversion of diatomite to porous Si/C composites as promising anode materials for lithium-ion batteries'. *Journal of Power Sources.* 2012;**219**:29–35.

[110] Liu N., Lu Z., Zhao J., *et al.* 'A pomegranate-inspired nanoscale design for large-volume-change lithium battery anodes'. *Nature Nanotechnology.* 2014;**9**(3):187–92.

[111] Terranova M.L., Orlanducci S., Tamburri E., Guglielmotti V., Rossi M. 'Si/C hybrid nanostructures for Li-ion anodes: an overview'. *Journal of Power Sources.* 2014;**246**:167–77.

[112] Timmons A., Todd A.D.W., Mead S.D., *et al.* 'Studies of $Si_{[sub\ 1-x]}C_{[sub\ x]}$ electrode materials prepared by high-energy mechanical milling and combinatorial sputter deposition'. *Journal of the Electrochemical Society.* 2007;**154**(9):A865–74.

[113] Dahn J.R., Zheng T., Liu Y., Xue J.S. 'Mechanisms for lithium insertion in Carbonaceous materials'. *Science.* 1995;**270**(5236):590–3.

[114] Ko M., Chae S., Ma J., *et al.* 'Scalable synthesis of silicon-nanolayer-embedded graphite for high-energy lithium-ion batteries'. *Nature Energy.* 2016;**1**(9):16113.

[115] Yan Z., Obrovac M.N. 'Selecting inactive materials with low electrolyte reactivity for lithium-ion cells'. *Journal of Power Sources.* 2018;**397**:374–81.

[116] Peled E., Golodnitsky D., Ardel G. 'Advanced model for solid electrolyte interphase electrodes in liquid and polymer electrolytes'. *Journal of the Electrochemical Society.* 1997;**144**(8):L208–10.

[117] Peled E. 'The electrochemical behavior of alkali and alkaline earth metals in nonaqueous battery systems—the solid electrolyte interphase model'. *Journal of the Electrochemical Society.* 1979;**126**(12):2047–51.

[118] Blomgren G.E. 'Electrolytes for advanced batteries'. *Journal of Power Sources*. 1999;**81-82**:112–18.
[119] Yamaguchi A., Nakajima K., Fujishige Y., Fukushima Y., Nagamine M. 'Electrolyte and battery'. Japanese Patent No. JP4022889B2, Filed Feb. 12; 2004.
[120] Choi N.-S., Yew K.H., Lee K.Y., Sung M., Kim H., Kim S.-S. 'Effect of fluoroethylene carbonate additive on interfacial properties of silicon thin-film electrode'. *Journal of Power Sources*. 2006;**161**(2):1254–9.
[121] Leung K., Rempe S.B., Foster M.E., *et al.* 'Modeling electrochemical decomposition of fluoroethylene carbonate on silicon anode surfaces in lithium ion batteries'. *Journal of the Electrochemical Society*. 2014;**161**(3):A213–21.
[122] Jaumann T., Balach J., Langklotz U., *et al.* 'Lifetime vs. rate capability: understanding the role of FEC and Vc in high-energy Li-ion batteries with nano-silicon anodes'. *Energy Storage Materials*. 2017;**6**:26–35.
[123] Heiskanen S.K., Kim J., Lucht B.L. 'Generation and evolution of the solid electrolyte interphase of lithium-ion batteries'. *Joule*. 2019;**3**(10):2322–33.
[124] Ota H., Sakata Y., Inoue A., Yamaguchi S. 'Analysis of vinylene carbonate derived SEI layers on graphite anode'. *Journal of the Electrochemical Society*. 2004;**151**(10):A1659–69.
[125] Dalavi S., Guduru P., Lucht B.L. 'Performance enhancing electrolyte additives for lithium ion batteries with silicon anodes'. *Journal of the Electrochemical Society*. 2012;**159**(5):A642–6.
[126] Nguyen C.C., Lucht B.L. 'Improved cycling performance of Si nanoparticle anodes via incorporation of methylene ethylene carbonate'. *Electrochemistry Communications*. 2016;**66**:71–4.
[127] Nguyen C.C., Lucht B.L. 'Development of electrolytes for Si-graphite composite electrodes'. *Journal of the Electrochemical Society*. 2018;**165**(10):A2154–61.
[128] Krause L.J., Chevrier V.L., Jensen L.D., Brandt T. 'The effect of carbon dioxide on the cycle life and electrolyte stability of Li-ion full cells containing silicon alloy'. *Journal of the Electrochemical Society*. 2017;**164**(12):A2527–33.
[129] Gómez-Cámer J.L., Bünzli C., Hantel M.M., Poux T., Novák P. 'On the correlation between electrode expansion and cycling stability of graphite/Si electrodes for Li-ion batteries'. *Carbon*. 2016;**105**(9):42–51.
[130] Jeong G., Lee S.M., Choi N.S., Kim Y.-U., Lee C.K. 'Stabilizing dimensional changes in Si-based composite electrodes by controlling the electrode porosity: an in situ electrochemical dilatometric study'. *Electrochimica Acta*. 2011;**56**(14):5095–101.
[131] Wachtler M., Wagner M.R., Schmied M., Winter M., Besenhard J.O. 'The effect of the binder morphology on the cycling stability of Li–alloy composite electrodes'. *Journal of Electroanalytical Chemistry*. 2001;**510**(1–2):12–19.
[132] Jow T.R., Shacklette L.W., Maxfield M., Vernick D. 'The role of conductive polymers in alkali-metal secondary electrodes'. *Journal of the Electrochemical Society*. 1987;**134**(7):1730–3.

[133] Maxfield M., Jow T.R., Sewchok M.G., Shacklette L.W. 'Alloy/conducting-polymer composite electrodes: electrolytes, cathodes, and morphology'. *Journal of Power Sources*. 1989;**26**(1–2):93–102.

[134] Karkar Z., Guyomard D., Roué L., Lestriez B. 'A comparative study of poly-acrylic acid (PAA) and carboxymethyl cellulose (CMC) binders for Si-based electrodes'. *Electrochimica Acta*. 2017;**258**:453–66.

[135] Komaba S., Shimomura K., Yabuuchi N., Ozeki T., Yui H., Konno K. 'Study on polymer binders for high-capacity SiO negative electrode of Li-ion batteries'. *The Journal of Physical Chemistry C*. 2011;**115**(27):13487–95.

[136] Higgins T.M., Park S.-H., King P.J., *et al*. 'A commercial conducting polymer as both binder and conductive additive for silicon nanoparticle-based Lithium-ion battery negative electrodes'. *ACS Nano*. 2016;**10**(3):3702–13.

[137] Zhao H., Du A., Ling M., Battaglia V., Liu G. 'Conductive polymer binder for nano-silicon/graphite composite electrode in lithium-ion batteries towards a practical application'. *Electrochimica Acta*. 2016;**209**:159–62.

[138] Kusumoto Y., Fukui A., Nakamura H. 'Negative electrode for lithium secondary battery and lithium secondary battery'. Japanese Patent No. JP2001401286A, Filed Dec. 28; 2001.

[139] Krause L.J., Jensen L.D. 'Polyimide electrode binders'. U.S. Patent No. 7,972,725 B2, Filed Sept. 1; 2005.

[140] Choi N.-S., Yew K.H., Choi W.-U., Kim S.-S. 'Enhanced electrochemical properties of a Si-based anode using an electrochemically active polyamide imide binder'. *Journal of Power Sources*. 2008;**177**(2):590–4.

[141] Wilkes B.N., Brown Z.L., Krause L.J., Triemert M., Obrovac M.N. 'The electrochemical behavior of polyimide binders in Li and Na cells'. *Journal of the Electrochemical Society*. 2016;**163**(3):A364–72.

[142] Mazur S., Lugg P.S., Yarnitzky C. 'Electrochemistry of aromatic polyimides'. *Journal of the Electrochemical Society*. 1987;**134**(2):346–53.

[143] Yoon T., Chapman N., Nguyen C.C., Lucht B.L. 'Electrochemical reactivity of polyimide and feasibility as a conductive binder for silicon negative electrodes'. *Journal of Materials Science*. 2017;**52**(7):3613–21.

[144] Hatchard T.D., Bissonnette P., Obrovac M.N. 'Phenolic resin as an inexpensive high performance binder for Li-ion battery alloy negative electrodes'. *Journal of the Electrochemical Society*. 2016;**163**(9):A2035–9.

[145] Obrovac M.N. 'Si-alloy negative electrodes for Li-ion batteries'. *Current Opinion in Electrochemistry*. 2018;**9**:8–17.

[146] Hassan F.M., Chabot V., Elsayed A.R., Xiao X., Chen Z. 'Engineered Si electrode nanoarchitecture: a scalable postfabrication treatment for the production of next-generation Li-ion batteries'. *Nano Letters*. 2014;**14**(1):277–83.

[147] Zhang C., Yang F., Zhang D., *et al*. 'A binder-free Si-based anode for Li-ion batteries'. *RSC Advances*. 2015;**5**(21):15940–3.

[148] Hatchard T.D., Fielden R.A., Obrovac M.N. 'Sintered polymeric binders for Li-ion battery alloy anodes'. *Canadian Journal of Chemistry*. 2018;**96**(7):765–70.

[149] Koyama A., Yoshino A. 'Secondary battery negative electrode'. Japanese Patent Application No. JPH0574461A, Filed Sept. 13; 1991.
[150] Park J.-H., Hwang S.-S. 'Carboxymethyl cellulose-based binder material and lithium battery using the same'. Korean Patent Application No. KR100522698B1, filed Oct. 1; 2003.
[151] Li J., Lewis R.B., Dahn J.R. 'Sodium carboxymethyl cellulose'. *Electrochemical and Solid-State Letters*. 2007;**10**(2):A17–20.
[152] D.B L. 'Electrodes including polyacrylate binders and methods of making and using the same'. US Patent No. 7,875,388 B2, Filed Feb. 6; 2007.
[153] Magasinski A., Zdyrko B., Kovalenko I., *et al.* 'Toward efficient binders for Li-ion battery Si-based anodes: polyacrylic acid'. *ACS Applied Materials & Interfaces*. 2010;**2**(11):3004–10.
[154] Komaba S., Yabuuchi N., Ozeki T., *et al.* 'Comparative study of sodium polyacrylate and poly(vinylidene fluoride) as binders for high capacity Si–graphite composite negative electrodes in Li-ion batteries'. *The Journal of Physical Chemistry C*. 2012;**116**(1):1380–9.
[155] Hochgatterer N.S., Schweiger M.R., Koller S., *et al.* 'Silicon/graphite composite electrodes for high-capacity anodes: Influence of binder chemistry on cycling stability'. *Electrochemical and Solid-State Letters*. 2008;**11**(5):A76–80.
[156] Kovalenko I., Zdyrko B., Magasinski A., *et al.* 'A major constituent of brown algae for use in high-capacity Li-ion batteries'. *Science*. 2011;**334**(6052):75–9.
[157] Mochizuki T., Aoki S., Horiba T., *et al.* '"Natto" binder of poly-γ-glutamate enabling to enhance silicon/graphite composite electrode performance for lithium-ion batteries'. *ACS Sustainable Chemistry & Engineering*. 2017;**5**(7):6343–55.
[158] Nguyen C.C., Seo D.M., Chandrasiri K.W.D.K., Lucht B.L. 'Improved cycling performance of a Si nanoparticle anode utilizing citric acid as a surface-modifying agent'. *Langmuir*. 2017;**33**(37):9254–61.
[159] Komaba S., Ozeki T., Okushi K. 'Functional interface of polymer modified graphite anode'. *Journal of Power Sources*. 2009;**189**(1):197–203.
[160] Bridel J.-S., Azaïs T., Morcrette M., Tarascon J.-M., Larcher D. 'In situ observation and long-term reactivity of Si/C/CMC composites electrodes for Li-ion batteries'. *Journal of the Electrochemical Society*. 2011;**158**(6):A750–9.
[161] Buchholz F.L. (eds.) 'Preparation and structure of polyacrylates' in Brannon-Peppas L., Harland R.S. (eds.). *Absorbent Polymer Technology. Amsterdam: Elsevier Science*; 1990. pp. 23–44.
[162] Fujitani S., Yagi H., Sayama K., Yoshida T., Tarui H. 'New a-Si alloy thin film anode with self organized micro columnar structure'. 203rd Meeting of the Electrochemical Society, Paris, France 2003; Abstract 1152; 2003.

Chapter 2

Application of Zintl–Klemm rules to silicon-based LIB anodes

J. Vaughey[1], X. Li[1], B. Han[1], Y. Zhang[1], R. Uppulari[1], B. Key[1], F. Dogan[1], Saida S. Cora[2], and Niya Sa[2]

Silicon is an earth abundant element that has been at the core of numerous scientific advances ranging from microelectronics to photovoltaics to energy storage. In the area of energy storage it has been studied due to its combination of high theoretical capacity and low voltage that, when paired with high-capacity cathodes, is capable of generating a high-energy-density system capable of meeting next-generation energy-storage needs [1–3]. With this great promise comes important limitations that have yet to allow widespread introduction of silicon into lithium ion cells. These issues include the large (~350 percent) volume expansion that occurs going from elemental silicon to $Li_{15}Si_4$ (the room temperature endpoint in an electrochemical lithiation), electrolyte reactivity, solid electrolyte interphase (SEI) layer stability, the role of silicon purity, and parasitic side reactions that irreversibly remove active lithium from the cathode [4, 5]. This chapter addresses recent progress in understanding the reactivity of the silicon at various states of charge and how this reactivity manifests itself in an electrochemical cell and in the cycling stability of silicon and lithiated silicides.

2.1 Energy storage

Silicon is a main group element with the electronic configuration $1s^2 2s^2 2p^8 3s^2 3p^2$. In nature, silicon-containing minerals mainly exist as silica or silicates (or related species) that are created by stripping the four outer shell electrons to create the Si^{+4} cation. The compounds are among the most common in and on the Earth and are distributed across the globe [6]. Elemental silicon is generated from these starting materials in several ways but among the most common is the reduction of silica by

[1]Electrochemical Energy Storage Group, Chemical Sciences and Engineering Argonne National Laboratory Lemont, USA

[2]Department of Chemistry, University of Massachusetts Boston 100 William T Morrissey Blvd Boston, USA

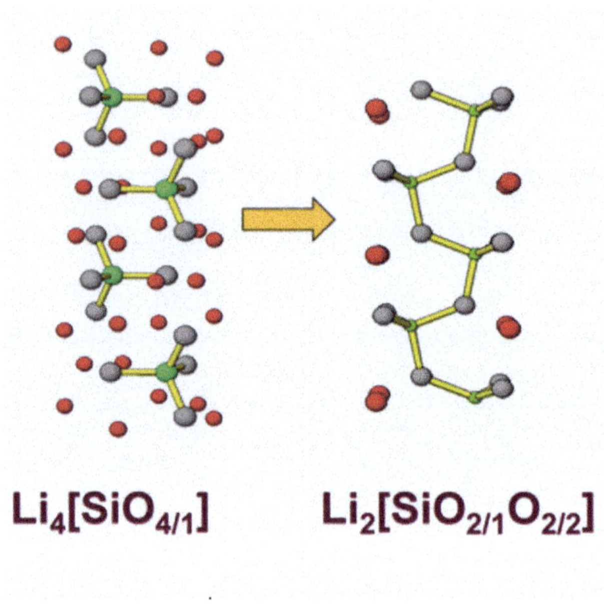

Figure 2.1 *The structural relationship between the anionic species in surface passivation materials: (a) silicate anion in Li_4SiO_4, (b) the condensed silicate chain in Li_2SiO_3. Produced by the protonation and dehydration of the species in (a)*

carbon (to form CO_2) at ~1 700 °C [7]. The silicon isolated can act as a feedstock for other purposes (i.e. steel) or be further processed and purified for more specialized applications [8]. Further purification is necessary for most applications where strict control over electronic conductivity is required [8]. In many of these cases, the silicon and its nanometer-scale passivation coating of silica are the product. For energy storage, the scale of usage and the various types of manufacturing lines and workspace environments it is used in or for necessitate passivated silicon. Significant effort has been placed on understanding how this passivation layer interacts with the charged anode, electrolytes, and the SEI layer [9–11].

In an electrochemical cell, the thin silica passivation coating has been found to be electrochemically stable but is chemically active [12]. Numerous published studies reporting results performed under various conditions have highlighted that rather than reduction to silicon with formation of lithia, the silica gets chemically converted to a series of lithium silicates derived from the gradual condensation of the kinetic product $Li_4[SiO_4]$ to the thermodynamic product Li_2SiO_3 [9–13]. Using Magic Angle Spinning-Nuclear Magnetic Resonance (MAS-NMR) ^{29}Si spectroscopy, the intermediate species condensation products, e.g. pyrosilicates, can be seen in the spectra as they gradually change concentrations moving toward the more condensed phases after exposure to low levels of water [13]. The condensation pathway is shown in Figure 2.1. Further spectroscopic studies on these interfacial species indicate that they gradually diffuse away from the interface into the surrounding SEI layer and either become too dilute to detect or possibly react with SEI components to form new silicate salts or silicon-containing species [13]. Within the cell, chemical reactions between the elemental silicon surface and transient water (or other oxygen containing species) may reform various silicates, but these should be pulled into the same mechanism of lithiation, condensation, and dispersion as the ones generated initially from the initial passivation layer [14]. Under the conditions used, the product of silicon oxidation is SiO_2 rather than the postulated suboxide "SiO" or "SiO_x" [15]. As a solid at room temperature and pressure, "SiO" manifests itself as a nanoscale mixture of SiO_2 and Si [16]. However, while silicates are redox inactive in an electrochemical cell, the underlying elemental silicon is a commonly studied and highly sought-after anode material for lithium ion batteries [1–3].

2.2 Zintl–Klemm rules

Based on the electronic structure noted in the begining of Section 2.1, most common silicon materials are derived from oxidation of silicon to a Si^{+4} cation; however, the electrode properties of silicon are derived from its ability to gradually fill the empty 3p orbitals and reduce the silicon to a silicide species. These reduced species, using crystallographic, theoretical, and spectroscopic analyses, have generated increased interest as synthesis and characterization methods have evolved to handle these very reactive species [17, 18]. The lithium-silicon phase diagram, shown in Figure 2.2, has six distinct phases that contain either unique $[Si]_x$ clusters or mixtures of constituent cluster species [17, 18]. The structure of many of these species can be understood through the application of Zintl–Klemm electron counting rules [17–20]. Figure 2.3 highlights the clusters seen in the various known Li–Si binary phases [17, 18, 21, 22]. The theoretical understanding of these materials, derived from Zintl–Klemm concepts, is that as the silicon is reduced, the number of Si–Si

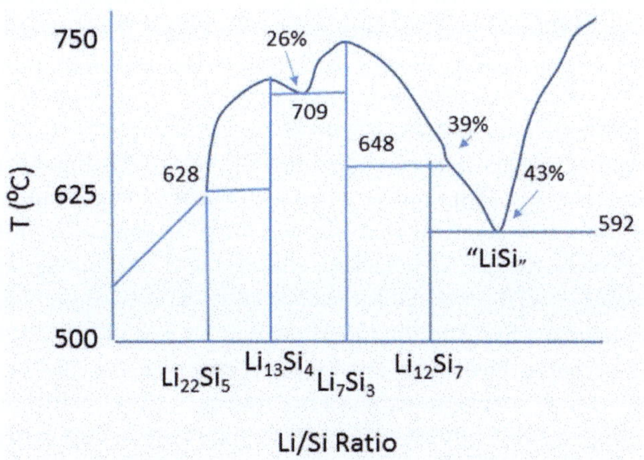

Figure 2.2 An approximation of the lithium–silicon binary phase diagram

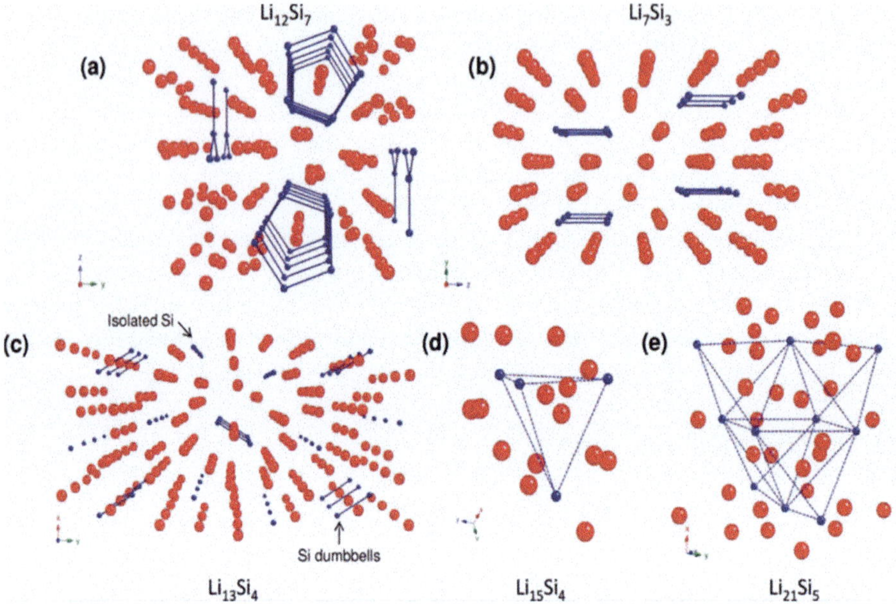

Figure 2.3 Structures of various binary lithium silicide phases [21]

bonds are broken and are replaced by silicon–anion-based lone pairs, as is commonly observed in the larger main group cations, i.e. $Sn^{+2} \rightarrow Sn^{+4}$, as the lone pair effect. To clarify, this reduced silicon chemistry is not unique to silicon but occurs for all of the main group elements [23]. The new silicide structures are built from various stable silicon clusters that meet the electron counting rules and crystallographically pack together with the added lithium cations to form the observed phases (synthesized at high temperatures, see Figure 2.2). At room temperature, the stoichiometries of a few of the electrochemically accessible lithium silicides are slightly different, reflecting the equilibriums present at room temperature. Notably LiSi and $Li_{22}Si_5$ are not observed, while $Li_{15}Si_4$ is now the endmember [24]. It should be noted that since this phase contains four fully reduced Si^{-4} anions but only 15 Li cations, it is technically electron-rich and not a closed shell species. MAS-NMR spectroscopic studies by Key *et al.*, focused on room temperature in situ lithium insertion reactions into this phase, noted the formation of a new metastable phase on their experimental timescale that cannot be quenched but was consistent with partially filling the cation vacancy in the structure [21, 22]. The spectra is shown in Figure 2.4. Other recent studies by Fassler *et al.* have noted the increased instability of these 15-4 phases by substitution of a divalent cation, such as Mg or Zn, for a Li into the structure to form, for example $Li_{14}MgSi_5$, an electron precise species [25].

For the intermediate species between Si and $Li_{15}Si_4$, the coordination of the silicon by other silicon atoms (or other main group elements in the case of mixed anion materials, e.g. Li[AlSi], $Na[GaSn_2]$) can be equated with the anions formal "oxidation state," which translates to state of charge in the electrochemical cell [17, 18, 26, 27].

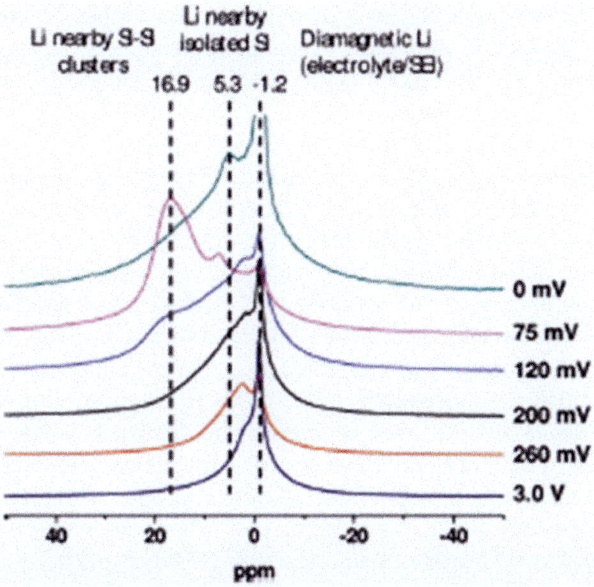

Figure 2.4 *Ex situ ^{7}Li MAS NMR spectra of battery samples stopped during the second discharge. Dashed lines indicate resonances with known lithium assignments [22].*

The forms of these anions in the structure are typically related to other elements with similar electron counts [17–20]. As an example, the Zintl phase NaSi is more accurately described as $Na_4[Si_4]$ as the structure is built of discreet $[Si_4]^{4-}$ packed anions. These discreet $[Si_4]^{4-}$ are isoelectronic to white phosphorus, $[P_4]$, an allotrope of elemental phosphorus. The structure of NaSi is shown in Figure 2.5 [28].

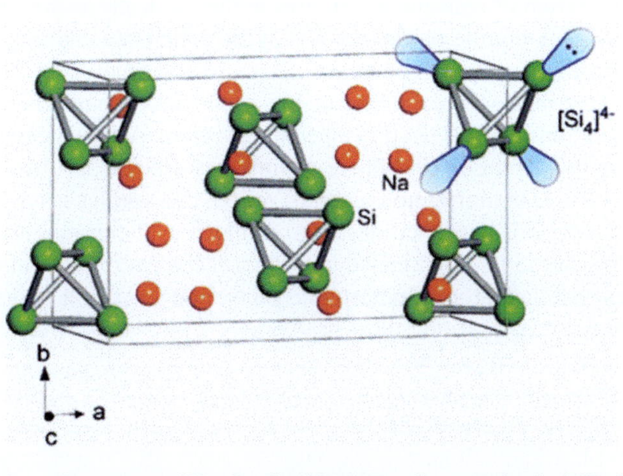

Figure 2.5 *Structure of Na_4Si_4, highlighting the tetrahedral $[Si_4]^{4-}$ cluster*

Within the lithium-silicon family of compounds, sometimes erroneously referred to as "alloys," several discrete clusters including pentamers, D_{3h} stars, and dimers have been reported that exist as a function of formal oxidation state (state of charge), packing requirements, and temperature [17, 18]. Most of the species are not observed in corresponding elemental structures where tetrahedra, helices, dimers, and isolated atoms are commonly observed [17–20]. In the field of energy storage, these various materials are created as a function of lithiation (state of charge) and in situ studies have shown that the species present reflect the weighted average of the possible species [21, 22]. Because the cells are typically cycled at room temperature, the kinetics of cluster formation are in part controlled by local issues including availability of reactive species and issues associated with the diffusion of highly charged species across the silicide electrode particle. This aspect of cell cycling has been studied and various mechanisms proposed [29]. In the MAS-NMR spectroscopic studies noted earlier, in situ cells of Li–Si were examined and compared to spectra of baseline samples of known lithium silicides, see Figure 2.4. At room temperature, the speciation of silicon anions changes as a function of state of charge and is reflective of low levels of homogeneity and interconversion [30]. These observations are manifested in several well-known aspects of silicon anode electrochemistry, notably the rapid formation of amorphous silicon phases in the electrode and the stability and crystallization of $Li_{15}Si_4$ at low voltage [24, 31]. The initial observation can be understood as related to the relatively large number of bonds that must be broken to form the Zintl cluster phases and the availability of a kinetic pathway that may involve a second (or third) reduction of the silicon atom before additional elemental silicon atoms are initially reduced. This lack of sample homogenity has been noted in ^{29}Si NMR studies the lithiation of silicon and noted extensive inhomogeneity on the initial cycles in the silicon species [30]. Under certain circumstances fully reduced silicon (Si^{-4}) was identified at the surface while elemental neutral silicon was detected in the bulk, with a gradient of partially reduced silicides between them. This state of charge-driven mixture of silicon anions and lack of long range order also lends itself to the formation of amorphous features in the electrode as the combination of clusters, and cations will be poorly packed in the solid state and may lack enough homogeneity to create a crystalline lattice. This situation changes at complete lithiation when the electrode eventually becomes a simpler mixture of lithium cations and fully reduced Si^{-4} anions [24]. Achieving some measure of homogeneity, the silicon components in the electrode crystallize to form the silicide salt $Li_{15}Si_4$ around 10 mV vs Li/Li^+ [24]. Due to this rapid crystallization, this species is typically avoided under conditions of cell testing because the amorphous silicide species is associated with enhancing cell failure possibly due to loss of the isotropic volume expansion characteristics postulated for the amorphous phase and formation of distinct crystallites and grain boundaries [31].

2.3 Structural considerations

The formation of these molecular cluster phases is in contrast to the commercially used graphitic carbon-based anodes. In those systems, the lithium intercalation is in conjunction with reduction of the graphitic layers but is not associated with bond breaking of the carbon host network, with formation of the charged anode material LiC_6 [32]. This maintenance of connectivity is advantageous for many purposes including maintaining electronic conductivity, ionic diffusion pathways due to the host lattice retaining its structure, and minimizing volume expansion as a function of state of charge. In the case of lithium silicide anodes, the necessity to maintain charge neutrality is done by silicon–silicon bond breaking and bond formation [17–20]. This charge compensation mechanism is thus tied to the need for these charged clusters to diffuse and shift Si–Si bonds to form the next species in the phase diagram [31]. Especially at room temperature, the diffusion can be hindered and slow as transient intermediate phases form in addition to the necessity for highly charged species to come together in bond-forming reactions on discharge [24]. As noted earlier, evidence for non-homogenity in the electrode has been reported with this slow diffusion mechanism as one of the probable causes [30]. With a diffusion-limited mechanism, increasing the temperature is useful, and reports indicate better kinetics and a more homogeneous local electrode structure as the temperature is raised [33]. Observationally, from Figure 2.2, in addition to faster interconversion of the clusters as a function of state of charge, new packing arrangements are noted that produce alternative phases not seen at room temperature, although they are built of the same subunits. One of the more notable examples is the isolation of $Li_{22}Si_5$ ($Li_{4.2}Si$), an overlithiated metallic Zintl phase at elevated temperatures. Since Zintl phases are driven to be closed shell systems, they typically are brittle (salt-like) and are poor electronic conductors. These highly lithiated and over-reduced materials, like $Li_{22}Si_5$, show metallic conductivity and it has been postulated that this results from the inclusion of un-ionized lithium atoms into lattice vacancies for better lattice packing adding unbound electrons to the material [34]. In contrast, as noted earlier, the end-member species seen electrochemically is the formally lithium deficient phase $Li_{15}Si_4$ ($Li_{3.75}Si$). While they contain the same general building units, the stoichiometries and electronic properties are different.

2.4 Binary Zintl phase reactivity

While these fully reduced species have been isolated and studied, a key aspect of their place in the electrochemical cell environment is their lack of a closed shell configuration [17, 18]. This makes these species reactive and unstable in the presence of organic species, including binders and electrolyte components. A recent study by Han *et al.* investigated the reactivity of various lithium silicides, notably the intermediate species Li_7Si_3, and showed using various methods that the formally Si^{-2} anion degraded the polyvinyldifluoride (PVDF) binder and was formally oxidized in the presence of many common electrolyte solvents [35]. In this case, the team was

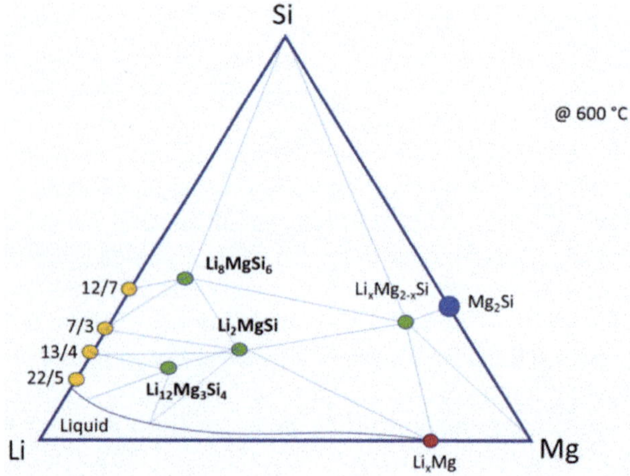

Figure 2.6 *An approximation of the ternary Li-Mg-Si phase diagram based on reported phases*

able to identify $Li_{12}Si_7$, the next most oxidized species in X-ray diffraction data and MAS-NMR spectroscopic studies. This is consistent with simple electron donation (reduction) to the solvent balanced by silicon oxidation. More reduced species, e.g. $Li_{15}Si_4$, are also noted to be very reactive as one might expect from the presence of the ionically large and isolated Si^{-4} anion.

2.5 Ternary Zintl phases

While these studies by Han *et al.* have highlighted the role of these silicide clusters in the electrochemical cell environment, the study was followed up with various attempts to make the charged species ($Li_{15}Si_4$) electron precise by addition of Mg cations to the composition. This is consistent with Fassler's observation that the substituted electron precise species $Li_{14}MgSi_4$ formed rapidly and looked to be much more stable under storage conditions [25]. Structurally the $Li_{14}MgSi_4$ phase is a simple packed structure and exists as a solid solution with the phase Li_2MgSi, also a Si^{-4} salt. The solid solution is written as $Li_{14.5-x}Mg_{0.5+x/2}Si_4$. In a follow-up study, Han *et al.* evaluated points in this solid solution for their reactivity against common electrolyte solvents and binders [36]. Specifically looking at the endmember composition, Li_2MgSi, no reactivity against common organic cell materials was observed and the Si^{-4} anion was not obviously oxidized based on spectroscopic studies. Looking at the ternary Li-Mg-Si phase diagram (Figure 2.6) the addition of magnesium to the low voltage 15-4 structure appears for the most part, to eliminate the ability to convert to a different reduced silicide cluster species. In the phase diagram, but for the presence of Li_8MgSi_6, the presence of magnesium inhibits electron transfer that allows formation of intermediate silicon cluster phases. The reactivity data are consistent that without the availability of formally Si^{-3}, Si^{-2}, or Si^{-1} phases (as seen in

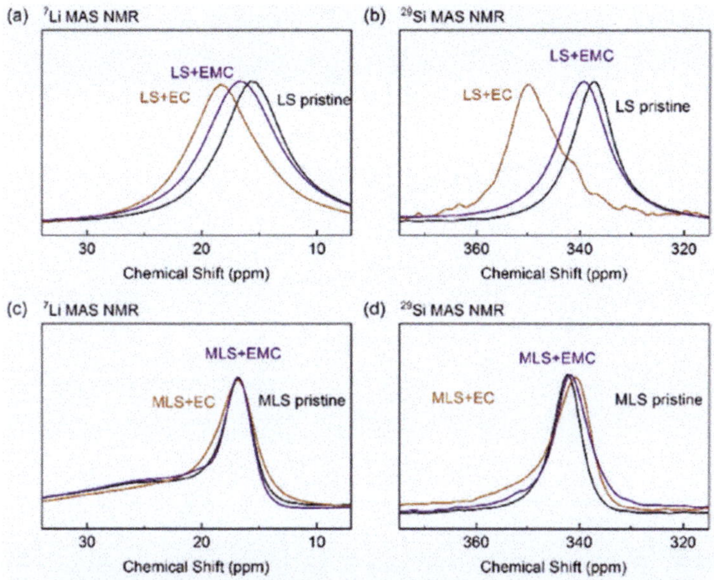

Figure 2.7 *(a) 7Li and (b) 29Si MAS NMR spectra of the pristine LS model compound and its mixtures with EMC and EC, respectively, with the volume ratio of 1:1. (c) 7Li and (d) 29Si MAS NMR spectra of the pristine MLS model compound and its mixtures with EMC and EC, respectively, with the volume ratio of 1:1. The mixture of LS + EC and MLS + EC has been heated up to 325 K and then cooled back to 298 K [37].*

the Li–Si binary system), the four electron oxidation to elemental silicon is too high a barrier and the magnesiated material is stable to the electrochemical cell components. A comparison of the reactivity of Li_7Si_3 and Li_2MgSi with electrolytes using MAS-NMR spectroscopy is shown in Figure 2.7.

In an electrochemical cell, this observation was experimentally recreated by the addition of small amounts of the salt Mg(TFSI)2 to the commonly used 1.2M $LiPF_6$ EMC/EC 3:7 lithium-ion battery (LIB) electrolyte [35]. Subsequent transmission electron microscopy (TEM), energy dispersive spectroscopy (EDS), MAS-NMR, and X-ray photoelectron spectroscopy (XPS) studies noted the presence at the surface of Mg^{+2} at an approximate stoichiometric ratio of $Li_{14.5}Mg_{0.5}Si_4$. With this ternary salt formation at the surface, baseline Si electrode samples were noted to have extended cycle and calendar life compared to electrolytes without the additive, see Figure 2.8. It was also noted that other divalent species, e.g. Ca^{+2}, had a similar effect on surface speciation.

Through an understanding of electron-transfer processes for the reduced silicon species, additives can be designed that alter the progression of electrode phases, moving the active electrodes from those available using the various redox active

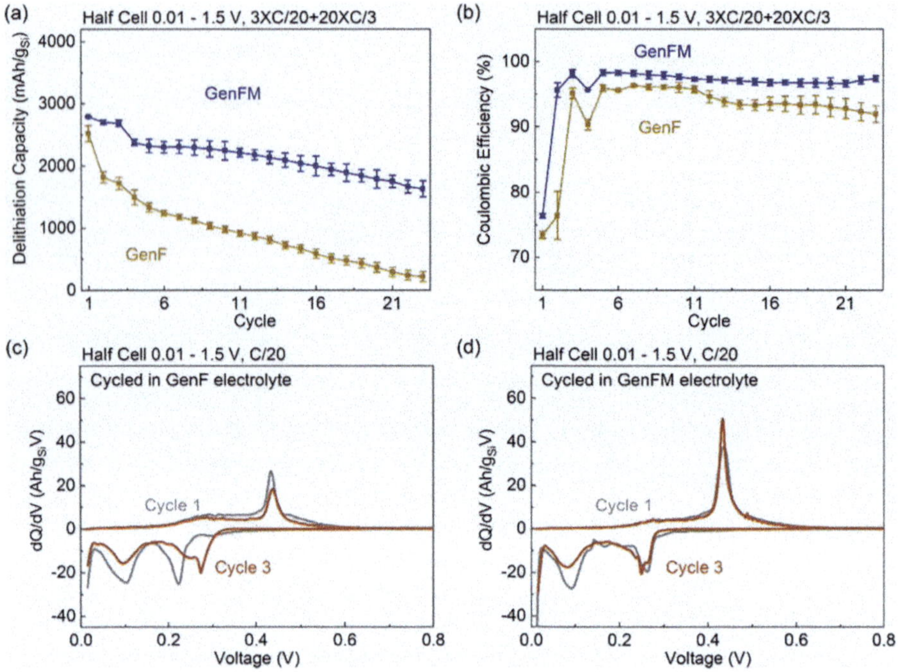

Figure 2.8 *Half-cell electrochemical test results on Si electrodes. The GenF and GenFM electrolytes were used in the electrochemical tests. The delithiation capacities and coulombic efficiencies are shown in panels (a) and (b), respectively. The differential capacities at the first and third formation cycles in GenF and GenFM electrolytes are shown in panels (c) and (d), respectively. The cells were cycled between 0.01 and 1.5 V, first at the rate of C/20 for three formation cycles, then at the rate of C/3 for aging cycles. The capacities are normalized by the mass of silicon. Error bars represent the standard deviations of at least three measurements for each sample [37].*

lithium silicides to the redox inactive ternary lithium magnesium silicides. In various studies, simple salts that can be added to the electrolyte translate to the formation of new surface ternary phases that, while redox inactive, are good lithium-ion conductors [36]. Alternative methods to achieve the desired phases have been attempted and studied, including synthesis of Mg_2Si-coated Si, and however the needed stoichiometric control of the ternary phases has not been achieved to date [38]. For the in situ formed phases, studies continue on the persistence of the desired ternary phases, their formation mechanism, and interactions with the SEI layer.

Table 2.1 Equilibrium emf for Si-based Zintl phase compounds at 415 °C and room temperature (RT)

	$Si/Li_{12}Si_7$ (V)	$Li_{12}Si_7/Li_7Si_3$ (V)	$Li_7Si_3/Li_{13}Si_4$ (V)	$Li_{13}Si_4/Li_{21}Si_5$ (V)
415 °C	332	288	158	44
Room temperature	582	520	428	300

2.6 Electrochemical characterization of the Si-based Zintl phase compounds

The electrochemical reaction of lithium with Si and the Si-based Zintl phase compounds is a complicated process, and it involves the conventional nucleation and growth of a new phase, amorphization, in addition to a charge transfer process, volume expansion, deformation, and grain refinement [1, 39]. The equilibrium electromotive force (emf) estimated from the formation energy for the Si-based Zintl phase compounds at 415 °C and room temperature are shown in Table 2.1 as reported by Wen and Huggins [40].

Crystalline silicon transforms to an amorphous Li_xSi (x≅3) above 60 mV during the first electrochemical lithiation cycle, coincidently around the composition where most of the silicon-silicon bonding has been eliminated in favor of discrete clusters and anions. The first electrochemical lithiation cycle exhibits a single flat plateau at ~0.1 V indicating a single two-phase addition reaction with a distinct plateau, rather than the multiphase reactions as predicted by the high-temperature equilibrium Li–Si phase diagram [39, 41]. For the electrochemical lithiation down to 60 mV, the amorphous Li_xSi (x≅3) is further reduced by lithium addition until the mixture reaches the composition $Li_{15}Si_4$. This composition represents y a relative uniform mixture of lithium cations and isolated silicon anions and is noted to rapidly crystallize with elimination of anion disorder . Such a phase transformation (an amorphous to crystalline phase) can proceed via a non-diffusional process with no change in composition. Once the phase transformation takes place, the entire amorphous particle will transform abruptly into a crystalline phase through the short-range atom rearrangement. The electrochemical delithiation depends on the cut-off lithiation potential. For fully lithiated $Li_{15}Si_4$ Zintl compounds, the delithiation electrochemical profile exhibits a single plateau at ~0.4 V ascribed from a two-phase reaction from the crystalline $Li_{15}Si_4$ to amorphous Li_xSi (x≅3), followed by a solid solution reaction from amorphous Li_xSi to amorphous silicon without a phase change [42, 43]. The over-potential is expected with an understanding of the mechanism that requires two $[Si]^{-4}$ anions to overcome an electrostatic barrier to dimerize to the next stage Zintl anion, $[Si_2]^{6-}$. It is notable that typical electrochemical lithiation or delithiation for silicon and the Si-based Zintl phase anodes exhibits amorphization of silicon. Such amorphization is due to the now nonhomogeneous mixture of silicon anions and the lowered free energy of the metastable amorphous Li–Si phase compared to the equilibrium crystalline phases. For instance, the initial lithiation at 0.1 V observed at RT for silicon experimentally is lower than the predicted lithiation voltage due to the large nucleation energy battier for the crystalline phase formation, while the formation of the disordered structure at a low packing density of the amorphous phase is more favorable. Another feature that is particularly interesting is the interfacial energy available for nucleation and growth of a new phase in this system, an important consideration for nanoscale materials [44].

The electrochemical cycling profiles for silicon particles or silicon films are closely related to particle size, depth of lithiation, and surface roughness of the film [45, 46]. Consequently, lithium insertion and extraction potentials are departed from the theoretical equilibrium predictions at room temperature. For instance, silicon particles at tens of micrometers or larger show a flat plateau for lithiation and delithiation. For silicon particles under 200 nm in diameter, a sloping plateau rather than a flat plateau with a smooth decay is seen for the lithiation electrochemical profile including the fully lithiated states [36]. Further reducing the particle under 20 nm results in a disappearance of the plateau at the first lithiation cycle with a sloping electrochemical cycling profile. The dQ/dV profile gives prominent lithiation peaks at 60 and 250 mV, and delithiation peaks at 300 and 500 mV. Structural changes corresponding to these voltages (from dQ/dV) are not fully understood as the small number of silicon atoms and the opportunity to reach a local equilibrium may result in a short-range order within the amorphous Li_xSi sample. Thicker silicon films exhibit flat plateau for delithiation while thinner film shows a sloping line profile [46].

2.7 Conclusions

In the field of silicon-based LIB electrodes, the fundamental understanding of the underlying energy storage mechanism using Zintl–Klemm electron counting rules has resulted in a new approach to study these technologically important materials. By identifying the electron imprecise nature of the charged species and their reactivity within the electrochemical cell, strategies to mitigate this reactivity resulted in the in situ synthesis of nanometer scale coatings that, while having little effect on the volume expansion of the electrode, do significantly reduce the reactivity of the surface to its environment. This advantage helps significantly reduce side reactions as the charged electrode reacts with the electrolyte over time, seen in both calendar life testing and cycle life extensions. In combination with other methodologies, e.g. fluoroethylene carbonate addition, nanostructured electrodes, advances in the utility and commercialization of silicon-based materials will require advances along multiple lines of attack and new approaches to solving identified problems.

Acknowledgements

The authors thank Brian Cunningham, Steven Boyd, and David Howell from the Vehicle Technologies Program, at the U.S. Department of Energy, Office of Energy Efficiency and Renewable Energy for their support. The submitted manuscript has been created by UChicago Argonne, LLC, Operator of Argonne National Laboratory ("Argonne"). Argonne, a US Department of Energy Office of Science laboratory, is operated under Contract No. DE-AC02-06CH11357. Financial support from research faculty start-up funds, the Joseph P. Healey Research Grant, and the Endowed Faculty Career Development Award from the University of Massachusetts Boston is acknowledged.

References

[1] Obrovac M.N., Chevrier V.L. 'Alloy negative electrodes for Li-ion batteries'. *Chemical Reviews*. 2014;**114**(23):11444–502.
[2] Obrovac M.N. 'Si-alloy negative electrodes for Li-ion batteries'. *Current Opinion in Electrochemistry*. 2018;**9**:8–17.
[3] Lee J.K., Oh C., Kim N., Hwang J.-Y., Sun Y.-K. 'Rational design of silicon-based composites for high-energy storage devices'. *Journal of Materials Chemistry A*. 2016;**4**(15):5366–84.
[4] Cao C., Shyam B., Wang J., Toney M.F., Steinrück H.-G. 'Shedding X-ray light on the interfacial electrochemistry of silicon anodes for Li-ion batteries'. *Accounts of Chemical Research*. 2019;**52**(9):2673–83.
[5] Huggins R.A., Nix W.D. 'Decrepitation model for capacity loss during cycling of alloys in rechargeable electrochemical systems'. *Ionics*. 2000;**6**(1-2):57–63.
[6] Holland T. 'Dependence of entropy on volume for silicate and oxide minerals – a review and a predictive model'. *American Mineralogist*. 1989;**74**:5–13.
[7] Xing Z., Lu J., Ji X. 'A brief review of metallothermic reduction reactions for materials preparation'. *Small Methods*. 2018;**2**(12):1800062.
[8] Jalali B., Fathpour S. 'Silicon photonics'. *Journal of Lightwave Technology*. 2006;**24**(12):4600–15.
[9] Sivonxay E., Aykol M., Persson K.A. 'The lithiation process and Li diffusion in amorphous SiO2 and Si from first-principles'. *Electrochimica Acta*. 2020;**331**:135344.
[10] Schroder K.W., Dylla A.G., Harris S.J., Webb L.J., Stevenson K.J. 'Role of surface oxides in the formation of solid-electrolyte interphases at silicon electrodes for lithium-ion batteries'. *ACS Applied Materials & Interfaces*. 2014;**6**(23):21510–24.
[11] Coyle J.E., Brumbach M.T., Veith G.M., Apblett C.A. 'Investigating the chemical reactivity of lithium silicate model SEI layers'. *The Journal of Physical Chemistry C*. 2020;**124**(15):8153–61.

[12] Philippe B., Dedryvère R., Allouche J., *et al.* 'Nanosilicon electrodes for lithium-ion batteries: interfacial mechanisms studied by hard and soft X-ray photoelectron spectroscopy'. *Chemistry of Materials*. 2012;**24**(6):1107–15.

[13] Hubaud A.A., Yang Z., Schroeder D.J., Dogan F., Trahey L., Vaughey J.T. 'Interfacial study of the role of SiO2 on Si anodes using electrochemical quartz crystal microbalance'. *Journal of Power Sources*. 2015;**282**:639–44.

[14] Abel P.R., Lin Y.-M., Celio H., Heller A., Mullins C.B. 'Improving the stability of nanostructured silicon thin film lithium-ion battery anodes through their controlled oxidation'. *ACS Nano*. 2012;**6**(3):2506–16.

[15] Gunduz D.C., Tankut A., Sedani S., Karaman M. 'Crystallization and phase separation mechanism of silicon oxide thin films fabricated via e-beam evaporation of silicon monoxide'. *Physica Status Solidi*. 2015;**12**(9–11):1229–35.

[16] Hirata A., Kohara S., Asada T., *et al.* 'Atomic-scale disproportionation in amorphous silicon monoxide'. *Nature Communications*. 2016;**7**(1).

[17] Nesper R. 'Structure and chemical bonding in zintl-phases containing lithium'. *Progress in Solid State Chemistry*. 1990;**20**(1):1–45.

[18] van der Marel C., Vinke G.J.B., van der Lugt W., Vinke W., Lugt vander. 'The phase diagram of the system lithium-silicon'. *Solid State Communications*. 1985;**54**(11):917–9.

[19] Corbett J.D. 'Polyatomic Zintl anions of the post-transition elements'. *Chemical Reviews*. 1985;**85**(5):383–97.

[20] Corbett J.D. 'Polyanionic clusters and networks of the early P-element metals in the solid state: beyond the Zintl boundary'. *Angewandte Chemie International Edition*. 2000;**39**(4):670–90.

[21] Key B., Bhattacharyya R., Morcrette M., Seznéc V., Tarascon J.-M., Grey C.P. 'Real-time NMR investigations of structural changes in silicon electrodes for lithium-ion batteries'. *Journal of the American Chemical Society*. 2009;**131**(26):9239–49.

[22] Key B., Morcrette M., Tarascon J.-M., Grey C.P. 'Pair distribution function analysis and solid state NMR studies of silicon electrodes for lithium ion batteries: understanding the (de)lithiation mechanisms'. *Journal of the American Chemical Society*. 2011;**133**(3):503–12.

[23] Wang F., Miller G.J. 'Revisiting the Zintl-Klemm concept: alkali metal trielides'. *Inorganic Chemistry*. 2011;**50**(16):7625–36.

[24] Obrovac M.N., Christensen L. 'Structural changes in silicon anodes during lithium insertion/extraction'. *Electrochemical and Solid-State Letters*. 2004;**7**(5):A93–6.

[25] Baran V., van Wüllen L., Fässler T.F. 'Substitution of lithium for magnesium, zinc, and aluminum in Li15 Si4: crystal structures, thermodynamic properties, as well as (6) Li and (7) Li NMR spectroscopy of Li15 Si4 and Li15-x Mx Si4 (M=Mg, Zn, and Al)'. *Chemistry*. 2016;**22**(19):6598–609.

[26] Fleischauer M.D., Obrovac M.N., Dahn J.R. 'Al–Si thin-film negative electrodes for Li-Ion batteries'. *Journal of The Electrochemical Society*. 2008;**155**(11):A851–4.

[27] Vaughey J.T., Corbett J.D. 'Synthesis and structure of $NaGaSn_2$: a Zintl phase with a helical framework structure'. *Journal of the American Chemical Society*. 1996;**118**(48):12098–103.
[28] Yamanaka S. 'Silicon clathrates and carbon analogs: high pressure synthesis, structure, and superconductivity'. *Dalton Transactions*. 2010;**39**(8):1901–15.
[29] John Wen C., Huggins R.A. 'Chemical diffusion in intermediate phases in the lithium silicon system'. *Journal of Solid State Chemistry*. 1981;**37**:271–8.
[30] Oumellal Y., Delpuech N., Mazouzi D., *et al.* 'The failure mechanism of nano-sized Si-based negative electrodes for lithium ion batteries'. *Journal of Materials Chemistry*. 2011;**21**(17):6201.
[31] Netz A., Huggins R.A. 'Amorphous silicon formed in situ as negative electrode reactant in lithium cells'. *Solid State Ionics*. 2004;**175**(1-4):215–19.
[32] Persson K., Sethuraman V.A., Hardwick L.J., *et al.* 'Lithium diffusion in graphitic carbon'. *The Journal of Physical Chemistry Letters*. 2010;**1**(8):1176–80.
[33] Piernas-Muñoz M.J., Trask S.E., Dunlop A.R., Lee E., Bloom I. 'Effect of temperature on silicon-based anodes for lithium-ion batteries'. *Journal of Power Sources*. 2019;**441**:227080–9.
[34] Lupu C., Mao J.-G., Rabalais J.W., Guloy A.M., Richardson J.W. 'X-ray and neutron diffraction studies on Li4.4Sn'. *Inorganic Chemistry*. 2003;**42**(12):3765–71.
[35] Han B., Piernas-Muñoz M.J., Dogan F., *et al.* 'Probing the reaction between PVDF and LiPAA vs Li_7Si_3: investigation of binder stability for Si anodes'. *Journal of the Electrochemical Society*. 2019;**166**(12):A2396–402.
[36] Dogan F., Sanjeewa L.D., Hwu S.-J., Vaughey J.T. 'Electrodeposited copper foams as substrates for thin film silicon electrodes'. *Solid State Ionics*. 2016;**288**:204–6.
[37] Han B., Liao C., Dogan F., *et al.* 'Using mixed salt electrolytes to stabilize silicon anodes for lithium-ion batteries via in situ formation of Li-M-Si ternaries (M = Mg, Zn, al, Ca)'. *ACS Applied Materials & Interfaces*. 2019;**11**(33):29780–90.
[38] Song S.W., Striebel K.A., Reade R.P., Roberts G.A., Cairns E.J. 'Electrochemical studies of nanoncrystalline Mg2Si thin film electrodes prepared by pulsed laser deposition'. *Journal of the Electrochemical Society*. 2003;**150**(1):A121–7.
[39] Zhang W.-J. 'Lithium insertion/extraction mechanism in alloy anodes for lithium-ion batteries'. *Journal of Power Sources*. 2011;**196**(3):877–85.
[40] Wen C.J., Huggins R.A. 'Chemical diffusion in intermediate phases in the lithium-silicon system'. *Journal of Solid State Chemistry*. 1981;**37**(3):271–8.
[41] Jin Y., Zhu B., Lu Z., Liu N., Zhu J. 'Challenges and recent progress in the development of Si anodes for lithium-ion battery'. *Advanced Energy Materials*. 2017;**7**(23):1700715.
[42] Li J., Dahn J.R. 'An in situ X-ray diffraction study of the reaction of Li with crystalline Si'. *Journal of the Electrochemical Society*. 2007;**154**(3):A156.
[43] Chevrier V.L., Dahn J.R. 'First principles model of amorphous silicon lithiation'. *Journal of the Electrochemical Society*. 2009;**156**(6):A454–8.

[44] Limthongkul P., Jang Y.-I., Dudney N.J., Chiang Y.-M. 'Electrochemically-driven solid-state amorphization in lithium–metal anodes'. *Journal of Power Sources*. 2003;**119–121**:604–9.
[45] Hatchard T.D., Dahn J.R. 'In situ XRD and electrochemical study of the reaction of lithium with amorphous silicon'. *Journal of the Electrochemical Society*. 2004;**151**(6):A838–42.
[46] Kim J.-B., Lim S.-H., Lee S.-M. 'Structural change in Si phase of Fe/Si multilayer thin-film anodes during Li insertion/extraction reaction'. *Journal of the Electrochemical Society*. 2006;**153**(3):A455–8.

Chapter 3

Electrochemistry of silicon

Ji Chen[1] and Chunsheng Wang[1]

3.1 Introduction

Among the many alloying anode candidates for lithium-ion batteries (LIBs), Si has been regarded as one of the most prospective anodes because of its outstanding specific capacity, easy availability in the earth's crust, and the well-developed manufacturing techniques from industry. Si anode has been studied by researchers for more than three decades [1, 2]. At room temperature, the lithiation of Si is electrochemically feature-rich and dependent on many factors including but not limited to crystalline structure, particle size, and electrode structure. At high temperatures around 400 °C, electrochemical studies demonstrated the formation of binary Li–Si crystalline phases, which basically matched the Li–Si phase diagram [2, 3]. As reported by Wilson and Dahn [4], room-temperature Si electrodes exhibited quite different voltage profiles from the high-temperature results. In this chapter, the electrochemistry of Si, as well as the associated phase changes and atomic structural and environmental evolution, will be discussed in detail.

3.2 High-temperature Li–Si phases

At elevated temperatures, the formation of $Li_{12}Si_7$, Li_7Si_3, $Li_{13}Si_4$, and $Li_{21}Si_5$ (Figure 3.1a–e) has been confirmed by detailed crystallography studies [2]. Electrochemically, lithium reacts with silicon following the equilibrium binary Li–Si phase diagram (verified by high-temperature coulometric titration technique with "Al, Li-Al(s)|LiC1-KCl(eut.)|Li-Si(s)" cell), with multiple two-phase structural transitions, evidenced by distinct voltage plateaus. It is worthwhile noticing that there has been different stoichiometry between the $Li_{22}Si_5$ and the $Li_{21}Si_5$ phases in the literature. In most cases, there is no particular need to differentiate the two phases [7].

[1]University of Maryland, Maryland

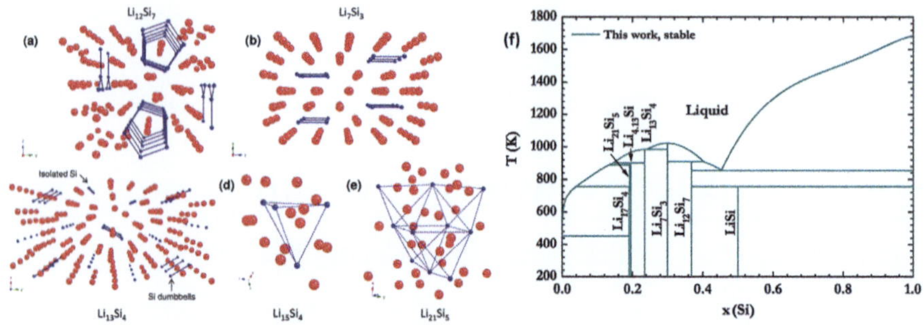

Figure 3.1 *Structural representations of (a) $Li_{12}Si_7$, 13 Li, and 9 Si crystallographic sites (Si–Si distances in rings and stars, 2.35–2.40 Å); (b) Li_7Si_3, 3 Li, and 1 Si sites (Si–Si dumbbell distances, 2.332 Å); (c) $Li_{13}Si_4$, 8 Li, and 2 Si sites (Si–Si dumbbells, 2.383 Å and isolated silicon ions); (d) $Li_{15}Si_4$, 2 Li, and 1 Si sites (isolated silicon ions, Si–Si distances >4.5 Å), and (e) $Li_{21}Si_5$, 16 Li, and 4 Si sites (Si–Si distances >4.6 Å) (Li and Si atoms are shown in red and blue, respectively; blue lines: Si–Si bonds; blue dashed lines: closest Si–Si contacts >3.0 Å)[5]. (f) Phase diagrams of the Li–Si system[6].*

3.3 Room-temperature electrochemical (de)lithiation of Si and corresponding phase changes

3.3.1 Difference between room- and high-temperature Si electrochemistry

Although silicon has the highest achievable (at high temperatures) theoretical specific capacity (~4 200 mAh/g for $Li_{22}Si_5$) among alloy-type lithium anode materials, there is still much effort needed to finally successfully apply it in commercial lithium-ion cells.

At room temperature, for micro-sized crystalline silicon powders, the half-cell discharge curve (Figure 3.2a) features only a single plateau between 1.0 and 0.0 V. In fact, silicon and lithium do not form any of the four high-temperature phases during room-temperature electrochemical lithiation but instead undergo a phase transition from a crystalline phase to an amorphous lithiated phase, even at a very slow rate of C/100 or C/200 [8]. This amorphous lithiated silicon phase then crystallizes at the end of lithiation, forming a metastable $Li_{15}Si_4$ phase. This stoichiometry determines that the theoretical capacity of Si is 3 579 mAh/g, not widely reported 4 200 mAh/g, which corresponds to the $Li_{22}Si_5$ phase. While many previous reports show the initial lithiation capacity exceeding 3 579 mAh/g, at room temperature, the extra capacity is mainly attributed to lithiation of surface oxides and/or solid electrolyte interface (SEI) formation.

Typical charge/discharge profile of −325 mesh polycrystalline Si cycled at a rate of C/100 (calculated based on 3 579 mAh/g capacity) is shown in Figure 3.2a.

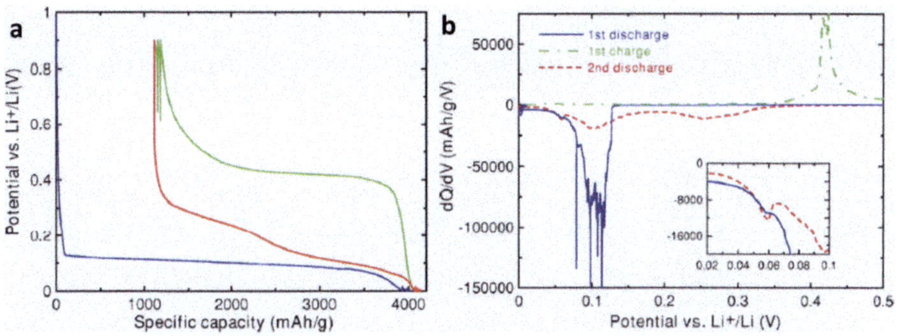

Figure 3.2 *(a) Potential vs specific capacity profile for the in situ X-ray diffraction (XRD) cell. The test cell was cycled at C/100 between 0.005 and 0.9 V using a signature discharge and charge method as described in the experimental section. (b) Differential capacity vs potential plot for the in situ XRD cell cycled at C/100 calculated from the data shown in (a). The inset focuses on the plot at low potential [8].*

During the first discharge (indicating lithiation in Li||Si half cells), a slightly sloping plateau is observed between 0.12 and 0.08 V. Then, the profile exhibits a small plateau near the end of the first lithiation when the full capacity is almost reached. During the first delithiaton, the potential increases sharply to about 0.45 V at only ~5 percent of the delithiation capacity. This sharp increase is followed by a single flat plateau, suggestive of a two-phase region, then an upward slope, indicative of a single-phase region. The Si electrode exhibited a first-cycle discharge capacity of 4 032 mAh/g, including an irreversible capacity loss (1 117 mAh/g). The large irreversible capacity loss is typical for Si electrodes independent of Si particle size. The near-zero potential brings substantial challenge for electrolyte, which decomposes and contributes much to the irreversible capacity. Other contributions presumably come from electrically disconnected particles due to the severe volume shrinkage upon delithiation, as well as Li atoms kinetically "trapped" in the electrically connected particles and thermodynamically irreversible lithiation of surface oxides. The second discharge profile is more feature-rich compared with the first discharge, showing three sloping plateaus. Figure 3.2b shows the corresponding differential capacity vs potential plot derived from profiles in Figure 3.2a, amplifying the characteristics observed in the voltage curves (enlarged low-potential data shown in the inset panel). The large sharp peaks appeared during the first charge/discharge indicate both processes undergo two-phase transition. The second discharge differs distinctly with the first one, although both are Si lithiation processes, indicating different phase-related processes. A small ~60 mV peak can be observed at the inset panel on both curves, which corresponds to the formation of $Li_{15}Si_4$ phase and will be discussed in detail later, also consistent with other reports [9, 10].

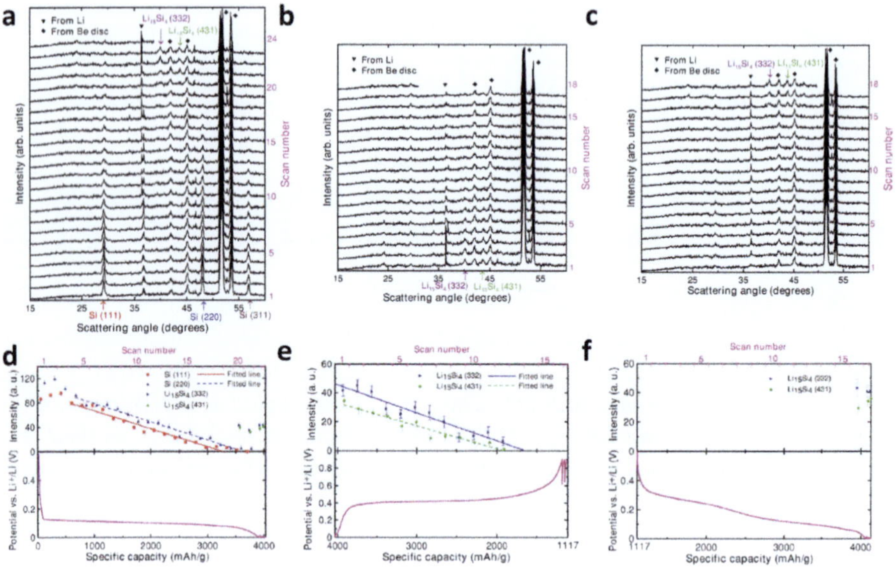

Figure 3.3 *(a–c) Plots of in situ XRD patterns collected during the cycling of the test cell. The cycling rate was C/100. (a) XRD scans for the first discharge, (b) XRD scans for the first charge, and (c) XRD scans for the second discharge. The numbers shown to the right indicate the scan numbers. (d–f) Plots of the intensities of different Bragg peaks during the cycling of the test cell. The cycling rate was C/100. (d) during the first discharge, (e) during the first charge, and (f) during the second discharge. The top panel in each figure shows the Bragg peak intensity as a function of capacity. The bottom panel shows the potential as a function of capacity [8].*

3.3.2 Structural evolution during the first lithiation and delithiation

The phase transition information was obtained by in situ/operando XRD with a cell cycled at C/100. Figure 3.3a–c shows the in situ collected XRD patterns during the first charge/discharge and second discharge, respectively. Technically, the background peaks include 42°, 46°, 52°, and 53° from the Be window, and the peak at ~37° from Li foil as counter/reference electrode. The scan number increases with discharge or charge time/capacity, from the bottom to the top of each panel. Figure 3.3a indicates that the peak intensities of crystalline Si(111), (220), and (311) decrease during the first lithiation. Since scan 19, the Si peaks completely disappear from the XRD patterns. The small residual Si(111) peak after scan 19 is from the test cell parts, in accordance with other reports [9–12]. Although no clearly visible diffraction peaks are observed from amorphous lithium silicide in Figure 3.3a, the flat plateau during the first lithiation in Figure 3.2a substantially supports the existence of this lithium silicide with particular composition in the two-phase region. Overall,

Table 3.1 $Li_{15}Si_4$ *crystalline structure [9]*

	Phase		$Li_{15}Si_4$	
	Space group		$I\bar{4}3d$	
	Z		4	
	A (Å)		10.6852(9)	
	Bragg R		1.68	
Atom	**Site**	*x*	*y*	*z*
Li(1)	12a	0.375	0	0.25
Li(2)	48e	0.118(1)	0.156(1)	0.961(1)
Si	16c	0.4588(2)	0.4588(2)	0.4588(2)

Figure 3.3a suggests that the lithiation of crystalline Si produces amorphous lithiated Si by the reaction (3.1):

$$\mathrm{Si} + x\mathrm{Li} \rightarrow \left(1 - \frac{x}{y}\right)\mathrm{Si} + \frac{x}{y}a - \mathrm{Li}_y\mathrm{Si} \tag{3.1}$$

The value of y in a-Li_ySi for the amorphous phase composition needs to be determined. This amorphous lithium silicide phase coexists with crystalline Si during the two-phase region before the final crystallization at ~60 mV indicated by the sudden appearance of a new set of peaks ($Li_{15}Si_4$) in the last several scans (Figure 3.3a). The corresponding slope change in voltage profile also verifies this two-phase process. First observed and indexed by Obrovac and Christensen using Rietveld refinement [9], this $Li_{15}Si_4$ crystalline phase formed at the end of discharge [9, 10]. $Li_{15}Si_4$ has been shown to be similar in structure to $Cu_{15}Si_4$ and $Li_{15}Ge_4$ [9, 10], in particular with similar lattice constants compared to those of $Li_{15}Ge_4$. In this compound, all the Si atoms are crystallographically equivalent, and each of them has 12 neighboring Li (Table 3.1). While Obrovac and Christensen identified the stoichiometry of $Li_{15}Si_4$, Key *et al.* [5] and Li *et al.* [8] utilized in situ nuclear magnetic resonance (NMR) and XRD, respectively, to demonstrate that this $Li_{15}Si_4$ phase can accommodate a small range of additional Li.

Electrochemically, this $Li_{15}Si_4$ phase forms once the capacity of Si electrode reaches 3 560 ± 100 mAh/g. This metastable phase has only been observed via room-temperature electrochemical lithiation of Si. The sudden crystallization of a-Li_ySi to $Li_{15}Si_4$ during lithiation is not a typical feature often seen among alloy anodes, as crystalline phases are generally reached through two-phase processes. In fact, $Li_{15}Si_4$ is the lowest in Li content among the phases having completely isolated Si atoms. Energetically, the Li content lower than $Li_{15}Si_4$ crystalline phases are unattainable during lithiation due to high barriers to Si rearrangement, even though their final configurations possess lower energy. Once Li fully isolated the Si atoms, the high Li mobility and structural similarity in a-Li_ySi and $Li_{15}Si_4$ promote the system to achieve a lower-energy $Li_{15}Si_4$ state at room temperature [13–15].

No observable crystalline Si peaks recovered in the XRD patterns acquired during the first charge. Upon delithiation, the $Li_{15}Si_4$ peaks intensities gradually decrease (Figure 3.3b). This indicates that the delithiation of crystalline $Li_{15}Si_4$ ($Li_{3.75}Si$) phase results in an amorphous Li_zSi phase in the coexisting two-phase reaction (3.2). The two-phase ($Li_{15}Si_4$ and a-Li_zSi) region is indicated by the flat plateau in the charge potential curve (Figure 3.2a) and the sharp peak in the dQ/dV curve (Figure 3.2b) during the first charge. The value of z in a-Li_zSi for the amorphous phase coexisting with $Li_{15}Si_4$ phase during delithiation needs to be determined.

$$Li_{3.75}Si - xLi \rightarrow \left(1 - \frac{x}{(3.75-z)}\right) Li_{3.75}Si + \frac{x}{(3.75-z)} a - Li_zSi \tag{3.2}$$

Figure 3.3c shows the $Li_{15}Si_4$ phase reformed during the second discharge at about 60 mV. This potential value is also well supported by the dQ/dV data from Figure 3.2b. Therefore, restricting the cutoff potential to those values higher than 60 mV can avoid the formation of the crystalline $Li_{15}Si_4$ phase, which will be discussed later. A least-squares fitting program was applied to fit every and each of the Si and $Li_{15}Si_4$ Bragg diffraction peaks (Si(111), Si(220), $Li_{15}Si_4$(332), and $Li_{15}Si_4$(431)) in all of the XRD patterns in Figure 3.3a–c with pseudo-Voigt functions, and the result indicates relatively small error bars. Figure 3.3d–f shows the intensity of the Si and $Li_{15}Si_4$ Bragg peaks as a function of the charge/discharge capacity of the cell cycled at C/100 during the first discharge/charge and second discharge, respectively.

During the first discharge (Figure 3.3d), the Si(111) and Si(220) peak intensities decrease linearly with cell capacity, except for the initial data points, which is anticipated based on (3.1). The Si-related peak intensities reached 0 at about 3 350 mAh/g. Taking (3.1) into consideration, $y = 3.5 \pm 0.2$ is determined for the stoichiometry of a-Li_ySi coexisting with crystalline Si during the first discharge. The reason why the Si peak intensities remained constant during the beginning part of the discharge is still unclear, but may partially due to the lithiation of surface oxide layer and initial formation of the SEI, which will be discussed at the end of this chapter. The next obvious change occurs as the potential drops to 60 mV, a set of $Li_{15}Si_4$ Bragg peaks appears during the crystallization of amorphous a-Li_ySi phase. As discussed before, this new crystal phase can exist over a range of lithiation capacity, so here we denote it $Li_{15\pm w}Si_4$. During the first charge (Figure 3.3e), the $Li_{15}Si_4$(332) and $Li_{15}Si_4$(431) peak intensities decreased linearly with cell charge capacity as anticipated based on (3.2). The two-phase ($Li_{15}Si_4$ and amorphous Si) region (flat plateau) ends when the $Li_{15}Si_4$ intensity hits zero, and thus the value of z in a-Li_zSi phase that coexists with $Li_{15}Si_4$ can be determined to be 2.0. It should be noted that this z value is subject to a rather large error caused by the irreversible capacity, but the error can only lead to smaller z values. During the second discharge (Figure 3.3f), a-Li_xSi dominates above the potential of 60 mV and afterward replaced by $Li_{15}Si_4$ at potentials lower than 60 mV. The recrystallized $Li_{15}Si_4$ peak intensities are similar to those of the initially formed during the first discharge. Although there is no XRD evidence that *a*-Li_xSi experienced phase transitions during second discharge, the two sloping plateaus in the voltage profile indicate the possible phase changes before the final formation of $Li_{15}Si_4$ and will be discussed in Chapter 5, Section 5.2.

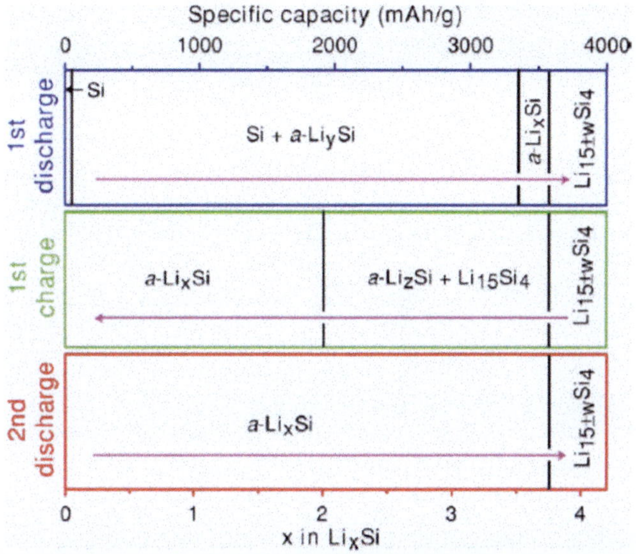

Figure 3.4 *Phase diagram describing the phases that form during the charge–discharge cycling of a Li/Si electrochemical cell between 0.005 and 0.9 V at room temperature. Discharge corresponds to the addition of Li to the electrode. In the figure and in the text, a-*Li_xSi *designates a single-phase region of variable composition,* Li_ySi *designates the fixed composition that coexists with c-Si during the first lithation, and* Li_zSi *designates the fixed composition that coexists with* $Li_{15}Si_4$ *during delithiation.* $Li_{15\pm w}Si_4$ *designates a variable stoichiometry single-phase region [8].*

The in situ XRD measurements (Figure 3.3d–f) were used to draw the phase diagram (Figure 3.4) that summarizes the phase transitions in the first several cycles (lower voltage cutoff below 60 mV) discussed above. There is a large hysteresis observed between charge and discharge if the final discharge product is $Li_{15}Si_4$. In other cases, if the discharge is cutoff to potentials above 60 mV, then there will not be $Li_{15}Si_4$ phase and the anode remains in the amorphous phase Li_xSi afterward with maximum $x = 3.5$.

After the initial charge/discharge cycle discussed above, typical potential profiles in the subsequent cycles are exhibited in Figure 3.5. As discussed above, the polycrystalline Si powder anode undergoes a phase transition from a crystalline to an amorphous lithiated $Li_{15}Si_4$ phase in the first cycle and never recrystallizes back to the original crystalline Si again. Actually, no matter whether starting with crystalline or amorphous Si, or even a Li-Si alloy phase, it will be converted to the same a-Si after the initial lithiation and complete delithiation [17]. So the second and afterward lithiation always starts from a-Si, no matter the initial Si structure. The bulk a-Si lithiation voltage curve in the beginning features two gently sloping

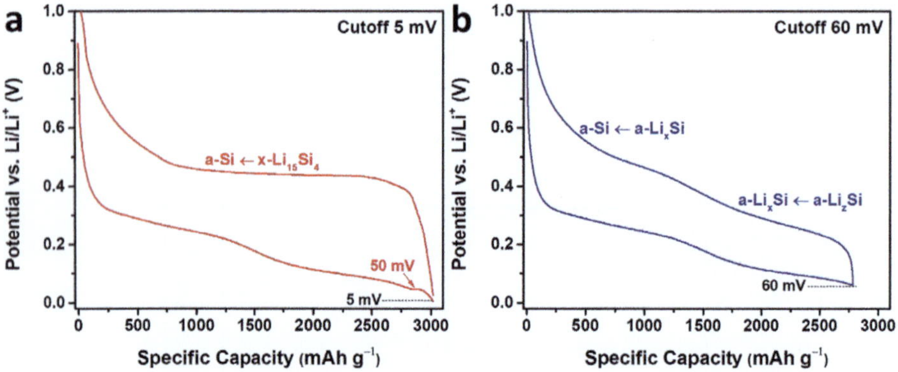

Figure 3.5 *Typical third cycle charge/discharge curves of −325 mesh polycrystalline Si powder anode with different cutoff voltage of (a) 5 mV and (b) 60 mV [16]*

quasi-plateaus. Computational simulations provided some insights into the features of Si lithiation [13] potential curves. It is indicated that the higher-voltage quasi-plateau corresponds to Li entering an Si-rich atomic environment where Si atoms mainly bond to other neighboring Si atoms. The lower-voltage quasi-plateau originates from Li entering Li-rich environments with Si atoms bonded mainly to Li neighbors. The overall sloping feature indicates a solid solution-type reaction. After the initial two quasi-plateaus, electrochemical behavior of the following lithiation and subsequent delithiation depends on the lower cutoff voltage applied at the end of lithiation. When the cutoff voltage is lower than 50 mV, a small voltage plateau can be observed near 50 mV vs Li/Li^+ in the lithiation voltage profile, which is characteristic of the formation of a crystalline $Li_{15}Si_4$ phase and indicated by arrow in Figure 3.5a. When the lithiation ends above ~60 mV, there will be no small plateau at the end of the lithiation (Figure 3.5b). For delithiation, with a lower cutoff below 50 mV, it will be identical to the above-described crystalline $Li_{15}Si_4$ delithiation (Figures 3.2a and 3.5a). In contrast, if the lithiation of Si is stopped above 60 mV, then the delithiation starts from a-$Li_{3.5}$Si and features two quasi-plateaus (Figure 3.5b) that are attributed to the reverse solid-solution process of the discharge plateaus [9, 17, 18].

3.4 Anisotropic expansion of Si during lithiation

From the discussions above, the whole picture of basic crystal structure transformation during Si lithiation/delithiation has been built. However, the crystal plane-specific lithiation behavior remains undiscovered. Special Si nanopillars were designed and fabricated to investigate structural and morphological evolution in crystalline Si during lithiation. These nanopillars are advantageous for this study because of the following three reasons. First, the fabrication technique of Si nanopillars with various axial crystalline orientations same as the Si substrate is mature;

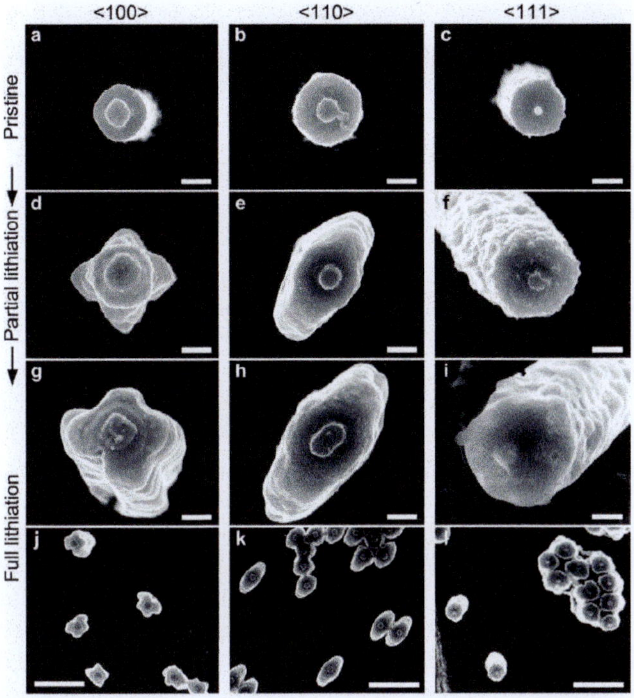

Figure 3.6 *Anisotropic lateral expansion of crystalline Si nanopillars with three different axial orientations (<100>, <110>, and <111>) upon lithiation. (a–l) Top-view SEM images of Si nanopillars of each crystal orientation and each lithiation state. The <100> axially oriented pillars are shown in the left column, <110> pillars are shown in the middle column, and <111> pillars are shown in the right column. The top row shows pristine pillars, the second row shows partially lithiated pillars held at 120 mV vs Li/Li^+, and the third row shows fully lithiated pillars held at 10 mV vs Li/Li^+. The images in the fourth row show low-magnification views of fully lithiated pillars of each axial orientation. Scale bars from (a) to (i) are 200 nm and from (j) to (l) are 2 μm [19].*

<100>, <110>, and <111> direction-oriented nanopillars are fabricated and used for the study. Second, the nanopillars are uniform in both width and height, and all stand vertically on a fixed substrate, which makes it simple for observing dimensional changes. Third, the crystalline orientations of the nanopillar can be easily determined during observations by checking the Si wafer substrate orientation.

Figure 3.6 shows the top view of expanding process of Si nanopillars with three different axial orientations (<100> on the left column, <110> on the middle column, and <111> on the right column) upon lithiation via SEM. First three rows of the figure represent different lithiation states of Si nanopillars, namely, top row for

pristine state (before lithiation), the second row for partial lithiation (voltage holding at 120 mV), and the third row for the full lithiation (10 mV). In the etching process, 600-nm silica nanospheres were used as etching mask, whereas the diameter of the resulting nanopillar is about 400 nm due to the side etching effect. Because of using nanospheres as the etching mask on the Si, the cross-sections for the Si nanopillars with different axial orientation are all circular, which is shown in Figure 3.6a–c. The size of the nanopillars with different orientations slightly varies because of the different etching rates of the lattice plane on the sidewalls. The buttons centered at the top of the pillars are where the silica nanospheres were physically attached during the etching. Upon lithiation, those Si nanopillars display anisotropic expansion. Even at the stage of partial lithiation (holding at 120 mV vs Li/Li^+), obvious anisotropic cross-sectional expansion is observed (Figure 3.6d–f), and the expansion behaviors of three nanopillars with different orientations are different. The <100> orientated nanopillars transformed into a cross shape (Figure 3.6d), <110> into an ellipse (Figure 3.6e), and <111> orientated one turns into hexagon (Figure 3.6f). As the lithiation proceeds further to 10 mV vs Li/Li^+, the anisotropic expansion of the nanopillars continues, resulting in a more significant cross and elliptical shape of the <100> and <110> nanopillars (Figure 3.6g–i). For the <111> nanopillar in the full lithiation stage, the shape becomes more rounded from the hexagonal shape (Figure 3.6i) due to the large volume expansion. Last row of the figure (Figure 3.6j–l) displays the low magnification view of multiple pillars at the stage of full lithiation, and anisotropic expansion behaviors that are consistent with single pillars (Figure 3.6j–i) are observed. When in close contact with each other, nanopillars will undergo a much more complicated shape change because of constrain by each other. This behavior will not be discussed here. Figure 3.6 implies, at specific crystallographic direction of the pristine Si lattice, the expansion is much greater.

Figure 3.7 shows the schematic explanation of the different cross-sectional expansion of silicon nanopillars upon lithiation. It is believed fast diffusion along the <110> direction and plastic deformation of the amorphous Li–Si alloy play a major role in those anisotropic expansions. In the pristine Si, <110> crystallographic direction has larger channels compared with <100> or <111> direction. Thus, the energy barrier for lithium ion diffusion along <110> direction should be the lowest. Upon lithiation, Li ions will undergo fastest diffusion into the Si crystal from the nanopillar sidewalls along the <110> directions (green zones in Figure 3.7), while much less diffusion will happen along the other lateral directions (light gray zones in Figure 3.7). Figure 3.7a–c shows the differences of <110> channels as well as Li diffusion on the sidewalls of different, orientated nanopillars. As the lithiation proceeds, amorphization of the Si crystal will happen first at the (110) surfaces because of the highest Li concentration there (fast diffusion rate), which is indicated as the blue zone in Figure 3.7d–f. Amorphous Li–Si alloy is much weaker than the crystalline Si in terms of yield strength, such that amorphous Li–Si alloy will be pushed away from the crystalline|amorphous two-phase interface once it begins to expand in volume. The amorphous region keeps growing during lithiation, and the previously deformed Li–Si alloy is pushed out to a higher extent. Due to differences in <110> channels on the side walls, unique cross-sectional shapes start

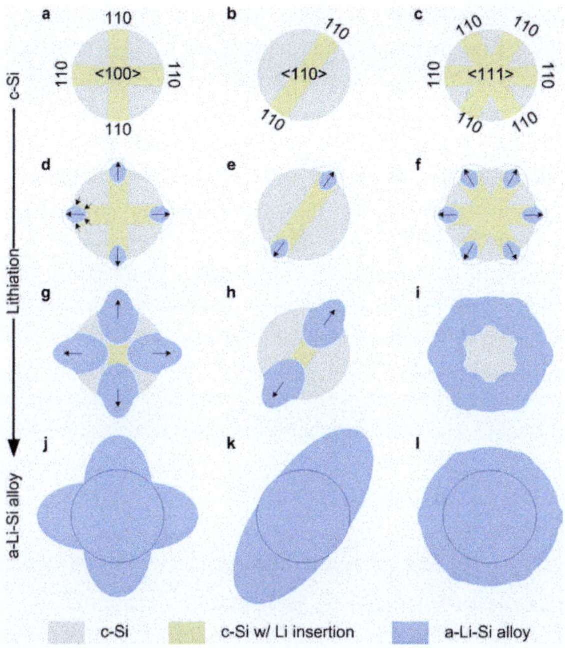

Figure 3.7 *Schematic explaining anisotropic expansion of Si nanopillars. (a–c) Li diffuses along the <110> lateral directions in each crystalline pillar. Green indicates the zones into which Li first diffuses. (d–f) Amorphization (blue) begins at the surface of the pillar along the <110> directions due to the relatively high concentration of Li. The amorphized material is constrained by the surrounding crystalline regions and deforms. (g,h) With further lithiation, the <100> and <110> pillars still have crystalline regions remaining in the core and near the edges, and this directs the amorphous regions to continue expanding along the <110> directions. (i) In the <111>-oriented pillar, the outer amorphous regions merge earlier in the lithiation process. (j,k) At full lithiation, the <100> and <110> axially oriented pillars expand to a cross shape and an ellipse as the amorphous regions are further displaced outward. (l) The <111> pillar expands relatively isotropically [19].*

to appear on the <100>, <110>, and <111> pillars, as shown in the SEM images (Figure 3.6d–f). The amorphous regions continue to grow along the <110> directions at the final stage of the lithiation. However, for the <100> and <110> axially oriented nanopillars, due to less <110> channels on their sidewalls, there are still crystalline regions stretching from the core of the nanopillar to the edges along other directions (not <110> shown in Figure 3.7g,h). These directions remain mostly unlithiated and their mechanically rigid structure does not expand. Thus, for the <110> and <100> oriented pillars, the LiSi phase is expanded along the <110> lateral directions, while little expansion occurs in the other directions because of their slow rate (Figure 3.7j,k). For the nanopillars orientated in <111> direction, there are six very close <110> channels so amorphous regions grown from these directions can merge earlier in the lithiation process (Figure 3.7i). Then the amorphous region will fully cover the entire sidewalls of the nanopillar, leading to a lack of driving force for anisotropic deformation along the <110> direction, as confirmed by the SEM image in Figure 3.6i. Based on the above analysis, the final cross-sectional shapes are displayed in Figure 3.7j–l.

To sum up, as the lithiation initiated, we suggest that Li migrates into the c-Si lattice through <110> ion diffusion channels and results in the collapse of some {111} planes, simultaneously breaking Si-Si bonds; this leads to the obvious decrease in height of the nanopillars with <111> and <100> axial orientations. Meanwhile, the Li-Si amorphous phase starts to form and grow at the surface of the <110> ion channels (the highest Li concentration), generating the cross, ellipse, and hexagonal-shaped cross-sections of the <100>, <110>, and <111> pillars, respectively. These findings provide us basic understanding of the crystal plane-specific structural evolution of Si lithiation.

3.5 Lithiation mechanism in Si

3.5.1 Atomic mechanism of Li insertion into crystalline Si

While in situ SEM characterization provides the crystal plane-specific insights, in situ TEM offers an atomic-resolution image for the Si lithiation process. It has been confirmed that the lithiation (crystalline-to-amorphous phase transition) kinetics of a single-crystal silicon matrix were found to vary greatly on different facets [19–21], with the slowest lithiation kinetics on the {111} surface and the fastest on the {110} surface [21]. A lithiation process of a single crystal Si nanowire is shown in Figure 3.8. The unlithiated Si nanowire has a growth direction of <111> with six {112} faces exposed (Figure 3.8a,b). An amorphous Li_xSi alloy shell is formed upon lithiation, which is indicated by the merge of the gray contrast (Figure 3.8c,d). The amorphous/crystalline interface (ACI) is also observed and is found to have ~1 nm thickness and a unique striped contrast (Figure 3.8e–g). At the same time, a darker layer consisting of a-Li_xSi and Li_2O is formed at the outermost surface of the lithiated wire due the lithiation of SiO_x layer on the surface of the pristine Si nanowire (Figure 3.8c,e). The figures view the nanowire along the [1 í 0] direction and it is found that the ACI is parallel to the Si (íí 2) plane in most of cases. As the lithiation proceeds, the ACI gradually move toward the center of the wire. During this process, the Si core shrinks progressively while the a-Li_xSi undergoes dramatic expansion (about 280 percent expansion in volume). It is interesting that the thickness of ACI remained nearly constant (Figure 3.8e,g), while it gradually flattened as indicated by the gradual disappearance of the bump with respect to the baseline (Figure 3.8g). Because of lack of planes with small d-spacing parallel to <111>, <111>-oriented silicon nanowires grown by the chemical vapor deposition method usually have sawtooth facets on their sidewalls [23]. At the beginning of lithiation, ACI was

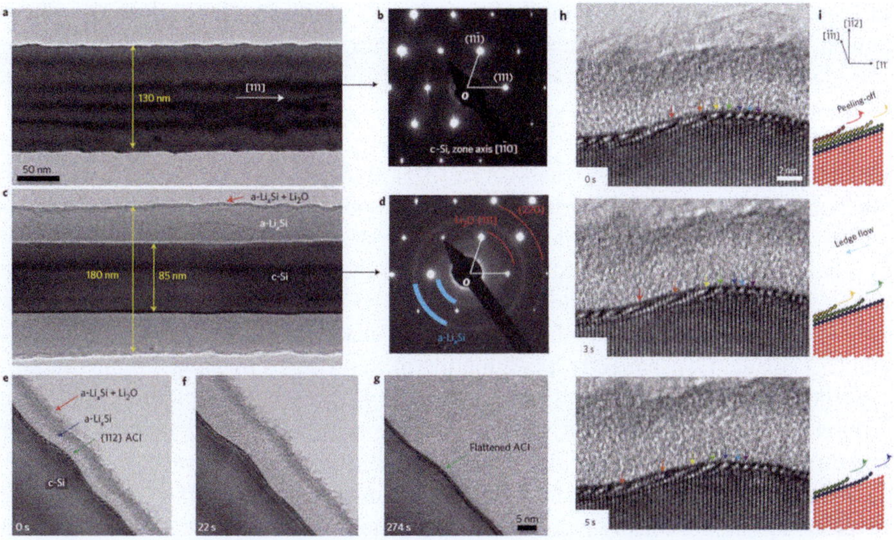

Figure 3.8 *Progressive migration of the sharp ACI during solid-state amorphization (lithiation) of a crystalline <111>-oriented silicon nanowire. (a) Pristine silicon nanowire with rough sidewalls due to faceting. (b) Electron diffraction pattern from the pristine single-crystal silicon nanowire. (c) Partially lithiated silicon nanowire with an a-Li$_x$Si layer surrounding the c-Si core. The ACI is sharp and flat. (d) Corresponding electron diffraction pattern indicating the amorphization process. (e–g) Time-lapse images showing migration of the {112} ACI during lithiation. The ACI is 1 nm thick and did not change in thickness, but became smoother during progressive migration. Ledge mechanism of lithiation in c-Si. (h) High-resolution TEM image sequences showing the lithiation process by lateral ledge flow (marked by colored arrows) in the ACI and simultaneous peeling-off of the {111} planes over 5 s. The ACI is ~3–4 atomic layers thick, and the atomic layers are distorted and different from both the regular lattice in the bottom c-Si and the mottled contrast in the top a-Li$_x$Si layer. (d) Schematic showing the peeling-off of the (111) atomic layers by continuous ledge flow in the planes during lithiation [22].*

formed nearly parallel to the sidewall, leading to similar surface morphology with pristine-faceted surface. However, the facets on the ACI gradually disappear during the lithiation (Figure 3.8c), which is caused by the different interfacial mobility of ACI at different orientations.

Figure 3.8h,i shows the atomic structure evolution of the ACI and a ledge mechanism upon Si lithiation was revealed. Ledges of successive inclined (íí 1) atomic planes are marked by the colored arrows in Figure 3.8h. The red dot arrays in Figure 3.8i stands for pristine lattice projected along the [1 í 0] direction of the silicon wire core. The high-resolution TEM images (Figure 3.8h) reveal a highly distorted atomic arrangement with clearly resolvable periodic feature (a short array of bright spots) in the ACI. The ACI also displays a very unique striped contrast, indicating a short-range ordered structure, which is different from highly ordered silicon lattice core or totally disordered Li_xSi amorphous shell. A ledge-mediated mechanism of solid-state amorphization was proposed based on the dynamic observation of migrating ACI. Lithium atoms are first inserted into the spaces between two neighboring (íí 1) plane layers, leading to the expanding of the spacing. The spacing expansion drives the peeling of the top (íí 1) layers. Silicon atoms from the ledges gradually migrate into the a-Li_xSi phase as the ledges flow along the interfacial (íí 1) plane. Consequently, the lithiation of Si proceeds by atomic removal from the silicon surface (Si atoms alloyed into the a-Li_xSi shell) during ledge migrations. All evidences support that the {111} facets are peeled in a layer-by-layer manner and through lateral ledge flow in each layer, as illustrated in Figure 3.8i.

The origin of the fact that {112} ACI moves faster than the {111} ACI is revealed by the atomically resolved TEM images. The terminology is clarified as follows: "{111} atomic facet" means an individual short facet in a long ACI, and "{112} or {111} ACI" means a long interface that has both the {112} or {111} planes, and may consist of many short {111} atomic facets/ledges, thus having finite, atomic-scale roughness. A high-resolution TEM image of a {112} ACI at the initial stage of lithiation is presented on Figure 3.9a, in which a ~15 nm-wide long (íí 1) facet can be seen. The inset shows a magnified TEM image near a {112} ACI, the atomic structure of which is reconstructed in Figure 3.9c. A typical flattened {112} ACI is shown in Figure 3.9b, generated after the large {111} facets have been removed by the peeling process during lithiation. It was found the real {112} surface (Figure 3.9d) is not flat and atomically smooth. Instead, on the {112} surface are a series of inclined {111} atomics facets. Each of those {111} facets consist of four {111} bilayers and has width of around 1 nm (or $3 \times d_{111}$). According to geometry, the diamond-cubic silicon lattice determines the translational periodicity within the {112} surface. As shown in Figure 3.9d, the periodicity in the horizontal <111> direction is the diagonal length of the <001> × <011> rectangular box. Therefore, to build a {112} surface with {111} facets, four {111} bilayers (blue segments in the inset of Figure 3.9a) are the smallest repeating units. Due to the relatively small repeating unit (consists of four {111} bilayers), the density of ledges on the {112} ACI is high. A high density of ledges provides more heterogeneous sites for the Si-Li reaction, leading to a high mobility of the {112} ACIs. The {110} ACIs were also found to have high density of inclined {111} atomic facets, thus displaying very similar high mobility with {112} ACIs. Figure 3.9a also shows that there is one atomically flat {111} ACI initially between two atomically rough {112} ACIs. This {111} ACI was, however, not stable and diminished during the lithiation, as shown in Figure 3.9b. The above observations with atomic resolution also suggest an efficient solid-state amorphization kinetic mechanism at room temperature [24], illustrated in Figure 3.9e,f. Due to the high energy needed to remove a silicon atom from c-Si surface with low index, which contributed to a large portion of the

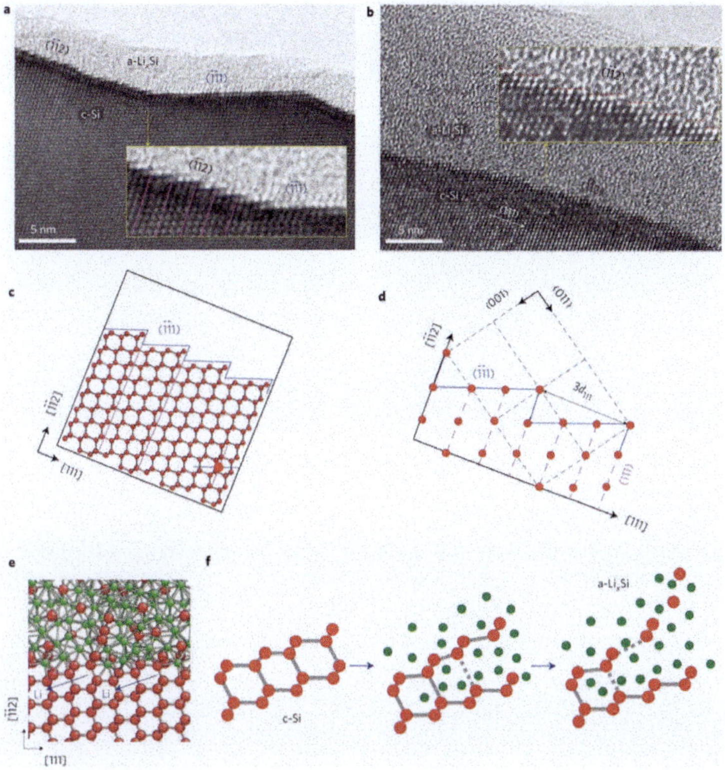

Figure 3.9 *Atomically resolved structure of the ACI. (a) High-resolution TEM image at the early stage of lithiation, showing the {112} ACI with a long smooth {111} facet in the middle and a series of short, inclined {111} atomic facets on the two sides. (b) The {112} ACI in the late stage of lithiation, with short {111} atomic facets surviving. (c) Reconstructed atomic structure of a {112} ACI based on the inset image in (a) showing that an exact {112} ACI actually consists of a series of periodic {111} atomic facets. (d) Geometrical origin of the periodicity (3d111) of {111} atomic facets. Red dots reveal the repeating units in the face-centered cubic lattice; each dot contains two silicon atoms in the adjacent {111} atomic sheets, as indicated by the big red dot in (c). (e) Relaxed atomic structure of a {112} ACI from molecular orbital theory calculations , showing the high local concentration of lithium atoms (green) that invade the exposed {112} surface of the c-Si (red). (f) Proposed mechanism of solid-state amorphization based on TEM observations: silicon atoms break away from the c-Si by a mechanism with low energy barriers, in which several lithium atoms enclose a single silicon or a Si–Si pair located near the edge of {111} atomic facets, weaken the Si–Si covalent bonding, and cause the easy dissociation of c-Si [22].*

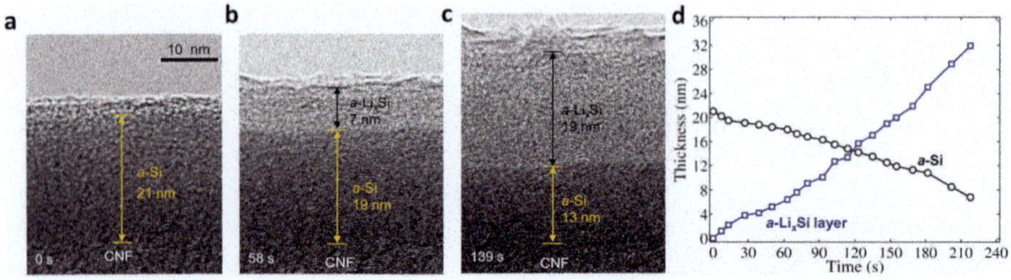

Figure 3.10 *(a−c) Time-elapse high-resolution TEM images showing the migration of a phase boundary between a-Si and a-Li_xSi. The a-Li_xSi layer in (c) grew to 19 nm thick and had a lighter contrast than the underneath a-Si, whose thickness decreased from 21 (shown in (a)) to 13 nm. (d) Thickness vs time in a-Si and a-Li_xSi [28].*

vacancy formation energy in c-Si (2.4–3.5 eV) [25], breaking the bond in the c-Si lattice is usually difficult, which makes the kinetics of c-silicon dissociation very slow at room temperature. Near the ACI (Figure 3.9e), however, because of the high local concentration of lithium in the lithium-rich a-Li_xSi phase, the interfacial interaction between lithium atoms and Si atom/Si–Si pair is promoted at the edge of {111} atomic facets [5, 26]. Previous study shows Si–Si bond can be weakened by high concentration of metallic atoms (lithium as in here) through the electron transfer process[25]. A weakened Si–Si bond promotes the dissociation of silicon atoms from the c-Si at ACI (Figure 3.9f), leading to an increased amorphization rate at room temperature. This high mobility of ACIs are desirable for high rate operation of batteries. Other than that, applying voltage and doping is also found to play roles in increasing the rate of electron/ion transport and interfacial reaction, thus enhancing the amorphization rate [27]. Though the key to fast solid-state amorphization in c-Si is the high concentration of lithium atoms near the ledge of {111} atomic facets of ACIs.

3.5.2 Atomic mechanism of Li insertion into amorphous Si

Figure 3.10a–c present the dynamic propagation of the amorphous–amorphous interface (AAI). The pristine a-Si layer on the CNF is about 21 nm (Figure 3.10a). Once lithiation starts, Li^+ ions began to consume the Si atoms to form an a-Li_xSi alloy layer. Meanwhile, a sharp phase boundary was formed between the a-Si and a-Li_xSi layer. During the propagation of AAI, the thickness of the lithiated layer was profoundly increased and the a-Si layer was gradually depleted (Figure 3.10b,c). The thickness change of a-Si layer and a-Li_xSi layer vs time is plotted on Figure 3.10d. The propagation velocity of AAI was found to be nearly constant (0.06 nm/s), implying the rate-limiting step is the lithiation reaction at the phase boundary [29]. The average value of x in the a-Li_xSi phase was estimated to be around 2.5 based on the volume expansion calculation assuming cylindrical symmetry. This x = 2.5 corresponds to volume expansion of ~170 percent [30]. Li distribution in the a-Li_xSi layer cannot be resolved from TEM contrast. Thus, digital image correlation (DIC) analysis was performed to quantitatively determine the spatial distribution of lithiation-induced strains from TEM images. DIC results found minor Li diffusion-induced strain in the a-Li_xSi layer but very high (~160 percent) local phase-transformation-induced strain in the region swept by the a-Si/a-Li_xSi phase boundary. It is evident that most of the lithiation-induced expansion happens at the a-Si/a-Li_xSi phase boundary and concentration gradient-driven lithium ion diffusion across the a-Li_xSi ($x \sim 2.5$) does not play an important role. Although electron beam irradiation can accelerate the lithiation sometimes, the reproducible observation of clear phase boundaries remained unchanged in all the TEM experiments (Figure 3.10), undoubtedly demonstrating that the two-phase mechanism dominates the electrochemical lithiation of a-Si. In situ TEM observation also shows that the AAI keeps its existence until all a-Si is consumed. After all a-Si is consumed, a second step of lithiation in a-Li_xSi ($x \sim 2.5$) proceeds, resulting in the final product of a-Li_xSi with higher Li content ($x \sim 3.75$) and a total volume expansion of ~280 percent.

No visible interface was observed during the second lithiation step. This lithiation step also proceeded so fast that it cannot be studied by current in situ imaging technique. And whether this lithiation occurs by a single-phase or two-phase reaction needs to be answered in the future study. In summary, in situ TEM observation in this section unravels the two-phase process of electrochemical lithiation in a-Si. The linear thickening of a-Li_xSi with time (Figure 3.10d) indicates that the reaction is rate-limited by the migration of AAI. This is the reason why the first step of lithiation undergoes a two-phase process rather than the single-phase one. The high energy required for breaking Si–Si bond, even in a-Si, results in a limited kinetics of Si dissociation from a-Si network at room temperature, similar to the case of c-Si discussed before.

In this section, we compared the similarity and difference of the lithiation behavior of c-Si and a-Si via in situ high-resolution TEM experiments. Both c-Si and a-Si undergo a two-phase lithiation mechanism of c-Si/a-Si to a-Li_xSi through the migration of a sharp ACI/AAI (Figures 3.9a and 3.11). While the migration of ACI is highly directional, promoted by layer-by-layer ledge flow at the c-Si surface [22], the migration of AAI is homogeneous because of the lack of the long-range order in a-Si. Nevertheless, compared with c-Si, though the atomic structure of a-Si is disordered, there is no big difference of local binding environment between c-Si and a-Si (Figures 3.9e and 3.11b). Considering that both a-Si and c-Si involve the strong Si-Si bonding with only small number of over-/mis-coordinated defects, the lithiation kinetics cannot be very good, with a-Si possibly better than c-Si. In both cases, the local Li^+ concentration near the phase boundary is assumed to be high in the a-Li_xSi phase. Thus no matter at the surface of c-Si or a-Si, the interfacial interaction was facilitated via the process of several Li atoms enclosing a single Si atom or Si–Si pair [5, 26], demonstrated in Figures 3.9f and 3.11c. Previous silicide study shows Si–Si bond can be weakened by high concentration of metallic atoms (lithium in our case) through the electron transfer process [25]. A weakened Si–Si bond then promotes the dissociation of silicon atoms from the a-Si at AAI (Figure 3.11c), leading to faster lithiation process at room temperature.

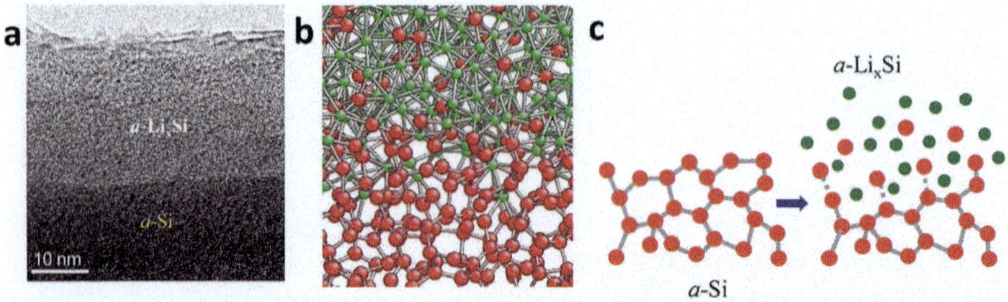

Figure 3.11 *(a) A high-resolution TEM image showing a sharp phase boundary between a-Si and a-Li$_x$Si (x ~ 2.5). (b) A snapshot from molecular dynamics simulation showing the atomic structures near the AAI. The a-Si consists of a continuous random network of Si atoms. The Li concentration is locally high near the surface of a-Si. (c) Schematics showing the proposed mechanism of dissolution of Si atoms from the a-Si surface into a-Li$_x$Si alloy [28].*

3.5.3 Localized phase transformation during lithiation and delithiation

With the information from in situ XRD, SEM, and TEM, the structural transformation of Si during first lithiation/delithiation cycle has been well studied. However, the (de)lithiation of Si after the first cycle undergoes a pathway between a-Si and a-Li$_x$Si, making it impossible to characterize the structural changes with conventional methods such as XRD. Especially, the short-range order of the amorphous materials a-Si and a-Li$_x$Si remains unclear, impeding attempts to optimize Si electrode performance based on structural evolutions.

Clare P. Grey *et al.* have demonstrated that in situ ^{7}Li NMR experiments can be utilized to monitor changes that cannot be readily studied by ex situ methods. To deconvolute the complicated Li NMR shifts with specific local Li environments in Li–Si alloys, they first obtained the standard crystalline phases $Li_{12}Si_7$, Li_7Si_3, $Li_{13}Si_4$, $Li_{15}Si_4$, and $Li_{21}Si_5$ (Figure 3.12a), and collected the corresponding ^{7}Li MAS NMR spectra (Figure 3.12b). These Li NMR shifts and associated specific Li chemical environments and silicon clusters types were then utilized to assist the determination of the structural evolutions that occur upon discharge of the Si–Li half cells.

3.5.3.1 Possible lithium silicide phases and their characteristics

The XRD patterns and corresponding profile fittings of all model Li–Si compounds are shown in Figure 3.12a. The result indicated that most of the as-synthesized compounds were of high impurity, except in $Li_{12}Si_7$, there are small portion of unreacted Si in $Li_{15}Si_4$, and very small portion of impurities (such as Li_2Si). $Li_{13}Si_4$, on the other hand, contains several kinds of impurities, including $Li_{21}Si_5$, $Li_{12}Si_7$, and a weaker unknown impurity phase(s).

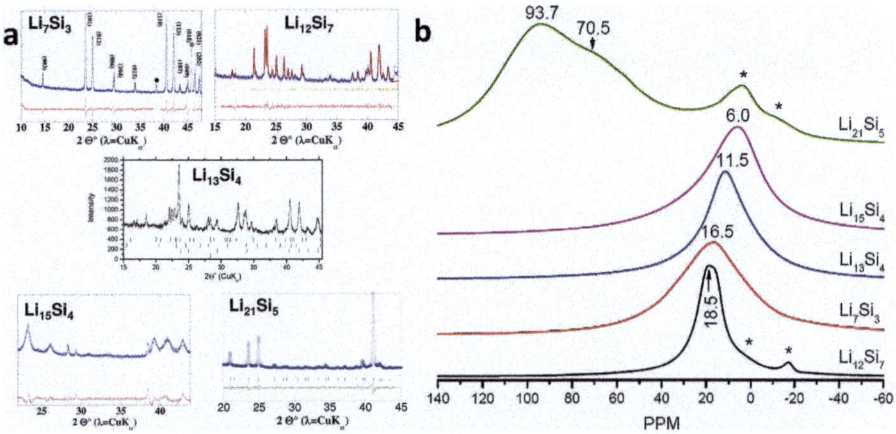

Figure 3.12 *(a) XRD patterns of lithium silicide intermetallic compounds. Fittings of four of the patterns, performed using the Fullprof software in profile-matching mode, are shown; the difference between the experimental data and the fit is shown underneath each pattern. The tick marks represent the reflections for respective structure models and impurity phases. The second set of tick marks in $Li_{15}Si_4$ and $Li_{12}Si_7$ represents Si and Li_2Si. For $Li_{13}Si_4$, tick marks are shown for the $Li_{13}Si_4$ (top; major phase), $Li_{21}Si_5$ (top middle), Li_7Si_3 (bottom middle), and Si (bottom); the remaining unindexed peaks correspond to unknown phases. * = Be; filled circle = Al. (b) ^7Li MAS NMR spectra of $Li_{12}Si_7$, Li_7Si_3, $Li_{13}Si_4$, $Li_{15}Si_4$, and $Li_{21}Si_5$ at a spinning speed of 38 kHz. Impurity phases are marked with asterisks (*) [5].*

The ^7Li MAS NMR spectra of the model compounds with increasing lithium content, $Li_{12}Si_7$, Li_7Si_3, $Li_{13}Si_4$, $Li_{15}Si_4$, and $Li_{21}Si_5$ are shown in Figure 3.12b. These materials lie between the Zintl phases and intermetallic compounds [31]. The feature of ^7Li MAS NMR spectrum (Figure 3.12b, top) of the most Li-rich phase, $Li_{21}Si_5$, is distinct: $Li_{21}Si_5$ is the only metallic phase reported among the known thermodynamic phases, whereas all the others are semiconductors [32–36]. The large and broad chemical shifts of overlapping resonances (at ~100–60 ppm) seen from this material originated from the Knight shift, which reflects the density of states at the Fermi level of the Li nucleus. The weaker resonances (~5 ppm) are resulted from Li in more shielded and/or less metallic/semiconducting environments and are assigned to impurity phases. This hypothesis is confirmed by performing the presaturation experiments by irradiating either the 70–94 ppm or the 4–0 ppm groups of peaks with a weak radio frequency field. The result suggests that they should not be ascribed to the same phase. In fact, Grey *et al.* did pay special attention to ensure the minimum oxygen contamination, if there is any. To ensure reproducibility, they collected the spectra of the model Li-Si compounds ≥2 occasions. Experimentally, only the $Li_{21}Si_5$ spectra exhibited some hint for possible contamination/decomposition of the sample, possibly caused by the highest lithiation degree thus highest reduction activity. All the other Li–Si compounds with lower Li content possess quite different electronic properties, showing resonances at much lower frequencies than $Li_{21}Si_5$ (Figure 3.12b). An obvious trend is clearly observed for these shifts, moving to lower frequencies as the Li content increases and the shielding (electron density) on the Li^+ increases. The least lithiated phase $Li_{12}Si_7$ contains planar Si_5 rings as well as trigonal-planar "star type" Si_4 clusters [32, 33], with the surrounding Li^+ give rise to ~18.5 ppm resonances. The minor impurities (presumably amorphous phases) lead to peaks at 2 and −17.6 ppm. Li_7Si_3 contains Si_2 dumbbells, with decreased average partial charges on the Li^+ [33, 35], resulting in a shift of the center of this spectrum down to 16.5 ppm. $Li_{13}Si_4$ contains both Si_2 dumbbells and isolated Si^{4-} ions, and the further shift of the NMR signal to even lower frequencies (11.5 ppm) is consistent with the increased Li content. Possibly, the combined effect of slow Li^+ motion, residual ^7Li dipolar coupling, and disorder impedes collection of well-resolved signals from the three crystallographic sites. The $Li_{15}Si_4$ NMR signal is not influenced by the large Knight shifts in the case of $Li_{21}Si_5$ spectrum. Instead, the uniformly distributed Si ions in the Li matrix give rise to a broad resonance at 6.0 ppm, which indicates the most shielded Li environment(s). In summary, Li–Si phases containing Si–Si bonds in the form of rings and/or clusters and/or dumbbells have the most deshielded Li environments, with peaks found around 16–18 ppm, whereas the isolated silicon ions provide more shielded environments of Li, generating resonance at around 6.0 ppm. Li–Si compounds with intermediate lithiation degrees exhibited shifts between those two, reflecting combined types of structural motifs. The above relations between Li–Si phases and chemical shifts will be used to assist elucidation of the atomic-level structural changes during electrochemical lithiation of Si.

Through meticulous ex situ ^7Li NMR studies and pair distribution function (PDF) analysis of X-ray data [26], Grey *et al.* successfully elucidated the mechanism of Si lithiation/delithiation and provided an illustration as shown in Figure 3.13. The investigation mainly covers the first discharge/charge, the second discharge, and the charge following a partial discharge (before $Li_{15}Si_4$).

3.5.3.2 Phase transformation during first lithiation

Based on the PDF and NMR results, a lithiation/delithiation model for c-Si has been proposed, to elucidate the electrochemistry/atomic environmental evolution observed during the first discharge and the following charge/discharge cycles. This model mainly consists of four steps. Lithiation of c-Si starts and proceeds with Si-Si bond breakage from the surface, generating both lithiated and isolated Si anions and small lithiated Si clusters until the complete consumption of bulk c-Si and its conversion to amorphous Li-Si compounds. Subsequently, further lithiation breaks the remaining Si clusters to form mostly fully lithiated and isolated Si environments. The principle behind this mechanism is (i) the high energy barrier for breaking up the c-Si matrix, suggesting that it is more kinetically favorable for lithium to react with silicon clusters meanwhile forming fully isolated Si ions or smaller clusters,

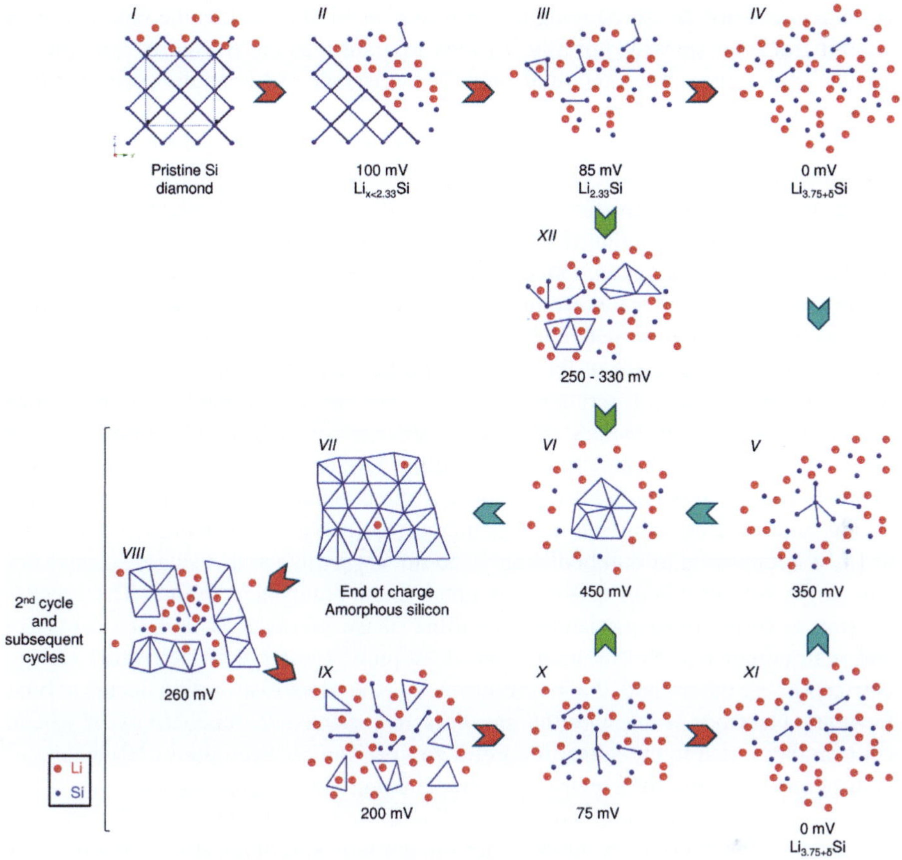

Figure 3.13 *Illustration of mechanisms by which silicon is lithiated and delithiated. I to IV correspond to lithiation of crystalline silicon discharged down to 0 mV. Red arrows denote discharge steps, whereas blue and green arrows denote charge steps. V to VIII correspond to the delithiation of fully lithiated silicon, and VIII to XI correspond to the relithiation of amorphous silicon to form fully lithiated silicon for cycle #x (x > 1) [26].*

than removing Si from the unreacted framework. (ii) Various Si clusters and anions are formed in the Si/Li-Si matrix, redistribution of them to form crystalline phase anticipated based on Li:Si atomic ratio of the amorphous mixture is energetically unfavorable, since the energy required for Si–Si bond breakage and/or rearrangement of the clusters is too high. Only when all of the Si clusters are further lithiated to form isolated Si atoms (<50 mV) is it possible for Li–Si amorphous phase to crystallize.

Generally, lithiation is believed to proceed from or close to the surface of Si particles, where Li^+ migrate through the SEI before reaching elemental Si. The large volume expansion accompanying the lithiation of Si also prefers the reaction front to start from the surface. Initially, Li ions migrate into the Si matrix through the surface sites, generating interstitial sites. The existence of Li and the corresponding increase in electron density of the silicon network will weaken the Si framework surrounding Li, ultimately resulting in breakage of Si–Si bond and the formation of small negatively charged Si clusters surrounded by Li^+ (Figure 3.13, II). Given the fact that (1) the large activation energy for breaking Si–Si bond, especially for c-Si, and (2) the Li insertion induced significant volume expansion, the following positive feedback can be anticipated. Once a fraction of Si framework breaks into isolated Si clusters, it will kinetically promote the continuous break up of these as formed Si clusters, rather than to lithiate and break the strong Si–Si bond originally existed in the framework. The result of the competition between the two processes (reacting with the framework vs reacting with the clusters) are determined by both kinetics and volume expansion factors, while both prevent the Li–Si system from reaching thermodynamic equilibrium (those crystalline phases with lower Li content than $Li_{15}Si_4$) even if the potential driving force is removed and the system is relaxed for a long time. The experimental results indicate that those intermediate phases such as Li_7Si_3 (consisting of dumbbells only) do not crystallize at this stage, because not any single type of silicon cluster dominates in the amorphous region. The crystallization to form single particular crystalline phase involves Si–Si bond breakage and rearrangements (Si migration), which requires too large an activation energy barrier. On the other hand, the inhomogeneous Li content distribution occurs at both local and electrode level, with the unsatisfactory electronic conductivity of silicon contributing much to the latter. A careful comparison between the Li NMR data of amorphous Li_xSi and the spectra from model compounds again indicates that the Si clusters are relatively small, likely consisting of dumbbells, stars, and also trimers/small rings. The continuous lithiation terminates when all of the silicon atoms of the tetrahedron matrix are reacted via Si–Si bond breakage (Figure 3.13, III, $Li_{2.33}Si$). Then the amorphous Li–Si phase crystallizes to form $Li_{15+\delta}Si_4$ phase (at about 50 mV), because at this end point the amorphous phase is largely dominated by isolated Si anions with only a few Si clusters left (Figure 3.13, IV). It is speculated that one reason why the metastable $Li_{15}Si_4$ phase rather than the thermodynamic phase $Li_{21}Si_5$ dominates the final lithiation product at room temperature is that this $Li_{15}Si_4$ phase can to some extent accommodate defects (such as Si–Si dumbbells) more readily than $Li_{21}Si_5$.

3.5.3.3 Phase transformation during first delithiation

The delithiation mechanism of the essentially fully lithiated phase starts from a few nuclei, which either come from the remaining incomplete lithiation phases or are newly formed upon delithiation. The amorphous (delithiated) Si phase grow directly from these nuclei without obvious formation of any intermediate structures/phases with multiple small clusters. The final amorphous Si phase formed at the end of delithiation contains Si tetrahedra, but there is no long-range order, even no order beyond the second silicon coordination shell.

The mechanism of delithiation is shown in Figure 3.13 (IV to VIII). Different from the lithiation of c-Si, delithiation of $Li_{15}Si_4$ could, in theory, start from both the surfaces and from the bulk attributing to the high lithium diffusivity in the lithiated phases (evidence for fast Li diffusion in Li_xSi phase comes from NMR spectra) [26]. Delithiation of $Li_{15}Si_4$ should produce Si clusters if it follows the thermodynamic pathway, and these are not seen yet (by NMR experiments). It seems that once a nucleation site (a small Si cluster) is formed, this serves as starting point for subsequent growth to an a-Si matrix. Growth is promoted by the fast Li^+ diffusion in the Li_xSi phase, outgoing from the growing Si cluster/a-Si domain. Continuous growth from a newly generated Si cluster (or an existing defect that remains unlithiated at the end of the first discharge) is favored over nucleation to form a new Si cluster, because the latter would require two highly negatively charged (4−) Si anions in the $Li_{15}Si_4$ to migrate closer enough to combine. In sharp contrast, migration of a Si anion (4−) to a growing pre-existing Si cluster, which has a much lower formal charge (per Si atom), should proceed more energetically favorable (Figure 3.13, V to VI). Thus, the Li_xSi phase with isolated Si ions is gradually delithiated to form an a-Si phase. A number of Li ions remain in a-Si inevitably; these are ascribed to superficial Li on larger a-Si domains or Li that coordinated to much larger Si clusters.

3.5.3.4 Phase transformation during second and following lithiation

Upon lithiation, the amorphous silicon matrix can be lithiated homogeneously, without preferred crystallographic directions, so the whole framework can be partially lithiated at higher voltages with a significantly reduced overpotential than that required to break the c-Si matrix, partially breaking down the a-Si network. The lower-voltage electrochemical process is attributed to the further breakdown of the lithiated silicon clusters, and the amorphous phase recrystallizes at the end of the lithiation, reaching the fully lithiated phase.

Specifically, the lithiation of a-Si (Figure 3.13, from VIII to XI) proceeds by breaking down the a-Si matrix, forming Si clusters. The a-Si matrix is more open than the c-Si framework and consequently can facilitate lithium penetration, resulting in essentially complete lithiation of the whole a-Si matrix, forming Si clusters at the end of the higher-voltage process (Figure 3.13, IX). Successively, these Si clusters are further lithiated and broken down (Figure 3.13, X). At the end of the second full discharge (0 mV vs Li/Li^+), recrystallization of the fully lithiated $Li_{15+\delta}Si_4$ phase occurs (Figure 3.13, XI).

3.5.3.5 Phase transformation during delithiation of intermediate lithiated phases

For intermediate (de)lithiated phases, the ones that contain both Si clusters and lithiated Si are delithiated; these clusters appear to serve as nucleation sites, and the system retraces a similar electrochemical pathway to that seen on lithiation. Differently, if the lithiation proceeds completely to the final form of $Li_{15}Si_4$, the phase with all equivalent fully isolated Si anions, there are seldom Si nucleation sites, and delithiation initiates with the growth of a few Si clusters from which larger Si domains form and eventually the a-Si phase is reached.

The different delithiation mechanism that follows partial discharge to 85 mV is mainly attributed to the presence of more Si clusters compared with the state at 0 mV. The remaining Si clusters serve as nucleation sites within a-Li_xSi phase, and these clusters grow during the lower-voltage charging (delithiation) process, via the inclusion of isolated Si atoms as well as possibly by fusing with some of other smaller clusters (Figure 3.13, green arrows, from III to XII to VII/from X to VI to VII). During higher-voltage delithiation, the small clusters fuse together, forming larger a-Si domains as further Li extraction proceeds. It should be noted that the ^{7}Li NMR spectra following the partial delithiation are very similar to those seen following the partial lithiation of the a-Si phase, the existence of a larger numbers of Si nuclei allows the delithiation to follow a pathway that is the reversal to the process during discharge.

3.5.4 Effect of surface oxide on electrochemical reactions

In the previous sections we briefly went through the basic phase change during Si initial lithiation/delithiation cycles, as well as discussed, the crystal plane-specific lithiation and local Si structural evolution during lithiation/delihitation process. These are essentially focusing on the elemental Si reacting with Li. However, it should also be noted Si has a strong affinity to O, so there always is a layer of oxide on the surface of either crystalline or amorphous Si. One should not neglect the electrochemistry of silicon oxide (SiO_x) when thinking about Si. Kristina Edström *et al.* investigated the evolution of SiO_x|Si interface, including the lithiation of the SiO_x, and subsequent Si alloying process, as well as SEI layer formation, associated with the first lithiation/delithiation cycle of Si/C/CMC composite anodes in Li||Si half-cells. This is accomplished by nondestructive depth-resolved photoelectron spectroscopy (PES) analysis by varying the energy of the photons (X-ray), different from the traditional destructive Ar ion sputtering. Specifically, the Si electrodes were studied at two different synchrotron facilities equipped with either soft X-ray PES (photon energy, hν = 100–800 eV) or hard X-ray PES (hν = 2 000 – 7 000 eV) in addition to the use of in-house XPS (Al Kα = 1 486.6 eV).

Figure 3.14 demonstrates Si 2p XPS spectra of the pristine Si electrode obtained by PES with various photon energies (hν). The classification of PES is as follows: Low energies (100–800 eV) indicate soft X-ray PES (MaxIV Lab. synchrotron), middle energy (1 486.6 eV) indicates in-house XPS, and high energies (2 000 – 7 000 eV) indicate hard X-ray PES (Bessy II synchrotron). The spectra show a first Si 2p peak (consist of 2p3/2 and 2p1/2 because of spin–orbit coupling) assigned to bulk elemental Si (~99.5 eV, in gray) and the second one assigned to SiO_x (~103.5 eV, red). Significant changes in peak intensity ratios were observed as the photon energy changes. In fact, the increase of the photon energy induces the increase of the kinetic energy of the photoelectrons. As the kinetic energy further determines the inelastic mean free path of the photoelectrons [38], the difference in photon energy results in a significant change of the analysis depth in PES. Nanosized Si powders

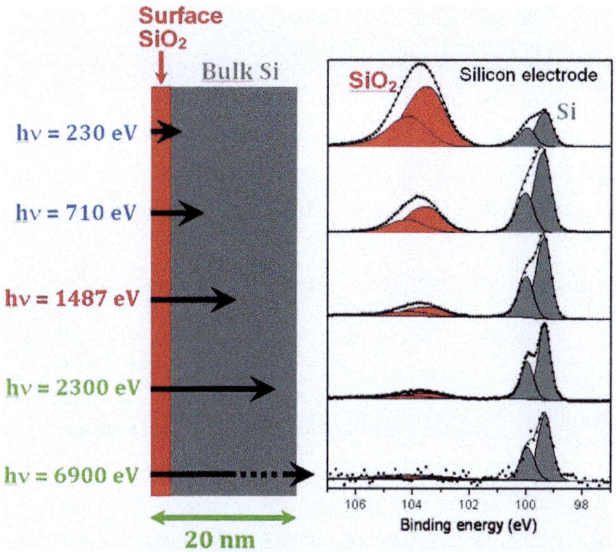

Figure 3.14 *Influence of the photon energy on the Si 2p spectrum of the pristine silicon electrode. Schematic view of the evolution of the analysis depth (arrow) as a function of the photon energy hν [37].*

(~50 nm average diameter) covered by a thin SiO_2 layer were used in the PES analysis. Figure 3.14 shows the various photon energy provide us the ability to focus on the extreme surface of the particles or, on the other end, to collect more signal from the bulk. This designed experimental approach can obtain information about the depth-specific distribution of chemical species from the samples and avoids the traditional argon-ion sputtering technique prevalently utilized for depth-profiling, which is destructive for samples especially those Li–ion battery electrodes with SEI layer. It should be noted that the increase of the photon energy also leads to decreased PES cross-sections. It is also worth noting that the surface oxide layer may be covalently bound with the CMC binder, which has been proposed as the origin of the effectiveness of this binder by several authors [39, 40]. Unfortunately, it is relatively difficult to monitor the change in these bonds or even verify their existence by XPS Si spectra since it does not obviously modify the local chemical environment of Si.

3.5.5 Species evolution during lithiation/delithiation of Si

Figure 3.15 exhibits the Si 2p spectra of the Si/C/CMC anode during the initial discharge/charge cycle and their analysis depth dependence obtained by changing the photon energy (hν). At the beginning of polarization to 0.5 V vs Li/Li^+, no changes were observed compared with the pristine electrode. The Si 2p SiO_x/Si ratio also remains unchanged. At this point, the lithiation of silicon has not initiated yet. The first observable changes occur after discharge to 0.1 V, with two new components

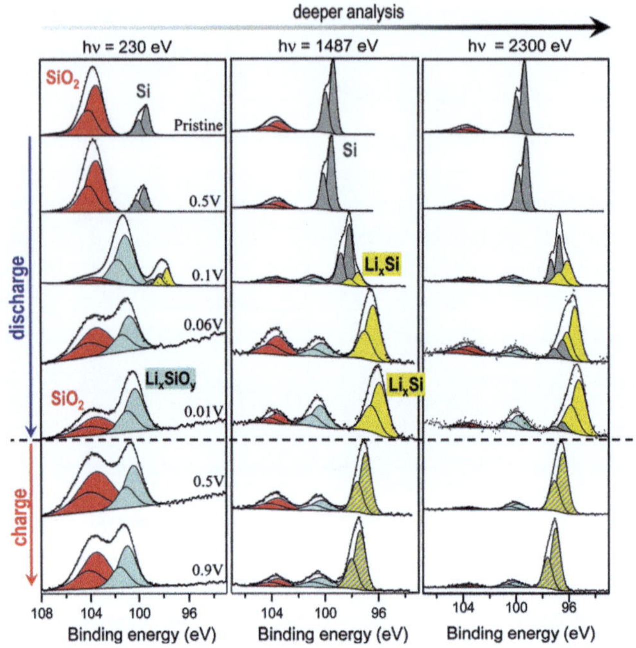

Figure 3.15 *Si 2p spectra of the Si/C/CMC composite electrodes upon the first discharge/charge cycle. Evolution as a function of the photon energy [37].*

appearing. The first one is associated with Li_xSi alloy, i.e. Si 2p3/2 at 97.5 eV (yellow in Figure 3.15). The spectra obtained when the electrode was discharged to 0.1 V showed initial sign of Si lithiation (yellow low-binding energy components in the Si spectra). This is a little bit different from a previous XRD study by Dahn *et al.* [8] and the detailed ^{7}Li NMR study of Si lithiation/delithiation by Key *et al.* [5], in which these authors show that the lithiation has not started yet at the exact beginning of the ~0.1 V plateau but begins to take place later in that plateau. This minor difference can be elucidated by the fact that XRD and NMR probes much more information from the whole sample with only a minor contribution from the surface, especially for those microsized samples. The PES results here reflect that the Li–Si alloy formation has just initiated at the surface of the Si particles at the starting point of the ~0.1 V plateau. The second component (Si 2p3/2 at 100–101 eV, blue in Figure 3.15) appearing at 0.1 V is attributed to a new phase containing Si. Comparison of spectra collected with three different penetration depth (photon energies 230, 1 487, and 2 300 eV) indicates that the intensity maximum occurs in the case with the smallest analysis depth. Thus this component should be located at the outermost surface of the particles. This surface phase is presumably assigned to Li_xSiO_y and will be discussed in detail later. At this stage, a shift of the Si component (gray in Figure 3.15) toward low binding energy is also noticed in all the Si 2p spectra. This shift could at first be incorrectly interpreted as the lithiation-induced binding energy decrease. However, the unchanged full width at half maximum of this component compared with respect to the pristine silicon material clearly reflects that this shift should be explained by a differential charging effect. Experimentally, the emission of photoelectrons leaves the surface positively charged during the XPS experiment. For electronic conductors, this positive charge can be easily compensated by the electrical connection between the sample and a common ground with the instrument. However, when a sample consists of mixed conducting and insulating compounds (like LIB composite electrodes), it is often observed that the signal shift differs from each type of compounds (a.k.a. differential charging effect). In particular, this is typical when the surface of the sample is modified by a passivation film such as SEI in LIB electrodes [41, 42]. The charging effect can partly be attributed to local charge accumulation due to poor electronic conductivity in the insulating material. Consequently, the gray shifted component is speculated to be Si that remains unlithiated. As further discharge continues (0.06 and 0.01 V vs Li/Li^+), the elemental Si component in Si 2p spectra disappears, with the exception for the largest analysis depth (2 300 eV), which indicates that the lithiation process is nearly complete. The ~0.1 V plateau corresponding to this stage is associated with an amorphization (from c-Si to a-Li_xSi) process, as reported previously [9, 10]. The Li_xSi component is shifted continuously to lower binding energies as lithiation proceeds, which is presumably explained by the increase of the lithium content x in Li_xSi but also might be caused by a differential charging effect particular for this electronically conductive phase. Additionally, when lowest analysis depth (230 eV) was used, there is no observable Li_xSi alloy component, showing that the lithiated core of the particles is covered by a thin shell of SiO_2 and/or Li_xSiO_y phases. It is noted that the SiO_2 component seems to abnormally increase in the spectra from 0.1 to 0.06 V. Meanwhile, the atomic percentage of Si observed at the surface drops from 5.9 to 0.9 percent, and consequently the absolute amount of SiO_2 decreases. After charging back to 0.5 and 0.9 V vs Li/Li^+, no significant changes associated with SiO_2 and Li_xSiO_y surface phases are observed. A gradually increasing binding energy of the Li_xSi component reflects the gradual decrease of the lithiation degree in the alloy. Different from discharge process, no separate Si or Li_xSi peaks are observed. This result does not agree well with that obtained by Key *et al.* [26] proposed by ^7Li NMR and X-ray PDF analyses, which indicated directly the formation of a-Si upon charge. Indeed, this difference originates from the different particle size and thus different delithiation behavior. In fact, Key *et al.* have worked with −325 mesh particles (<44 μm), while the Si particle size is ~50 nm in the PES study. Nanoparticles are good for their cycling stability, but it is rather difficult to maintain good electrical contact during delithiation-induced contraction. So incomplete and nonuniform delithiation is observed in this case [43], as supported by the obvious spectra broadening of the Li_xSi component.

An important finding of the PES study is the plausible interpretation of the Si containing component related to the additional surface phase appearing at 100−101 eV (blue in Figure 3.15). This component was firstly tentatively assigned as SiO_x or Li_xSiO_y phases. For comparison, SiO powder (Aldrich, <44 μm, a mixture of SiO_2 and Si) was analyzed, but the Si 2p spectra of this mixture mainly exhibited the

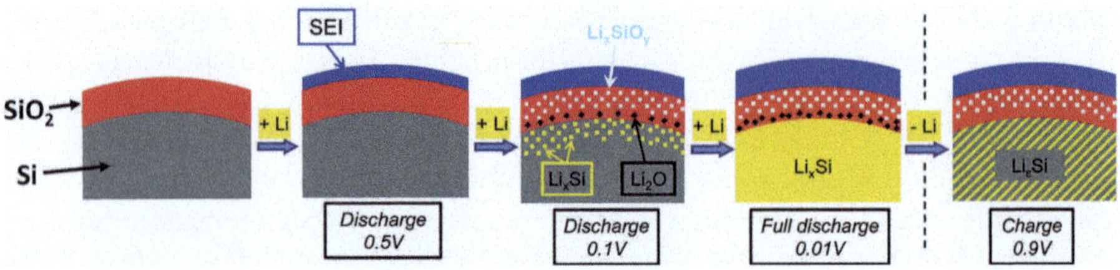

Figure 3.16 *Schematic view of the mechanisms occurring at the surface of the silicon nanoparticles. Formation of the SEI at the beginning of discharge. Formation of the Li–Si alloy upon further discharge, together with Li_2O and Li_xSiO_y interfacial phases. Partial reversibility upon charge [37].*

signature of 101.8 eV, which is still far from that observed for the surface species. Meticulous PES study of SiO_x interfacial layer between SiO_2 and Si has been reported previously [44, 45]. It is observed that the Si 2p component corresponding to SiO_x species is relatively weak compared to SiO_2. In addition, the SiO_x signal usually includes multiple components with shifting oxidation states (Si^{4+}, Si^{3+}, Si^{2+}, and Si^{+}); thus, the associated binding energies are within a range of as much as 4 eV. On the contrary, the blue component appeared in a limited binding energy range (Figure 3.15), not likely from SiO_x layer. Thus the more plausible assignment is Li_xSiO_y phase formed upon the electrochemical lithiation of the surface oxide layer. Li_2SiO_3 and Li_4SiO_4 were successfully synthesized and used for standard compound. The obtained Si 2p3/2 binding energies for Li_2SiO_3 and Li_4SiO_4 were 102.3 and 100.9 eV, respectively. The binding energy of the blue component in Figure 3.15 matches well with that of Li_4SiO_4. Consequently, it is tentatively concluded that Li_4SiO_4 appears to be the phase generated at the surface of the silicon upon lithiation of surface silicon oxides.

Figure 3.16 schematically summarizes the electrochemical lithiation/delithiation mechanisms near the surface of Si, with the inclusion of the evolution of surface oxide layer. As the polarization proceeds to 0.5 V vs Li/Li^+, Li^+ has not reacted with either SiO_2 or Si, but already a thin SEI layer has formed on the surface. The properties and composition of the SEI is not the subject of this chapter and will be discussed in another chapter.

Upon further discharge, the SEI layer becomes thicker at the ~0.1 V plateau (vs Li/Li^+), meanwhile lithiation of Si initiates, forming Li–Si alloy. Lithiation of the surface SiO_2 layer results in the formation of the above-discussed Li_4SiO_4 as well as Li_2O. At full discharge state of ~0.01 V vs Li/Li^+, the lithiation should be completed; however, dynamic hindrance may lead to some unreacted silicon remaining as the core of the Li_xSi particle. The amount of Li_2O increases as lithiation proceeds continuously, reaching its maximum amount at the end of discharge; however, there are still unreacted SiO_2 remaining on the surface. The SEI thickness also increased to its maximum at full discharge, with its composition not significantly changed. After charging back to ~0.9 V vs Li/Li^+, most of Li^+ have been extracted, with only limited amount remaining alloyed with Si ($Li_{\varepsilon}Si$ with $\varepsilon \rightarrow 0$). The SEI thickness decreased slightly, accompanying the disappearance of Li_2O.

3.6 Outlook

Learning from those seminal works, we now have a clear picture of the phase evolutions and localized atomic structural changes associated with the Si lithiation/delithiation process. This knowledge can guide us on the right path to the successful utilization of Si anodes. For instance, it is suggested to limit the cutoff potential above the crystallization of a-Li_xSi to avoid the two-phase delithiation of c-$Li_{15}Si_4$ process, which is more challenging than the delithiation of a-Li_xSi. The existence of surface oxide and its reduction products Li_xSiO_y should also be considered when designing binder or SEI, because mostly those determine interfacial bonding. It is believed that stable cycling of Si-based anodes with high coulombic efficiencies will be realized in the near future with dedicated effort in electrode and/or electrolyte design.

References

[1] Lai San-Cheng., Lai S.-C. 'Solid lithium-silicon electrode'. *Journal of the Electrochemical Society*. 1976;**123**(8):1196–7.

[2] Wen C.J., Huggins R.A. 'Chemical diffusion in intermediate phases in the lithium-silicon system'. *Journal of Solid State Chemistry*. 1981;**37**(3):271–8.

[3] Sharma R.A., Seefurth R.N. 'Thermodynamic properties of the lithium-silicon system'. *Journal of the Electrochemical Society*. 1976;**123**(12):1763–8.

[4] Wilson A.M., Dahn J.R. 'Lithium insertion in carbons containing nanodispersed silicon'. *Journal of the Electrochemical Society*. 1995;**142**(2):326–32.

[5] Key B., Bhattacharyya R., Morcrette M., Seznéc V., Tarascon J.-M., Grey C.P. 'Real-time NMR investigations of structural changes in silicon electrodes for lithium-ion batteries'. *Journal of the American Chemical Society*. 2009;**131**(26):9239–49.

[6] Braga M.H., Dębski A., Gąsior W. 'Li–Si phase diagram: enthalpy of mixing, thermodynamic stability, and coherent assessment'. *Journal of Alloys and Compounds*. 2014;**616**(8):581–93.

[7] Chevrier V.L., Zwanziger J.W., Dahn J.R. 'First principles study of Li–Si crystalline phases: charge transfer, electronic structure, and lattice vibrations'. *Journal of Alloys and Compounds*. 2010;**496**(1-2):25–36.

[8] Li J., Dahn J.R. 'An in situ X-ray diffraction study of the reaction of Li with crystalline Si'. *Journal of the Electrochemical Society*. 2007;**154**(3):A156–61.

[9] Obrovac M.N., Christensen L. 'Structural changes in silicon anodes during lithium insertion/extraction'. *Electrochemical and Solid-State Letters*. 2004;**7**(5):A93–6.

[10] Hatchard T.D., Dahn J.R. 'In situ XRD and electrochemical study of the reaction of lithium with amorphous silicon'. *Journal of the Electrochemical Society*. 2004;**151**(6):A838–42.

[11] Limthongkul P., Jang Y.-I., Dudney N.J., Chiang Y.-M. 'Electrochemically-driven solid-state amorphization in lithium-silicon alloys and implications for lithium storage'. *Acta Materialia*. 2003;**51**(4):1103–13.
[12] Limthongkul P., Jang Y.-I., Dudney N.J., Chiang Y.-M. 'Electrochemically-driven solid-state amorphization in lithium–metal anodes'. *Journal of Power Sources*. 2003;**119**:604–9.
[13] Chevrier V.L., Dahn J.R. 'First principles studies of disordered lithiated silicon'. *Journal of the Electrochemical Society*. 2010;**157**(4):A392–8.
[14] Chan M.K.Y., Wolverton C., Greeley J.P. 'First principles simulations of the electrochemical lithiation and delithiation of faceted crystalline silicon'. *Journal of the American Chemical Society*. 2012;**134**(35):14362–74.
[15] Gu M., Wang Z., Connell J.G., *et al*. 'Electronic origin for the phase transition from amorphous Li(x)Si to crystalline Li15Si4'. *ACS Nano*. 2013;**7**(7):6303–9.
[16] Chen J., Fan X., Li Q., *et al*. 'Electrolyte design for LiF-rich solid–electrolyte interfaces to enable high-performance microsized alloy anodes for batteries'. *Nature Energy*. 2020;**5**(5):386–97.
[17] Obrovac M.N., Krause L.J. 'Reversible cycling of crystalline silicon powder'. *Journal of the Electrochemical Society*. 2007;**154**(2):A103–8.
[18] Li J., Smith A., Sanderson R.J., Hatchard T.D., Dunlap R.A., Dahn J.R. 'In situ ^{119}Sn Mössbauer effect study of the reaction of lithium with Si using a Sn probe'. *Journal of the Electrochemical Society*. 2009;**156**(4):A283–8.
[19] Lee S.W., McDowell M.T., Choi J.W., Cui Y. 'Anomalous shape changes of silicon nanopillars by electrochemical lithiation'. *Nano Letters*. 2011;**11**(7):3034–9.
[20] Liu X.H., Zheng H., Zhong L., *et al*. 'Anisotropic swelling and fracture of silicon nanowires during lithiation'. *Nano Letters*. 2011;**11**(8):3312–18.
[21] Goldman J.L., Long B.R., Gewirth A.A., Nuzzo R.G. 'Strain anisotropies and self-limiting capacities in single-crystalline 3D silicon microstructures: models for high energy density lithium-ion battery anodes'. *Advanced Functional Materials*. 2011;**21**(13):2412–22.
[22] Liu X.H., Wang J.W., Huang S., *et al*. 'In situ atomic-scale imaging of electrochemical lithiation in silicon'. *Nature Nanotechnology*. 2012;**7**(11):749–56.
[23] Ross F.M., Tersoff J., Reuter M.C. 'Sawtooth faceting in silicon nanowires'. *Physical Review Letters*. 2005;**95**(14):146104.
[24] Johnson W. 'Thermodynamic and kinetic aspects of the crystal to glass transformation in metallic materials'. *Progress in Materials Science*. 1986;**30**(2):81–134.
[25] Tu K.N. 'Selective growth of metal-rich silicide of near-noble metals'. *Applied Physics Letters*. 1975;**27**(4):221–4.
[26] Key B., Morcrette M., Tarascon J.-M., Grey C.P. 'Pair distribution function analysis and solid state NMR studies of silicon electrodes for lithium ion batteries: understanding the (de)lithiation mechanisms'. *Journal of the American Chemical Society*. 2011;**133**(3):503–12.
[27] Liu X.H., Zhang L.Q., Zhong L., *et al*. 'Ultrafast electrochemical lithiation of individual Si nanowire anodes'. *Nano Letters*. 2011;**11**(6):2251–8.

[28] Wang J.W., He Y., Fan F., *et al.* 'Two-phase electrochemical lithiation in amorphous silicon'. *Nano Letters*. 2013;**13**(2):709–15.

[29] Weichert K., Sigle W., van Aken P.A., *et al.* 'Phase boundary propagation in large LiFePO4 single crystals on delithiation'. *Journal of the American Chemical Society*. 2012;**134**(6):2988–92.

[30] Huang S., Zhu T. 'Atomistic mechanisms of lithium insertion in amorphous silicon'. *Journal of Power Sources*. 2011;**196**(7):3664–8.

[31] Chan C.K., Peng H., Liu G., McIlwrath K., Zhang X.F., Huggins R.A. 'High-performance lithium battery anodes using silicon nanowires. materials for sustainable energy: a collection of peer-reviewed research and review articles from nature publishing group: world scientific'. 2011:187–91.

[32] Nesper R. 'Structure and chemical bonding in Zintl-phases containing lithium'. *Progress in Solid State Chemistry*. 1990;**20**(1):1–45.

[33] von Schnering H.G., Nesper R., Curda J., Tebbe K.-F. 'Li12Si7, eine Verbindung MIT trigonal-planarem Si4-Cluster und planaren Si5-Ringen'. *Angewandte Chemie*. 1980;**92**(12):1070–1070.

[34] Nesper R., von Schnering H.G., Curda J. 'Li12Si7, eine Verbindung MIT trigonal-planaren Si4-Clustern und isometrischen Si5-Ringen'. *Chemische Berichte*. 1986;**119**(12):3576–90.

[35] Klemm W., Struck M. 'Notiz über die Verbindungen zwischen lithium und Silicium'. *Zeitschrift für anorganische und allgemeine Chemie*. 1955;**278**(3-4):117–21.

[36] Nesper R., von Schnering H.G. 'Li21Si5, a Zintl phase as well as a Hume-Rothery phase'. *Journal of Solid State Chemistry*. 1987;**70**(1):48–57.

[37] Philippe B., Dedryvère R., Allouche J., *et al.* 'Nanosilicon electrodes for lithium-ion batteries: interfacial mechanisms studied by hard and soft X-ray photoelectron spectroscopy'. *Chemistry of Materials*. 2012;**24**(6):1107–15.

[38] Seah M.P. 'Data compilations: their use to improve measurement certainty in surface analysis by Aes and XPS'. *Surface and Interface Analysis*. 1986;**9**(2):85–98.

[39] Lestriez B., Bahri S., Sandu I., Roué L., Guyomard D. 'On the binding mechanism of CMC in Si negative electrodes for Li-ion batteries'. *Electrochemistry Communications*. 2007;**9**(12):2801–6.

[40] Hochgatterer N.S., Schweiger M.R., Koller S., *et al.* 'Silicon/graphite composite electrodes for high-capacity anodes: influence of binder chemistry on cycling stability'. *Electrochemical and Solid-State Letters*. 2008;**11**(5):A76–80.

[41] Ehinon K.K.D., Naille S., Dedryvère R., Lippens P.-E., Jumas J.-C., Gonbeau D. 'Ni $_3$ Sn $_4$ electrodes for Li-ion batteries: Li–Sn alloying process and electrode/electrolyte interface phenomena'. *Chemistry of Materials*. 2008;**20**(16):5388–98.

[42] Dedryvere R., Foix D., Franger S., Patoux S., Daniel L., Gonbeau D. 'Electrode/electrolyte interface reactivity in high-voltage spinel LiMn1. 6Ni0. 4O4/Li4Ti5O12 lithium-ion battery'. *The Journal of Physical Chemistry C*. 2010;**114**(24):10999–1008.

[43] Oumellal Y., Delpuech N., Mazouzi D., *et al.* 'The failure mechanism of nano-sized Si-based negative electrodes for lithium ion batteries'. *Journal of Materials Chemistry*. 2011;**21**(17):6201–8.

[44] Himpsel F., McFeely F., Taleb-Ibrahimi A., Yarmoff J., Hollinger G. 'Microscopic structure of the SiO_2/Si interface'. *Physical Review. B*. 1988;**38**(9):6084.

[45] Grunthaner P.J., Hecht M.H., Grunthaner F.J., Johnson N.M. 'The localization and crystallographic dependence of Si suboxide species at the SiO_2/Si interface'. *Journal of Applied Physics*. 1987;**61**(2):629–38.

Chapter 4
Electrolytes used in silicon anodes

Zhangxing Shi[1], Sisi Jiang[1], and Lu Zhang[1,2]

As the most versatile and attractive energy storage system, lithium-ion batteries (LIBs) continue to draw vast attention because of their high energy density, low self-discharge property, nearly zero-memory effect, high open-circuit voltage, and long lifespan [1–4]. While LIBs promoted worldwide development of the portable electronic devices that have profoundly changed our daily lives and remain the most viable choice for electric vehicles (EVs) and hybrid EVs, their limited capacities (no higher than 200 mAh/g) hinder wider transportation applications. Silicon (Si) emerges as one of the most promising next-generation anode materials by providing impressive gravimetric (3 579 mAh/g, based on $Li_{15}Si_4$) and volumetric (9 786 mAh/cm^3, based on the initial volume of Si) capacities, as well as high abundance, low cost, negligible toxicity, high safety, and low operating potential (0.2–0.4 V vs. Li/Li^+) [5–7]. However, such extraordinarily high capacity comes with a catch, that is Si anodes suffer up to a 400 percent volume expansion upon lithiation [7], which results in particle pulverization, loss of conductivity, and an unstable electrode-electrolyte interface, essentially leading to fast capacity fading. While approaches have been actively engaged to mitigate this issue, including engineering Si particles and surrounding components, significant efforts are still needed to tackle this important barrier.

In this chapter, we focus on a crucial and indispensable component, the electrolyte, of LIBs, and summarize and discuss what have been done to mitigate the inferior performance from the intrinsic volume changes of Si anode. The electrolyte is typically composed of a lithium salt as the solute and carbonate solvents as the medium; and the main function is to provide ionic conductivity to enable electrochemical reactions. Therefore, the performance of LIBs, evaluated by rate, cycling life, and safety, is greatly affected by the properties of electrolytes, such as ionic conductivity, compatibility, stability, and flammability [8]. Importantly, the so-called compatibility and electrochemical stability, especially on the anode side, depend mostly on the formation of kinetic barriers, which usually happens at the interfaces during the first formation cycle. In the case of Si anodes, this process is

[1]Chemical Sciences and Engineering Division, Argonne National Laboratory, USA
[2]Joint Center for Energy Storage Research, Argonne National Laboratory, USA

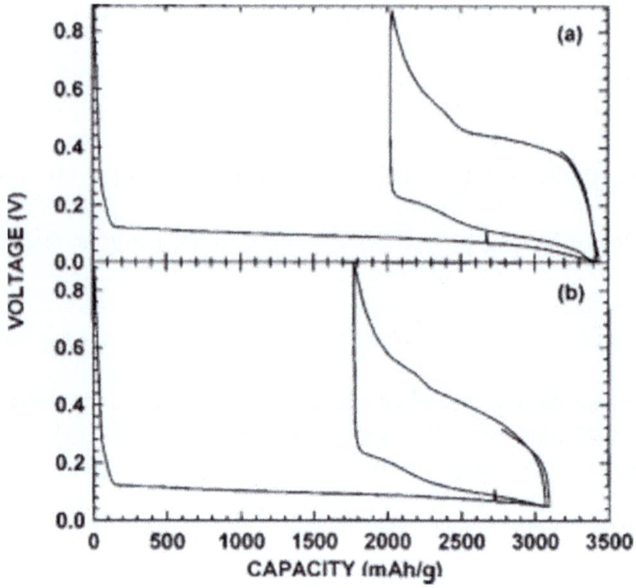

Figure 4.1 Voltage profiles of crystalline Si half-cells with lower cut-off voltages at (a) 0 V and (b) 50 mV. Reprinted with permission from [5] Copyright 2004 Electrochemical Society.

particularly unusual for the following reasons. First, the first lithiation of Si is different from the one of graphite and involves phase transition. Figure 4.1a–b shows the voltage curves of Li-Si half-cells with lower cut-off at 0 V and 50 mV, respectively. Both cells show a single flat plateau during the first lithiation, indicating crystalline Si becoming amorphous and a single two-phase region. Moreover, the crystalline/amorphous conversion is dependent on the lower cut-off potentials. The amorphous Si converts to crystalline $Li_{15}Si_4$ again if exposed to voltages of less than 50 mV, while Si restricted to above a 50-mV voltage stays completely amorphous. Because two-phase regions have been associated with high internal stresses, leading to particle cracking, poor electrical contact, and capacity fading, cycling Si above 50 mV should benefit the performance [5, 6]. Second, as lithiation of Si occurs, not only does the voltage go down but the volume of Si electrode material starts to expand. This scenario is very damaging because in addition to the very negative voltage where nearly all electrolyte is thermodynamically unstable, the expansion of Si particles creates more surface area, which accelerates the decomposition of electrolytes, leading to excess formation of the solid electrolyte interface (SEI) layer and to low initial capacity and Coulombic efficiency [9]. In addition, the repeated lithiation/delithiation process may crack this SEI layer, leading to even more growth. Given all of those challenges, the electrolyte for Si anodes has to meet even more stringent requirements. This chapter will provide an overview of the electrolyte used in Si-based LIBs.

4.1 Electrolytes based on aprotic solutions

Aprotic organic solutions of one or more Li salts are widely used as electrolytes in LIBs, which consist of Li salts, aprotic organic solvents, and functional additives. Naturally, these electrolytes have been directly applied in Si-based LIBs as they share some common desirable properties for a typical energy storage device, including high ionic conductivity, good chemical and electrochemical stability, and the ability to form effective SEI [10]. On the other hand, the common drawbacks of such electrolytes are also retained, including flammability of the organic solvents, moisture sensitivity and decomposition of lithium salts, and an unstable SEI layer. The SEI layer is one particular challenge due to the more severe surface cracking that results from the aforementioned drastic volume expansion of Si anodes. Therefore, re-examination of the electrolyte components seems to be a very necessary activity, and re-engineering the formulation of electrolytes tailored to the Si anode has been a very active topic.

4.1.1 Lithium salts

As a key component in organic liquid electrolytes, Li salts provide a source of Li ions that are essential to enabling the ionic conductivity of the electrolytes. The choice of Li salts is based on the following requirements: (1) high solubility in organic electrolyte solvents; (2) good mobility of Li ions in the electrolyte media; (3) compatibility or inertness against solvents, electrode materials, separator, and cell parts; (4) sufficient electrochemical stability within the operating window of LIBs; and (5) good thermal stability for high-temperature applications. However, only a handful of Li salts able to meet these requirements have been utilized for electrolyte formulation in LIBs, and even fewer Li salts were investigated in the electrolytes for Si-based LIBs (see Figure 4.2).

Lithium hexafluorophosphate ($LiPF_6$) is the salt used most often in common LIB electrolytes; it also manages to work with Si anodes. The complex anion (PF_6^-) has a well-distributed negative charge, which makes $LiPF_6$ more soluble in low dielectric solvents than other lithium salts, such as lithium fluoride (LiF), lithium chloride (LiCl), lithium oxide (Li_2O), lithium carbonate (Li_2CO_3), etc. $LiPF_6$ is considered inert toward electrolyte solvents, electrode materials, and other cell parts. However, $LiPF_6$ is very sensitive to moisture (even at the order of the parts per million (ppm) level) and undergoes decomposition to generate phosphorus pentafluoride (PF_5),

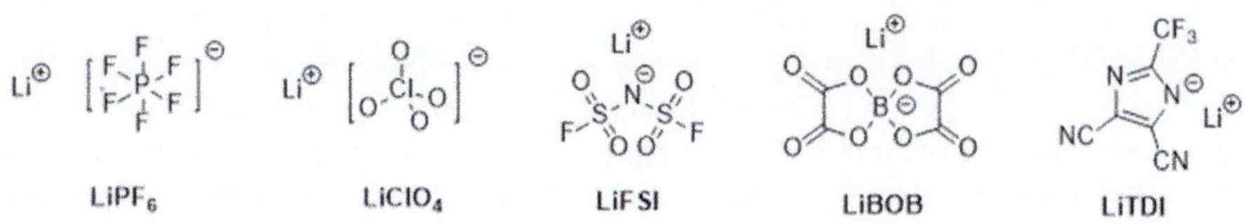

Figure 4.2 Representative examples of lithium salts used in liquid electrolytes for Si-based electrodes

hydrofluoric acid (HF), and phosphoryl fluoride ($O{=}PF_3$), which are highly reactive and corrosive [8]. The thermal stability of $LiPF_6$ in solution is also relatively low. For example, the decomposition temperature of $LiPF_6$ in an ethylene carbonate/dimethyl carbonate (EC/DMC) mixture is around 80 °C, which hinders its application with high-temperature electrolytes [11]. Hence, several Li salts have been studied to replace $LiPF_6$ or to be mixed with $LiPF_6$ to provide better performance; these include lithium perchlorate ($LiClO_4$) [12], lithium bis(fluorosulfonyl)imide (LiFSI), lithium bis(oxalato)borate (LiBOB), and lithium 4,5-dicyano-2-(trifluoromethyl) imidazolide (LiTDI). Although $LiPF_6$ is not the perfect choice for any given property and even possesses several drawbacks, because of its well-balanced combination of desired properties, $LiPF_6$ is still the salt of choice in most Si LIB studies [8].

$LiClO_4$ is another salt commonly used in liquid electrolytes in conventional LIBs. It has good solubility in carbonate solvents, high conductivity (9.0 mS/cm in EC/DMC at 20 °C), and great stability against oxidative decomposition at the cathode (up to 5.1 V on a spinel cathode surface in EC/DMC) [13]. Compared to $LiPF_6$, $LiClO_4$ has two primary advantages: (1) it is less hygroscopic and more stable against trace amounts of moisture, and (2) the SEI formed with $LiClO_4$ electrolytes has lower impedance than that formed with the $LiPF_6$ electrolytes [14]. However, $LiClO_4$ has rarely been utilized as an electrolyte solute for Si-based electrodes; only a few examples of $LiClO_4$ electrolytes have been reported for the cycling performance and SEI studies of Si-based electrodes [12, 15, 16]. The major concern is that $LiClO_4$, as a strong oxidant resulting from the high oxidation state of chlorine (VII), can cause severe side reactions, especially at high temperature and high current charge [8]. Lucht *et al.* studied the effects of $LiClO_4$ and $LiPF_6$ in EC:diethyl carbonate (DEC) (1:1) on Si electrodes [17]. As shown in Figure 4.3, there is no obvious difference in cycling performance between the two salts in the early stage of cycling tests; however, the long-term capacity retention of the $LiClO_4$ electrolyte is much lower than that of the $LiPF_6$ electrolyte.

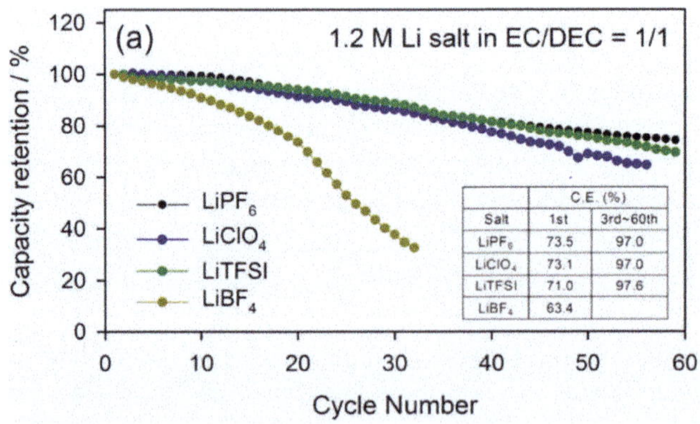

Figure 4.3 *Capacity retention levels of Si electrodes cycled with various electrolytes. Reprinted with permission from [17]. Copyright 2017 Electrochemical Society.*

Imide Li salts, such as bis(trifluoromethanesulfonyl)imide (LiTFSI) and LiFSI, have attracted increased attention. These nitrogen-based anions rely on the electron-withdrawing effect of the N-substituents to reduce the electron density, facilitating the dissociation of the corresponding salts. Compared to the well-studied LiTFSI, LiFSI replaces the fluorinated alkyl with F, which results in more promising properties, including higher solubility, better ionic conductivity, and greater thermal stability up to 180 °C [18]. However, the practical application of LiFSI electrolytes was hindered by the corrosion of the aluminum current collector that happens at high potential [10]. In 2013, Edström and colleagues reported improved cycling performance of nanosilicon electrodes using LiFSI instead of $LiPF_6$ in Li-Si half-cells [19]. The X-ray photoelectron spectroscopy (XPS) analysis showed that LiFSI passivates the electrode surface and prevents the fluorination process of the surface of Si particles (from silicon dioxide or silica (SiO_2) to SiO_xF_y) during long-term cycling, which helps maintain the favorable interactions between binders and active materials. In 2015, Abraham *et al.* studied the cycling performance of pouch cells (*c.* 20 mAh) with a $Li_{1.03}(Ni_{0.5}Co_{0.2}Mn_{0.3})_{0.97}O_2$ (NCM 523) cathode, Si-graphite anode, and LiFSI-containing electrolyte [20]. The results showed that cells containing only LiFSI suffer rapid capacity fading, whereas cells with LiFSI-fluoroethylene carbonate (FEC) and LiFSI-vinylene carbonate (VC) showed significant improvements in cycling impedance and capacity retention.

LiBOB is another salt that is built on a weakly associating anion [21]. While it provides good solubility and conductivity, LiBOB is more frequently considered as an electrolyte additive as LiBOB can prompt a sturdier SEI layer [22]. However, LiBOB has been studied more recently as an electrolyte solute in liquid electrolytes for Si-based electrodes. Yu and colleagues examined the effects of the mixed electrolytes—consisting of LiBOB and $LiPF_6$ or $LiClO_4$—on the cycling performance of Si/graphite/disordered carbon (Si/G/DC) composite electrodes [16]. By varying the ratio of the salts, the influences of LiBOB were systematically investigated by constant current charge–discharge and electrochemical impedance spectra (EIS) techniques. The results showed that better electrochemical performance of the Si/G/DC composite electrodes was achieved by using electrolytes containing mixed salts compared to a single-salt electrolyte. Among the tested compositions, the Si/G/DC electrode with 0.5 M LiBOB/0.38 M LiPF6 electrolyte mixture containing VC exhibited a high reversible capacity (512 mAh/g) and excellent capacity retention (*c.* 84 percent after 220 cycles).

LiTDI is a promising electrolyte solute with great electrochemical stability (up to 4.6 V) and excellent thermal stability (up to 250 °C) [11]. LiTDI solutions have a higher Li^+ transference number but lower ionic conductivity than $LiPF_6$ solutions at the same concentration. The LiTDI electrolyte has been investigated for Si-based electrodes and showed better electrochemical performance than a conventional $LiPF_6$ electrolyte. Younesi and colleagues utilized the LiTDI electrolyte for nanosilicon composite electrodes with FEC or VC [23]. The results demonstrated that the Si-based electrode can maintain a capacity of 1200 mAh/g for more than 100 cycles. Unlike $LiPF_6$ electrolytes, the LiTDI electrolyte and its decomposition products showed no side reactions with the Si nanoparticles (SiNPs). It was also

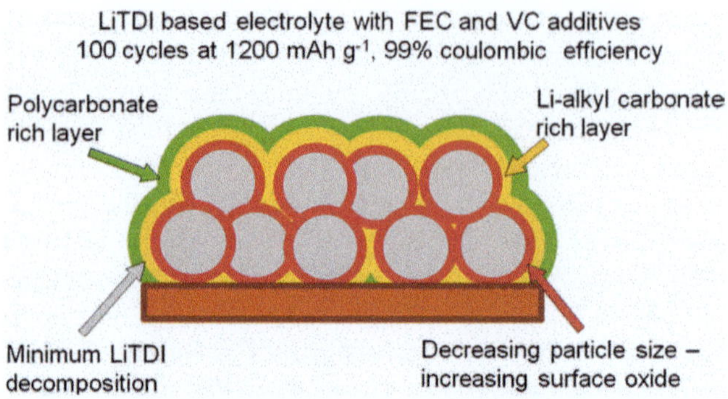

Figure 4.4 SEI formation at the surface of Si nanoparticles upon cycling with LiTDI electrolytes. Reprinted with permission from [23]. Copyright 2016 American Chemical Society.

suggested that the resulting SEI containing polycarbonates and the high LiF content formed in the LiTDI electrolytes was favorable for long-term cycling performance (Figure 4.4).

4.1.2 Aprotic organic solvents

In Si-based LIBs, the strongly reducing anodes and the strongly oxidizing cathodes prevent the use of any solvents with labile protons. In order to fulfill the basic requirements of electrolytes, an ideal aprotic solvent should possess the following properties: (1) a high dielectric constant to dissolve a sufficient amount of the Li salts; (2) low viscosity for facile ion transport; (3) good chemical and electrochemical stability to stay inert during the cell cycling; (4) a low melting point and a high boiling point to accommodate a wide operational temperature window; and (5) the ability to form a stable SEI for long-term cycling [8]. These multifaceted requirements make it impractical for a solo solvent system to meet all of these criteria, and thus a mixture of two or more solvents is generally used instead. A wide range of aprotic solvents has been investigated for Si-based anodes, which can be categorized as carbonates, ethers, and other organic solvents.

4.1.2.1 Carbonates

The majority of electrolytes studied for Si-based LIBs in previous reports are $LiPF_6$ dissolved in a mixture of cyclic alkyl carbonates and one or more linear alkyl carbonates. When mixed together, a synergistic effect can occur, including high anodic stability, high solubility of Li salts, and low viscosity to promote ion conductivity. The common cyclic alkyl carbonates are EC, propylene carbonate (PC), FEC, and VC. The common linear carbonates include DMC, diethyl carbonate (DEC), and ethyl methyl carbonate (EMC). While the state-of-the-art electrolyte in LIBs, $LiPF_6$ in EC/EMC mixtures, does not seem to work well with Si-based LIBs, a PC-based electrolyte has been investigated with promising performance. Lu *et al.* studied the

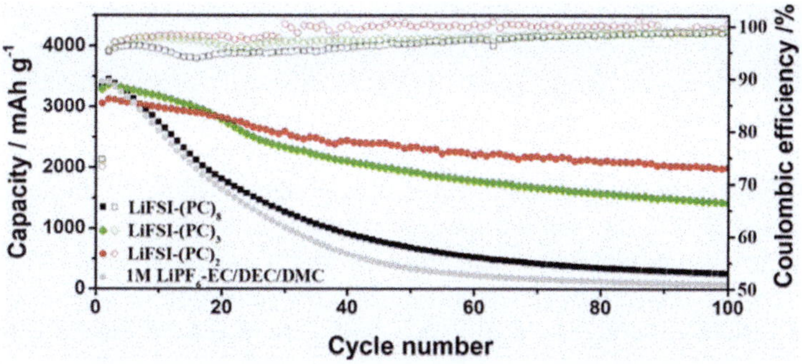

Figure 4.5 *Delithiation capacities of LiFSI-PC electrolytes with various concentrations. Reprinted with permission from [24]. Copyright 2018 Wiley-VCH Verlag GmbH & Co. KGaA, Weinheim.*

cycling performance of SiNPs in a concentrated electrolyte consisting of LiFSI and PC with a 1:2 molar ratio (LiFSI-$(PC)_2$) [24]. The SiNP electrode showed a significantly improved cycling performance with an initial delithiation capacity of *c.* 3 000 mAh/g and a capacity retention of 67 percent after 100 cycles, which is about a 6.8 times higher capacity than the cells with the dilute LIFSI-PC electrolyte (Figure 4.5). Raman spectra revealed that most of the PC solvent and FSI anions are complexed by Li^+ to form a specific solution structure like a fluid polymeric network. The reduction of the FSI anions played an important role because of the increased concentration of contact ion pairs or aggregates, which contribute to the formation of a more mechanically robust and chemically stable complex SEI layer. The complex SEI layer can effectively suppress the morphology evolution of Si particles and self-limit the excessive growth, which mitigates the crack propagation of the Si electrode and the deterioration of the kinetics [24]. Alpas and colleagues found that no cracks were formed on the Si surface of Si-aluminum electrodes at a low voltage scan rate of 0.50 mV/s when cycled in PC-based electrolytes [25]. However, cracks were observed on the Si surface when cycled in EC-based electrolytes, and the extent of the cracking was arrested by the surrounding aluminum matrix and subsequently closed during delithiation. Lithiation at a high scan rate (5.00 mV/s) increased the probability of crack formation; however, only a few cracks formed, and these were closed during delithiation in PC-based electrolytes. The results suggested that composite Si-aluminum anodes could provide durable electrodes for LIBs, especially when using a PC-based electrolyte.

While FEC has been more frequently used as an electrolyte additive for Si-based electrodes, it also can serve as a good electrolyte solvent or co-solvent. Liang and colleagues investigated FEC-based electrolytes consisting of 50 wt% FEC as the co-solvent and compared them with the electrolyte using FEC as the additive (10 wt%) for a freestanding Si-carbon nanotube composite electrode [26]. The EC-free, FEC-based electrolyte achieved higher specific capacity and better capacity retention during long-term cycling. After 500 cycles, the cells maintained a remarkable capacity retention when using DEC-FEC (1:1 wt./wt.), which is 88 and 60 percent higher than the cells using EC-DEC-FEC (45:45:10 wt./wt./wt.) and EC-FEC (1:1 wt./wt.), respectively. Through scanning electron microscopy coupled with energy dispersive X-ray (SEM-EDX) and XPS analyses, a possible reaction route for the formation of fluorinated semi-carbonates and polyolefins from FEC is proposed. The inferior cell performance related to the EC-containing electrolytes is likely attributable to the formation of more polyolefins, which do not favor Li ion transport. Somorjai *et al.* studied the electrolyte interface on Si surfaces by sum frequency generation vibrational spectroscopy and X-ray reflectivity [27, 28]. The cyclability of Si anodes in LIBs is affected by the reduction of the electrolyte on the anode surface to produce a stable SEI layer. One of the key steps for the major improvement of LIBs is unraveling the SEI's structure-related diffusion properties since charge and discharge rates of LIBs are diffusion-limited. The authors combined two surface-sensitive techniques to explore the first monolayer on the Si surface and probe the first several layers of electrolyte, respectively. Four electrolytes consisting of 1 M $LiClO_4$ in EC, FEC, and their mixtures (e.g., EC/FEC = 1/1) were investigated. The results suggested that the addition of FEC to the EC solution causes the first monolayer to rearrange itself more perpendicular to the anode surface, whereas subsequent layers are less affected and tend to maintain their, on average, surface-parallel arrangements. Feng *et al.* reported stable Si-based anode cycling in a fluorinated carbonate electrolyte with high salt-to-solvent ratios [29]. The authors developed a safer high salt-to-solvent electrolyte consisting of a mixture of nonflammable solvents, di-2,2,2-trifluoroethyl carbonate (TFEC), and FEC. The results showed that this electrolyte could not only enhance the cycling stability for the Si nanoparticle anode but also solve the safety issues of conventional carbonate solvents. A high initial reversible capacity of 2644 mAh/g and a low capacity fading rate (only 0.064 percent per cycle) after 300 cycles were achieved when using 3.5 M LiFSI in FEC-TFEC solvents (Figure 4.6). However, the authors did not differentiate the impacts from LiFSI and fluorinated carbonates.

4.1.2.2 Ethers

Ethers are proposed as alternative electrolyte solvents because of their low viscosity and resultant high ionic conductivity, especially for cell systems using a lithium metal anode. The commonly used ether-based electrolyte solvents include dimethoxyethane (DME) and 1,3-dioxolane (DOL). However, ether-based solvents tend toward oxidative decomposition at relatively low potentials, leading to severe capacity fade. In a full-cell battery setup, ethers are more easily decomposed on highly catalytic metal oxide cathodes (below 4.0 V). Thus, the use of ether-based solvents or co-solvents for LIBs is limited compared to carbonate solvents. Tobishima *et al.* studied the influence of glyme-based nonaqueous electrolyte solutions for Si-based anodes [30]. Poly(ethyleneglycol) dimethyl ethers ($CH_3O(CH_2CH_2O)_nCH_3$, n = 1, 2, 3, and 4) are generally known as "glymes." When mixing glymes with1 M $LiPF_6$-EC/methylene ethylene carbonate (MEC) (30:70 in volume ratio), the discharge capacity became higher for Li/Si-C cells. The cycle life of Li/Si-C cells was improved with glyme-based electrolytes. The

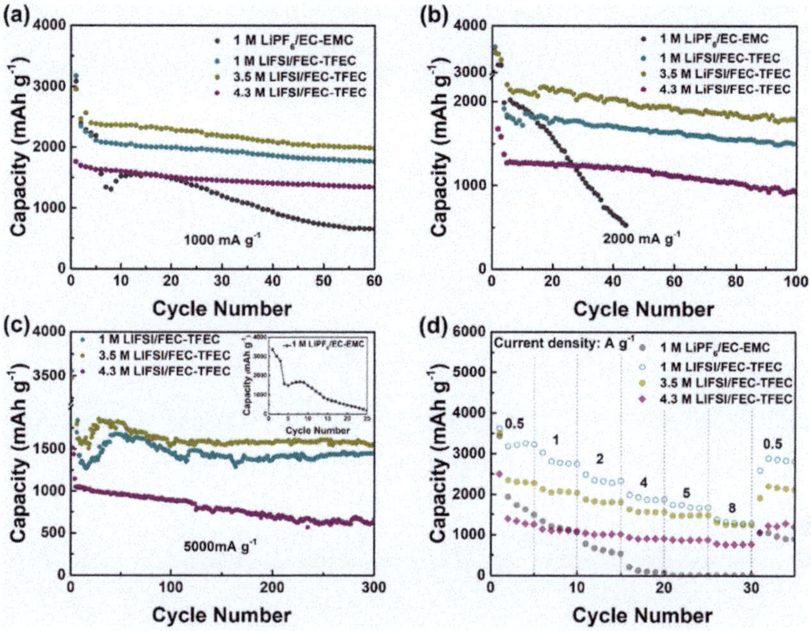

Figure 4.6 *Electrochemical performance of the Si-based anodes in different electrolytes (0.01–3 V). Cycling capability at (a) 1 000 mA/g, (b) 2 000 mA/g, (c) 5 000 mA/g, and (d) the rate property. Reprinted with permission from [29]. Copyright 2019 American Chemical Society.*

primary reason for the degradation of the charge–discharge cycling life of Li/Si-C cells is a physical breakdown and loss of an electronic conductive path of the Si electrode. This phenomenon resulted from a large volume change in electrode driven by the expansion and shrinking of Si during the charge–discharge process of lithium in Si. In addition, the reactivity of electrolyte solutions toward lithium also contributed to the degradation. Glymes-based electrolytes have much higher stability against lithium and form a thin carbon layer that help to suppress the physical breakdown of Si–C. Aurbach *et al.* reported an ether-based electrolyte (LiTFSI-lithium nitrate ($LiNO_3$)/DOL) for Si-based electrodes [31]. Cells using this electrolyte (containing $LiNO_3$) showed much higher capacity (1 275 mAh/g) than cells using the LiTFSI/DOL electrolyte (no $LiNO_3$) (575 mAh/g) after 1 000 cycles at 6 C rate. The role of $LiNO_3$ in DOL-DME solvents in the electrochemical performance and interfacial characteristics of the Si electrode were evaluated. The electrolyte created a protective SEI layer composed of the nitro compound (RNO_2) species as an essential stabilizing compound. The protective surface layer inhibited the decomposition of LiTFSI but seemed not to inhibit the decomposition of DOL. The authors proposed that the $LiNO_3$ initiated a polymerization of DOL via NO_2 radicals to form flexible polymeric RNO_2 compounds. The proposed mechanisms of electrolyte decomposition and SEI formation are

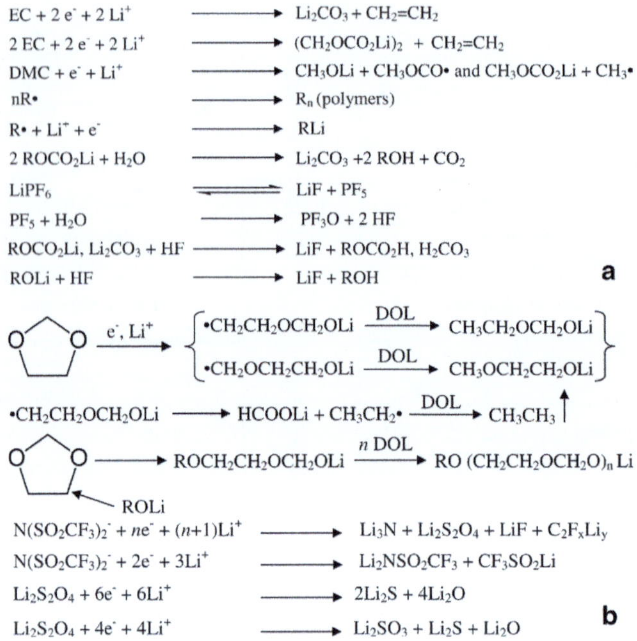

Figure 4.7 *Reduction reactions of $LiPF_6$/EC–DMC (a) and LiTFSI/DOL solutions (b). Reprinted with permission from [31]. Copyright 2012 American Chemical Society.*

illustrated in Figure 4.7. Utsugi and colleagues reported the use of fluorinated ethers as electrolytes for silicon oxide electrodes [32]. The authors evaluated the cycling performance of Si-based electrodes using 1,1,2,2-tetrafluoroethyl 2,2,3,3-tetrafluoropopyl ether. The cells maintained a capacity retention of 67 percent after 500 cycles, which is much higher than conventional carbonate electrolytes.

4.1.2.3 Other aprotic solvents

Other than carbonates and ethers, phosphorus-based and sulfone-based solvents have also been studied as electrolyte solvents in Si LIBs. These nonflammable solvents can eliminate the possible fire hazard of conventional flammable solvents. Lucht *et al.* developed the flame-retardant co-solvents triphenyl phosphate (TPP) and dimethyl methyl phosphonate (DMMP) mixed with a conventional $LiPF_6$-EC/EMC electrolyte [33]. The standard electrolyte showed poor cycling performance that lost 50 percent of the initial capacity for the Si/Li half-cells during 50 cycles, whereas electrolytes containing TPP and DMMP with LiBOB showed improved cycling efficiency and capacity retention. However, adding those viscous phosphorus solvents leads to poor ion conductivity and wettability that may hinder their practical application. To reduce the viscosity of the electrolyte, a series of binary systems including carbonate or ether mixtures have been investigated, and improved ionic conductivity was achieved. In 2015, Utsugi *et al.* utilized a triethyl-phosphate (TEP)–FEC electrolyte for a silicon oxide anode to improve its safety and energy density [34]. The cells showed a capacity retention (78 percent) after 250 cycles comparable to that of a conventional electrolyte, such as an EC/DEC electrolyte (capacity retention of 79 percent).

Sulfones are potential electrolyte solvents because of their high dielectric constant, low flammability, and excellent anodic stability on various cathodes. They have attracted attention as possible electrolyte solvents for high-potential cathodes. Wang and colleagues reported that the cycling performance of the $LiNi_{0.5}Mn_{1.5}O_4$. (LNMO) electrode in a sulfone-based electrolyte was significantly better than LNMO in an EC-based electrolyte [35]. The electrolyte containing sulfone/EMC mixed solvents was also used in SiO_x-C/LNMO pouch cells to investigate the cycling performance, and then was compared with that of EC-based cells. The sulfone-based electrolyte showed superior antioxidative ability at high voltage and stable cycling of the SiO_x-C anode.

4.1.3 Electrolyte additives

As an effective and economical way to implement a certain function to the electrolyte, electrolyte additives have been regarded as the "third" essential component of electrolyte formulation. Typically, an additive should not make up more than 10 percent of the volume of the whole electrolyte so that the added functionality can be achieved without sacrificing the bulk properties of the electrolyte. During the past two decades, additives have been extensively studied to address one or more issues related to the practical use of Si-based electrodes, and many efforts have been devoted to the following topics: (1) enhancing the mechanical and chemical stability of the SEI layer; (2) improving thermal stability or expanding the electrochemical window; and (3) addressing safety concerns regarding the battery [10]. Figure 4.8 summarizes some representative examples of electrolyte additives that have been studies in literature.

As continuous SEI growth has been regarded as one primary cause for the rapid capacity fade of Si-based electrodes, development of additives that can generate a stable SEI layer is of crucial importance. As mentioned in Section 4.1.2.1, VC and FEC are the two common carbonates that can serve as co-solvents, but more frequently they are studied as effective additives. For an electrolyte system based on carbonate solvents, VC serves as a sacrificial additive that can be reduced prior to other carbonates and afford an SEI layer that is impermeable to electrolytes, thus preventing further decomposition. The effect of VC is evidenced by controllable interfacial impedance during several initial electrochemical cycles [36]. FEC emerges as another effective SEI-forming additive for Si-based LIB systems. The effects of FEC on the cycling performance and the associated mechanism have been studied extensively. A series of studies based on different characterization techniques, including Fourier transform infrared (FTIR) spectroscopy, XPS [37, 38], mass spectrometry, nuclear magnetic resonance (NMR) [39], microscopies [40], etc., reveal that the

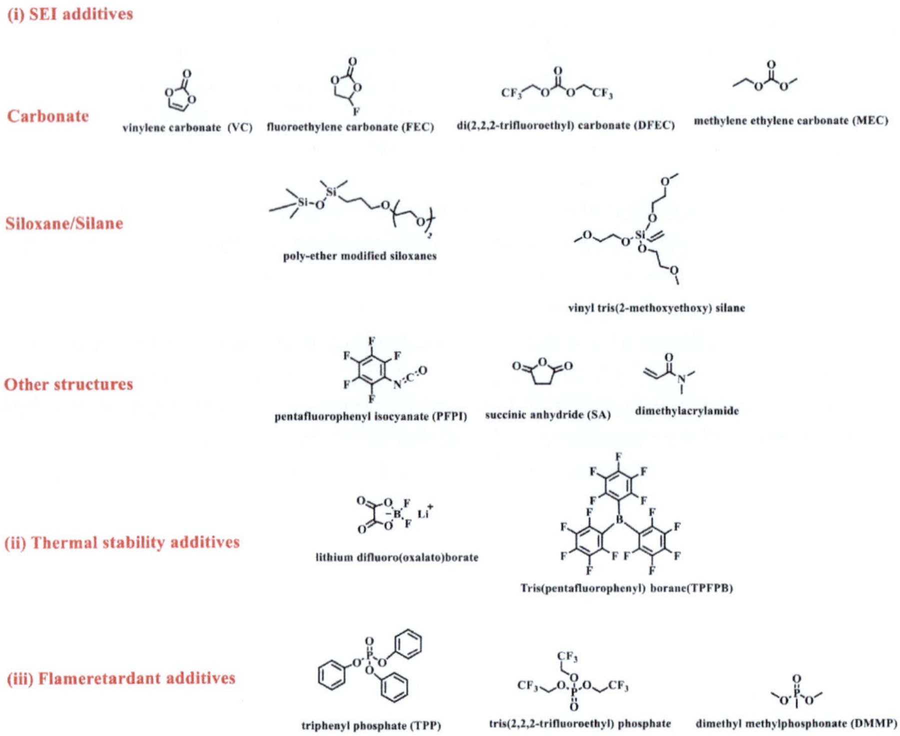

Figure 4.8 Chemical structures of representative electrolyte additives for the Si anode-based LIBs

presence of FEC leads to the formation of a less porous and more stable SEI layer on Si-based anodes. The main components of this SEI layer are poly (FEC), LiF, Li_2CO_3, and lithium alkyl carbonate, among others, which suppress the electrolyte decomposition. A comparative study between FEC- and VC-containing electrolytes shows that the impedance build-up on the anode surface was lower for FEC than it was for VC possibly owing to the difference in SEI composition. The authors also pointed out that the amount of FEC present in the electrolyte matters when it comes to electrochemical performance. An electrolyte of 10 wt% FEC appears to have the best combination of reasonable cost, low impedance, and high capacity retention [37, 41–43]. An ab initio study indicated that the FEC might be more effective on the Si surface as it prompts the formation of LiF that could glue the SEI components via aggregates [42]. Grey and coworkers utilized a combination of solution and solid state NMR to probe the changes from addition of FEC [43]. They found the soluble poly(ethylene oxide)-like linear oligomeric electrolyte breakdown products that are observed after cycling in the standard EC-based electrolyte are suppressed in the presence of 10 vol% FEC additive. Based on NMR analysis, some possible reaction routes were proposed in Figure 4.9. Although the exact mechanistic role of FEC is still not fully understood, the improvements to cycling performance may

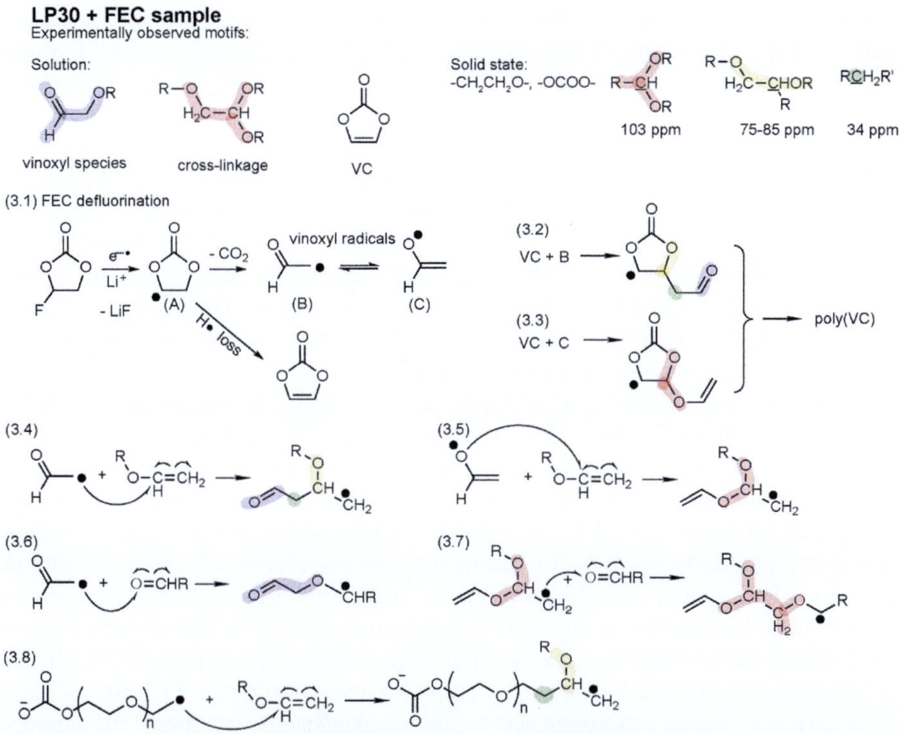

Figure 4.9 *Possible reaction schemes consistent with the chemical signatures detected by solution and solid-state NMR for the LP30 (1 M $LiPF_6$ in EC/DMC = 50/50, v/v) + FEC sample: (3.1) Reduction of FEC, (3.2 and 3.3) examples of reactions between VC and vinoxyl radicals as the initial steps for radical polymerization to form poly(VC), (3.4 and 3.5) reactions between alkene termination and vinoxyl radicals, (3.6) possible reactions between vinoxyl radical and aldehyde species, (3.7) reaction between secondary radical formed in (3.5) and aldehyde Species, forming branched structures, and (3.8) possible reaction between alkene termination with reduced EC Intermediate. Reprinted with permission from [43]. Copyright 2017 American Chemical Society.*

be attributed to twofold: (1) FEC forms polymeric carbonates, which constitute an essential part of a flexible and resilient SEI layer on Si-based anodes that helps to accommodate the volume change [44]; and (2) faster secondary reactions of carbonate species are found in FEC containing electrolytes that contribute to the SEI formation triggered [45].

Due to the promising performance of FEC, other modified carbonate-based compounds have been extensively studied as electrolyte additives for Si anodes. Di(2,2,2-trifluoroethyl) carbonate (DFDEC) was coupled with FEC as a binary additive to stabilize the SEI layer on the Si/graphite composite anode. By charactering the composition of SEI, the role of DFDEC was found to be to promote the inorganic species, such as Li_2CO_3, LiF, and $OPF_{3-y}(OR)y/Li_xPF_yO_z$, thereby thickening the SEI layer. With the right amount of DFDEC (1 wt%) and FEC (10 wt%), the binary additives can deliver an improved initial Coulombic efficiency of 84 percent, a high discharge capacity of 742–601 mAh/g, and 81 percent capacity retention at the 50th cycle [46]. MEC is another example that has been evaluated as an electrolyte additive for Si anodes. The Si electrodes cycled with 10 percent MEC-containing electrolyte exhibit improved capacity retention after 100 cycles compared to a standard electrolyte (73 vs. 46 percent), which may be attributed to preformation of poly(MEC) on the electrode surface that not only stabilizes the interface but also alleviates cracking of the electrode [47].

Siloxane- and silane-based additives are one family of interesting molecules as they also contain Si atoms that may have stronger interactions with Si anodes. Tobishima and colleagues studied a series of polyether modified siloxanes as electrolyte additives [48]. They found that the addition of polyether modified siloxane to the electrolyte of 1 M $LiPF_6$ in EC/EMC (volume ratio = 3:7) not only increased the cycling efficiencies but also improved the impedance of the Si–C/lithium cobalt oxide ($LiCoO_2$) cells. In addition, the thermal stability of the electrolyte was improved as its thermal decomposition temperature shifted to higher temperatures [48]. Zhang and colleagues reported vinyl tris(2-methoxyethoxy) silane (VTMS) as an effective electrolyte additives for Si/graphite composite anodes. Five wt% VTMS in a carbonate-based electrolyte led to improved capacity retention. The authors believed such improvement can be attributed to the formation of a more flexible SEI containing Li_2CO_3, LiF, and organosilicone [49]. Ryu and colleagues investigated the possible mechanism of alkoxy silane additives on Si anodes. By using an electrochemical quartz crystal microbalance (EQCM) combined with various electrochemical techniques, the authors discovered that the presence of additives in the electrolyte suppressed the mass accumulation onto the Si electrode, implying the irreversible electrolyte reductions were reduced [50].

Several other families of additives have also been investigated toward making the SEI layer sturdier. Isocyanate compounds, such as pentafluorophenyl isocyanate (PFPI), were evaluated in $Si/LiNi_{1/3}Mn_{1/3}Co_{1/3}O_2$ (NMC) full cells, and an improved performance was observed, which was comparable to the cells using VC or FEC. Besides reductive decomposition, the oxidation of PFPI on the cathode was also observed, leading to decreased Coulombic efficiencies compared to those of cells using VC and FEC. The improved performance may result from the significantly reduced consumption of active lithium by decreasing the parasitic side reactions on the anode side [51]. Zhang *et al.* reported the performance improvement of Si/graphite composite nanofiber anodes by utilizing a succinic anhydride (SA) additive. The authors believed that SA helps to preserve the integrity of the electrode structure [52]. Dimethylacrylamide (DMAA) is another effective film-forming additive for Si anodes. By adding 2.5 wt% DMAA to the FEC-containing electrolyte, the cycling performance of the Si half-cells was enhanced, with a retention of 80 percent after 500 cycles [53]. The authors believed that adding DMAA enables the SEI formation on the Si anode at a higher reductive potential and thus helps to mitigate the decomposition of the electrolyte solvents. Tris(pentafluorophenyl) borane (TPFPB) is a strong Lewis acid that can serve as an anion receptor. Park and colleagues put TPFPB in Si LIBs and observed significant enhancement in cycling performance. TPFPB was added into the electrolyte consisting of 1 M $LiClO_4$ in EC/DEC (1/1 vol) and enabled improved properties by forming stable SEI layers and suppressing surface pulverization [54]. Thermal stability is another desirable feature for LIBs. Lithium difluoro(oxalato)borate (LiDFOB) is an effective additive for LIBs [55], and a recent study showed that LiDFOB is also beneficial to Si anodes. LiDFOB was evaluated in full cells using a Si-Gr composite anode and a lithium cobalt oxide (LCO)/lithium NMC oxide blended cathode. As expected, the LiDFOB-added electrolyte improves the discharge capacity retention of Si-Gr/LCO-NMC full cells at 45 °C and the storage performance of full cells at 60 °C as a result of the formation of a LiDFOB-derived SEI on the Si-C anode and the LCO-NMC cathode [56].

Finally, because of the increased safety concerns regarding thermal runaway of LIBs, efforts have been conducted to remedy the flammability of the electrolyte by using flame-retardant additives (FRAs). The most popular FRAs are molecules based on organophosphorus compounds, and some examples include tris(2,2,2-trifluoroethyl) phosphate and dimethyl methylphosphonate [57, 58]. Recently, TPP was evaluated as a FRA for Si-Gr anodes. However, while 15 vol% TPP and FEC were used, the electrolytes still outperformed the carbonate baseline electrolytes[59]. Table 4.1 summarizes the cell performance measures realized by various Si-based electrodes in electrolytes with different additives.

In general, using electrolyte additives is an effective and economical way to tune the cell system toward improved properties, including cycling performance, thermal stability, and safety. An effective electrolyte additive for Si anodes should be able to participate the initial interfacial reactions that occurs on the anode so that it can modify the SEI with desirable features, such as enhanced flexibility or mechanical modulus, to accommodate the stress caused by drastic volume change. Although some improvements have been achieved by using additives, the current state-of-the-art electrolyte is still far from meeting the battery life required by practical applications, and there is still significant work ahead before an optimized electrolyte system for a Si anode is identified. It should be pointed out that the majority of these improvements were achieved in half-cells only, where Li metal serves as the anode, and they might not be transferable to LIB full cells.

4.2 Electrolytes based on ionic liquid solutions

Conventional liquid electrolytes as discussed in the Section 4.1 are based on organic aprotic solvents. While these organic solvents have many desirable features, such as low viscosity, high conductivity, a wide electrochemical window, and versatile and tunable functionality, their intrinsic flammability and volatility still raise

Table 4.1 Cell performance measures of Si-based electrodes in electrolytes with different additives

Item	Cell configuration	Additives	Electrolyte	Reversible capacity (mAh/g)	1st cycle Coulombic efficiency (%)	Cycle	Capacity retention (%)	Ref.
1	Li/Si film	VC (2 wt%)	$LiPF_6$-EC/EMC	2000	72.5	200	100	[36]
		Control		2000	61.1		25	
2	Li/Si	FEC (10 wt%)	$LiPF_6$-EC/DEC	3200	89	50	83	[37]
		Control		3000	90		33	
3	Li/Si-graphite	DFDEC, FEC	$LiPF_6$-EC/EMC	750		50	81	[46]
		Control		750		50	69	
4	Li/Si	MEC		3400		100	73	[47]
		Control					47	
5	Si/C	VTMS (5 wt%)	$LiPF_6$-EC/DEC	630	84.8	50	96	[49]
				555	82.1		83	
6	Si/NMC111	PFPI (2 wt %)	LiPF6-EC/DEC	122	73	100	45	[51]
		Control		131	62		12	
7	Li/Si-C nanofiber	SA (3 wt%)	$LiPF_6$-EC/DEC	970	71.5	50	83	[52]
		Control		950	67.3		63	
8	Li/Si	DMAA (2.5 wt%)	$LiPF_6$-EC/DMC/DEC/FEC	1950	80	500	80	[53]
		Control		1522	77		62	
9	Si-C/LCO-NMC	LiDFOB (1 wt%)	LiPF6-EC/DMC/DEC	N/A	N/A	50	92	[56]
		Control					87	
10	Li/Si film	TPFPB (2 wt%)	$LiClO_4$-EC/DEC	2050	59.8	100	87	[60]
		Control		2000	59.8		49	

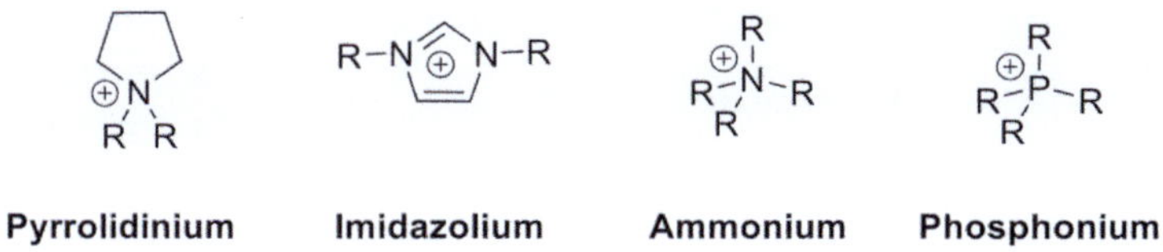

Figure 4.10 *Structure of representative cations in room-temperature ionic liquids*

serious safety concerns as they have been associated with many safety incidents. Room-temperature ionic liquids (RTILs, i.e., molten salts at room temperature under ambient pressure) have been actively studied as safer alternative solvents for electrolytes because of their unique features, including negligible vapor pressure, incombustibility, and excellent thermal and chemical stability, as well as synthesis feasibility, intrinsic high conductivity, and wide electrochemical window [10]. Finding appropriate cations and anions is important to developing RTILs. Regarding anions, the choice is relatively limited owing to the required chemical and electrochemical stability. The most promising ones are bis(fluorosulfonyl)imide (FSI) and bis(trifluoromethanesulfonyl)imide (TFSI). On the other hand, RTILs with plenty of cations have been studied for Si-based anodes, including pyrrolidiniums, imidazoliums, ammoniums, and phosphoniums (see Figure 4.10). Because RTILs cannot be constructed with small Li^+, despite the high conductivity, additional Li salts are still needed to formulate RTIL-based electrolytes for LIBs.

Pyrrolidinium is one of the promising and popular cations that is in constant use in constructing RTILs. When coupled with FSI or TFSI anions, the resultant RTILs can exhibit many promising features, including low viscosity, high ionic conductivity, and great cathodic stability. In 2012, Song *et al.* reported stabilized cycling performance of the $SiO_{1.3}$ anode using an RTIL-based electrolyte of 1 M LiTFSI in *N*-methyl-*N*-propylpyrrolidinium bis(trifluoromethylsulfonyl)imide (Py_{13}TFSI) [61]. The $SiO_{1.3}$ electrodes were cycled between 0.1 V and 1.5 V vs. Li/Li^+ in both a Py_{13}TFSI electrolyte and a Gen 2 electrolyte (1 M $LiPF_6$/EC:EMC). The reductive decomposition of Py_{13} and TFSI ions occurs at 0.5–0.3 V and effectively passivates the electrode surface with a stable SEI layer, leading to suppressed change in particle morphology and enhanced cycling performance. As shown in Figure 4.11, cells using a LiTFSI/Py_{13}TFSI electrolyte showed a capacity retention of 88 percent and average capacity around 950 mAh/g after 200 cycles, whereas cells using $LiPF_6$/EC:EMC lost the majority of their capacity during the same cycling procedure. The RTIL electrolyte also provided higher Coulombic efficiency that was maintained at 97 percent. Attenuated total reflectance-FTIR (ATR-FTIR) spectroscopy and XPS analysis on the cycled electrodes showed that the SEI layer formed in the RTIL electrolyte contains Py_{13} ions, TFSI ions, and their decomposition products. Notably, the electrode surface cycled in the LiTFSI/Py_{13}TFSI electrolyte remained dense and flat. In contrast, the electrode surface cycled in the Gen 2 electrolyte became rough and porous with aggregated particles. In 2013, Sakaguchi *et al.* also investigated the electrolyte of Py_{13}-TFSI and Py_{13}-FSI for lanthanum silicide ($LaSi_2$)/Si composite

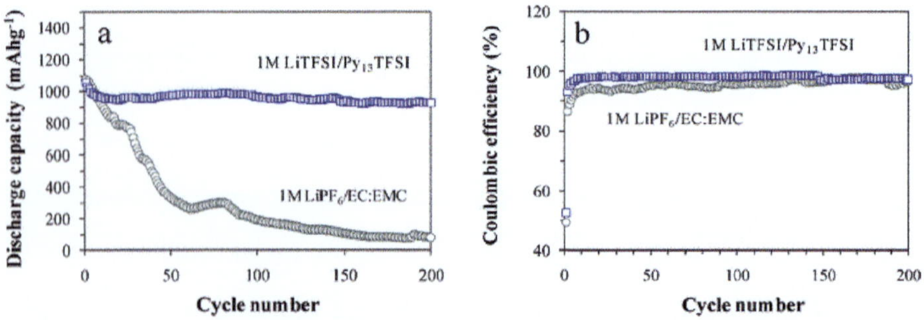

Figure 4.11 *Plots of (a) discharge capacity and (b) Coulombic efficiency vs. the cycle number of* $SiO_{1.3}$ *electrodes in lithium half-cells. Reprinted with permission from [61]. Copyright 2012 Royal Society of Chemistry.*

anodes [62]. The $LaSi_2$/Si thick-film electrodes in the Py_{13}-FSI electrolyte outperformed other electrolytes, including one that was carbonate based. The reversible capacity at the 250th cycle and its retention were 800 mAh/g and 80 percent, respectively. The excellent performance is attributed to the higher stability of Py_{13} cations against cathodic decomposition and an easier dissolution of Li ions and FSI anions at the electrode/electrolyte interface. Bund *et al.* studied the cycling performance of a nonporous Si anode with both carbonate-based ($LiPF_6$ in EC/DMC) and RTIL-based electrolytes (LiTFSI in *N*-butyl-*N*-methyl-pyrrolidinium bis(trifluoromethyl sulfonyl)imide) [63]. The nanoporous Si electrode showed high reversibility when cycled in the $LiPF_6$ EC/DMC electrolyte, whereas the Si electrode cycled in the RTIL electrolyte showed reduced capacity and reversibility of the Li ion exchange. The latter results were caused by the high viscosity of the RTIL and ineffective cathodic passivation of the Si substrate in the RTIL-based electrolyte. SEM imaging showed minor morphological changes due to the large volume change during Li insertion.

In 2015, Lee and colleagues demonstrated a stable Si-RTIL interface for next-generation LIBs [64]. By combining a high-performance Si electrode with an RTIL electrolyte (LiFSI in Py_{13}-FSI), the authors developed a highly energy-dense lithium-ion cell with an impressively long cycling life. The cells were able to maintain more than 75 percent of the initial capacity after 500 cycles (Figure 4.12). Such high performance is enabled by a stable half-cell Coulombic efficiency of 99.97 percent, averaged over the first 200 cycles. The detailed characterization revealed new insights into the previously convoluted mechanisms of the SEI formation on Si-based electrodes. The theoretical simulation to model the interface and microstructural-compositional analyses confirmed the location and constitution of various interfacial components.

Appetecchi and colleagues reported another example of a LiTFSI and Py_{13}-TFSI electrolyte for Si nanowire anodes. The authors conducted thorough electrochemical characterizations using cyclic voltammetry, cycling tests, and impedance

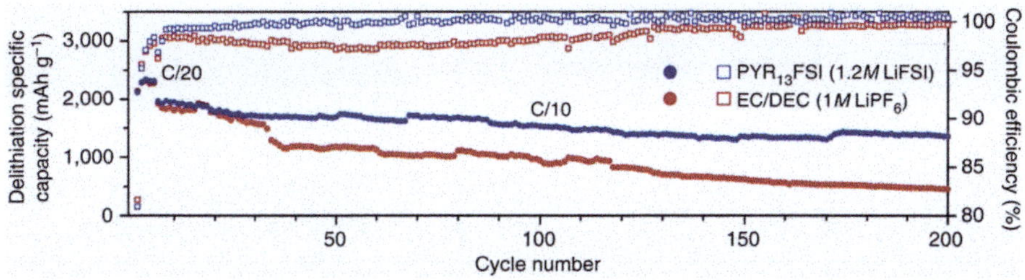

Figure 4.12 Specific capacities and Coulombic efficiencies of the SiNP electrode in RTIL and EC/DEC electrolytes. Reprinted with permission from [64]. Copyright 2015 Springer Nature.

spectroscopy [65]. The results indicated that the LiTFSI and Py_{13}-TFSI electrolyte was very compatible with Si anodes, promoting a highly reversible lithiation/delithiation process that led to a high reversible capacity of 2200 mAh/g and long cycling life with capacity retention of 50 percent after 500 cycles. The stable, long-term cycling performance was related to the transformation of the nanowires into durable porous network structures of Si nanoligaments, which was facilitated by the electrolyte formulation. Zhan and colleagues developed a fluorine-substituted RTIL as an electrolyte solvent for Si anodes. The one fluorine atom introduced into *N*-(2-fluoroethyl)-*N*-methylpyrrolidinium bis(trifluoromethanesulfonyl)imide ($Py_{1,2F}$-TFSI) seems to significantly affect the cycling performance for Si anodes [66]. The cells with $Py_{1,2F}$-TFSI-containing electrolytes delivered much better cycling performance levels than those containing FEC or fluorine-free ionic liquids. Sakaguchi's group studied the volume expansion of Si-based anodes in conventional carbonate electrolytes and RTIL electrolyte (Py_{13}-TFSI) [67]. The authors reported that combining a Py_{13}-TFSI-containing electrolyte and a charge capacity limit of 1 000 mAh/g significantly suppressed the volume expansion of the Si anode and improved the life cycle (Figure 4.13) [67]. In contrast, the Si layer expanded significantly in an organic electrolyte even with the charge capacity limit and even in an ionic-liquid electrolyte without the limit. The authors demonstrated that the homogeneously distributed Si lithiation–delithiation, phase-transition control from the Si to Li-rich Li–Si alloy phases, formation of a surface film with structural and/or mechanical stability, and faster Li^+ diffusion all contribute to suppressing Si volume expansion. Tran-Van *et al.* demonstrated improved cycling performance of binder-free microporous Si electrodes using RTIL electrolytes [68]. The freestanding microporous Si layer on a copper current collector was cycled using a LiTFSI/Py_{13}-FSI electrolyte. The electrode performance levels were determined under constant charge conditions (500, 750, and, 1 000 mAh/g) and exhibited significantly higher stability for more than 1 800 charge and discharge cycles.

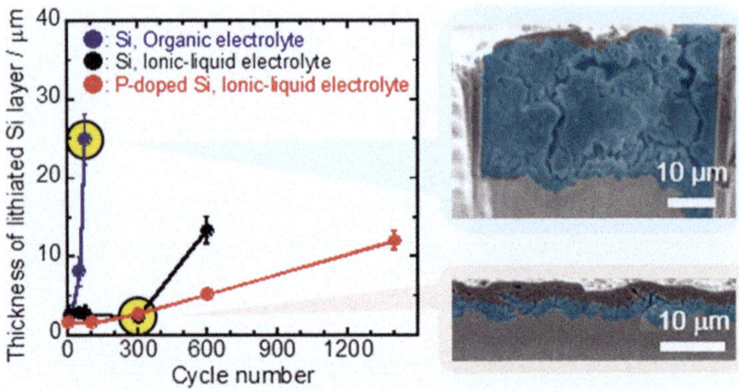

Figure 4.13 *Suppression of Si volume expansion with an RTIL electrolyte. Reprinted with permission from [67]. Copyright 2019 American Chemical Society.*

4.3 Solid-state electrolytes

While the liquid electrolyte dominates the chemistry for commercial LIBs, solid-state electrolytes (SSEs) have attracted considerable attention because of their potential safety, energy density, and cycle-life benefits. The increased safety emerges from the absence of the flammable liquid electrolytes typically employed in Li-ion cells. Adoption of SSEs could also lead to significant increases in energy density, as SSEs may enable energy-dense materials, including Si: using solid electrolytes may help mitigate continuous SEI growth. In general, a successful SSE should possess at least two characteristics: (1) high bulk ionic conductivity and fast interfacial transport of Li ions at both the battery anode and cathode; and (2) good mechanical stability and chemical inertness during extended battery cycling [69].

4.3.1 Ceramic-based SSEs

Ceramic materials are promising candidates for SSEs as they can have high ionic conductivity when compatible ions are doped, among which lithium silicon phosphorus oxynitride (LiSiPON) [70], Li_2S-P_2S_3 [71], and $Li_xLa_yZr_zTa_mO_n$ (LLZTO) [72] are some representative examples that have been extensively studied for Si-based anodes. Similar to all solid-state batteries, the interfacial contact between Si anodes and the solid electrolytes is of crucial importance as the interfacial resistance is largely dictating the performance. One of the factors that affects the interfacial contact is the thickness of the electrodes. Li and colleagues investigated the compatibility of Si anodes and Ta-doped $Li_{6.4}La_3Zr_{1.4}Ta_{0.6}O_{12}$ (LLZTO) solid electrolytes and the stability of the Si anode [73]. The Li/LLZTO/Si cells with Si layers thinner than 180 nm exhibit good cycling performance with a capacity retention over 85 percent after 100 cycles (Figure 4.14). As the Si layer's thickness is increased to larger than 300 nm, the capacity retention becomes 77 percent after 100 cycles. When the thickness is close to 900 nm, the cells can barely cycle for a limited number

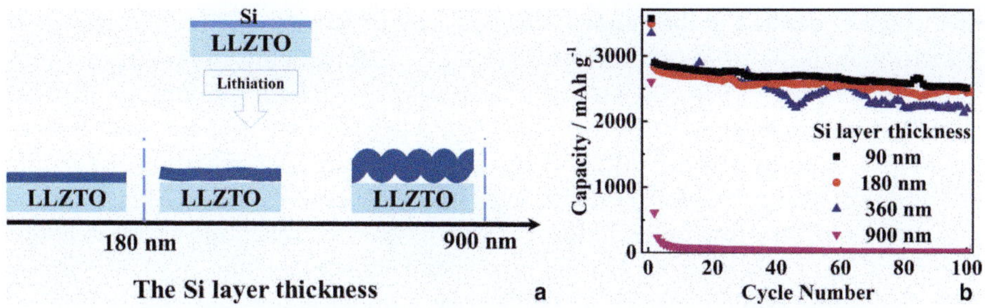

Figure 4.14 *Thickness-dependent electrochemical performance of Si-based anode with LLZTO. Reprinted and revised with permission from [73]. Copyright 2017 American Chemical Society.*

of times as a result of the destructive volume change at the interfaces [73]. Many strategies have been proposed to mitigate the interfical contact issues. One angle is to engineer the nanostructured Si with different morphologies, such as a nanosheet, nanowire, or porous structure, to accommodate the volume changes. For instance, a ball-milled porous Si composite anode exhibited capacity retention of greater than 90 percent after up to 100 cycles when made into all-solid-state cells by pairing it with Li_3PS_4 [74]. Another strategy is to apply special treatments to the electrode/electrolyte interface to alleviate the interfacial impedance. For instance, to reduce the amount of resistance, argo plasma etching was used to pretreat the LLZTO electrolyte surface before it was used in Si/LLZTO/Li all-solid-state cells. The cells achieve an initial capacity of about 2 700 mAh/g and were stabilized at 1 200 mAh/g after prolonged cycles [72].

4.3.2 Polymer-based SSEs or solid-state polymer electrolytes

Polymers are another promising candidates for use in constructing SSEs, and solid-state polymer electrolytes (SPEs) are attractive from a range of perspectives, including their superior mechanical toughness, lower cost, lightweight characteristics, and compatibility with large-scale, roll-to-roll manufacturing processes [69]. Polyethylene glycol (PEO) is one of the most studied polymers for SPEs for Si anodes. Kobayashi and colleagues combined the SiO/graphite anode with a PEO-based SPE by precoating the electrode with the SPE. A polymer battery system consisting of SiO/graphite paired with poly(ethylene oxide/2-(2-methoxyethoxy)ethyl was reported to be fabricated with a combination of precoating of the electrode and overcoating of the SPE. The reversible capacity of the [SiO/graphite | SPE | Li] cell delivered promising cycling performance with a reversible capacity of more than 1 000 mAh/g and a capacity retention of 72 percent at the 100th cycle [75]. Park *et al.* reported a Li-polymer battery system composed of Li^+ conducting, PEO-functionalized SiNPs as active anode materials and nanostructured block copolymers (poly(styrene-b-ethylene oxide)) as solid electrolytes. The PEO chains introduced onto the surface of the SiNPs can prevent particle aggregation of SiNPs during cycling and facilate Li^+ transport to the active centers in the anodes. The self-assembly nature of block copolymer electrolytes helps to restrain the formation of macroscopic ionic clusters during Li insertion/desertion. Two nonvolatile additives (both nonionic plasticizer dimethyl phthalate (DMP) and an ionic liquid (EMIm)(BF_4)) were incorporated into SPEs, and remarkably different cycling performances were observed. While the addition of (EMIm)(BF_4) yielded high initial capacity (2 380 mAh/g), the cell suffered from very low reversibility because of the side reactions between active materials and ionic species. On the other hand, the cells using the DMP-doped SPE had a reversible capacity of 1 850 mAh/g under aggressive test profiles of 80 °C and a voltage window of 0–4.5 V [76].

While both ceramic and polymer materials show promising merits, blending them together seems a necessary approach to achieving further improvements. Guo and colleagues developed an SPE by combining a polymer, poly(propylene carbonate) (PPC), and a ceremic, LLZTO ($Li_{6.4}La_3Zr_{1.4}Ta_{0.6}O_{12}$), through ball milling. After depositing Si on the SPE, a flexible interface between the Si anodes and the SPE was constructed. The interfacial conductivity at room temperature was very high around 10^{-4} S/cm, and the fabricated Si/SPE/Li cell showed excellent rate and cycle performance [77]. Chen *et al.* reported a Li-conducting, PEO-based SSE with 40 vol% ceramic particles (LICGC). The room-temperature conductivity also reached approximately 10^{-4} S/cm with the incorporation of carbonate plasticizers [78].

4.3.3 Gel polymer electrolyte

In contrast to SPEs, where Li salts are dissolved and solvated by the high-molecular-weight polymer chains, in gel polymer electrolytes (GPEs), the liquid electrolyte is immobilized in a polymeric matrix. GPEs combine the advantages of both the liquid and solid components, and have attracted increasing attention because of their "dual nature" ability to function as both electrolytes and the separator. These advantages of both the liquid and solid components include high ionic conductivity and good interfacial properties from the liquid phase and good mechanical properties from the solid component. Various polymer systems, including PEO, poly(methyl methacrylate) (PMMA), poly(vinylidene fluoride-hexafluoropropylene) (PVdF-HFP), have been studied as polymer matrices in GPEs. Li *et al.* demonstrated the infiltration of a GPE into Si-coated, vertically aligned carbon nanofibers (SiVACNFs), a high-capacity-three-dimensional (3D) nanostructured anode, and its electrochemical characterization. The GPE was prepared by swelling the copolymer, poly(vinylidene fluoride-hexafluoropropylene) PVdf-HFP, in the liquid electrolyte of LiTFSI/EC-DMC. The prepared GPE showed high conductivity of ~2.2×10^{-3} S/cm and sufficient cationic transference number at ambient temperature, along with good thermal and mechanical stabilities. A half-cell cycling of this fabricated electrode showed a very high capacity of 3 450 mAh/g at C/10 and 1 732 mAh/g at 1 C [79]. An *in situ* coupling strategy was investigated by Park *et al.* to combine the chemically cross-linked, triacrylate-based GPE with a nanostructured Si anode (Figure 4.15).

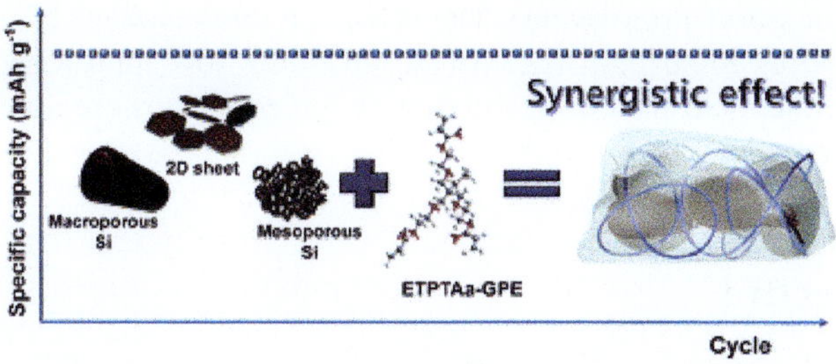

Figure 4.15 *Schematic illustration showing the synergistic coupling of various silicon anode materials (mesoporous, micro-sized macroporous, 2D sheet) and cross-linked ETPTA polymer-mediated GPEs. Reprinted with permission from [80]. Copyright 2016, The Royal Society of Chemistry.*

The mesoporous Si combined with GPE showed significant improvement in cycling performance, with a specific capacity of 2 000 mAh/g and a capacity retention of 92 percent after 100 cycles. The GPE also enabled a remarkable suppression (1.5–2 times less than the liquid electrolytes) in volume expansion of the Si anodes during cycling [80].

4.3.4 Materials other than electrolytes used

Owing to their promising features, SSEs or PSEs have also been explored for other functionalities for Si LIBs. Ci *et al.* utilized LPS ($Li_7P_3S_{11}$), a highly conductive SSE material, as a coating agent for SiNPs to mitigate the expansion issue at the particle level. A thin LPS layer was coated onto the Si nanoparticle surface to form a core-shell structure by a facile liquid-phase *in situ* reaction. As a result, the fabricated cells delivered improved cycling performance, possibly resulting from the better-protected SiNPs [81]. Qiu *et al.* designed an inward growth of SEI by filling a PEO-based conductive SPE into a hollow-structured Si anode. The as-prepared composite delivers a high specific capacity of more than 2 100 mAh/g and a long-term cycle stability with a reversible capacity of 1 350 mAh/g over 500 cycles. The improvement was believed to be related to the prevention of excessive SEI growth by filling the internal space [82].

Overall, SSEs or SPEs are appealing alternatives to the conventional organic electrolytes for Si-based anodes as they provide much improved safety control by eliminating flammable organic solvents and contributing enhanced mechanical strength, which can be used to mitigate the expansion issue. However, for real applications in Si LIBs, many hurdles are still present. Despite the progress in improving ionic conductivity at operating temperature, tuning mechanical strength to preserve electrode integrity, and stabilizing SEI formation, more efforts need to be conducted to improve the interfacial resistance and the electrochemical window. While no candidates can satisfy all of these desirable requirements, combining polymer-inorganic ceramic materials may lead to finding a balanced composite with better promise. As for SPEs, the most studied PEO-based electrolytes still possess intrinsic drawbacks, including low ionic conductivity and narrow electrochemical windows, and new polymer designs are needed to further improve performance.

4.4 Perspective

Electrolytes are traditionally considered to be of less influence than the electrode materials in LIBs. However, researchers have become more aware of how profoundly the chemical composition of electrolytes affects the cell performance in many aspects. The electrolyte systems have to be tailored to specific cell chemistries so that the proper interfacial reactions can be tuned to afford a kinetically stabilized cell system and facilitate Li^+ shuttling between the cathode and anodes. With the emergence of Si-based anodes, reformulation of the electrolytes has to be carried out so they are compatible with the new surface chemistry.

Si-based anodes have huge potential as a high-capacity anode material for LIBs. However, the same reaction that makes Si a high-capacity anode also results in drastic volume changes upon alloying with Li, which poses many challenges and has hindered their practical application. Although many efforts have been made to address those challenges through the design and optimization of Si-based material structures, the inevitable expansion is always cracking the electrode/electrolyte interface. Therefore, the electrolytes for Si anodes have to be tailored to adapt to this continuously changing interface to maintain the fragile stability. Despite the more stringent requirements, carbonate-based electrolytes still hold the highest promise for enabling Si anodes. The addition of FEC is by far the most effective way to improve the cycling performance of most Si anodes.

Other electrolytes that have been studied for Si anodes have also been discussed, including ionic liquids, SSEs, SPEs, and GPEs. Although many of those studies reported improved performance, the reported results are still far from meeting the requirements for commercialization. For instance, many studies are using a low mass loading or are based on special processing of the electrode, to achieve better performance, which are not feasible for a commercial device.

In order to enable high performance of Si LIBs, much more work needs to be performed, including the optimization of Si-based electrode structures and development of tailored electrolyte systems. For future electrolyte development, understanding SEI formation on the expanding surface of Si anodes is of top priority, which may lead to construction of a sturdier interface between Si anodes and electrolytes. Electrolyte additives will continue to be one of the most effective approaches toward this end, with the addition of binary or multiple components being even more effective. In addition, new electrolyte systems that can accommodate the SEI cracking may be worth exploring. Solvents that have better reductive stabilities, such as ethers, may have a chance to bypass the SEI issue due to their excellent thermodynamic stability.

References

[1] Tarascon J.M., Armand M. 'Issues and challenges facing rechargeable lithium batteries'. *Nature*. 2001;**414**(6861):359–67.

[2] Armand M., Tarascon J.-M. 'Building better batteries'. *Nature*. 2008;**451**(7179):652–7.

[3] Goodenough J.B., Park K.-S. 'The Li-ion rechargeable battery: a perspective'. *Journal of the American Chemical Society*. 2013;**135**(4):1167–76.

[4] Larcher D., Tarascon J.-M. 'Towards greener and more sustainable batteries for electrical energy storage'. *Nature Chemistry*. 2015;**7**(1):19–29.

[5] Obrovac M.N., Christensen L. 'Structural changes in silicon anodes during lithium insertion/extraction'. *Electrochemical and Solid-State Letters*. 2004;**7**(5):A93–6.

[6] Obrovac M.N., Krause L.J. 'Reversible cycling of crystalline silicon powder'. *Journal of the Electrochemical Society*. 2007;**154**(2):A103–8.

[7] Zhu B., Wang X., Yao P., Li J., Zhu J. 'Towards high energy density lithium battery anodes: silicon and lithium'. *Chemical Science*. 2019;**10**(30):7132–48.

[8] Xu K. 'Nonaqueous liquid electrolytes for lithium-based rechargeable batteries'. *Chemical Reviews*. 2004;**104**(10):4303–418.

[9] Shi F., Ross P.N., Somorjai G.A., Komvopoulos K. 'The Chemistry of Electrolyte Reduction on Silicon Electrodes Revealed by *in Situ* ATR-FTIR Spectroscopy'. *The Journal of Physical Chemistry C*. 2017;**121**(27):14476–83.

[10] Xu K. 'Electrolytes and interphases in Li-ion batteries and beyond'. *Chemical Reviews*. 2014;**114**(23):11503–618.

[11] Xu Z., Yang J., Li H., Nuli Y., Wang J. 'Electrolytes for advanced lithium ion batteries using silicon-based anodes'. *Journal of Materials Chemistry A*. 2019;**7**(16):9432–46.

[12] Ohara S., Suzuki J., Sekine K., Takamura T. 'Li insertion/extraction reaction at a Si film evaporated on a Ni foil'. *Journal of Power Sources*. 2003;**119–121**:591–6.

[13] Tarascon J., Guyomard D. 'New electrolyte compositions stable over the 0 to 5 V voltage range and compatible with the Li1+xMn2O4/carbon Li-ion cells'. *Solid State Ionics*. 1994;**69**(3):293–305.

[14] Aurbach D., Zaban A., Schechter A., Ein-Eli Y., Zinigrad E., Markovsky B. 'The study of electrolyte solutions based on ethylene and diethyl carbonates for rechargeable Li batteries: I'. *Li Metal Anodes*. 1995;**142**(9):2873–82.

[15] Liu Y., Matsumura T., Imanishi N., Hirano A., Ichikawa T., Takeda Y. 'Preparation and characterization of Si/C composite coated with polyaniline as novel anodes for Li-Ion batteries'. *Electrochemical and Solid-State Letters*. 2005;**8**(11):A599–602.

[16] Li M.-Q., Qu M.-Z., He X.-Y., Yu Z.-L. 'Effects of electrolytes on the electrochemical performance of Si/graphite/disordered carbon composite anode for lithium-ion batteries'. *Electrochimica acta*. 2009;**54**(19):4506–13.

[17] Yoon T., Chapman N., Seo D.M., Lucht B.L. 'Lithium salt effects on silicon electrode performance and solid electrolyte interphase (SEI) structure, role of solution structure on sei formation'. *Journal of the Electrochemical Society*. 2017;**164**(9):A2082–8.

[18] Kang S.-J., Park K., Park S.-H., Lee H. 'Unraveling the role of LiFSI electrolyte in the superior performance of graphite anodes for Li-ion batteries'. *Electrochimica Acta*. 2018;**259**:949–54.

[19] Philippe B., Dedryvère R., Gorgoi M., Rensmo H., Gonbeau D., Edström K. 'Improved performances of nanosilicon electrodes using the salt LiFSI: a photoelectron spectroscopy study'. *Journal of the American Chemical Society*. 2013;**135**(26):9829–42.

[20] Trask S.E., Pupek K.Z., Gilbert J.A., *et al.* 'Performance of full cells containing carbonate-based LiFSI electrolytes and silicon-graphite negative electrodes'. *Journal of the Electrochemical Society*. 2016;**163**(3):A345–50.

[21] Xu W., Angell C.A. 'Weakly coordinating anions, and the exceptional conductivity of their nonaqueous solutions'. *Electrochemical and Solid-State Letters*. 2001;**4**(1):E1–4.

[22] Lee S.J., Han J.-G., Park I. 'Effect of lithium bis(oxalato)borate additive on electrochemical performance of $Li_{1.17}Ni_{0.17}Mn_{0.5}Co_{0.17}O_2$ cathodes for lithium-Ion batteries'. *Journal of the Electrochemical Society*. 2014;**161**(14):A2012–9.

[23] Lindgren F., Xu C., Niedzicki L., *et al.* 'Sei formation and interfacial stability of a Si electrode in a LiTDI-Salt based electrolyte with FEC and Vc additives for Li-Ion batteries'. *ACS Applied Materials & Interfaces*. 2016;**8**(24):15758–66.

[24] Chang Z.-H., Wang J.-T., Wu Z.-H., Gao M., Wu S.-J., Lu S.-G. 'The electrochemical performance of silicon nanoparticles in concentrated electrolyte'. *The Journal is ChemSusChem*. 2018;**11**(11):1787–96.

[25] Bhattacharya S., Alpas A.T. 'Effects of electrolyte composition and voltage scan rate on cyclic performance of silicon-aluminum anodes for lithium-ion cells'. *The Journal is ECS Transactions*. 2017;**75**(23):1–15.

[26] Yao K., Zheng J.P., Liang R. 'Ethylene carbonate-free fluoroethylene carbonate-based electrolyte works better for freestanding si-based composite paper anodes for Li-ion batteries'. *Journal of Power Sources*. 2018;**381**(2):164–70.

[27] Horowitz Y., Han H.-L., Soto F.A., Ralston W.T., Balbuena P.B., Somorjai G.A. 'Fluoroethylene carbonate as a directing agent in amorphous silicon Anodes: electrolyte interface structure probed by sum frequency vibrational spectroscopy and ab initio molecular dynamics'. *Nano Letters*. 2018;**18**(2):1145–51.

[28] Horowitz Y., Steinrück H.-G., Han H.-L., *et al.* 'Fluoroethylene carbonate induces ordered electrolyte interface on silicon and sapphire surfaces as revealed by sum frequency generation vibrational spectroscopy and X-ray reflectivity'. *Nano Letters*. 2018;**18**(3):2105–11.

[29] Zeng G., An Y., Xiong S., Feng J. 'Nonflammable fluorinated carbonate electrolyte with high salt-to-solvent ratios enables stable silicon-based anode for

next-generation lithium-ion batteries'. *ACS Applied Materials & Interfaces*. 2019;**11**(26):23229–35.

[30] Inose T., Watanabe D., Morimoto H., Tobishima S.-I. 'Influence of glyme-based nonaqueous electrolyte solutions on electrochemical properties of si-based anodes for rechargeable lithium cells'. *Journal of Power Sources*. 2006;**162**(2):1297–303.

[31] Etacheri V., Geiger U., Gofer Y., *et al*. 'Exceptional electrochemical performance of Si-nanowires in 1,3-dioxolane solutions: a surface chemical investigation'. *Langmuir*. 2012;**28**(14):6175–84.

[32] Noguchi T., Hasegawa T., Yamauchi H., Yamazaki I., Utsugi K. 'Effect of using fluorinated ether and sulfone as electrolyte solvents for lithium ion batteries with lithium-rich layered cathodes and silicon oxide anodes'. *ECS Transactions*. 2017;**80**(10):291–303.

[33] Dunn R.P., Nadimpalli S.P.V., Guduru P., Lucht B.L. 'Flame retardant co-solvent incorporation into lithium-ion coin cells with thin-film Si anodes'. *Journal of the Electrochemical Society*. 2014;**161**(1):A176–82.

[34] Matsumoto K., Inoue K., Utsugi K. 'A highly safe battery with a non-flammable triethyl-phosphate-based electrolyte'. *Journal of Power Sources*. 2015;**273**:954–8.

[35] Li C., Chen Y., Chen Y., Lao B., Qi L., Wang H. 'Stability analysis for 5V high energy density pouch batteries of Si anode and SL/EMC electrolytes'. *Journal of Alloys and Compounds*. 2019;**773**(8):105–11.

[36] Chen L., Wang K., Xie X., Xie J. 'Enhancing electrochemical performance of silicon film anode by vinylene carbonate electrolyte additive'. *Electrochemical and Solid-State Letters*. 2006;**9**(11):A512–15.

[37] Nguyen C.C., Lucht B.L. 'Comparative study of fluoroethylene carbonate and vinylene carbonate for silicon anodes in lithium ion batteries'. *Journal of the Electrochemical Society*. 2014;**161**(12):A1933–8.

[38] Dalavi S., Guduru P., Lucht B.L. 'Performance enhancing electrolyte additives for lithium ion batteries with silicon anodes'. *Journal of the Electrochemical Society*. 2012;**159**(5):A642–6.

[39] Jin Y., Kneusels N.-J.H., Marbella L.E., *et al*. 'Understanding fluoroethylene carbonate and vinylene carbonate based electrolytes for Si anodes in lithium ion batteries with NMR spectroscopy'. *Journal of the American Chemical Society*. 2018;**140**(31):9854–67.

[40] Choi N.-S., Yew K.H., Lee K.Y., Sung M., Kim H., Kim S.-S. 'Effect of fluoroethylene carbonate additive on interfacial properties of silicon thin-film electrode'. *Journal of Power Sources*. 2006;**161**(2):1254–9.

[41] Uchida S., Yamagata M., Ishikawa M. 'Effect of electrolyte additives on non-nano-Si negative electrodes prepared with polyimide binder'. *Journal of the Electrochemical Society*. 2015;**162**(3):A406–12.

[42] Okuno Y., Ushirogata K., Sodeyama K., Tateyama Y. 'Decomposition of the fluoroethylene carbonate additive and the glue effect of lithium fluoride products for the solid electrolyte interphase: an ab initio study'. *Physical Chemistry Chemical Physics*. 2016;**18**(12):8643–53.

[43] Jin Y., Kneusels N.-J.H., Magusin P.C.M.M., *et al.* 'Identifying the structural basis for the increased stability of the solid electrolyte interphase formed on silicon with the additive fluoroethylene carbonate'. *Journal of the American Chemical Society*. 2017;**139**(42):14992–5004.

[44] Huang W., Wang J., Braun M.R., *et al.* 'Dynamic structure and chemistry of the silicon solid-electrolyte interphase visualized by cryogenic electron microscopy'. *Matter*. 2019;**1**(5):1232–45.

[45] Markevich E., Fridman K., Sharabi R., *et al.* 'Amorphous columnar silicon Anodes for advanced high voltage lithium ion full cells: dominant factors governing cycling performance'. *Journal of the Electrochemical Society*. 2013;**160**(10):A1824–33.

[46] Jo H., Kim J., Nguyen D.-T., *et al.* 'Stabilizing the solid electrolyte interphase layer and cycling performance of silicon–graphite battery anode by using a binary additive of fluorinated carbonates'. *The Journal of Physical Chemistry C*. 2016;**120**(39):22466–75.

[47] Nguyen C.C., Lucht B.L. 'Improved cycling performance of Si nanoparticle anodes via incorporation of methylene ethylene carbonate'. *Electrochemistry Communications*. 2016;**66**:71–4.

[48] Inose T., Tada S., Morimoto H., Tobishima S.-ichi. 'Poly-ether modified siloxanes as electrolyte additives for rechargeable lithium cells'. *Journal of Power Sources*. 2006;**161**(1):550–9.

[49] Wang J., Zhang L., Zhang H. 'Effects of electrolyte additive on the electrochemical performance of Si/C anode for lithium-ion batteries'. *Ionics*. 2018;**24**(11):3691–8.

[50] Ryu Y.-G., Lee S., Mah S., *et al.* 'Electrochemical behaviors of silicon electrode in lithium salt solution containing alkoxy silane additives'. *Journal of the Electrochemical Society*. 2008;**155**(8):A583–9.

[51] Nölle R., Achazi A.J., Kaghazchi P., Winter M., Placke T. 'Pentafluorophenyl isocyanate as an effective electrolyte additive for improved performance of silicon-based lithium-ion full cells'. *ACS Applied Materials & Interfaces*. 2018;**10**(33):28187–98.

[52] Li Y., Xu G., Yao Y., *et al.* 'Improvement of cyclability of silicon-containing carbon nanofiber anodes for lithium-ion batteries by employing succinic anhydride as an electrolyte additive'. *Journal of Solid State Electrochemistry*. 2013;**17**(5):1393–9.

[53] Zhu G., Yang S., Wang Y., Qu Q., Zheng H. 'Dimethylacrylamide, a novel electrolyte additive, can improve the electrochemical performances of silicon anodes in lithium-ion batteries'. *RSC Advances*. 2019;**9**(1):435–43.

[54] Han G.-B., Lee J.-N., Choi J.W., Park J.-K. 'Tris(pentafluorophenyl) borane as an electrolyte additive for high performance silicon thin film electrodes in lithium ion batteries'. *Electrochimica Acta*. 2011;**56**(24):8997–9003.

[55] Liu J., Chen Z., Busking S., Amine K. 'Lithium difluoro(oxalato)borate as a functional additive for lithium-ion batteries'. *Electrochemistry Communications*. 2007;**9**(3):475–9.

[56] Lee S.J., Han J.-G., Lee Y., *et al.* 'A bi-functional lithium difluoro(oxalato) borate additive for lithium cobalt oxide/lithium nickel manganese cobalt oxide cathodes and silicon/graphite anodes in lithium-ion batteries at elevated temperatures'. *Electrochimica Acta*. 2014;**137**:1–8.

[57] Xiang H.F., Xu H.Y., Wang Z.Z., Chen C.H. 'Dimethyl methylphosphonate (DMMP) as an efficient flame retardant additive for the lithium-ion battery electrolytes'. *Journal of Power Sources*. 2007;**173**(1):562–4.

[58] Zhang S.S., Xu K., Jow T.R. 'Tris(2,2,2-trifluoroethyl) phosphite as a co-solvent for nonflammable electrolytes in Li-ion batteries'. *Journal of Power Sources*. 2003;**113**(1):166–72.

[59] Smart M.C., Krause F.C., Hwang C., *et al.* 'Electrolytes with improved safety developed for high specific energy Li-ion cells with si-based anodes'. *ECS Transactions*. 2013;**50**(26):365–74.

[60] Yang Z., Zhang J., Kintner-Meyer M.C.W., *et al.* 'Electrochemical energy storage for green grid'. *Chemical Reviews*. 2011;**111**(5):3577–613.

[61] Song J.-W., Nguyen C.C., Song S.-W. 'Stabilized cycling performance of silicon oxide anode in ionic liquid electrolyte for rechargeable lithium batteries'. *RSC Advances*. 2012;**2**(5):2003–9.

[62] Usui H., Shimizu M., Sakaguchi H. 'Applicability of ionic liquid electrolytes to LaSi2/Si composite thick-film anodes in Li-ion battery'. *Journal of power sources*. 2013;**235**:29–35.

[63] Ivanov S., Vlaic C.A., Du S., Wang D., Schaaf P., Bund A. 'Electrochemical performance of nanoporous Si as anode for lithium ion batteries in alkyl carbonate and ionic liquid-based electrolytes'. *Journal of applied electrochemistry*. 2014;**44**(1):159–68.

[64] Molina Piper D., Evans T., Leung K., *et al.* 'Stable silicon-ionic liquid interface for next-generation lithium-ion batteries'. *Nature communications*. 2015;**6**:6230.

[65] Kim G.-T., Kennedy T., Brandon M., *et al.* 'Behavior of germanium and silicon nanowire anodes with ionic liquid electrolytes'. *ACS nano*. 2017;**11**(6):5933–43.

[66] Men F., Yang Y., Shang Y., *et al.* 'Fluorine-substituted ionic liquid for Si anode in Li-ion battery'. *Journal of power sources*. 2018;**401**:354–61.

[67] Domi Y., Usui H., Yamaguchi K., Yodoya S., Sakaguchi H. 'Silicon-based anodes with long cycle life for lithium-ion batteries achieved by significant suppression of their volume expansion in ionic-liquid electrolyte'. *ACS applied materials & interfaces*. 2019;**11**(3):2950–60.

[68] Luais E., Ghamouss F., Sakai J., Defforge T., Gautier G., Tran-Van F. 'Improved cycling performances of binder-free macroporous silicon Li-ion negative electrodes using room temperature ionic liquid electrolyte'. *Journal of Solid State Electrochemistry*. 2019;**23**(3):937–41.

[69] Zhao Q., Liu X., Stalin S., Khan K., Archer L.A. 'Solid-state polymer electrolytes with in-built fast interfacial transport for secondary lithium batteries'. *Nature energy*. 2019;**4**(5):365–73.

[70] Lee S.-J., Baik H.-K., Lee S.-M. 'An all-solid-state thin film battery using LISIPON electrolyte and Si–V negative electrode films'. *Electrochemistry communications*. 2003;**5**(1):32–5.

[71] Cervera R.B., Suzuki N., Ohnishi T., *et al*. 'High performance silicon-based anodes in solid-state lithium batteries'. *Energy & Environmental Science*. 2014;**7**(2):662–6.

[72] Ferraresi G., El Kazzi M., Czornomaz L., Tsai C.-L., Uhlenbruck S., Villevieille C. 'Electrochemical performance of all-solid-state Li-ion batteries based on garnet electrolyte using silicon as a model electrode'. *ACS energy letters*. 2018;**3**(4):1006–12.

[73] Chen C., Li Q., Li Y., Cui Z., Guo X., Li H. 'Sustainable interfaces between Si anodes and garnet electrolytes for room-temperature solid-state batteries'. *ACS applied materials & interfaces*. 2018;**10**(2):2185–90.

[74] Okuno R., Yamamoto M., Terauchi Y., Takahashi M. 'Stable cyclability of porous Si anode applied for sulfide-based all-solid-state batteries'. *Energy Procedia*. 2019;**156**:183–6.

[75] Kobayashi Y., Seki S., Mita Y., *et al*. 'High reversible capacities of graphite and SiO/graphite with solvent-free solid polymer electrolyte for lithium-ion batteries'. *Journal of power sources*. 2008;**185**(1):542–8.

[76] Choi I., Ahn H., Park M.J. 'Enhanced performance in lithium–polymer batteries using surface-functionalized Si nanoparticle anodes and self-assembled block copolymer electrolytes'. *Macromolecules*. 2011;**44**(18):7327–34.

[77] Huo H., Sun J., Chen C., *et al*. 'Flexible interfaces between Si anodes and composite electrolytes consisting of poly(propylene carbonates) and garnets for solid-state batteries'. *Journal of power sources*. 2018;**383**:150–6.

[78] Chen X.C., Liu X., Samuthira Pandian A., *et al*. 'Determining and minimizing resistance for ion transport at the polymer/ceramic electrolyte interface'. *ACS energy letters*. 2019;**4**(5):1080–5.

[79] Pandey G.P., Klankowski S.A., Li Y., *et al*. 'Effective infiltration of gel polymer electrolyte into silicon-coated vertically aligned carbon nanofibers as anodes for solid-state lithium-ion batteries'. *ACS applied materials & interfaces*. 2015;**7**(37):20909–18.

[80] Bok T., Cho S.-J., Choi S., *et al*. 'An effective coupling of nanostructured Si and gel polymer electrolytes for high-performance lithium-ion battery anodes'. *RSC advances*. 2016;**6**(9):6960–6.

[81] Xu X., Ai Q., Pan L., *et al*. 'Li7P3S11 solid electrolyte coating silicon for high-performance lithium-ion batteries'. *Electrochimica Acta*. 2018;**276**(45):325–32.

[82] Ma T., Yu X., Cheng X., Li H., Zhu W., Qiu X. 'Confined solid electrolyte interphase growth space with solid polymer electrolyte in hollow structured silicon anode for Li-ion batteries'. *ACS applied materials & interfaces*. 2017;**9**(15):13247–54.

Chapter 5

Interfacial chemistry on silicon anode

Guang Yang[1] and Jagjit Nanda[1]

5.1 Introduction

This chapter is mainly focused on the surface chemistry and structure of the silicon (Si) anode. Because of its ultrahigh theoretical lithium (Li) storage capacity (3 579 mAh/g for $Li_{15}Si_4$)[1], Si is considered one of the most promising alternatives to graphite anodes for Li-ion batteries. Unlike graphite anode, Si anode does intercalate Li cations upon lithiation. Instead, Si alloys with Li, resulting in a drastic structural disruption (e.g., amorphization of crystal Si (c-Si) as a starting material). This disruption leads to up to a 300 percent volumetric change [2] and structural morphological evolution, resulting in an intrinsically unstable solid-electrolyte interphase (SEI) on the Si anode surface, further capacity fading, and limited cycle life.

It is commonly accepted that the successful operation of almost all Li battery systems relies largely on a stable SEI layer [3]. For example, graphite-based anodes currently dominate the Li-ion battery market largely because graphite is capable of forming a self-limiting SEI [4]. Thus, it is critically important to develop an understanding of the underlying chemistry, structures, and properties of the SEI on a Si anode to ensure a reasonable time frame for putting Si-based anodes in practical batteries that meet the current US Department of Energy (DOE) goals of >375 Wh/kg, >750 Wh/L, and the capability of operating for >1 000 cycles with >80 percent capacity retention over 10 years.

This chapter specifically tackles the foundational understanding of the chemistry and structure of the SEI on Si anodes. Its primary goal is to give the readership a broad overview of the SEI formation mechanism, the reductive decomposition reaction paths of commonly seen carbonate electrolyte components and additives, Si SEI chemistry and morphological characterization tools, and the current understanding of SEI models on a Si surface.

[1]Chemical Sciences Division, Oak Ridge National Laboratory, United States

5.2 SEI formation mechanisms

Most widely used Li-ion battery electrolytes contain small aprotic molecules, such as ethylene carbonate (EC), ethyl-methyl carbonate (EMC), and diethyl carbonate (DEC), with a Li salt dissolved in these solvent molecules. These electrolyte components experience decomposition at the electrode/electrolyte interface when the redox potential of the electrodes falls out of the electrochemical stability window of the electrolyte, resulting in the formation of a passivation (or SEI) layer on the electrode surface. The early SEI formation mechanism was schematically illustrated by Goodenough and Kim [5] (Figure 5.1). When the Fermi energy of the anode (μ_A) becomes higher than that of the lowest unoccupied molecular orbital (LUMO) of the electrolyte, electrolyte reduction decomposition occurs. Similarly, oxidation reduction of the electrolyte will occur if the cathode has a Fermi energy (μ_C) lower than that of the highest occupied molecular orbital (HOMO) of the cathode. The stability window of the electrolyte is defined as the difference between the energy of the LUMO and the HOMO, denoted as E_g. The reduction stability of the electrolyte determines the nature of SEI formation on a Si surface. Most carbonate-based electrolytes have a reduction potential value close to 1 V vs Li^+/Li, as shown in Figure 5.2. For example, the so-called GenII electrolyte composed of 1 M $LiPF_6$ in EC–EMC (3:7 wt%) starts to undergo reduction decomposition at 0.9 V. The lowest reduction decomposition potential listed in Figure 5.2 is for a pure EC solvent. However, it is impractical to use a pure solvent as the electrolyte because it will provide no movable charge carriers. Regardless, a Si anode would have to be held

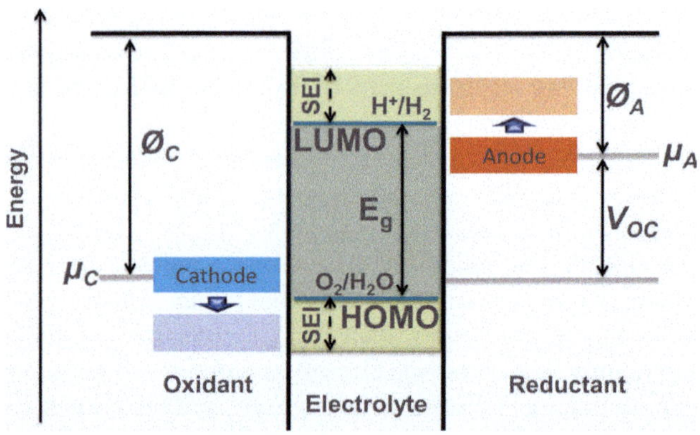

Figure 5.1 *Schematic illustration of the open-circuit energy diagram of an electrolyte. Φ_A and Φ_C represent the work functions of the anode and cathode, respectively. E_g is the electrochemical stability window for the electrolyte. μ_A and μ_C are the redox potential of the anode and cathode, respectively. Under electron-reduction or electro-oxidation conditions, the SEI forms at the anode or cathode surface. (Reprinted with ACS permission) [5].*

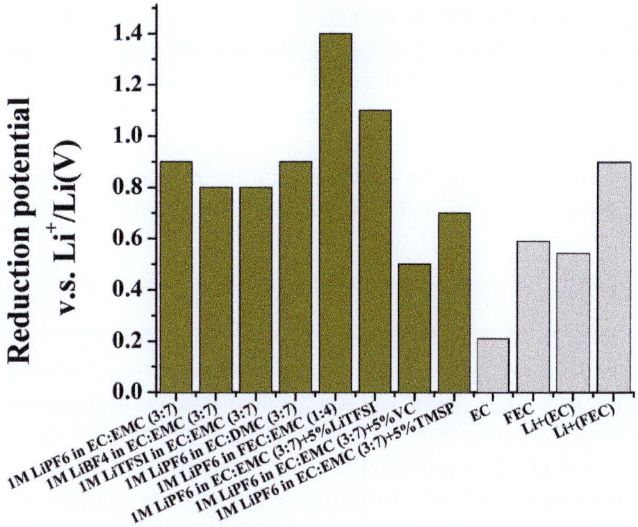

Figure 5.2 *The experimental reduction potentials of some carbonate electrolytes [6] (in gold) and density functional theory (DFT)—calculated values (in gray) of solvents and the Li^+-solvent ion complex [7]*

at below 50 mV vs Li^+/Li to approach a full lithiation state [1, 8], more than an order of magnitude lower than the reduction potential of the electrolytes. Therefore, reduction decomposition of the electrolyte on a Si anode surface is inevitable. In this context, a stable SEI to conformally cover the Si anode surface to protect the electrolytes from repeated decomposition upon charge/discharge cycling is mandatory for successful development of Si anodes.

5.3 The effect of electrolyte solvent and salt on Si SEI formation

Because the reduction decomposition of electrolyte components on the anode surface directly contributes to SEI formation, the electrolyte constituents play a decisive role in shaping SEI chemistry and structure. Traditionally, a Li-ion battery electrolyte consists of a binary organic carbonate solvent mixture with a Li salt dissolved in it [9]. The binary mixture is usually composed of a high-dielectric-constant solvent such as EC to better dissociate the Li salt, and a low-viscosity ingredient such as dimethyl carbonate (DMC), DEC, or EMC to promote Li^+ transport.

The scheme described in Figure 5.1 considers only the basic thermal dynamic properties of either a solvent or a Li salt. The dissolution of the Li salt in the electrolyte solvent leads to strong Li^+-solvent interactions, which ultimately alter the reduction potential [10]. This process is shown in Figure 5.2, in which the coordination of Li^+ with EC causes the reduction potential to increase from 0.2 V to 0.6 V vs. Li^+/Li [7]. As Xu summarized in his seminal review [9], three factors that affect SEI formation at the anode/electrolyte interface—the understanding of which remains elusive—are (1) which solvent nucleophilic sites interact with Li^+, (2) the number of solvent molecules Li^+ can accommodate in its solvation sheath, and (3) the preference of Li^+ in coordinating with any solvent.

Without including exhaustive detail regarding how coordination chemistry affects the SEI on a Si surface, this section focuses on the basic reduction mechanisms of several commonly used electrolyte components for Si anodes.

5.3.1 Reduction of solvent molecules

5.3.1.1 Reduction of the cyclic carbonate—ethylene carbonate

The EC molecule has probably been the most intensively studied to explain its reduction mechanism. Since 1964, EC has been considered as an electrolyte cosolvent for improving the dissociation of salts in electrolyte solutions owing to its high dielectric constant [11]. The significant role of EC reduction in building a stable SEI on graphitic carbonaceous anodes, to eliminate the graphite exfoliation unveiled by Fong *et al.* [12], resulted in the development of a practical electrolyte formula for the Li-ion industry [9]. The mechanism of Si lithiation, as discussed in Section 5.2, is different from that of graphite lithiation. And it has been demonstrated by many studies that commonly used carbonate-based electrolytes in commercial Li-ion batteries cannot stabilize Si anodes well [13]. Regardless, EC is still the cyclic carbonate cosolvent most commonly used to formulate electrolytes for Si anodes. Therefore, a proper understanding of its reduction paths at the anode/electrolyte interface is meaningful for next-generation electrolyte development.

As confirmed by Wang *et al.* [14], after the formation of $Li^+(EC)_n$ by coordination with Li^+ cations, EC molecules undergo a two-step reduction process: they are initially reduced to an ion-pair intermediate that undergoes homolytic C–O bond cleavage to produce a radical anion coordinated with Li^+, followed by different termination pathways of the radical anion. Wang *et al.* also indicated that almost all the products of EC decomposition may have been observed experimentally, including dilithium butylene dicarbonate (Li_2BDC), Li ethylene dicarbonate (Li_2EDC), $LiO(CH_2)_2CO_2(CH_2)_2OCO_2Li$, $Li(CH_2)_2OCO_2Li$, and Li_2CO_3. It was found that the formation of Li_2BDC is highly favored thermodynamically, but that Li_2BDC is highly soluble in carbonate solvents. Therefore, compared with Li_2BDC, Li_2EDC dominates in the SEI. It is worth noting that most SEI reaction mechanisms discussed in the literature have been deduced by experimental observation [10]. Quantum chemistry and molecular dynamics (MD) simulations play a key role in exploring SEI reaction mechanisms, especially when intermediate species cannot be directly measured in the reaction paths.

Two feasible EC reduction paths are shown in Figure 5.3, although the final chemical structures are alike [18]. One-electron reduction can occur at high potential, owing to insufficient electrons with low electrical resistance, and two-electron transfer can occur at low potential [19, 20]. When the anode cycles from high to low potential, a passivation film is generated on the anode surface by one-electron reduction. At low potential, "compact" Li compounds can be generated either from

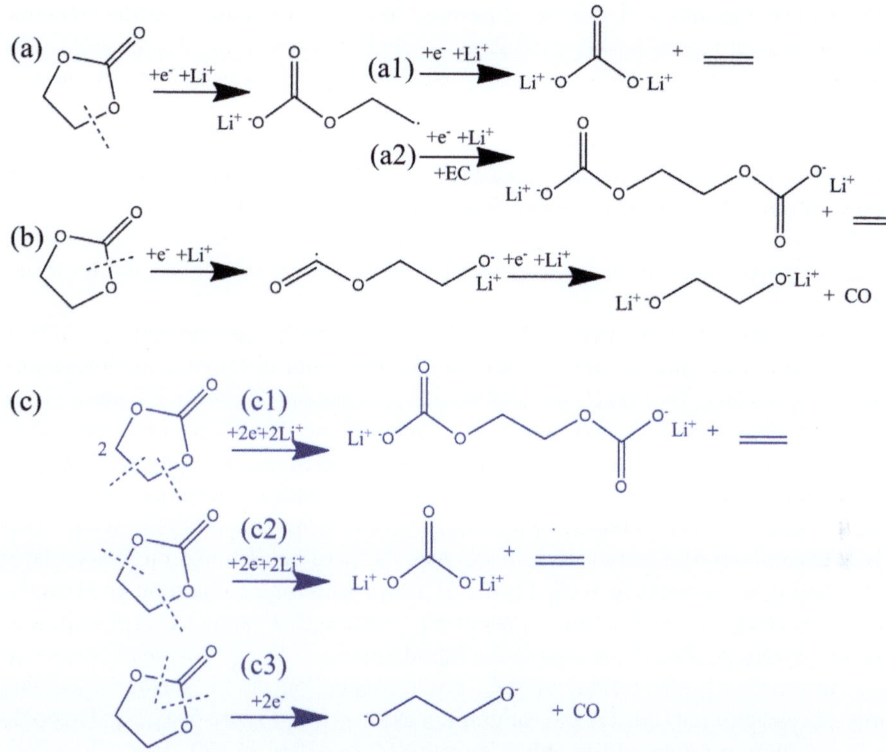

Figure 5.3 *EC decomposition paths at the anode/electrolyte interface for (a) and (b) one-electron reduction and (c) two-electron reduction. Reaction path a1 is from Aurbach et al. [15], a2 is from Wang et al. [14], b is from Xh [16], and c is from Leung and Budzien [17].*

continuous reduction of a preexisting Li-containing passivation layer on the anode surface or directly by complete reduction of Li compounds in the electrolyte driven by high energy at low potential [18]. A different concept assumes that either a slight or no passivation layer preoccupies the anode surface at high potential, thereby providing a high electron transfer rate favoring the multi-electron reduction route.

Taking the charge transfer rate into account at the anode/electrolyte interface, the reduction kinetics of EC should be considered, rather than just the reduction thermodynamics. Li and Balbuena clarified that a two-electron transfer reaction could be ten times faster than single-electron transfer. Therefore, the rate-limiting step is single-electron reduction [21].

Finally, since the electrolyte components affect the SEI thickness and structure on the anode surface, it is easy to understand that the constituent in the electrolyte determines the EC reduction path on the anode surface. For example, more Li_2CO_3 is formed when the EC concentration is low (reaction path a1 or c2 in Figure 5.3), whereas the formation of LiEDC is favored at a high EC concentration (reaction path a2 or c1 in Figure 5.3) [14, 21–23].

Almost all the EC reduction paths listed in Figure 5.3 were initially derived from graphitic carbonaceous anodes. Based on these mechanisms, EC reduction on Si anodes has been specified.

Using the reactive force field on a Si surface, Yun and Han *et al.* [24] found that one-electron reduction of EC following reaction paths a and b in Figure 5.3 could be thermodynamically favorable because of the small energy barrier. The reduction mechanisms of EC on lithiated Si anodes are dependent on the lithiation stage of the anode.

Balbuena and coworkers [25] used ab initio molecular dynamics (AIMD) simulations to explore EC reduction on a Si anode with different lithiation stages. On a LiSi surface, two reaction mechanisms independent of surface features were observed, the most common one being cleavage of the C(carbonyl carbon)–O single bond based on the two-electron reduction process, yielding the $O(C_2H_4)OCO^{2-}$ adsorbed species. A less common reduction path involves the breaking of two C(ether carbon)–O bonds sequentially, which generates a C_2H_4 + CO_3^{2-} pair [26]. The common reduction path of EC was found to be common on less-lithiated $LiSi_2$ and $LiSi_4$ surfaces as well. For Li-Si alloys with high Li loading and reactivity, multi-electron reduction of EC occurred, yielding CO^{2-} and $O(C_2H_4)O^{2-}$ on a Si surface or $O(C_2H_4)OCO_{2-}$ anions in the liquid phase.

Anionic polymerization of EC could occur on a Si surface, yielding poly(ethylene oxide) (PEO)-like oligomers as shown in Figure 5.4 [27]. This path is initiated by a nucleophilic attack on the EC, resulting in CO_2 loss. The ethylene oxide radical anion ($CH_2CH_2O^-$) formed as a result attacks the EC to initiate anionic polymerization. Jin and Grey *et al.* [28] later confirmed this reaction mechanism [28] on a Si anode using nuclear magnetic resonance (NMR). The linear PEO chain formed as a result seemed to be unstable and easily solubilize back into the electrolyte.

Note that the same types of compounds may coordinate with cations flowing in different conformations in the SEI (Figure 5.5), complicating the interpretation of the SEI chemical information further [29].

In summary, there are multiple reduction paths for EC on the anode surface. Upon coordination with a Li^+ cation, the reduction potential of the EC is elevated. The coordination of EC-Li^+ affects the EC reduction decomposition. EC decomposition may require multiple steps, and it can proceed through one-electron, two-electron, or multi-electron reduction paths. These various paths could lead to similar SEI species eventually. EC reduction is kinetics dependent, with the final

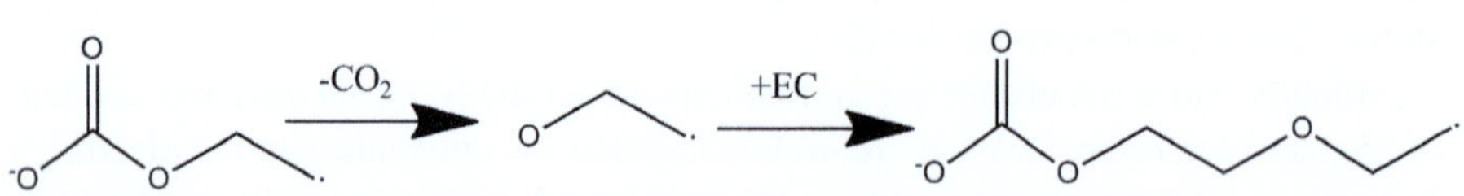

Figure 5.4 Formation of a PEO-like oligomer by anionic polymerization [27]

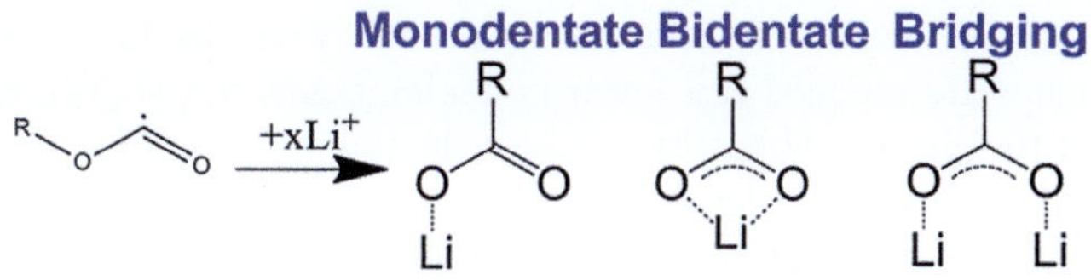

Figure 5.5 *The formation mechanism of various carboxylate conformations by interim ester radicals from EC reduction (Reprinted with Elsevier permission) [29]*

decomposition products relying on the Si surface passivation status and the lithiation state of the Si anode.

5.3.1.2 Reduction of linear carbonates

Linear carbonates were introduced to electrolytes because of their low boiling points, low viscosity, low dielectric constant, and ability to homogeneously blend with cyclic carbonates to increase the ion conductivity [9]. Commonly used linear carbonates in Li-ion batteries include DEC [8, 30, 31], DMC [32], EMC [33, 34], and others. The reduction paths of the linear carbonates are summarized in Figure 5.6. For a more comprehensive summary of linear carbonate reduction paths, readers may resort to studies by An *et al.* and Yan *et al.* [18, 20].

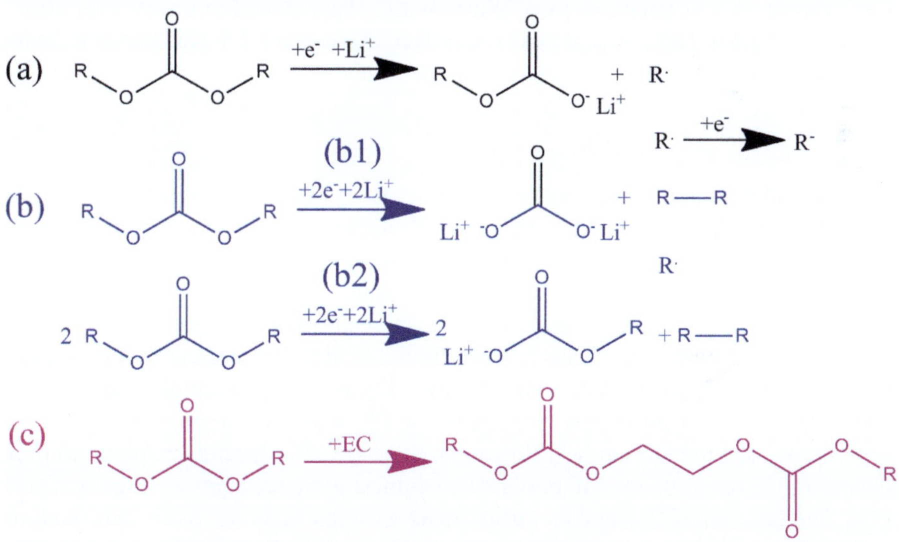

Figure 5.6 *Linear carbonate reduction paths following (a) one-electron and (b) two-electron mechanisms. (c) Combination reduction of the linear and cyclic carbonates. Reaction path a (up) is from [35], a (down) is from [36], b1 is from [37], b2 is from [38], and c is from [20].*

While cyclic EC carbonate decomposes at >0.8 V vs. the Li^+/Li reference, the linear carbonates are reduced at a lower potential, below 0.5 V [10]. Each solvent component in the electrolyte has its unique reduction potential. For a pure solvent, the first electron reduction order is EC > PC > VC (vinylene carbonate) > DMC > EMC > DEC [39]. Upon coordination with Li^+ to form the solvent-Li^+ complex, the order of the first electron reduction was found to be VC > PC ~ EC ~ DMC (cis-trans) > DMC (cis-cis) [6].

Similar to cyclic carbonate reductions, linear carbonate reduction may follow a single-electron path, shown in Figure 5.6a, or a two-electron reduction mechanism (Figure 5.6b). The former involves a step-wise reaction in which the single-electron reduction occurs first to a linear carbonate, resulting in an unstable radial intermediate and a Li compound, such as $R(OCO_2)Li$. The radical intermediate then goes through a second single-electron reduction path to form a more stable compound. The two-electron reduction involving one linear carbonate results in the formation of Li_2CO_3 (path b1), whereas two linear carbonates are reduced to form two $R(OCO_2)Li$ molecules (path b2). It has been noted that on a graphite anode, before the net electron transfers between the electrode and the electrolyte, a small amount of $LiPF_6$ salt could decompose to $PF5^-$ and LiF, initiating electrophilic reactions of PF_5 with EC or DMC molecules to produce insoluble $(CH_2CH_2O)_2$ and $(CH_2OCOOCH_3)_2$ (Figure 5.6c) [20]. The occurrence of such a reduction reaction on a Si anode has been observed by Shi and coworkers [40]. Using depth-resolved, in situ attenuated total reflectance–Fourier transform infrared (ATR-FTIR) spectroscopy, they found that the reduction mechanisms of a commonly used electrolyte consisting of EC and DEC varied on a Si anode surface. Whereas EC was reduced to Li ethylene dicarbonate (Li_2EDC), DEC was selectively reduced to diethyl 2,5-dioxahexane dicarboxylate on the surface of the native Si-oxide film.

5.3.2 *Reduction of lithium salt*

It was found in numerous studies related to Si SEI that new fluorinated species were formed on the top layer of a Si surface. Recalling the chemical formulas of the commonly used carbonate-based electrolyte components (solvents and salts), the major source of the fluorine in the SEI was the $LiPF_6$ salt.

It is known that $LiPF_6$ is extremely sensitive to trace amounts of moisture in electrolytes and commonly leads to hydrofluoric acid (HF) formation following the reaction paths (a1) and (a2) in Figure 5.7 [41, 42], which are eventually in dynamic equilibrium status. Continuous reaction of the PF_5 with the carbonate solvent molecules may lead to reaction (a1) shifting to the right. The further liberation of protons from HF can produce LiF based on an autocatalytic mechanism (Figure 5.7a3) [43]. The generated HF can also further react with the Si oxide, likely contributing to form favorable interactions between the binder and the Si [46]. HF also will react with any Li_2O that might exist on the lithiated Si surface to produce LiF and H_2O [47]. The reaction of HF with Li_2CO_3 results in LiF and $ROCO_2H$ [7].

The decomposed reactive Lewis acid product, POF_3, may react with EC to produce CO_2 and $OPF_2OCH_2CH_2F$ (Figure 5.7b). Continued reaction produces capped

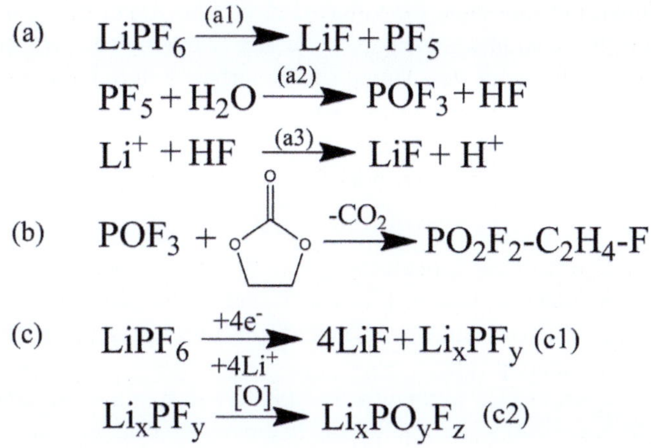

Figure 5.7 Possible decomposition paths of $LiPF_6$ on an anode surface. Reaction path a1 is from [41], a2 is from [42], a3 is from [43], b is from [44], and c is from [45]

oligoethylene oxides [44]. The primary phosphorus-containing species in the SEI layer may also result from the direct reduction of $LiPF_6$ to form a Li_xPF_y type of compound (Figure 5.7c). Part of the Li_xPF_y may react further with water or oxygen to generate $Li_xPF_yO_z$ [45].

Generally, salt reduction in the carbonate electrolyte leads to negative impacts on the cell performance. First, it results in a decrease in the salt concentration and hence solvated ions in the electrolyte; these may reduce the ionic conductivity of the electrolyte, and polarization effects may arise to increase the cell overpotential. Second, decomposition byproducts such as LiF and HF are detrimental to the overall cell performance. For example, LiF has poor Li^+ conductivity and consequently blocks Li^+ transport between the electrolyte and the anode. HF reacts with SiO_x and several other intermediate decomposition products to generate undesirable SEI species. Gaseous byproducts such as CO_2 result in increasing internal pressure during battery operation. The salt decomposition reaction at the Si/electrolyte interface can be mitigated by the introduction of additives such as VC and fluoroethylene carbonate (FEC) to the electrolyte.

5.4 The effect of electrolyte additives on Si SEI formation

An effective method of improving the SEI stability and quality is adding a small amount of a passivation film formation additive to the electrolytes. The general rule for introducing an additive to the electrolyte is that the additive must not alter the main functionality of the electrolyte, and it must be consumed at the anode/electrolyte interface to modify the SEI chemistry. Several studies have confirmed that commonly used carbonate electrolytes such as 1.2M $LiPF_6$ in EC:EMC 3:7 (wt%) (GenII) alone are incapable of stabilizing the Si surface. Thus, an SEI-forming additive is almost indispensable for Si-based electrodes. Usually, the onsite reduction potential of these additives is higher than that of other electrolyte components, and they are thus capable of passivating the Si surface before the other electrolyte components decompose. Electrolyte additives can include carbonate solvents, other solvent molecules, Li salts, and gases.

5.4.1 Carbonate solvent additive—FEC

FEC is probably the most efficient electrolyte additive for stabilizing a Si SEI, as has been validated by intensive studies [10, 18]. It was shown that the addition of less than 10 wt% FEC drastically improved the charge/discharge stability and prolonged the cycling life of a battery [8].

The initial report of FEC as an effective additive to stabilize the cycling performance of a 200 nm Si thin film anode was delivered by Choi and coworkers [48]. They found that the addition of 3 wt% FEC to 1.3 M $LiPF_6$ in EC:DEC (3:7 in volume) resulted in a smoother SEI surface on the Si anode, and the discharge capacity retention increased after 80 galvanostatic cycles from <70 percent without FEC to around 90 percent with FEC. The onsite reduction of the FEC additive, as reflected by the differential capacity plot, occurred at 1.5 V, 0.15 V higher than for EC in an FEC-free electrolyte. The preferable reduction decomposition of FEC at the Si/electrolyte interface led to the formation of an LiF- and Si-F-enriched SEI layer, based on X-ray photoelectron spectroscopy (XPS) analysis. Those compounds are more stable than EC-SEI derivatives such as linear alkyl carbonates (Si $OCH_2CH_2OCO_2Li$, $SiCH_2CH_2OCO_2Li$, or $R(OCO_2Li)_2$) because of their higher bonding energy [48].

The role and the reaction paths of LiF in the SEI layer derived from the FEC additive triggered considerable interest because the improved performance of the Si anodes was always associated with the presence of LiF as a decomposition product [49–51]. The generation of LiF can result from the direct defluorination of FEC (Figure 5.8a) or from the defluorination of intermittent species during several ring-opening reactions of FEC (Figure 5.8b and 5.8c). Since LiF has been shown to be more kinetically stable under reduction than other SEI species, it may better stabilize the anode when it approaches the Si surface [53]. Unlike organic SEI species, inorganic Li^+ forms ionic bonds to inorganic species such as LiF. Therefore, the enrichment of LiF facilitates Li^+ ion transport across the SEI layer based on hopping-type transport mechanisms, as demonstrated by simulation [54].

Note that despite the profound discovery of FEC as a useful additive for Si anodes in a few pioneering studies, the exact mechanism of FEC decomposition remains elusive. One would agree that the presence of only inorganic compounds derived from FEC reduction is insufficient to stabilize a Si anode, because they are too stiff to buffer the severe volumetric change that results from repeated (de) lithiation of the Si anode. Therefore, FEC reduction is thought to generate polymeric organic species that are capable of providing a certain elasticity to cushion the Si volumetric change upon (de)lithiation.

Nakai *et al.* [55] used XPS and time-of-flight secondary ion mass spectrometry (ToF-SIMS) to investigate the detailed characteristics of an FEC-derived SEI

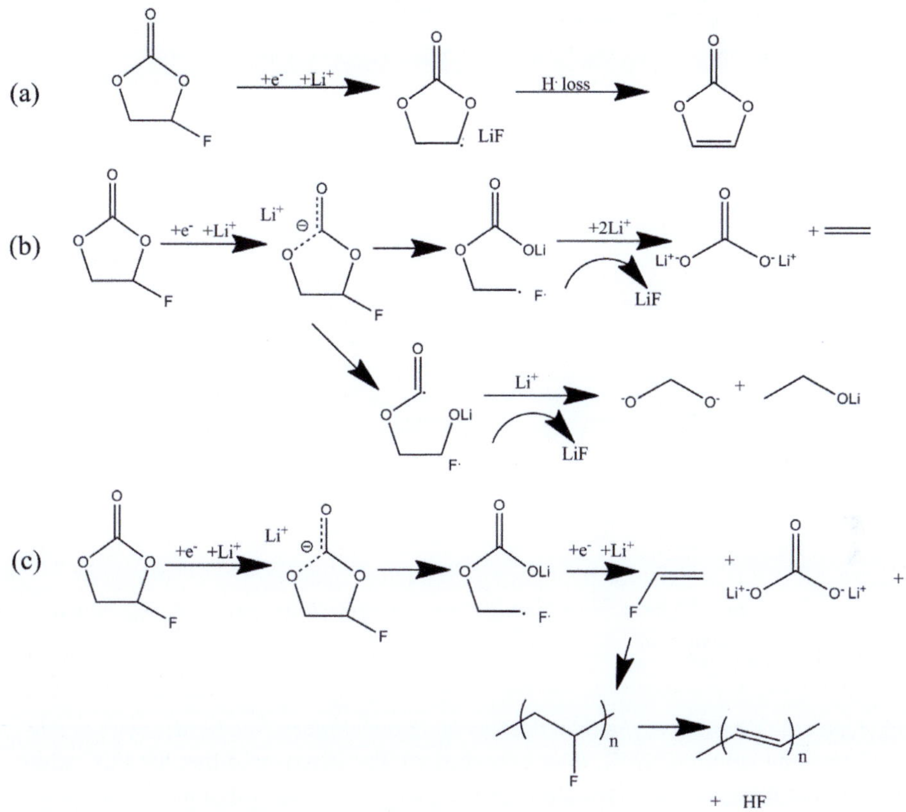

Figure 5.8 *FEC reduction decomposition paths and products. Reaction path a is from [28], b is from [8], and c is from [52].*

on a 2-μm-thick Si thin film electrode cycled in 1 mol/L of $LiPF_6$ in FEC:DEC (1:1 v/v). In addition to the enrichment of fluorine, they observed the formation of a polyene compound from the reduction of FEC. The possible reaction path is shown in Figure 5.8c. Later, Jin and Grey *et al.* [56] conducted NMR studies using a ^{13}C-labeled FEC additive. The ^{13}C NMR study of the electrolyte after cycling indicated that the oxygen-deficient polymers contained primarily alkyl chains instead of unsaturated carbons, leaving the existence of the polyene an open question.

Aurbach *et al.* [57] explored SEI chemistry by cycling Si nanowires in 1 M $LiPF_6$ in DMC and EC (1:1 w/w) solutions. FTIR and XPS results reflected that with the addition of 10 wt% of FEC, the SEI compositions were enriched in polymeric or oligomeric species as a result of VC polymerization (resulting from FEC decomposition, as in the reaction shown in Figure 5.9a). Those FEC-derived species caused the SEI to be thinner than its counterpart derived from an FEC-free electrolyte, which decreased the charge transfer resistance at the Si surface.

Combining ^{13}C isotope labeling and 1D/2D solid-state NMR, Jin and Grey *et al.* [28] found that the addition of FEC resulted in the formation of insoluble branched

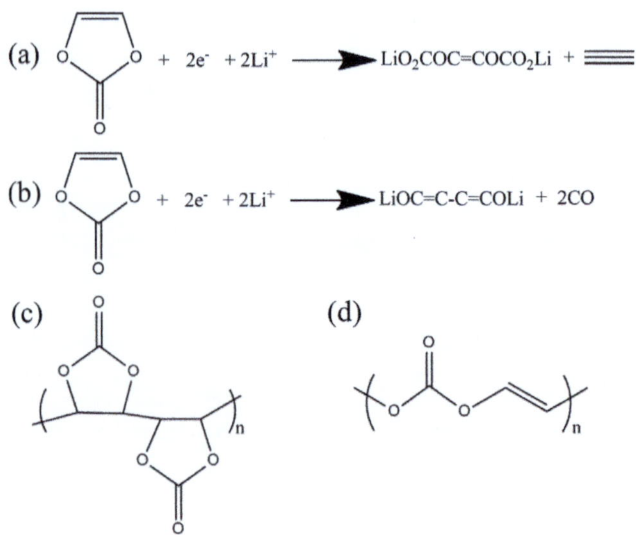

Figure 5.9 *(a) and (b) VC two-electron reduction decomposition paths [58] and two possible VC reduction polymerization products; (c) radical polymer and (d) linear polymer [59]*

ethylene oxide-based polymers in the SEI layer on a binder-free Si nanowire anode. Facilitated by solution NMR, their experiments also discovered that the FEC additive helped mitigate the dissolution of the PEO-like linear oligomeric electrolyte breakdown products from EC reduction. The promoted insolubility of FEC derivatives was thought to be due to cross-linked polymerization, as observed by a higher ^{13}C shift at 103 ppm than the shift of the aliphatic units at 34 ppm.

The exact reaction mechanism for FEC reduction initiated on a Si surface is not yet thoroughly understood. In an effort to shed light on it, Hou and Persson *et al.* [7] employed classical MD simulations coupled with first-principles quantum mechanics to unveil the detailed solvation structure and reactivity of $LiPF_6$ in EC/FEC. Despite its lower concentration, FEC was found to significantly alter the solvation structure and reduction behavior of the electrolyte in two ways: (1) FEC more than doubled the portion of the contact-ion pairs (CIPs) at the electrode/electrolyte interface, and (2) FEC participated in the formation of the primary solvation shell. As a consequence of the alteration of the Si surface coordination chemistry by FEC, the onsite reduction potential was increased to 0.9 V, agreeing well with several experimental observations [8]. The increased CIP ratio due to FEC assisted in the formation of LiF because of the promoted PF_5^- and consequent HF formation.

Based on these discussions, the functions of an FEC additive in increasing the quality of the SEI on a Si anode and further promoting the electrochemical performance of an Li-Si battery can be summarized as follows. (1) Before SEI formation, FEC promotes the concentration of CIPs and contributes to primary coordination shell formation. (2) FEC facilitates earlier formation of the passivation layer at a higher reduction potential compared with other electrolyte components. (3) The enrichment of inorganic species, such as LiF, close to the Si anode surface increases the electrochemical stability and the ion transport across the SEI layer. (4) The metastable intermediates of the FEC reduction facilitate alternative reaction pathways to EC reduction, leading to preferable polymerization reactions to form polyene, polycarbonate, and branched PEO oligomers. The resultant polymeric species in SEI can accommodate the volumetric change in the Si anode, thereby promoting the cycling life of the Li-Si battery.

5.4.2 Carbonate solvent additive—VC

VC has been used in Li-ion batteries to increase the coulombic efficiency (CE) and thermal stability of graphite anodes and is considered to be the most effective additive for enhancing the lifetime and thermal stability of graphite anodes [60, 61]. This additive was first tested by Aurbach *et al.* [62] on a graphite anode in 1 M $LiAsF_6$ in an EC and DMC (1:1) electrolyte. The VC addition was found to promote cell cyclability and mitigate the irreversible capacity. As reflected by spectroscopic studies, the surface chemistry of the anode featured poly (alkyl) Li-carbonate species, which were thought to suppress both solvent and salt anion reduction.

Ota and Yamaguchi *et al.* [58] performed an exhaustive study of the surface chemistry of a graphite anode derived from VC-containing electrolytes. Although the reduction mechanism of the VC was not investigated directly on a Si anode, the findings should benefit the understanding of the surface chemistry of Si when a VC additive is used. Based on detailed analysis via NMR, FTIR, and XPS, they found that the reduction decomposition of VC on the graphite surface generated both polymeric organic species—including poly(VC), the VC oligomer, the ring opening polymer of VC, and polyacetylene—and inorganic compounds containing vinylene dicarbonate $(CHOCO_2Li)_2$, Li divinylene dicarbonate $(CH{=}CHOCO_2Li)_2$, Li divinylene dialkoxide $(CH{=}CHOLi)_2$, and Li carboxylate (RCOOLi). The passivation layer formed by VC decomposition also mitigated the reductive gassing issue brought about by EC reduction.

As proposed by Ota and Yamaguchi *et al.*, the formation of Li vinylene dicarbonate $(CHOCO_2Li)_2$ stemmed from the ring-opening reaction of VC under a two-electron reduction path, as shown in Figure 5.9a. Two-electron reduction of VC could also generate acetylene. Since acetylene is metastable, and gassing of CO was observed by gas chromatography, the final product should be Li divinylene dialkoxide $(CH{=}CHOLi)_2$ with the reaction path shown in Figure 5.9b. Although the final polymeric reduction products of the VC were analyzed and unveiled by 2D ^{1}H-^{13}C NMR, the authors did not specify the possible reaction paths leading to the generation of these polymeric species.

Looking at the chemical formula of VC in Figure 5.9, VC polymerization could proceed in two ways: via C=C double bond breakage, resulting in a repetition of EC units (Figure 5.9c), or via a ring-opening reaction to form linear chains of polycarbonates in which the C=C double bonds pertain (Figure 5.9d) [59]. Wang and Balbuena *et al.* [63] used DFT calculations to elucidate the reduction mechanism of VC as a solvent additive. From a thermodynamic standpoint, based on which the formation of a linear polymer (Figure 5.9d) has a lower energy barrier than the formation of a radical polymer (Figure 5.9c), they concluded that the formation of a linear polymer (Figure 5.9d) is more probable. Ouatani at Gonbeau *et al.* [59] combined ab initio simulations with the XPS valence spectra of both the passivation film on a graphite surface and synthetically made VC-polymeric derivatives—the radical poly(VC) shown in Figure 5.9c and a ring-opening polymerized linear product shown in Figure 5.9d. Interestingly, they concluded that the radical poly(VC) was the main VC-derived polymeric product in the SEI, contrary to the conclusion drawn by Wang *et al.* [63].

Xie and coworkers [64] initiated a study of applying a VC additive to stabilize a Si anode. Adding 1 wt% of VC to 1 M $LiPF_6$/EC/DMC (1:1 vol), with a 150 nm Si thin film as a model anode, stabilized the CE value after 10 cycles at close to 100 percent; in comparison, the VC-free electrolyte had a CE value of 96 percent. Based on scanning electron microscopy (SEM) micrographs, the Si surface became smoother and more uniform when it was cycled with a VC additive. Probably their most interesting finding regarding the Si surface chemistry was that LiF drastically decreased with the addition of VC. This result indicates that passivation on a Si surface by VC reduction may inhibit the decomposition of $LiPF_6$ to form LiF, as the salt in this case was the only fluorine source. An FEC additive is an additional source of fluorine and thus enriches the LiF in the SEI layer, with the reduction paths shown in Figure 5.8. The stabilization mechanism of the VC additive should differ from the stabilization mechanism shown for FEC.

In fact, the merits of LiF for stabilizing the Si anode/electrolyte interfacial region remain an open question. Etacheri *et al.* [57] found polycarbonates as decomposition products of FEC. The role of LiF in enhancing reversibility was less emphasized. Rather, the mitigated internal resistance on the Si anode was not attributed to the better ionic conductivity of LiF, as discussed in Section 5.4.1, but rather to the elastic thin polymeric film derived from VC decomposition, agreeing with a recent study by Jaumann *et al.* [65]. A later study by Jaumann and Giebeler *et al.* [61] detailed the benefits provided by FEC and VC with a nano-Si anode derived from $HSiCl_3$ and 1 M $LiPF_6$ in EC/DMC (1:1) (LP30) as a reference electrolyte. They found that an FEC additive slowed the SEI growth rate but was incapable of completely suppressing SEI growth. The continuous SEI growth resulted in a thick passivation film with delamination of the Si anode from the current collector. The defects resulting from LiF nanocrystals embedded in the passivation film were believed to be responsible for a defective SEI layer on the Si anode. With the addition of VC, the growth of the SEI was close to being completely suppressed. A defect-free and denser SEI layer derived from VC reduction provided the benefits of longer cycling life and better capacity retention under deep-discharge conditions. Thus, FEC provided better Li^+ conductivity and high-power applications at the cost of a shorter battery life, whereas VC increased the chemical homogeneity of the SEI and consequently promoted battery reversibility and cycle life.

Regardless, there seems to be a consensus that both VC and FEC additives lead to the formation of a flexible polymeric SEI species on Si anodes, regardless of the type of Si. The polymeric feature makes the SEI capable of accommodating the large volume changes of Si during lithiation. The question remains open whether the characteristics of the polymeric components derive from both FEC and VC. A vast majority of the literature acknowledged the existence of the polycarbonate when FEC and VC were used as electrolyte additives [18]. Recently, Jin and Grey *et al.* [66] noted in a study that the Si anode (Si nanowire in their case) surface was enriched in highly cross-linked PEO-type SEI components. They used magic-angle spinning solid-state NMR spectroscopy to characterize the SEIs generated on binder-free Si nanowire anodes from pure FEC and VC electrolytes and from an LP30 (1 M $LiPF_6$ in EC/DMC (1:1)) electrolyte as a reference. Their ^{13}C NMR experiments with both ^{13}C-enriched and natural-abundance electrolytes revealed that the organic electrochemical decomposition products of FEC and VC mainly consisted of cross-linked PEO and alkyl chains, with carbonate and carboxylate units forming 15 percent of the carbon content of the organic SEI. As noted by the authors, the resulting highly cross-linked PEO had three potential merits for stabilizing a Si anode: (1) it reduced solvent penetration and swelling of the SEI, helping to prevent further solvent reduction; (2) it accommodated the volumetric change in the Si, and (3) it might possibly have improved Li^+ conductivities compared with the linear PEO found in EC-based electrolytes [67]. The similarity of the polymeric reduction products in SEIs derived from FEC and VC may stem from that the fact that FEC could be defluorinated to form VC and vinxoyl radicals, as indicated by solution NMR [28].

In fact, Soto *et al.* [68], using first-principles calculations, validated that a more stable SEI was a benefit of the polymeric feature resulting from the VC/FEC reduction [68] (Figure 5.10). Focusing on the electrochemical stability of the decomposition products on the outer SEI layer, they noted that without VC/FEC additives, the outer SEI layer was prone to additional attacks by radicals, causing the SEI to continue to grow. Their findings supported other work indicating that VC/FEC additives result in fast polymerization on an anode surface, forming a more stable SEI and leading to controlled SEI growth, as shown in Figure 5.10.

It has been generally accepted that the carbonate-based electrolytes cannot effectively stabilize Si anode surface. It is partly because of that the carbonate-based electrolytes have been developed for graphite anode. Therefore, alternatives to carbonate-based electrolytes for Li-Si batteries are sought after. Recently, our team demonstrated that ether-based electrolytes could ensure a conformal SEI on Si and keep the Si "fracture-free". Benchmarking against the optimal, commonly used carbonate electrolyte with the FEC additive, the Si anode cycled in the ether-based electrolyte a reduced early parasitic current (by 62.5%) and interfacial resistance (by 72.8%), while cell capacity retention is promoted by >7% over the course of 110 cycles. The mechanism behind the more conformal SEI on Si cycled in ether-based electrolyte is that the reduction composition of the ether solvent forms a more elastic polyether SEI layer, thereby capable of accommodating several volumetric change of the Si.[69]

In summary, both FEC and VC additives have drawn significant interest for their potential to promote SEI stability for Si anodes. Although FEC has been given much more attention, recent studies have documented the merits of VC. The reduction

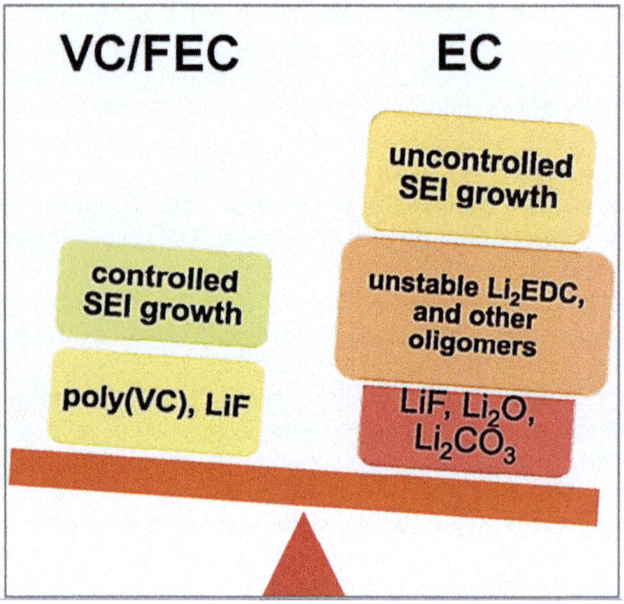

Figure 5.10 *A proposed principle for more stable SEI formation with FEC and EC electrolyte additives based on insights obtained from comparative AIMD calculations. EC reduction tends to form unstable SEI species such as Li_2EDC and soluble oligomers, leading to an unstable outer SEI layer prone to intermittent radical attack. Effective additives such as VC and FEC help form a polymeric passivation layer, resulting in more controllable SEI growth [68]. (Reprinted with ACS permission).*

decomposition of small amounts of FEC and VC additives occurs before the decomposition of other electrolyte components and eventually alters the SEI chemistry. The advantage of inorganic components in the SEI, such as LiF, is still under debate. However, the polymeric decomposition species derived from FEC and VC reduction are considered beneficial because they accommodate severe volumetric change in Si anodes upon cycling, thereby improving CEs and capacity retention in Si anodes for Li-ion batteries.

5.5 SEI characterization techniques

The compositional and morphological complexity of the SEI necessitates multimodal characterization to obtain a comprehensive understanding. Therefore, many techniques have been employed to study the SEI on Si anodes, including its physical and chemical structure, morphology, composition, electrical and electrochemical properties, and formation and evolution dynamics. This section aims to give readers an overview of the experimental means developed to explore different properties of the SEI. In general, post-mortem characterization methods are the most popular. However, cell disassembly and Si anode clearing after charge/discharge cycling inevitably alter the SEI structure and chemistry, thereby introducing uncertainty and imprecision in unveiling the SEI properties. Nondestructive strategies are thus desirable, ideally implemented under in operando conditions to study SEI growth on the Si anode during cell electrochemical cycling. However, investigation of SEI growth under realistic operation conditions is challenging.

5.5.1 Morphological methods

5.5.1.1 Optical microscopy

The most readily accessible microscopic technique used to observe the surface morphology of the Si anode is the optical microscope. It is also the oldest, the simplest, and probably the most underexplored method of all the microscopies [70]. Bhattacharya *et al.* [71] employed a model Si anode composed of micron-size (<10 μm) Si particles embedded in a ductile aluminum matrix electron collector; they cycled the anodes in EC- and PC-based electrolytes at different voltage scanning rates and cut-off voltages. The change in the surface topography was explored using a 3D surface optical profilometer. It was found that the Si anode was coated with deposits stemming from electrolyte reduction and SEI formation. The instability of the SEI due to the chemo-mechanical fracture of the Si upon continuous lithiation/delithiation was also captured by 3D optical profilometry imaging.

5.5.1.2 Electron microscopy

Although optical microscopy is easily accessible and inexpensive, its spatial resolution is inherently restricted by the visible light diffraction limit (Abbe's law), which makes it capable of obtaining only topographical information from the surface SEI layer in the submicron-micron range.

To overcome this inherent limitation, high-energy electron beams, which have much shorter wavelengths than visible light, are used. SEM focuses an electron beam onto the surface of a sample and rasters it over a desired range on the sample surface. The interactions between the surface atoms and the electrons are dependent on the atom types, and they produce various signals that convey topographical information for the sample surface. Two advantages of SEM compared with optical microscopy are (1) it enables high spatial resolution and (2) it can map surface elements concurrently with energy-dispersive X-ray spectroscopy (EDX) [70].

Numerous studies have employed SEM as an effective tool for postmortem analysis of surface SEI and Si anode topography. For example, the fracture failure mechanisms of a single-c-Si (100) anode were explored by Shi *et al.* using SEM, as shown in Figure 5.11 [72]. As electrochemical cycling proceeded, the loss of active Si increased as a result of lithiation depth. Simultaneously, the surface morphology of the Si (100) anode continuously evolves, as is clearly demonstrated in the top-view SEM micrographs. Crack pattern development is seen to be dominant and to grow in two orthogonal directions, resulting in isolated small squares that form a

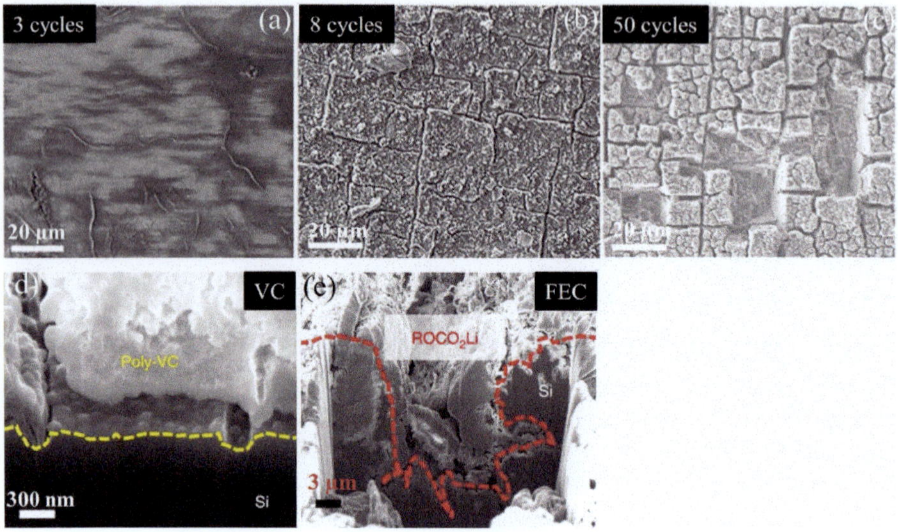

Figure 5.11 *Top view of SEM images of Si (100) electrode surface morphology obtained from (a) 3, (b) 8, and (c) 50 cycles between 2.0 and 0.01 V at a scan rate of 0.1 mV s^{-1}, with a Li metal counter electrode. Cross-sectional SEM micrographs of Si (100) cycled in EC/DEC/ Li hexafluorophosphate ($LiPF_6$) electrolyte with (d) 5 percent VC and (e) FEC additives. The cross-sectional area was produced by focused ion beam. (Reprinted with permission from Nature) [72].*

regular repetitive pattern on the electrode surface. In an effort to mitigate surface fracturing using different additives, the authors found that poly-VC from VC reduction prior to the reduction of the other electrolyte components aggregated and filled in the vertical surface cracks, as indicated by the focused-ion-beam cross-sectional SEM image (Figure 5.11d), thereby promoting crack-face bridging and inhibiting crack propagation. The crack depth reflected by the cross-sectional SEM image after 30 cycles is merely 100 nm. In contrast, ROCO2Li salt as a major product of FEC additive decomposition coated the SI crack edge preferentially, altering the perpendicular direction of cracks and significantly decelerating the crack growth [73]. In this context, VC and FEC stabilized the Si anode surface following two different mechanisms. Similar effects of additives on the morphologies of Si nanoparticle anodes were studied by Nguyen *et al.* using SEM top-view micrographs [73].

Cheng *et al.* [74] investigated crack patterns generated during electrochemical cycling of amorphous Si (a-Si) thin film electrodes using SEM. They found that cracks generated in thick films were straight, with a few sharp directional changes and large "islands." In thinner films, cracks showed more wiggles and the average cracked area was smaller. Based on these observations, two strategies were used to mitigate fracturing of Si thin film anodes: (1) pattern the Si islands to be smaller than the average crack size and (2) control the critical thickness of the Si film to remain below the level at which the material would crack, thereby retaining the SEI integrity.

SEM operation requires a high-vacuum sample environment. Since most carbonate electrolytes are volatile, it is almost impossible to observe the morphological changes of Si anodes in operando using common electrolytes for Li-ion batteries. This limitation can be avoided by using a high-boiling-point ionic liquid (IL) electrolyte. Chen *et al.* [75] designed a coin cell with an IL electrolyte to enable the simultaneous collection of electrochemical data and time-resolved SEM images. A Si-polyimide binder electrode was used as a model anode. The researchers found that the Si morphology changed drastically and cracks formed during the first lithiation process, whereas the thickness variation in the Si composite layer was moderated in subsequent cycles. Despite the success of the in operando SEM experimental design enabled by the IL electrolyte, such a configuration does not allow observation of the SEI morphology change in operando with most commonly used carbonate electrolytes.

The electron beam employed in transmission electron microscopy (TEM) has a significantly higher accelerating voltage than the SEM beam, enabling the beam to pass through the sample to create a projection. Compared with SEM, TEM enables even higher spatial resolution of the sample, down to the 0.1 nm scale [76]. This high resolution is extremely useful for observing the morphology of a nanoscale tin SEI. In addition, the SEI chemistry can be reflected by electron energy loss spectroscopy (EELS) simultaneously.

Molecular layer deposition (MLD) coating of the Si particle surface has been found effective in enhancing the cyclic performance in Li-Si batteries. However, it is challenging to understand the fundamental mechanism of the coating functions because of its ultrathin nature. Wang *et al.* [77] applied in situ TEM and EELS mapping to probe in detail the structural and chemical evolution of both coated and uncoated Si particles upon lithiation/delithiation. It was found that the native oxide layer converted to crystalline Li_2O islands, increasing the impedance on the particle and lowering the CE drastically. In construct, an alucone MLD coating effectively removed the native oxide layer and therefore mitigated side reactions and detrimental effects of the native oxide on the Si particle. Consequently, the CE and capacity retention were improved.

TEM usually produces a higher electron beam dose than other electron microscopy techniques, which may damage the SEI layer and result in beam damage artifacts [78]. Recently, Yi *et al.* [79] applied cryogenic (scanning) TEM (cryo-STEM) to stabilize the beam-sensitive and air-reactive SEI layer at liquid nitrogen temperature and to provide high-resolution structural and chemical characterization of the SEI. In that study, a Si nanowire was cycled in an EC/DEC + 10 percent FEC electrolyte against a Li metal counter electrode. The SEI on the Si nanowire in the charged state was protected by the low temperature of the cryogenic conditions. A high-resolution TEM micrograph indicates that a crystalline Si core is surrounded by an amorphous Li_xSi shell (Figure 5.12a). Closer inspection of the SEI on the outer shell of the litigated Si indicates that the SEI is in a totally amorphous state (Figure 5.12b), different from the state of its FEC-free counterpart. The compositional chemistry and

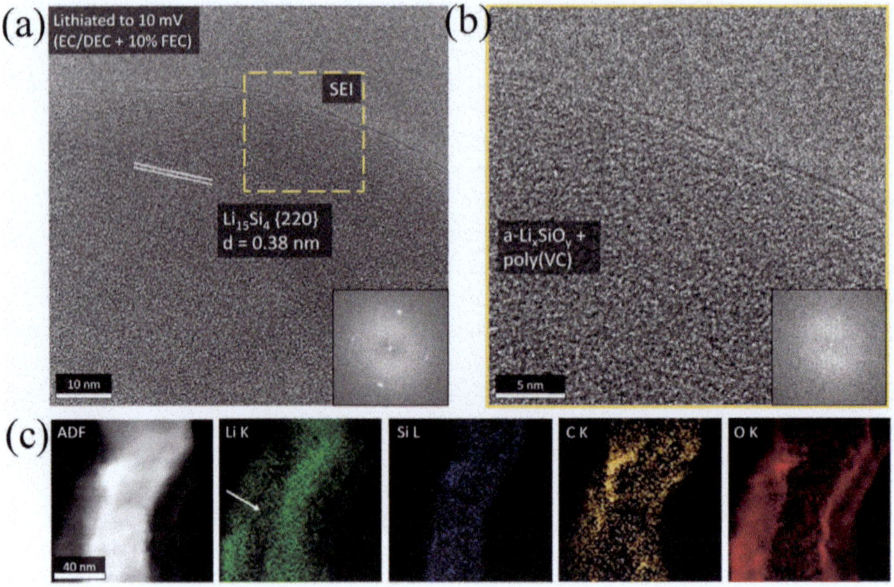

Figure 5.12 *(a) High-resolution TEM and fast Fourier transform (FFT) of a Si nanowire fully lithiated to 10 mV. (b) Magnified high-resolution TEM and FFT of the SEI. (c) STEM EELS mapping of the Si nanowire lithiated to 10 mV. The white arrow indicates the low Li intensity layer. All images were acquired under cryogenic conditions (Reprinted with ACS permission) [79]*

the chemical distribution were further explored by EELS mapping (Figure 5.12c), which indicates the existence of lithiated SiO_x and Li_xSiO_y, along with an amorphous outer layer different from that observed from EC decomposition.

5.5.1.3 Scanning probe microscopy

Atomic force microscopy (AFM), or scanning force microscopy, is a high-resolution microscopic technique in the scanning probe microscopy (SPM) category that is extremely useful for studying the surface morphology of the SEI on Si. Different from electron microscopy, which counts on the interactions of a high-energy electron beam with the specimen to gain topological information, SPM gathers surface morphology information by "feeling" the surface with a mechanical probe. Scanning the sample surface, piezoelectric elements convert the mechanical response of the AFM cantilever to electronic signals, thereby capturing the tiny and precise movements of the probe at nanoscale resolution [80]. Thus, the sample surface roughness/height can be precisely determined. Besides the SEI morphology, SPM can potentially map the mechanical, electrical, and electrochemical properties of the SEI, thereby providing a multimodal imaging system to visualize various intrinsic properties of the SEI.

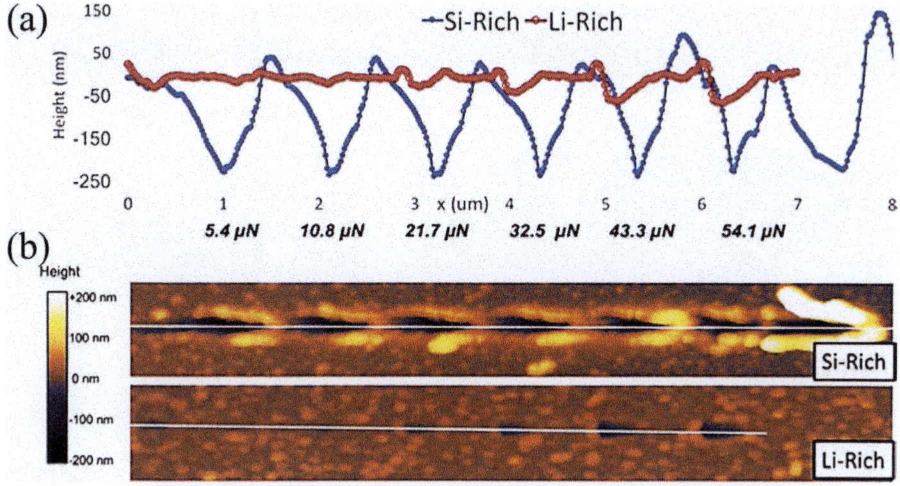

Figure 5.13 *(a)SPM indentation depths at different sites with increasing indentation forces aligned with each valley in the curve. (b) AFM images of indented film at both Si-rich and Li-rich regions. (Reprinted with ACS permission) [82].*

Becker *et al.* [81] used arrays of a-Si nanopillars as a model anode and electrochemically cycled it in 1 M Li trifluoromethanesulfonate in a propylene carbonate electrolyte. In situ AFM was used to qualitatively and quantitatively examine the morphology evolution of the nanopillars, including volume and height changes vs voltage in real time. It was found that the a-Si pillars did not crack in the initial cycle, but a permanent volume expansion was captured. During subsequent cycles, the a-Si roughened and deformed from the initial geometry, and eventually pillars with diameters >200 nm fractured. The size-dependent fracturing of a-Si suggested an important threshold size value for a-Si anode design to mitigate the possible impacts of mechanical instability on the SEI layer.

Besides passively sensing the surface morphology, SPM can also be used to indent the sample surface and measure the mechanical response of the SEI at different depths on the Si surface. Xu *et al.* [82] demonstrated that surface-sputtered Li_xSiO_y on a Si surface could be Li-rich and Si-rich along the depth direction. Ductile behavior was observed in the Li-rich layer, whereas the Si-rich layer showed brittle behavior. These evaluations were based on the indentation depth/load response, as shown in Figure 5.13a. At each depth of the surface Li_xSiO_y layer, the morphology also was collected (Figure 5.13b).

Recently, Caleb *et al.* [83] combined SPM with scanning spreading resistance microscopy (SSRM) imaging to measure synergistically the electronic resistivity and the morphology of the SEI formed on a model single-crystalline wafer Si anode (Figure 5.14). The electronic resistivity could be probed at each sampling point by a DC bias applied to the SPM probe of a metallic coating layer. The major finding of

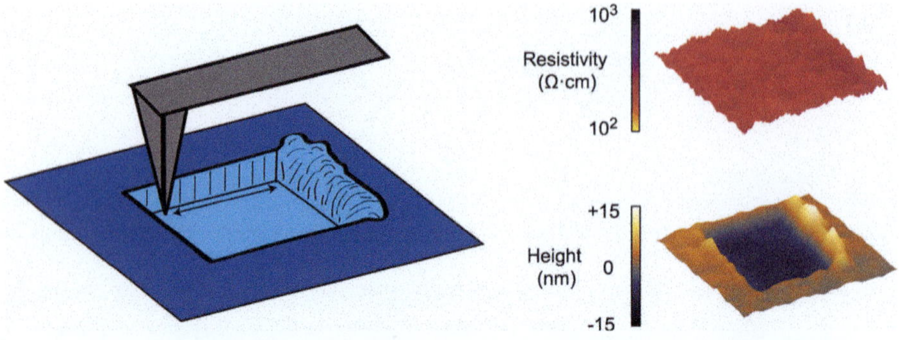

Figure 5.14 *Scheme of depth profile SSRM and its synergistic measurements of the electronic resistivity and the morphology of the Si SEI. (Reprinted with ACS permission) [83]*

their study was that the SEI had a uniform in-plane resistivity distribution on a given layer range but a highly inhomogeneous through-plane of the SEI layer.

To sum up this section, optical microscopy is easy to access and is so far largely underexplored for characterizing Si anodes. However, it has limited lateral spatial resolution because of the light diffraction limit. Electron microscopy, including SEM and TEM, break the spatial limitation and thus can provide morphological information from an SEI on Si at the nanoscale. When electron microscopy is used jointly with EDX (for SEM) and EELS (for TEM), the chemical distribution on a Si SEI can be visualized. SPM is another powerful tool that provides nanoscale morphological information for the SEI on Si. SPM is capable of probing not only the SEI morphology but also the local mechanical properties and electrical conductivity of the SEI.

5.5.2 Surface chemistry

Compositional information for SEI, and the chemical distribution both in-plane and through-plane of the Si anode, are central research interests. Surface analysis methods are techniques for discovering the chemical structure of an extremely shallow and thin area [84]. Multiple surface chemical analysis techniques with variable physical bases are therefore developed and utilized for the frontiers of current research on Si SEIs. Commonly used methods include Raman [29, 85–87] and FTIR spectroscopy [40, 88, 89], XPS [87, 90, 91], ToF-SIMS [91], EDX spectroscopy, and EELS [79]. The intrinsic difference among these many techniques lies in the signals used to irradiate and interact with the sample surface, including light, X-rays, neutrons, high-speed ions, and electrons.

5.5.2.1 Infrared spectroscopy

Infrared (IR) spectroscopy is a technique used to obtain an IR spectrum of the absorption or emission of a solid, liquid, or gas. The absorbance peaks within a narrow range of wavelengths result from the mutual interaction of the molecular

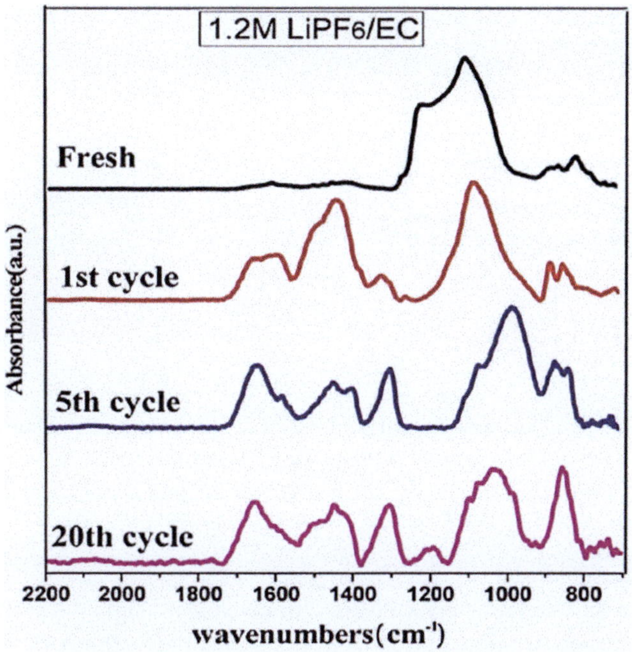

Figure 5.15 Infrared spectra of binder free-Si electrodes cycled with EC electrolyte. From top to bottom: fresh electrode, 1st, 5th, and 20th cycles. (Reprinted with ACS permission) [51].

vibrations and the inlet IR beam. And the reflection outlet IR beam carries useful chemical information. The term FTIR spectroscopy indicates that a mathematical process, Fourier transform, is required to convert the raw data into an actual spectrum [92].

FTIR is one of the most common spectroscopic techniques for Si SEI chemical identification. Figure 5.15 shows the IR spectra of binder-free model Si anodes cycled in $LiPF_6$/EC electrolyte after the 1st, 5th, and 20th cycles [51]. Multiple IR absorbance peaks appear for different Si samples carrying chemical information for either the pristine Si or the SEI at different cycles. Commonly seen SiO_x peaks are usually broad in the wavenumber range between 1 290 and 1 070 cm^{-1}. Li_2EDC was found to be the major product of the first cycle, with characteristic peaks at 1 652, 1 416, 1 305, 1 145, and 870 cm^{-1} [93]. The peaks centered at 1 420 and 870 cm^{-1} can be assigned to Li_2CO_3 and $LiPF_6$, respectively. As the number of cycles increases, IR peaks associated with Li_2CO_3 and O-Si-O absorptions decrease in intensity, whereas peaks associated with Li_2EDC increase in intensity, suggesting the increasing concentration of Li_2EDC as cycling is prolonged in the SEI.

At the ATR configuration, it is possible to turn FTIR into an in situ spectroscopic technique to study a Si SEI (see Figure 5.16). Shi *et al.* [40] initiated an effort to utilize depth-resolved, in situ ATR-FTIR to analyze the interface chemistry

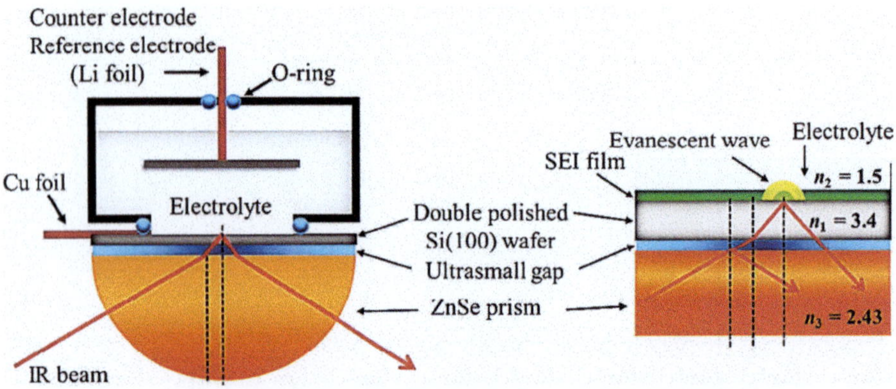

Figure 5.16 *Scheme of an in situ ATR-FTIR electrochemical cell setup. Single-c-Si (100) is used both as the IR transport window and as the working electrode. The ultrasmall air gap is much smaller than IR wavelengths(Reprinted with ACS permission) [40].*

of the SEI on Si electrodes upon cycling. In the study, single-c-Si (100) was used as the IR transport window and as a working electrode. The ATR configuration mitigated the signal probed from the bulk electrolyte, which would overwhelm the signal of the thin SEI. The composite ATR prism composing the 100 μm ultrathin Si and a semi-cylindrical ZnSe prism exhibited a wide energy transparency, allowing for penetration of evanescent waves into the sample. The incidence angle could be tuned continuously to adjust the IR penetration depth to the sample, encompassing the IR signal capture for SEI at the near surface of the Si anode. Such an in situ ATR-FTIR setup allows for synergistic recording of the surface chemical change of the Si anode and the electrochemical data (e.g., cyclic voltammogram), revealing the SEI formation and compositional change dependent on electrochemical cycling.

5.5.2.2 Confocal microscope Raman spectroscopy

Raman spectroscopy is another noninvasive chemical analysis technique based on the interactions between light and the chemical bonds of the analyte. Within the scattered light, a small portion (typically 0.0000001 percent) has a different energy, carrying useful information regarding the chemical structure of the analyte. This portion of inelastic scattered light is called "Raman scattering."

Raman spectroscopy is versatile in measuring multiple components on a Si anode, including the lattice mode of the Si (e.g., first order transverse optical, or TO, mode at 520 cm^{-1}) [94], the carbon D band and G band [85], and various Raman peaks in the fingerprint region assigned to the SEI [29]. The noninvasive nature of Raman spectroscopy allows for ease of in operando observation of the SEI growth and structural change of the Si anodes upon electrochemical cycling [95, 96].

Krause *et al.* [95] employed a Si nanowire anode with and without carbon coating to implement in situ Raman spectroscopy measurement (Figure 5.17). It was

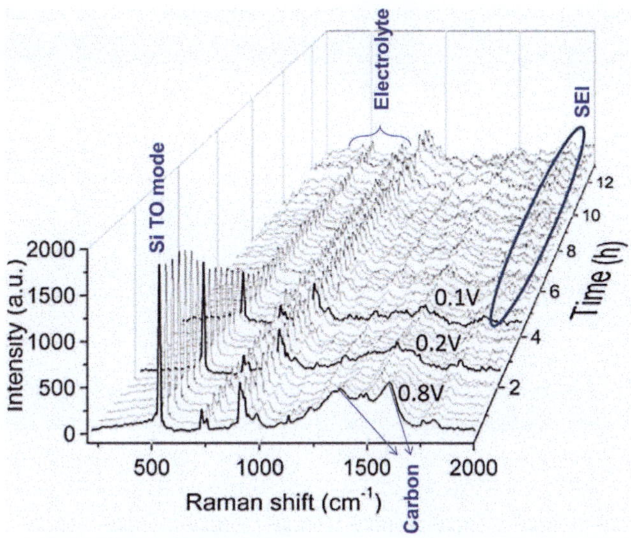

Figure 5.17 *Time-resolved Raman spectra collected form carbon-coated Si nanowires. A strong Si TO mode, vibrational modes of the carbon and electrolyte, and change over time as lithiation proceeds can be observed. (Reprinted with permission of JES) [95].*

found that during the first lithiation cycle, the Si TO mode centered at 520 cm^{-1} decreased in intensity, corresponding to the amorphization of the Si upon lithiation. A similar observation was made initially by Nanda *et al.* [96] during the discharge process of a Si/C composite anode. The intensity of all electrolyte-associated Raman peaks remained almost unchanged through the lithiation process, indicating that Raman scattering was contributed mainly from the bulk electrolyte. Regardless, a small peak at 1 859 cm^{-1} arose with diminishing Si TO mode as lithiation proceeded and was ascribed to SEI growth.

In addition to its noninvasive advantage, the combination of Raman spectroscopy with a motorized piezoelectric sample stage allows for quantifying and visualizing the relative abundance of the surface components in a composite Si anode. As shown in Figure 5.18, Rose *et al.* [85] performed Raman mapping on composite Si anodes composed of Si nanoparticles, carbon black, graphite, and Li polyacrylate binder. Each Raman mapping frame was acquired over 50 by 50 μm areas of the anodes in 1 by 1 μm steps. Using the Raman spectrum of each individual component (Figure 5.18a) as the Eigen vector for principal component analysis, the relative abundance of each component was quantified on the mapped region, as shown in Figure 5.18b. It was found, based on this technique, that the c-Si portion was greatly reduced after 100 galvanostatic cycles, whereas a-Si increased in abundance. However, even after 100 cycles, c-Si was still observed in a few spots, clearly showing that inhomogeneous electrochemically inactive regions were distributed on Si composite anodes.

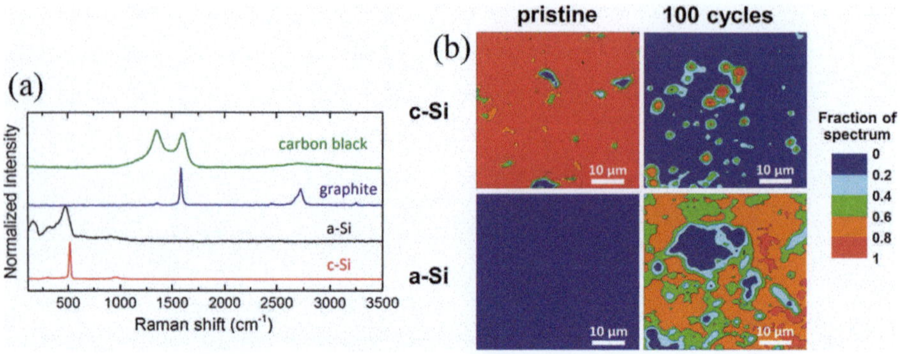

Figure 5.18 *(a) Raman spectra of the four primary components of cycled Si-graphite anodes: c-Si, a-Si, graphite, and carbon black. (b) Principal component analysis of the Raman mapping before/after 100 cycles. (Reprinted with ACS permission) [85].*

Confocal Raman microscopy is restricted by the light diffraction limit, which constrains its spatial resolution to the submicron-micron range [97]. The light diffraction limit can be overcome by tip-enhanced Raman spectroscopy (TERS). TERS is based on the plasmonic response of a metallic SPM tip. Focusing on the tip apex (Figure 5.19a), the Raman laser can produce surface plasmon resonance (SPR) to enhance the local electric field, thereby increasing the Raman cross section of the

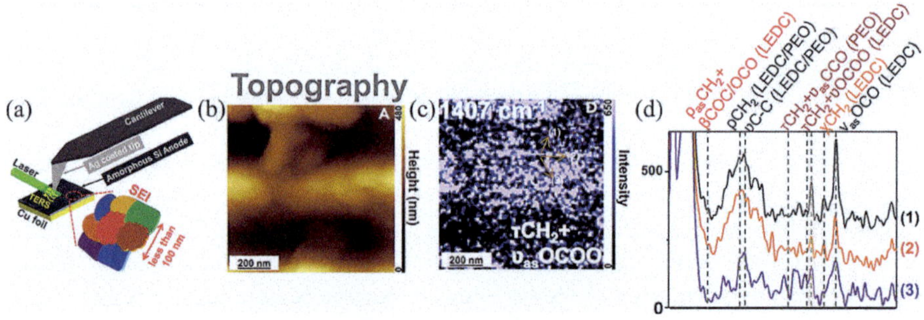

Figure 5.19 *(a) Schematic illustration of a TERS setup for measuring the SEI on the surface of cycled a-Si with nanoscale in-plane resolution. (b) AFM topography and (c) corresponding TERS chemical imaging of a 1 407 cm^{-1} peak, standing for –H_2 twisting and asymmetric –OCOO stretching bands. (d) Individual TERS spectra collected from the TERS mapping at a designated spot from (c). The 1 407 cm^{-1} peak stems from a moiety found in Li_2EDC. A 50 nm a-Si thin film anode was used and cycled in $LiPF_6$ (1 M) in EC+ DEC (1:1 by volume) electrolyte. (Reprinted with Elsevier permission) [29].*

analyte molecule [98]. TERS can detect single molecules and their conformational variations when the molecules are in a gap between the tip and a smooth metallic substrate (referred to as "gap-mode" TERS) [99]. Since the SPR region underneath the tip apex is on the scale of ~10 nm, despite a possibly larger tip diameter [29], the chemical distribution can be mapped with nanoscale spatial resolution by TERS. In addition, the topography probed by SPM (Figure 5.19b) can be synergistically linked to chemical mapping SPM (Figure 5.19c). Detailed analysis of the TERS spectra from spots that are nanometers apart shows clear variation in the peak intensity (Figure 5.19d), demonstrating the nanoscale heterogeneity of the SEI on the Si surface [99]. A similar SPR phenomenon was employed recently in surface-enhanced Raman spectroscopy (SERS) by Ha *et al.* [86] with roughened copper mesh used as the SPR media. The in situ SERS setup was found capable of probing the change in near-surface SEI components, including alkyl carboxylate species and Li propionate on an a-Si anode upon charge/discharge. In addition, SPR sensitivity can be improved by harnessing gap-mode SERS enabled by arrays of metallic nanoparticles [100–103].

5.5.2.3 X-ray photoelectron spectroscopy

XPS is based on the photoelectric effect that can identify the elements that exist on the surface (usually <10 nm) of a material [104]. The photoelectric effect is triggered by the irradiation of the sample surface by a beam of X-rays. XPS is an extremely useful surface-sensitive quantitative spectroscopic technique to identify elements, their chemical states, and the overall electronic structures and densities of the electronic states on the surfaces of the materials.

Figure 5.20 shows a set of typical XPS spectra collected from the Si anode surface [105]. The model Si anode used was a composite one consisting of graphite, Si, carbon black, and polyacrylic acid binder. As shown in Figure 5.20a, the Si 2p core level of the pristine Si anode consists of the native SiO_x (104 eV) and Si (100.1 eV). After cycling in the carbonate electrolyte composed of $LiPF_6$ (1.2 M) in EC:EMC

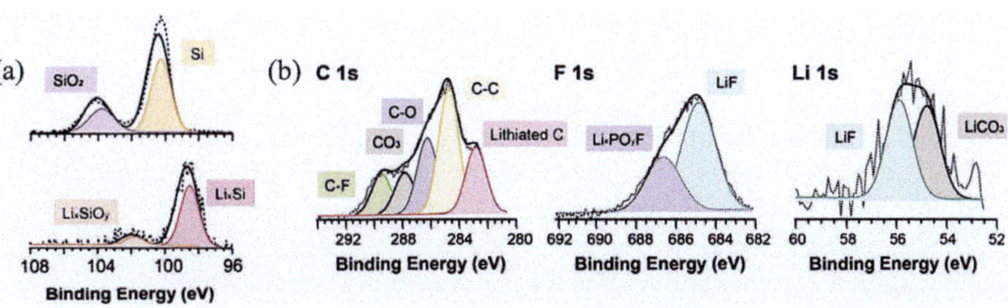

Figure 5.20 *XPS spectra of (a) Si 2p and (b) C 1s, F 1s, and Li 1s core levels. The spectrum in the upper panel of (a) was collected from a pristine Si anode, and the bottom spectrum is from Si after one cycle. Spectra in (b) were collected from a Si anode cycled 100 times. (Reprinted with Wiley permission) [105].*

Table 5.1 Assignment of XPS peaks of various core levels for SEI on Si

	Binding energy center (eV)	Assignment	Reference
C 1s	284.9	Aliphatic C–C–H	Schulz *et al.* [107]
	285.8	Ether C–O	Schulz *et al.* [108]
	286.9	Carbonate C*–O–COO	Schulz *et al.* [107]
	281.0	Carboxylate O–C=O	Schulz *et al.* [107]
	289.7	Li_2CO_3	Schulz *et al.* [107]
	290.8	Carbonate C*–O–**C**=OO	Becker *et al.* [109]
O 1s	531	Li_xSiO_y	Xu *et al.* [82]
	532.4	C=O	Beamson *et al.* [110]
	533.1	Ether C–O–C	Beamson *et al.* [110]
	533.9	$Li_xPO_yF_z$	Schulz *et al.* [107]
	534.1	C*–**O**–C=O(O)	Beamson *et al.* [110]
F 1s	684.7	LiF	Nie *et al.* [111], Schulz *et al.* [107]
	686.3	$Li_xPO_yF_z$	Schulz *et al.* [107]
		$Si\text{-}F_x$	Beamson *et al.* [110]
	688.7	LiP_xF_y	BS Parimalam *et al.* [45]
		Fluoroorganic	Beamson *et al.* [110], Schulz *et al.* [108]
Si 2p	102.2	Li_xSiO_y	B Philippe *et al.* [45]
		$Si\text{-}F_x$	Morar *et al.* [112]
	104	SiO_x	Nie *et al.* [111]

*Atom responsible for the assignment.

(3:7 by weight) with 10 wt% FEC for one cycle, peaks at around 102 and 99 eV replaced the SiO_x and Si peaks, representing Li silicate and lithiated Si, respectively [106]. After 100 continuous cycles between 1 and 0.01 V vs Li^+/Li, XPS core levels corresponding to components in the passivation layer emerged, including C 1s, F 1s, and Li 1s (Figure 5.20b). The deconvoluted peaks under each core level suggest the existence of the commonly seen species in an SEI on Si reduced from carbonate electrolytes, including Li_2EDC, Li_2CO_3, LiF, Li_xPO_yF, and organic polymeric compounds.

Table 5.1 includes some commonly seen XPS core levels related to the SEI species on a Si anode surface. The absolute values of the peak assignment tend to vary in the literature. The reasons for the inconsistency include (1) surface charge effect [113], (2) various chemical surroundings due to the SEI compositional complexity, and (3) different stoichiometry ratios for the same types of compounds (e.g., different x, y, and z values in $Li_xPO_yF_z$ from those literature).

In quantifying XPS spectra, assuming the number of electrons recorded is proportional to the atom number in a given state, it is possible to quantify the atomic percentage and the concentration of an associated functional group using XPS. Therefore, XPS can be a powerful tool for quantitative study of the concentration of the species in a Si SEI.

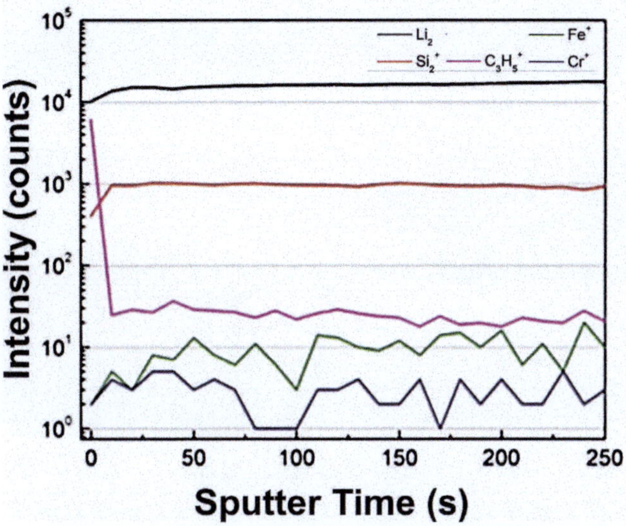

Figure 5.21 *ToF-SIMS depth profiles of a Li_3SiO_x thin film on an interdigitated platinum electrode. (Reprinted with ACS permission) [91]*

5.5.2.4 Time-of-flight secondary ion mass spectrometry

ToF-SIMS employs a focused, pulsed cesium or gallium particle beam to dislodge chemical species on the surface of a specimen. Particles produced closer to the site of impact tend to be dissociated ions (positive or negative). Secondary particles generated farther from the impact site tend to be molecular compounds, typically fragments of much larger organic macromolecules. After being dislodged, the particles are accelerated into a flight path on their way toward a detector. The mass of the particles is determined by measuring the exact time at which they reach the detector (i.e., the time-of-flight). The raster of the beam can create a mass of interest on a submicron lateral-resolution chemical image; and after the ion beam removes the surface layers, a depth profile at nanoscale resolution can be detected [114].

Based on the depth and the area scale of the scanned specimen, ToF-SIMS can be used to study the elemental/compound distribution across the full cross section of the electrode, or a few Si particles on or within the electrode. Han *et al.* deposited a model SEI compound, Li_3SiO_x, on an interdigitated platinum electrode. ToF-SIMS depth profiling was completed on this model system, as shown in Figure 5.21. Through post-acquisition retrospective data analysis, the distribution of different possible components across the Li_3SiO_x film was analyzed [91].

Bordes *et al.* [115] demonstrated that the Li^+ distribution in Li_xSi alloy particles could be mapped by ToF-SIMS chemical mapping and that it was heterogeneous at different lithiation stages. Figure 5.22 shows an SEM image of the cross section of a Si anode composed of micron-size Si particles and the ToF-SIMS chemical mapping of selected particles in the cross section for 25 percent lithiated electrodes. It is seen that some unlithiated Si is distributed in the core region, indicating that Li

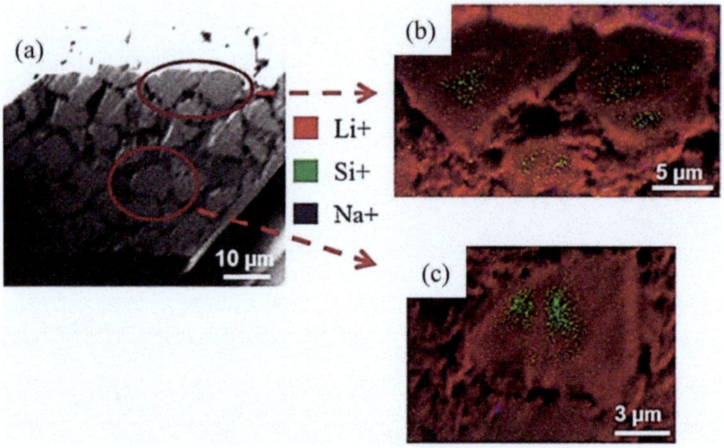

Figure 5.22 (a) SEM imaging of a full cross-section created by a focused ion beam for a 25 percent lithiated electrode and (b) and (c) ToF-SIMS chemical mapping of Li^+ (red), Si^+ (green), and Na^+ (blue) selected particles. (Reprinted with ACS permission) [115]

preferably distributed on the shell area in the Li_xSi particles. The uneven lithiation of the individual Si particles can be attributed to fracturing [115].

5.6 Summary

5.6.1 Silicon SEI models

Given the vital importance of the SEI in Li-ion batteries, a significant amount of effort has been invested over more than two decades to understanding the mechanism of SEI formation and quantify its composition [116]. To date, two models for SEI formation on Li-ion battery anodes are generally accepted: the mosaic-type polyhetero microphase model proposed by Peled *et al.* [116] and the multilayer model proposed by Aurbach *et al.* [117]. Although these models were proposed based on carbonaceous anodes, most of the SEI models proposed for use on Si anodes are derived from them. However, in considering the decomposition of electrolytes on a Si surface, one should bear in mind that the volumetric expansion of Si is >300 percent, more than 10 times that of its graphite anode counterpart. The SEI evolution on a Si surface should be synergistically considered with the morphological change of the Si anode.

Figure 5.23 presents a scheme of the SEI on a Si anode. Because early SEI formation on Si has non-passivating behavior [87], unblocked electron transfer across the electrode/electrolyte interface continuously decomposes the $LiPF_6$ to form inorganic compounds, including LiF, LiP_yF_z, and $Li_xPO_yF_z$. The multi-electron reduction of EC at a higher potential than the reduction of other electrolyte components results in the formation of Li_2EDC and Li_2CO_3. The reaction of Li^+ with the native SiO_2 layer results in the formation of Li silicate. Therefore, a Si SEI has a preferable distribution of inorganic species in the layer

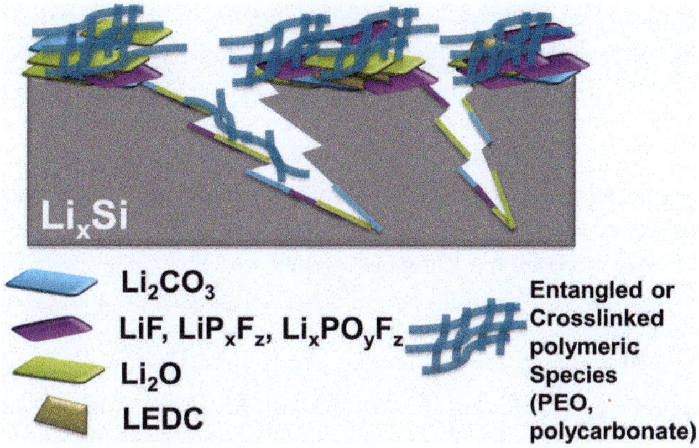

Figure 5.23 *Schematic of the SEI structure on a lithiated Si surface upon charge/discharge cycling*

in close proximity to the Si surface. The formation of inorganic SEI compounds on a Si surface during early galvanostatic cycles has been observed by several studies [42, 51, 87, 91, 118]. An inorganic-rich layer effectively blocks electron transfer, thereby mitigating the decomposition of salt and electrolyte components at early stages. Further reactions of the electrolyte components result in the formation of polymeric species, including PEO, polycarbonate, and polycarboxylate species [29, 66]. The elasticity of these polymeric species enables them to tolerate the volumetric change of Si upon continuous lithiation/delithiation, but only to an extent. The fracturing of Si generates new surfaces on the Si anode, bringing about new cycles of early reduction and continuous consumption of Li salt and electrolyte solvents.

The vast majority of the multimodal characterization techniques that allow probing of the morphological and chemical distributions of the SEI are ideal for submicron-micron length scales. New technologies are always desired for unveiling Si SEI properties at the nanoscale and even at atomic scale. A recent TERS study by our group indicated that within the sample plane, different components formed “nanomosaic” patches. Upon progressive galvanostatic cycling, the enrichment of different SEI species changed, but the nanomosaic structure of multiple species was retained. Therefore, a nanomosaic-multilayer model could best describe the SEI on Si surfaces.

Despite the great efforts made to obtain insightful understanding of the Si SEI, and the development of additives such as VC, FEC, and binary salts to form a more stable SEI layer on Si, much work remains to obtain a completely stabilized SEI to enable practical applications of Si in commercial Li-ion batteries.

5.6.2 Outlook

It has been demonstrated by numerous studies that carbonate-based electrolytes do not effectively passivate Si surfaces in the same way that they passivate conventional graphite anodes [13]. Therefore, new electrolyte formulas, novel cosolvents, additives, and binary or tertiary salts should be actively explored to form more robust SEI layers on Si surfaces

[118, 119]. The ideal SEI layer on Si should have the following properties: (1) high ionic conductivity, (2) high electron resistivity, (3) capability to coat Si surfaces conformally, (4) capability to tolerate in-plane volumetric changes in Si, and (5) chemical, electrochemical, and physical stability against Li_xSi (with varying x values) in electrolyte surroundings for the life span of the battery. Physical stability includes low solubility of SEI species, non-deformability of the SEI (e.g., resistance to swelling), and resistance to fracturing with Si volumetric expansion as electrochemical cycling progresses.

New SEI characterization techniques should be developed with more focus on operando-level operations. Multiple length-scale techniques should be developed that cover the range from the nanometer, submicron, and micron scales all the way up to the whole-cell scale. To accelerate the exploration and development of new solvents, additives, and salts, artificial intelligence combined with MD simulation and DFT should be used [120].

Acknowledgment

This work is supported by the US Department of Energy (DOE) Vehicle Technologies Office under the Silicon Electrolyte Interface Stabilization (SEISta) Consortium directed by Brian Cunningham and managed by Anthony Burrell. This manuscript has been authored by UT-Battelle LLC under Contract DE-AC05-00OR22725 with DOE. The US government retains and the publisher, by accepting the article for publication, acknowledges that the US government retains a nonexclusive, paid-up, irrevocable, worldwide license to publish or reproduce the published form of this manuscript, or allow other to do so, for US government purposes. DOE will provide public access to these results of federally sponsored research in accordance with the DOE Public Access Plan (http://energy.gov/downloads/doe-public-access-plan).

References

[1] Obrovac M.N., Christensen L. 'Structural changes in silicon anodes during lithium insertion/extraction'. *Electrochemical and Solid-State Letters*. 2004;**7**(5):A93

[2] Ashuri M., He Q., Shaw L.L. 'Silicon as a potential anode material for Li-ion batteries: where size, geometry and structure matter'. *Nanoscale*. 2016;**8**(1):74–103.

[3] Peled E. 'The electrochemical behavior of alkali and alkaline earth metals in nonaqueous battery systems—the solid electrolyte interphase model'. *Journal of the Electrochemical Society*. 1979;**126**(12):2047–51.

[4] Deng D. 'Li-ion batteries: basics, progress, and challenges'. *Energy Science & Engineering*. 2015;**3**(5):385–418.

[5] Goodenough J.B., Kim Y. 'Challenges for rechargeable Li batteries'. *Chemistry of Materials*. 2010;**22**(3):587–603.

[6] Delp S.A., Borodin O., Olguin M., Eisner C.G., Allen J.L., Jow T.R. 'Importance of reduction and oxidation stability of high voltage electrolytes and additives'. *Electrochimica Acta*. 2016;**209**:498–510.

[7] Hou T., Yang G., Rajput N.N., *et al.* 'The influence of FEC on the solvation structure and reduction reaction of LiPF6/EC electrolytes and its implication for solid electrolyte interphase formation'. *Nano Energy*. 2019;**64**:103881.
[8] Xu C., Lindgren F., Philippe B., *et al.* 'Improved performance of the silicon anode for Li-Ion batteries: understanding the surface modification mechanism of fluoroethylene carbonate as an effective electrolyte additive'. *Chemistry of Materials*. 2015;**27**(7):2591–9.
[9] Xu K. 'Nonaqueous liquid electrolytes for lithium-based rechargeable batteries'. *Chemical Reviews*. 2004;**104**(10):4303–418.
[10] Wang A., Kadam S., Li H., Shi S., Qi Y. 'Review on modeling of the anode solid electrolyte interphase (SEI) for lithium-ion batteries'. *npj Computational Materials*. 2018;**4**(1):1–26.
[11] Elliott W. Report No. 1, contract nas 3-6015 (N 65-11518). Sept. 1964. 1964
[12] Fong R., von Sacken U., Dahn J.R. 'Studies of lithium intercalation into carbons using nonaqueous electrochemical cells'. *Journal of the Electrochemical Society*. 1990;**137**(7):2009–13.
[13] Burrell A. *Silicon Electrolyte Interface Stabilization (SEISta) Third Quarter Progress Report*; 2018.
[14] Wang Y., Nakamura S., Ue M., Balbuena P.B. 'Theoretical studies to understand surface chemistry on carbon anodes for lithium-ion batteries: reduction mechanisms of ethylene carbonate'. *Journal of the American Chemical Society*. 2001;**123**(47):11708–18.
[15] Aurbach D., Markovsky B., Weissman I., Levi E., Ein-Eli Y. 'On the correlation between surface chemistry and performance of graphite negative electrodes for Li ion batteries'. *Electrochimica Acta*. 1999;**45**(1-2):67–86.
[16] Xu K. 'Whether EC and PC differ in interphasial chemistry on graphitic anode and how'. *Journal of the Electrochemical Society*. 2009;**156**(9):A751
[17] Leung K., Budzien J.L. 'Ab initio molecular dynamics simulations of the initial stages of solid-electrolyte interphase formation on lithium ion battery graphitic anodes'. *Physical Chemistry Chemical Physics*. 2010;**12**(25):6583–6.
[18] An S.J., Li J., Daniel C., Mohanty D., Nagpure S., Wood III D.L. 'The state of understanding of the lithium-ion-battery graphite solid electrolyte interphase (SEI) and its relationship to formation cycling'. *Carbon*. 2016;**105**:52–76.
[19] Lu P., Li C., Schneider E.W., Harris S.J. 'Chemistry, impedance, and morphology evolution in solid electrolyte interphase films during formation in lithium ion batteries'. *The Journal of Physical Chemistry C*. 2014;**118**(2):896–903.
[20] Yan J., Xia B.-J., Su Y.-C., Zhou X.-Z., Zhang J., Zhang X.-G. 'Phenomenologically modeling the formation and evolution of the solid electrolyte interface on the graphite electrode for lithium-ion batteries'. *Electrochimica Acta*. 2008;**53**(24):7069–78.
[21] Li T., Balbuena P.B. 'Theoretical studies of the reduction of ethylene carbonate'. *Chemical Physics Letters*. 2000;**317**(3–5):421–9.

[22] Aurbach D., Levi M.D., Levi E., Schechter A. 'Failure and stabilization mechanisms of graphite electrodes'. *The Journal of Physical Chemistry B*. 1997;**101**(12):2195–206.

[23] Aurbach D., Moshkovich M., Cohen Y., Schechter A. 'The study of surface film formation on noble-metal electrodes in alkyl carbonates/Li salt solutions, using simultaneous in situ AFM, EQCM, FTIR, and EIS'. *Langmuir*. 1999;**15**(8):2947–60.

[24] Yun K.-S., Pai S.J., Yeo B.C., Lee K.-R., Kim S.-J., Han S.S. 'Simulation protocol for prediction of a solid-electrolyte interphase on the silicon-based anodes of a lithium-ion battery: ReaxFF reactive force field'. *The Journal of Physical Chemistry Letters*. 2017;**8**(13):2812–18.

[25] Martinez de la Hoz J.M., Leung K., Balbuena P.B. 'Reduction mechanisms of ethylene carbonate on Si anodes of lithium-ion batteries: effects of degree of lithiation and nature of exposed surface'. *ACS Applied Materials & Interfaces*. 2013;**5**(24):13457–65.

[26] Gomez-Ballesteros J.L., Balbuena P.B. 'Reduction of electrolyte components on a coated Si anode of lithium-ion batteries'. *The Journal of Physical Chemistry Letters*. 2017;**8**(14):3404–8.

[27] Shkrob I.A., Zhu Y., Marin T.W., Abraham D. 'Reduction of carbonate electrolytes and the formation of solid-electrolyte interface (SEI) in lithium-ion batteries. 1. Spectroscopic observations of radical intermediates generated in one-electron reduction of carbonates'. *The Journal of Physical Chemistry C*. 2013;**117**(38):19255–69.

[28] Jin Y., Kneusels N.-J.H., Magusin P.C.M.M., *et al*. 'Identifying the structural basis for the increased stability of the solid electrolyte interphase formed on silicon with the additive fluoroethylene carbonate'. *Journal of the American Chemical Society*. 2017;**139**(42):14992–5004.

[29] Nanda J., Yang G., Hou T., *et al*. 'Unraveling the nanoscale heterogeneity of solid electrolyte interphase using tip-enhanced Raman spectroscopy'. *Joule*. 2019;**3**(8):2001–19.

[30] Nanda J., Yang G., Hou T., *et al*. 'Unraveling the nanoscale heterogeneity of solid electrolyte interphase using tip-enhanced Raman spectroscopy'. *Joule*. 2019;**3**(8):2001–19.

[31] Schroder K., Alvarado J., Yersak T.A., *et al*. 'The effect of fluoroethylene carbonate as an additive on the solid electrolyte interphase on silicon lithium-ion electrodes'. *Chemistry of Materials*. 2015;**27**(16):5531–42.

[32] Shen B.H., Wang S., Tenhaeff W.E. 'Ultrathin conformal polycyclosiloxane films to improve silicon cycling stability'. *Science Advances*. 2019;**5**(7):eaaw4856.

[33] Hays K.A., Ruther R.E., Kukay A.J., *et al*. 'What makes lithium substituted polyacrylic acid a better binder than polyacrylic acid for silicon-graphite composite anodes?' *Journal of Power Sources*. 2018;**384**:136–44.

[34] Schiele A., Breitung B., Hatsukade T., *et al*. 'The critical role of fluoroethylene carbonate in the gassing of silicon anodes for lithium-ion batteries'. *ACS Energy Letters*. 2017;**2**(10):2228–33.

[35] Eriksson T., Andersson A.M., Bishop A.G., Gejke C., Gustafsson T., Thomas J.O. 'Surface analysis of $LiMn_2O_4$ electrodes in carbonate-based electrolytes'. *Journal of the Electrochemical Society*. 2002;**149**(1):A69

[36] Aurbach D. 'Review of selected electrode–solution interactions which determine the performance of Li and Li ion batteries'. *Journal of Power Sources*. 2000;**89**(2):206–18.

[37] Radvanyi E., De Vito E., Porcher W., Jouanneau Si Larbi S, Larbi S.J.S. 'An XPS/AES comparative study of the surface behaviour of nano-silicon anodes for Li-ion batteries'. *Journal of Analytical Atomic Spectrometry*. 2014;**29**(6):1120–31.

[38] Leroy S., Blanchard F., Dedryvère R., *et al.* 'Surface film formation on a graphite electrode in Li-ion batteries: AFM and XPS study'. *Surface and Interface Analysis*. 2005;**37**(10):773–81.

[39] Tasaki K. 'Solvent decompositions and physical properties of decomposition compounds in Li-ion battery electrolytes studied by DFT calculations and molecular dynamics simulations'. *The Journal of Physical Chemistry B*. 2005;**109**(7):2920–33.

[40] Shi F., Ross P.N., Somorjai G.A., Komvopoulos K. 'The chemistry of electrolyte reduction on silicon electrodes revealed by *in situ* ATR-FTIR spectroscopy''. *The Journal of Physical Chemistry C*. 2017;**121**(27):14476–83.

[41] Aurbach D., Markovsky B., Shechter A., Ein-Eli Y., Cohen H. 'A comparative study of synthetic graphite and Li electrodes in electrolyte solutions based on ethylene carbonate-dimethyl carbonate mixtures'. *Journal of the Electrochemical Society*. 1996;**143**(12):3809–20.

[42] Philippe B., Dedryvère Rm., Gorgoi M., Rensmo Hk., Gonbeau D., Edström K. 'Role of the Lipf6 $salt_6$ salt for the long-term stability of silicon electrodes in Li-ion batteries – a photoelectron spectroscopy study'. *Chemistry of Materials*. 2013;**25**(3):394–404.

[43] Plakhotnyk A.V., Ernst L., Schmutzler R. 'Hydrolysis in the system LiPF6—propylene carbonate—dimethyl carbonate—H2O'. *Journal of Fluorine Chemistry*. 2005;**126**(1):27–31.

[44] Campion C.L., Li W., Lucht B.L. 'Thermal decomposition of $LiPF_6$-based electrolytes for lithium-ion batteries'. *Journal of the Electrochemical Society*. 2005;**152**(12):A2327

[45] Cao C., Abate I.I., Sivonxay E., *et al.* 'Solid electrolyte interphase on native oxide-terminated silicon anodes for Li-ion batteries'. *Joule*. 2019;**3**(3):pp.762–781.

[46] Munao D., van Erven J.W.M., Valvo M., Garcia-Tamayo E., Kelder E.M. 'Role of the binder on the failure mechanism of Si nano-composite electrodes for Li-ion batteries'. *Journal of Power Sources*. 2011;**196**(16):6695–702.

[47] Shiraishi S., Kanamura K., Takehara Z.-I. 'Study of the surface composition of highly smooth lithium deposited in various carbonate electrolytes containing HF'. *Langmuir*. 1997;**13**(13):3542–9.

[48] Choi N.-S., Yew K.H., Lee K.Y., Sung M., Kim H., Kim S.-S. 'Effect of fluoroethylene carbonate additive on interfacial properties of silicon thin-film electrode'. *Journal of Power Sources*. 2006;**161**(2):1254–9.

[49] Zhang S., He M., Su C.-C., Zhang Z. 'Advanced electrolyte/additive for lithium-ion batteries with silicon anode'. *Current Opinion in Chemical Engineering*. 2016;**13**:24–35.

[50] Tang J., Qu M., Wang M., Zhang M., Yan J., Mei Q. 'Line: large-scale information network embedding'. Proceedings of the 24th International Conference on World Wide Web. International World Wide Web Conferences Steering Committee; 2015.

[51] Nie M., Abraham D.P., Chen Y., Bose A., Lucht B.L. 'Silicon solid electrolyte interphase (SEI) of lithium ion battery characterized by microscopy and spectroscopy'. *The Journal of Physical Chemistry C*. 2013;**117**(26):13403–12.

[52] Nakai H., Kubota T., Kita A., Kawashima A. 'Investigation of the solid electrolyte interphase formed by fluoroethylene carbonate on Si electrodes'. *Journal of the Electrochemical Society*. 2011;**158**(7):A798.

[53] Peled E., Golodnttsky D., Ardel G., Menachem C., Bar Tow D., Eshkenazy V. 'The role of SEI in lithium and lithium ion batteries'. *MRS Proceedings*. 1995;**393**.

[54] Jorn R., Kumar R., Abraham D.P., Voth G.A. 'Atomistic modeling of the electrode–electrolyte interface in Li-ion energy storage systems: electrolyte structuring'. *The Journal of Physical Chemistry C*. 2013;**117**(8):3747–61.

[55] Nakai H., Kubota T., Kita A., Kawashima A. 'Investigation of the solid electrolyte interphase formed by fluoroethylene carbonate on Si electrodes'. *Journal of the Electrochemical Society*. 2011;**158**(7):A798–801.

[56] Deringer V.L., Bernstein N., Bartók A.P., *et al*. 'Realistic atomistic structure of amorphous silicon from machine-learning-driven molecular dynamics'. *The Journal of Physical Chemistry Letters*. 2018;**9**(11):2879–85.

[57] Etacheri V., Haik O., Goffer Y., *et al*. 'Effect of fluoroethylene carbonate (FEC) on the performance and surface chemistry of Si-nanowire Li-ion battery anodes'. *Langmuir*. 2012;**28**(1):965–76.

[58] Ota H., Sakata Y., Inoue A., Yamaguchi S. 'Analysis of vinylene carbonate derived SEI layers on graphite anode'. *Journal of the Electrochemical Society*. 2004;**151**(10):A1659–69.

[59] El Ouatani L., Dedryvère R., Siret C., *et al*. 'The effect of vinylene carbonate additive on surface film formation on both electrodes in Li-ion batteries'. *Journal of the Electrochemical Society*. 2009;**156**(2):A103–13.

[60] Xu K. 'Electrolytes and interphases in Li-ion batteries and beyond'. *Chemical Reviews*. 2014;**114**(23):11503–618.

[61] Jaumann T., Balach J., Langklotz U., *et al*. 'Lifetime vs. rate capability: understanding the role of FEC and VC in high-energy Li-ion batteries with nano-silicon anodes'. *Energy Storage Materials*. 2017;**6**:26–35.

[62] Aurbach D., Gamolsky K., Markovsky B., Gofer Y., Schmidt M., Heider U. 'On the use of vinylene carbonate (VC) as an additive to electrolyte solutions for Li-ion batteries'. *Electrochimica Acta*. 2002;**47**(9):1423–39.

[63] Wang Y., Nakamura S., Tasaki K., Balbuena P.B. 'Theoretical studies to understand surface chemistry on carbon anodes for lithium-ion batteries: how does vinylene carbonate play its role as an electrolyte additive?' *Journal of the American Chemical Society*. 2002;**124**(16):4408–21.

[64] Chen L., Wang K., Xie X., Xie J. 'Effect of vinylene carbonate (VC) as electrolyte additive on electrochemical performance of Si film anode for lithium ion batteries'. *Journal of Power Sources*. 2007;**174**(2):538–43.

[65] Jaumann T., Balach J., Klose M., *et al*. 'SEI-component formation on sub 5 nm sized silicon nanoparticles in Li-ion batteries: the role of electrode preparation, FEC addition and binders'. *Physical Chemistry Chemical Physics*. 2015;**17**(38):24956–67.

[66] Jin Y., Kneusels N.-J.H., Marbella L.E., *et al*. 'Understanding fluoroethylene carbonate and vinylene carbonate based electrolytes for Si anodes in lithium ion batteries with NMR spectroscopy'. *Journal of the American Chemical Society*. 2018;**140**(31):9854–67.

[67] Lehmann M.L., Yang G., Gilmer D., *et al*. 'Tailored crosslinking of poly (ethylene oxide) enables mechanical robustness and improved sodium-ion conductivity'. *Energy Storage Materials*. 2019;**21**:85–96.

[68] Soto F.A., Ma Y., Martinez de la Hoz J.M., Seminario J.M., Balbuena P.B. 'Formation and growth mechanisms of solid-electrolyte interphase layers in rechargeable batteries'. *Chemistry of Materials*. 2015;**27**(23):7990–8000.

[69] Yang G., Frisco S., Tao R., *et al*. 'Robust Solid/Electrolyte Interphase (Sei) Formation on Si Anodes Using Glyme-Based Electrolytes'. *ACS Energy Letters*; 2021. pp. 1684–93.

[70] Wu Y., Liu N. 'Visualizing battery reactions and processes by using in situ and in operando microscopies'. *Chem*. 2018;**4**(3):438–65.

[71] Bhattacharya S., Alpas A.T. 'Effects of electrolyte composition and voltage scan rate on cyclic performance of silicon-aluminum anodes for lithium-ion cells'. *Journal of the Electrochemical Society*. 2017;**164**(9):A2065–74.

[72] Shi F., Song Z., Ross P.N., Somorjai G.A., Ritchie R.O., Komvopoulos K. 'Failure mechanisms of single-crystal silicon electrodes in lithium-ion batteries'. *Nature Communications*. 2016;**7**(1):1–8.

[73] Nguyen C.C., Lucht B.L. 'Comparative study of fluoroethylene carbonate and vinylene carbonate for silicon anodes in lithium ion batteries'. *Journal of the Electrochemical Society*. 2014;**161**(12):A1933–8.

[74] Li J., Dozier A.K., Li Y., Yang F., Cheng Y.-T. 'Crack pattern formation in thin film lithium-ion battery electrodes'. *Journal of the Electrochemical Society*. 2011;**158**(6):A689.

[75] Chen C.-Y., Sawamura A., Tsuda T., Uchida S., Ishikawa M., Kuwabata S. 'Visualization of Si anode reactions in coin-type cells via operando scanning electron microscopy'. *ACS Applied Materials & Interfaces*. 2017;**9**(41):35511–5.

[76] Inkson B. 'Scanning electron microscopy (SEM) and transmission electron microscopy (TEM) for materials characterization'. *Materials Characterization Using Nondestructive Evaluation (NDE) Methods*. Elsevier; 2016.

[77] He Y., Piper D.M., Gu M., *et al.* '*In* situ transmission electron microscopy probing of native oxide and artificial layers on silicon nanoparticles for lithium ion batteries'. *ACS Nano*. 2014;**8**(11):11816–23.

[78] Wang X., Li Y., Meng Y.S. 'Cryogenic electron microscopy for characterizing and diagnosing batteries'. *Joule*. 2018;**2**(11):2225–34.

[79] Huang W., Wang J., Braun M.R., *et al.* 'Dynamic structure and chemistry of the silicon solid-electrolyte interphase visualized by cryogenic electron microscopy'. *Matter*. 2019;**1**(5):1232–45.

[80] Ohnesorge F., Binnig G. 'True atomic resolution by atomic force microscopy through repulsive and attractive forces'. *Science*. 1993;**260**(5113):1451–6.

[81] Becker C.R., Strawhecker K.E., McAllister Q.P., Lundgren C.A. 'In situ atomic force microscopy of lithiation and delithiation of silicon nanostructures for lithium ion batteries'. *ACS Nano*. 2013;**7**(10):9173–82.

[82] Xu Y., Stetson C., Wood K., *et al.* 'Mechanical properties and chemical reactivity of Li X Sio Y thin films'. *ACS Applied Materials & Interfaces*. 2018;**10**(44):38558–64.

[83] Stetson C., Yoon T., Coyle J., *et al.* 'Three-dimensional electronic resistivity mapping of solid electrolyte interphase on Si anode materials'. *Nano Energy*. 2019;**55**:477–85.

[84] Vickerman J.C., Gilmore I.S. *Surface Analysis: The Principal Techniques*. John Wiley & Sons; 2011.

[85] Ruther R.E., Hays K.A., An S.J., Li J., Wood D.L., Nanda J. 'Chemical evolution in silicon-graphite composite anodes investigated by vibrational spectroscopy'. *ACS Applied Materials & Interfaces*. 2018;**10**(22):18641–9.

[86] Ha Y., Tremolet de Villers B.J., Li Z., *et al.* 'Probing the evolution of surface chemistry at the silicon–electrolyte interphase via in situ surface-enhanced Raman spectroscopy'. *The Journal of Physical Chemistry Letters*. 2020;**11**(1):286–91.

[87] Yin Y., Arca E., Wang L., *et al.* 'Non-passivated silicon anode surface'. *ACS Applied Materials & Interfaces*. 2020.

[88] Dalavi S., Guduru P., Lucht B.L. 'Performance enhancing electrolyte additives for lithium ion batteries with silicon anodes'. *Journal of the Electrochemical Society*. 2012;**159**(5):A642–6.

[89] Pekarek R.T., Affolter A., Baranowski L.L., *et al.* 'Intrinsic chemical reactivity of solid-electrolyte interphase components in silicon–lithium alloy anode batteries probed by FTIR spectroscopy'. *Journal of Materials Chemistry A*. 2020;**8**(16):7897–906.

[90] Xu Y., Wood K., Coyle J., *et al.* 'Chemistry of electrolyte reduction on lithium silicide'. *The Journal of Physical Chemistry C*. 2019;**123**(21):13219–24.

[91] Han S.-D., Wood K.N., Stetson C., *et al.* 'Intrinsic properties of individual inorganic silicon-electrolyte interphase constituents'. *ACS Applied Materials & Interfaces*. 2019;**11**(50):46993–7002.

[92] Griffiths P.R., De Haseth J.A. *Fourier Transform Infrared Spectrometry*. John Wiley & Sons; 2007.

[93] Zhuang G.V., Xu K., Yang H., Jow T.R., Ross P.N. 'Lithium ethylene dicarbonate identified as the primary product of chemical and electrochemical reduction of EC in 1.2 M $LiPF_6$/EC:EMC electrolyte'. *The Journal of Physical Chemistry B*. 2005;**109**(37):17567–73.

[94] Parker J.H., Feldman D.W., Ashkin M. 'Raman scattering by silicon and germanium'. *Physical Review*. 1967;**155**(3):712–4

[95] Krause A., Tkacheva O., Omar A. 'In situ Raman spectroscopy on silicon nanowire anodes integrated in lithium ion batteries'. *Journal of the Electrochemical Society*. 2019;**166**(3):A5378–85.

[96] Nanda J., Datta M.K., Remillard J.T., O'Neill A., Kumta P.N. 'In situ Raman microscopy during discharge of a high capacity silicon–carbon composite Li-ion battery negative electrode'. *Electrochemistry Communications*. 2009;**11**(1):235–7.

[97] Yang G., Ivanov I.N., Ruther R.E., *et al.* 'Electrolyte solvation structure at solid-liquid interface probed by nanogap surface-enhanced Raman spectroscopy'. *ACS Nano*. 2018;**12**(10):10159–70.

[98] Stöckle R.M., Suh Y.D., Deckert V., Zenobi R. 'Nanoscale chemical analysis by tip-enhanced Raman spectroscopy'. *Chemical Physics Letters*. 2000;**318**(1-3):131–6.

[99] Zhang W., Yeo B.S., Schmid T., Zenobi R. 'Single molecule tip-enhanced Raman spectroscopy with silver tips'. *The Journal of Physical Chemistry C*. 2007;**111**(4):1733–8.

[100] Yang G., Nanda J., Wang B., Chen G., Hallinan D.T. 'Self-assembly of large gold nanoparticles for surface-enhanced Raman spectroscopy'. *ACS Applied Materials & Interfaces*. 2017;**9**(15):13457–70.

[101] Yang G., Hu L., Keiper T.D., Xiong P., Hallinan D.T. 'Gold nanoparticle monolayers with tunable optical and electrical properties'. *Langmuir*. 2016;**32**(16):4022–33.

[102] Yang G., Hallinan D.T. 'Self-assembly of large-scale crack-free gold nanoparticle films using a 'drain-to-deposit' strategy'. *Nanotechnology*. 2016;**27**(22):225604

[103] Yang G., Sacci R.L., Ivanov I.N. 'Probing electrolyte solvents at solid/liquid interface using gap-mode surface-enhanced Raman spectroscopy'. *Journal of the Electrochemical Society*. 2019;**166**(2):A178–87.

[104] Hüfner S. *Photoelectron Spectroscopy: Principles and Applications*. Springer Science & Business Media; 2013.

[105] Son S.-B., Cao L., Yoon T., *et al.* 'Interfacially induced cascading failure in graphite-silicon composite anodes'. *Advanced science*. 2019;**6**(3):1801007

[106] Philippe B., Dedryvère R., Allouche J., *et al.* 'Nanosilicon electrodes for lithium-ion batteries: interfacial mechanisms studied by hard and soft X-ray photoelectron spectroscopy'. *Chemistry of Materials*. 2012;**24**(6):1107–15.

[107] Schulz N., Hausbrand R., Wittich C., Dimesso L., Jaegermann W. 'XPS-surface analysis of SEI layers on Li-ion cathodes: Part II. SEI-composition and formation inside composite electrodes'. *Journal of the Electrochemical Society*. 2018;**165**(5):A833–46.

[108] Schulz N., Hausbrand R., Dimesso L., Jaegermann W. 'XPS-surface analysis of SEI layers on Li-ion cathodes: Part I. Investigation of initial surface chemistry'. *Journal of the Electrochemical Society*. 2018;**165**(5):A819–32.

[109] Becker D., Cherkashinin G., Hausbrand R., Jaegermann W. 'XPS study of diethyl carbonate adsorption on $LiCoO_2$ thin films'. *Solid State Ionics*. 2013;**230**(10):83–5.

[110] Briggs G.BaD. In Beamson G., Briggs D. (eds.) *High Resolution XPS of Organic Polymers: The Scienta Esca300 Database*. Chichester: John Wiley and Sons; 1992.

[111] Nie M., Chalasani D., Abraham D.P., Chen Y., Bose A., Lucht B.L. 'Lithium ion battery graphite solid electrolyte interphase revealed by microscopy and spectroscopy'. *The Journal of Physical Chemistry C*. 2013;**117**(3):1257–67.

[112] Morar J.F., McFeely F.R., Shinn N.D., Landgren G., Himpsel F.J. 'Synchrotron photoemission investigation: fluorine on silicon surfaces'. *Applied Physics Letters*. 1984;**45**(2):174–6.

[113] Wood K.N., Teeter G. 'Xps on Li-battery-related compounds: analysis of inorganic SEI phases and a methodology for charge correction'. *ACS Applied Energy Materials*. 2018;**1**(9):4493–504.

[114] Benninghoven A. 'Chemical analysis of inorganic and organic surfaces and thin films by static time-of-flight secondary ion mass spectrometry (ToF-SIMS'. *Angewandte Chemie International Edition in English*. 1994;**33**(10):1023–43.

[115] Bordes A., De Vito E., Haon C., Boulineau A., Montani A., Marcus P. 'Multiscale investigation of silicon anode Li insertion mechanisms by time-of-flight secondary ion mass spectrometer imaging performed on an in situ focused ion beam cross section'. *Chemistry of Materials*. 2016;**28**(5):1566–73.

[116] Peled E., Menkin S. 'Review—SEI: past, present and future'. *Journal of the Electrochemical Society*. 2017;**164**(7):A1703–19.

[117] Zaban A., Zinigrad E., Aurbach D. 'Impedance spectroscopy of Li electrodes. 4. A general simple model of the Li–solution interphase in polar aprotic systems'. *The Journal of Physical Chemistry*. 1996;**100**(8):3089–101.

[118] Chen J., Fan X., Li Q., *et al*. 'Electrolyte design for LiF-rich solid–electrolyte interfaces to enable high-performance microsized alloy anodes for batteries'. *Nature Energy*. 2020;**5**(5):386–97.

[119] Jia H., Zou L., Gao P., *et al*. 'High-performance silicon anodes enabled by nonflammable localized high-concentration electrolytes'. *Advanced Energy Materials*. 2019;**9**(31):1900784.

[120] Ye W., Chen C., Dwaraknath S., Jain A., Ong S.P., Persson K.A. 'Harnessing the materials project for machine-learning and accelerated discovery'. *MRS Bulletin*. 2018;**43**(9):664–9.

Chapter 6

Computational studies for understanding and developing silicon anodes

Fernando A. Soto[1], Diego E. Galvez-Aranda[1], Jorge M. Seminario[1], and Perla B. Balbuena[1]

6.1 Introduction

Energy storage technology is key for a rapid transition into a global automotive market dominated by electric vehicles (EVs). Therefore, it is necessary to develop revolutionary next-generation lithium-ion batteries (LiBs) that can sustain an increase in driving range for EVs. Graphite-based LiBs are quickly reaching their physical limit, but silicon (Si) is an ideal replacement due to its high Li storage capacity. Indeed, Si is one of the most promising anode materials in LiBs mainly due to their abundance and high theoretical gravimetric specific capacity (4200 mAh/g) [1]. Despite this increased capacity over carbonaceous materials, one of silicon's main roadblocks for its implementation as an anode material is the huge volume expansion and contraction that takes place upon cycling. Moreover, the work function of the Si anodes is approximately 0.37 V vs. Li^+/Li, and at these potentials most of the typical electrolyte solutions mainly composed of a salt in solution of carbonate solvents become easily reduced at the surface of the anode [2]. Therefore, during the normal operation of a battery, a multicomponent layer known as the solid-electrolyte interphase (SEI) layer forms as a result of the breaking down of molecular bonds of salts and solvent molecules making up the electrolyte solution. Ideally, this SEI layer should act like a sponge, blocking electrons and molecules, allowing passage to ions, and protecting the battery from degradation over time. Thus, designing this layer in a specific way is a promising solution to improve battery life.

It is noteworthy that the SEI layer formed on the graphite anode has been well studied in the last couple of decades. Peled *et al.* [3, 4] coined the term SEI layer in the 1970s and described it as a multilayered structure where the inner layer closer to the anode contains inorganic films mainly composed of lithium oxide (Li_2O), lithium fluoride (LiF), and lithium carbonate (Li_2CO_3). Meanwhile, the outer organic layer closer to the electrolyte solution was characterized as

[1]Department of Chemical Engineering, Texas A&M University, USA

composed of lithium ethylene dicarbonate (LEDC), lithium ethylene carbonate (LEC), lithium vinylene dicarbonate (LVDC), lithium butylene dicarbonate (LBDC), and nonconductive polymers [5, 6]. However, unveiling the multicomponent, multi-morphology character of the SEI and its specific functionalities and properties represents a challenge in developing new innovations for battery capacity and safety. Since many of these reactions take place at interface regions, they are often difficult to study effectively. Specifically, little information is known as to how the properties of the Si electrode affect the growth and stability of the SEI layer. Moreover, one of the main problems is given by the mechanical properties of the SEI-covered Si anode that expands during lithiation causing formation of cracks and that allow continuous SEI growth, anode pulverization, and eventual battery failure [2]. Currently, most of our knowledge about the SEI structure on Si anodes come from ex-situ measurements such as electron microscopy, infrared microscopy (IR), and X-ray photoelectron microscopy (XPS) [7].

Practical application of Si anodes is currently hampered by multiple challenges. The primary one is its tremendous volume change (~300 percent) upon full lithiation and the resultant expansion/shrinkage stress during lithiation/delithiation, which induces severe cracking and structural pulverization of Si, leading to a rapid capacity fading. This results in the formation of an unstable solid electrolyte interphase (SEI) on the Si surface, and causes lithium trapping in active Si material, consequently leading to irreversible fast capacity loss and low initial coulombic efficiency (CE). Material properties, electrode particle size, electrode morphology, and charging rate can modify the resulting mechanical degradation of the Si electrode.

Besides experimental studies, multiscale computer simulations contribute additional tools for gaining a deeper understanding of the chemical and mechanical processes that take place in particular at the battery interfacial regions, but also to describe the overall behavior of these devices. The purpose of this chapter is to discuss the recent computational studies of Si-based materials as anode electrodes for LiBs, including the initial stages of formation of the SEI layer and mechanical stresses and deformations associated with Si electrochemistry. Section 6.2 provides a summary of the computational methods and the properties that are investigated regarding interfacial and bulk properties of Si anodes.

6.2 Brief review of computational methodology

In recent decades, advances in computational power and integrated modeling software packages have promoted the integration of molecular modeling techniques into materials science. Numerical experimentation provides an understanding of complex material at low cost; hence, computational simulations provide much more significant value as screening tools and guidance for experiments [8]. With the increasing demand for materials with improved performance for energy applications, industrial and research communities have become increasingly involved in studying novel materials at the molecular level. The structure of materials is linked to the types of bonding between the constituent atoms (e.g., metallic, covalent, and ionic bonding).

Obtaining a fundamental description of bonding requires a calculation of the electronic distribution. The types of methods that yield such information are called electronic structure methods [9]. Electronic structure methods are the most accurate way to calculate the structure, energy, and properties of a molecule or group of molecules [10]. Due to the rapid growth of the number of density functional theory (DFT) methods, a complete overview of DFT becomes impossible. This chapter describes the essential elements of DFT-based methods.

In electronic structure codes such as the Vienna Ab Initio Simulation Package (VASP), the electronic wave functions are treated as plane-waves, and the core electrons are represented by project augmented wave (PAW) pseudopotentials [11]. For most of the simulations presented in this chapter, the calculations were performed with the Generalized Gradient Approximation (GGA) to the exchange-correlation functional of Perdew-Burke-Ernzerhof (PBE) [12]. The PBE functional remains superior for the energetics of covalent and non-covalent bonds [13]. Moreover, the mean absolute deviation of binding energies between plane-wave PBE and a Gaussian basis set rarely exceeds 1 kcal/mol [14]. It is important to mention that the use of pseudopotentials may change the equilibrium bond length by up to 0.05 Å. A self-consistent field method is applied to update the electronic density in order to reach the ground-state energy. For this minimization procedure, density mixing is used. Density mixing produces a new estimated electron density utilizing a linear combination of the input charge densities from a previous history of electron densities. Fast Fourier transform methods are used to transform various functions (i.e., wavefunctions and potentials) from real to reciprocal space and back.

In general, these are algebraic manipulations to solve the matrix problem efficiently. Typically, the Hamiltonian matrix is projected, approximated, preconditioned, or spanned in subspace. With geometry optimizations, errors arise due to the change of the reciprocal space. It is important to note that plane-waves are treated in reciprocal space. The solution for this is to fix the cut-off energy allowing the set of plane-waves to vary along the length of the reciprocal space. The partial occupancies of the wave functions are described by a Gaussian smearing method, typically set to 0.1 eV. This smearing method heats the material by including fictitious thermal energies. Regarding the time evolution of the materials, DFT-coupled Molecular Dynamics (DFT-MD) or ab-initio molecular dynamics (AIMD) simulations are employed to study the evolution of the systems at finite temperatures. The forces acting on the atoms are obtained through electronic structure calculations. In AIMD simulations, one of the biggest challenges is the computational cost. The long computational time required to calculate the forces that are acting on the ions limits the simulations to a few hundreds of atoms. There are various approaches to solving the dynamic trajectories in AIMD simulations. One of them (used in most of the applications discussed in this chapter) is Born-Oppenheimer molecular dynamics (BOMD), where the calculation of the forces acting on the atoms and the trajectory of the nuclei are treated separately. Therefore, for this method, the time step is determined by the dynamics of nuclei. That is, at each step, the forces are calculated from the energies given by the time-independent electronic Schrödinger equation for the fixed nuclei. Then, the nuclei move according to classical mechanics in the field given by the calculated forces. This is an expensive computational calculation, and several approximations can be made to minimize this expense. For instance, wave function extrapolation [15] and charge density extrapolation [16] methods between steps are applied to provide a reasonable starting configuration, therefore minimizing the time spent on the electronic minimization. To account for the state conditions of the process being studied, simulations are carried out in various ensembles . The micro-canonical NVE ensemble is used to investigate the dynamics of the systems with a fixed number of particles and volume in a constant-energy conformational space. Meanwhile, the canonical NVT ensemble is used to study the dynamics of systems under fixed temperature instead of energy. It is important to remark that at the start of the AIMD simulations, there will be large fluctuations of properties such as total energy and temperature. Also, it is important to determine when the simulation is equilibrated. The equilibration is achieved by looking at the convergence of properties, depending on the ensemble, such as energies, velocities, etc. Moreover, in order to generate an NVT ensemble, a modification must be introduced in order to account for the effect of the thermal bath that keeps the system at a constant temperature. For most of the computational work presented in this chapter, the Nosé-Hoover chain thermostat is employed [17]. The Nosé-Hoover chain thermostat technique assures ergodic sampling of phase space even for the harmonic oscillator. In this thermostat, degrees of freedom are added to the system to mimic a reservoir representing the properties of the desired NVT ensemble. The ergodicity is assured by coupling the original thermostat with several thermostats.

In addition to simulations with periodic boundary conditions, ab initio simulations are also carried out in molecular clusters by solving the time-independent Schrodinger equation in the Born-Oppenheimer approximation, where the orbitals are represented by Gaussian functions. The advantage of this type of simulations is that much accurate electronic structure theories can be used beyond density functional theory. However, the number of atoms that can be treated in these systems becomes reduced as the level of theory is improved. In any case, it is a great method for identifying accurate chemistry details and some applications are also discussed in this chapter.

The described time evolution can also be obtained with classical MD simulations [18]. In these simulations, Newton's second law of motion is solved for the atomic motion (same as in AIMD) with the difference that the interatomic forces are described by effective force fields. These effective force fields are mathematical functions that describe the field exerted by the electrons without their explicit inclusion. Significant efforts from the research community have been invested in using these types of methods in the battery field [19–27]. Particularly interesting was the development of simulations able to represent the most important events in the battery at the nanoscale. These "nanobattery" simulations [28–31] introduced new insights coming from the interactions among the various battery components. Applications to batteries featuring Si anodes are discussed in this chapter. It is important to remark that the key challenge of the classical MD simulations is the use of the correct force field. Thus, this is one of the areas where researchers have invested significant time and effort. In particular, reactive force fields [32] have proved to be extremely useful to capture (at least qualitatively) phenomena such as SEI nucleation and growth [33].

Finally, coarse-grained methods are the additional group of tools that make the link between the microscopic and macroscopic behavior [34, 35]. Among them, kinetic Monte Carlo methods [36] have been used to address longer time scales and larger systems in various applications of the catalysis and battery materials fields, among others [37–41]. In these methods, first a list of physically significant events including for example bulk and surface diffusion, adsorption, reaction, complexation, are described with a microscopic rate equation, usually following Arrhenius law. Obviously, the description of these microscopic events requires information from ab initio simulations or experiments. Classical MD simulations are also a good source of data for interactions among components at the nanoscale, such as sintering of nanostructures or complexation events such as those expected to arise during SEI layer formation and growth.

6.3 Si anode properties characterized by computational methods

6.3.1 Si anode lithiation

Si nanowires (SiNWs) are promising anode materials for LiBs. SiNWs exhibit excellent properties for high-flux Li-ion insertion due to its efficient electron transport and high surface area. It is noteworthy that the details of the local high-flux Li-insertion are buried, making it challenging to extract a fundamental understanding from the experimental analysis. This fundamental understanding gains more relevance if one considers that the Si electrode experiences a large volume expansion of ~300 percent during electrochemical lithiation [42, 43]. The lithiation of Si results in known crystalline Li_xSi phases, including LiSi (x = 1), $Li_{12}Si_7$ (x = 1.71), Li_7Si_3 (x = 2.33), $Li_{13}Si_4$ (x = 3.25), $Li_{15}Si_4$ (x = 3.75), $Li_{21}Si_5$ (x = 4.2), and $Li_{22}Si_5$ (x = 4.4) [44, 45]. Shenoy *et al.* [46] performed AIMD simulations for the LiSi phase with the aim of providing insights into the intermixing kinetics and the motion of Li and Si in crystalline and amorphous phases of Si. Using Si slabs in their amorphous and crystalline phase as well as bulk Si, the MD simulations reveal that the surface Li atoms gradually move inside the Si by breaking the Si–Si bonds and allowing the Si phase to expand and accommodate them. The formation of Si-Si clusters occur during the first discharge and atomic Si appears in the highly lithiated phases (i.e., $Li_{15}Si_4$ and $Li_{22}Si_4$) with the caveat that the simulations were run at a high temperature of 1 200 K. A key finding is that Li diffuses almost ten times faster in amorphous than in crystalline Si. Interestingly, the $Li_{15}Si_4$ phase is a metastable crystalline phase, opening the possibility of obtaining novel meta-stable structures under extreme conditions. Indeed, Romero and coworkers [47] reported a new set of crystal structures for Li–Si binary compounds (Li_9Si_4, Li_5Si_2, Li_3Si, Li_7Si_2, Li_4Si, Li_5Si, and Li_6Si) obtained by using the minima hopping method [48].

From a thermodynamic standpoint, very recently, Chan *et al.* [49] observed that the lithiation process in Si is predicted to follow a path (crystalline-L1) in which the structure resembles the formation of crystalline phases, while the delithiation of crystalline-Li_{15}/Si_4 can occur via multiple paths due to the formation of high-energy-state amorphous-Li_xSi alloys. Meanwhile, Pizzagalli *et al.* [50] utilized DFT calculations and AIMD simulations to propose the following lithiation procedure: initially, Li atoms are found to prefer staying on the surface of the amorphous Si nanocluster. As the lithiation continues, a saturated state is reached, at which the nanocluster surface is fully covered. Then, it becomes energetically favorable for one Li atom to penetrate the amorphous Si nanocluster core. The authors attribute that the driving force for this transition is the increased electrostatic repulsion as Li-ions get closer to each other. In a related ab initio DFT work [51], the authors demonstrated that inserting Li atoms into the crystalline SiNWs leads to a step-like process, where each Si shell is sequentially lithiated from the outside. Meanwhile, during delithiation, the step-like behavior observed in the lithiation cycle disappears and the Li-ions are distributed throughout the cross-section of the SiNWs. Thus, more efforts should be made to improve the understanding of the delithiation process of Si anodes as this step could be key to prevent the failure mechanism of the anode. Moreover, the electrolyte effects at the interface between the electrode and the electrolyte should be included to determine the details by which Li ions abandon the electrode surface and travel through interphases before reaching the electrolyte phase. For a review of first principles approaches to simulate lithiation in Si electrodes see the work of Zhang, Cui, and coworkers [52].

Galvez-Aranda *et al.* focused on the first charging events of a pristine Si anode using MD simulations, developing a nano-battery model able to perform a set of predictions and post-dictions and defining a testbed for the design of new materials [29]. The charging process is performed by applying a full voltage using a step function to the nanobattery model. At the beginning of the simulation, the Si crystal had a volume of 23 749 Å [3]. Due to the arrival of Li atoms, the Si anode increases its volume as Li-ions insert into the crystal yielding a volume 316 percent larger when the anode reaches its full lithiated state: $Li_{3.5}Si$. This large inhomogeneous volume expansion occurring because of the coexistence of different phases and the crystallographic orientation-dependent lithiation in crystalline materials is expected to be very detrimental for device operation. The simulations revealed that as the concentration of lithium increases during the charging process of the nanobattery model, the volume increases linearly. The linear volume increase has also been reported in several experiments, such as the one performed by Jerliu *et al.* [53]. The temperature on the simulation never exceeds 360 K during the charging process. Controlling the temperature is needed to avoid overheating. Another finding during the simulation of charging the Si anode of the nanobattery is the formation of sharp interfaces between the rich and poor Li phases. This is due to the fact that the lithiation of Si follows a two-phase mechanism; therefore, a lithiation-protocol is implemented in the MD simulations. Nanostructured electrodes of Si are often fabricated with crystalline Si. During operation of a Li-ion battery, the crystalline Si electrode is amorphized after the first cycle of charge and discharge. To avoid the stress field induced

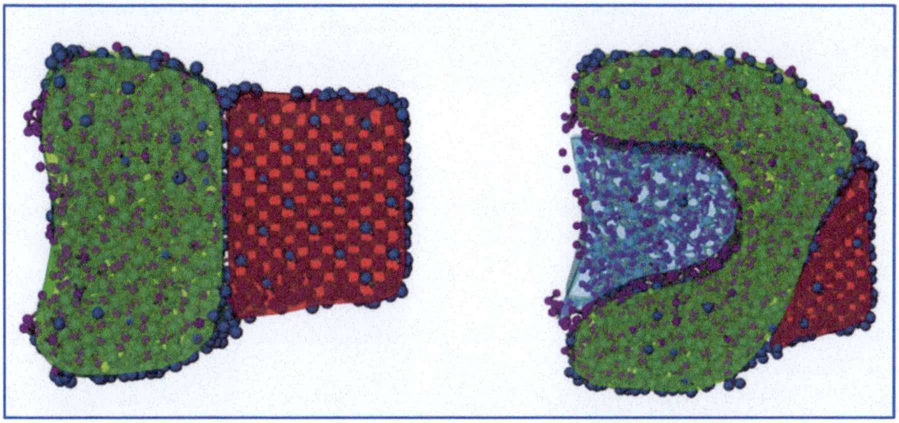

Figure 6.1 *Two-phase process of electrochemical lithiation of silicon in the nanobattery model. Left: At 80 ps, Li_1Si (green) and pristine-Si (red) is observed. Right: As charging continues at 525 ps, other concentrations are detected: $Li_{6.2}Si$ (cyan), Li_1Si (green), and pristine-Si (red). Adapted from [35]. Copyright © by Elsevier, with permission.*

by this phase transition, amorphous Si is often used as the electrode in experiments. Lithium alloying in a crystalline Si particle features the pristine original Si phase, or in general, a lithium-poor phase, separated from the alloyed regions containing Li, which in turn, can be separated in several phases of Li_xSi depending on the state of charge (SOC). During lithiation of the Si anode in the nanobattery model, various Li_xSi_y phases can be obtained before achieving the fully lithiated state of Si: $Li_{15}Si_4$. The Li_xSi_y phases observed during the charging are not necessarily thermodynamically stable; in most cases, they are amorphous alloys that are metastable structures (Figure 6.1) that exist before reaching the crystallographic $Li_{15}Si_4$ structure. Phase separations result in severe strain inhomogeneity. Most of the volume expansion occurs on the Li-rich phase near the particle surface since the Li-poor phase near the particle core is relatively unexpanded. Consequently, large tensile stresses are formed at the particle surface leading to severe mechanical damage close to the crystalline particle surface during lithiation. Conversely, amorphous Si exhibits relatively small damage because of smoother lithium concentration variation throughout the particle volume, owing to single-phase diffusion behavior. The nanobattery model proved to be good to emulate and understand the mechanical behavior during the first charging process of a Si anode. In this work, we focused on the study of the anode during the first charging process using the cathode only as a source of Li-ions.

MD calculations using reactive force field (ReaxFF) showed to be useful to characterize the lithiation process of Si electrodes being able to reproduce energy barriers of Li diffusion from DFT calculations in both crystalline and amorphous Si [54]. Diffusion mechanisms regarding the Li insertion in Si electrodes can be determined using these kinds of models. ReaxFF-MD model showed that during

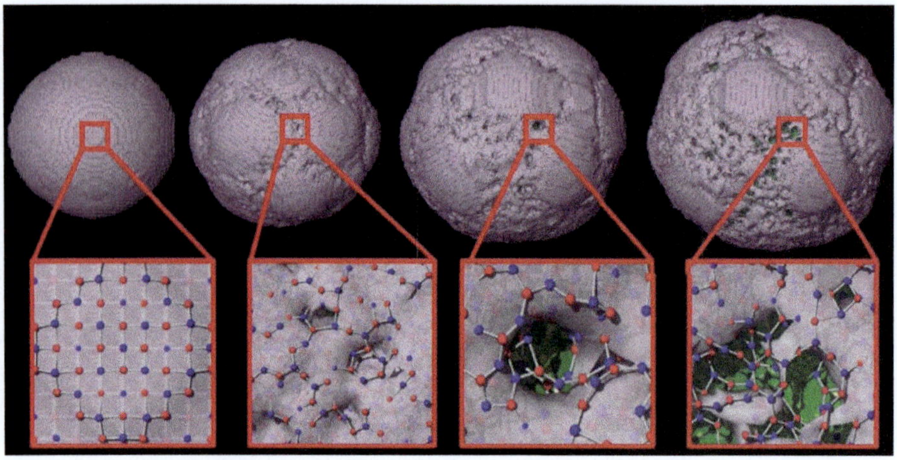

Figure 6.2 Stress concentration and crack nucleation of Si NPs. Grain structure and hole formation during volume expansion. Li_xSi (green) and LiF SEI. Li (blue) and F (red). Adapted from [55]. Copyright © Electrochemical Society, with permission.

lithiation of crystalline Si, the electrode changes into amorphous Li_xSi phases due to the simultaneous breaking of Si–Si bonds as a result of the stress generated between Si atoms. This lithiation results in an anisotropic volume expansion leading to a stress concentration and crack nucleation (Figure 6.2).

Moreover, the Li deposition and possibility of dendrite growth on Si anodes covered by a model LiF SEI layer were investigated with classical MD simulations in a nanobattery setting that includes both electrodes and the electrolyte layer [31]. This work contributed to understanding the mechanism of dendrite growth by simulating possible anode charge distribution protocols. The simulation was set as a nanobattery, where plating is characterized on an already cracked solid electrolyte interphase. The ionic motion took place under the application of an external field representing the battery charging. This study clarified that the drift velocity of the Li-ions in a typical commercial Li-ion battery is significantly slow: ~1 mm/h. This slow motion determines that dendrite growth would take place in a few hours. But an important point is that, after a Li-ion arrives at an active anode site, it takes an extremely short time, less than 1 ps, to become reduced. Thus, in order to emulate this behavior, this large difference in time-scales needs to be accounted for. When the MD simulation of the ionic transport is performed at much larger drift velocities, useful results can be obtained at reasonable computational expenses. For example, physical and chemical conditions before the ion deposition process are assumed to be known, and conditions that do not significantly influence the growth process are ignored. MD simulations were conducted on a pre-lithiated Si anode with a Li:Si ratio of 21:5. The cation deposition process was analyzed by testing various surface charge distributions in a nanosized square representing a crack of the solid electrolyte interphase, where the electrolyte solution comes into direct contact with the LiSi alloy anode (Figure 6.3). Depending on the selected charge distributions on the anode surface, Li nucleation and growth are observed during the simulation under an external applied field. It was concluded that strong changes in charge distributions taking place on the surface of the crack induce dendritic growth. At constant surface charge, the rate of growth is enhanced with respect to that when the charge varies with the height of the growing dendrite.

Several conclusions are drawn from the results presented in Section 6.3.1. Lithium intercalates crystalline Si and amorphous Si structures leading to various phases of crystalline Si and amorphous Si that are rich in Li content. Consequently, there has been considerable interest in this material as an anode electrode for LiBs and considerable effort in studying the delithiation mechanisms in order to improve the cycling capability during battery operation. Despite considerable work done over the years, much research still needs to be done to answer the question: How does the lithiated Si interact with the electrolyte solution? In Section 6.3.2, the focus is directed toward the initial electrochemical reduction of solvent molecules when in contact with a lithiated Si cluster.

6.3.2 *Si clusters: stability, lithiation, and interaction with electrolyte components at low degrees of lithiation*

Taking advantage of the smaller volume changes with the reduction of the particle size, in the past decade, research groups have used DFT methods to study the structure and reactivity patterns of small-sized Si clusters. Sporea *et al.* [56] used the Gaussian 98 code [57] to understand the interaction between Si clusters and alkali atoms, natural population analysis was performed. The charge of the alkali metal atoms (M) in Si_nM clusters was calculated to be near to one, increasing monotonically with n from about 0.7–0.8 u.a. for SiM to 0.85–0.95 for Si_6M. The charge on the metal atom is higher for potassium-doped clusters than those for Li or Na-doped systems since the ionization potential of potassium atoms are lower than those of Li and Na atoms. In a related study, Sporea *et al.* [58] reported a computational study of the possibility to form alkali-encapsulating Si clusters $A@Si_nA@Si_n$ with n = 10–20. The authors reported that the alkali-encapsulating Si clusters $A@Si_nA@Si_n$ are found to be stable but lying much higher in energy as compared to surface-bound alkali isomers. Yang *et al.* [59] used the Gaussian 03 computational chemistry package [57] to investigate the electron affinities and dissociation energies of small Si–Li clusters. The groundstate geometry showed that the Li atoms bound to the neutral Si_nLi cluster. The predicted electron affinities were 1.87 eV for Si_2Li, 2.06 eV for Si_3Li, 2.01 eV for Si_4Li, 2.61 eV for Si_5Li, 2.36 eV for Si_6Li, 2.21 eV for Si_7Li, and 3.18 eV for Si_8Li. The dissociation energies of Li atom from Si_nLi are predicted to be 2.75 eV for Si_2Li, 2.41 eV for Si_3Li, 1.99 eV for Si_4Li, 2.50 eV for Si_5Li, 2.07 eV for Si_6Li, 1.43 eV for Si_7Li, and 2.10 eV for Si_8Li. Merino *et al.* [60] used DFT calculations with a B3LYP functional [61, 62] to explore the Si_6Li_6 potential energy surface and the calculations showed that the most stable form of the Si_6Li_6 cluster is not planar; instead, it is a tridimensional structure formed by a Si tetrahedron interacting

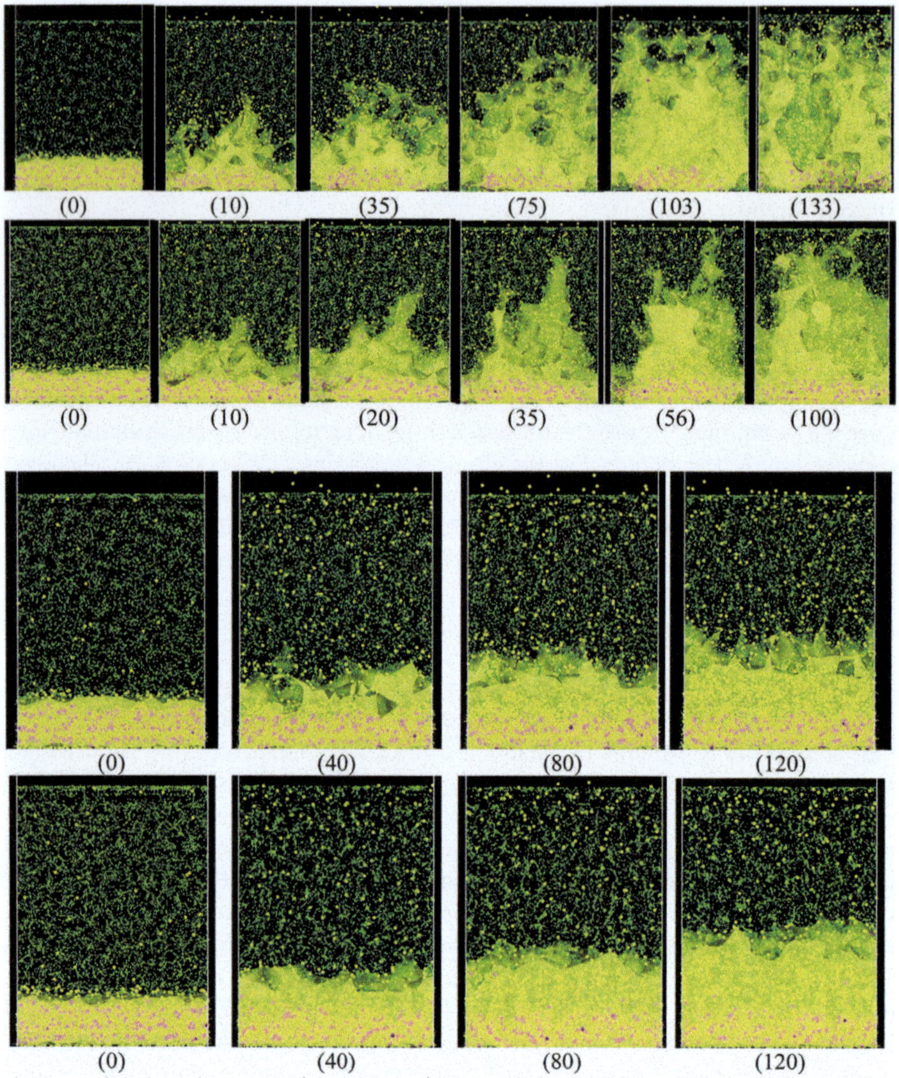

Figure 6.3 *Time dependence (ps) of dendritic growth. Top (first three rows): at constant surface charge of decreasing magnitude (−1e, −.54e, −88μe). The time where the dendrite growth reaches the positive electrode is 103 and 56 ps in the first two cases. Bottom (last row): at variable surface charge that depends on the vertical height of the dendritic growth. Li (yellow), Si (pink), EC and PF_6^- (green). Reproduced with permission from [31]. Copyright © Royal Society of Chemistry.*

with a Si dimer through four Li atoms. In an interesting study, Bae [63] used DFT and Monte Carlo (MC) calculations to study geometry and electronic properties of neutral and charged Si_nO_n (n = 1–7) clusters. Regarding the electronic properties of these clusters, the author reported that the odd number neutral Si oxide clusters are more stable than even number neutral silicon oxide clusters, while cationic and anionic Si_4O_4 clusters are the least stable as determined from second energy difference calculations.

More recently, empirical potentials and DFT methods have been used to study small- and medium-sized Si clusters [64–66]. Heydariyan *et al.* [67] used the evolutionary methodology with the USPEX (Universal Structure Predictor: Evolutionary Xtallography) algorithm [68] to search for low-energy Si_n (n = 8–80) structures. The low-energy structures were further optimized with density functional tight binding (DFTB) and DFT calculations. The authors classified the optimized Si clusters into two families: (a) for n = 8–26, where the structures include magic-number clusters (Si_6, Si_9, and Si_{10}) and other sub-units of the Si bulk diamond such as the puckered-hexagonal-ring S_6 cluster; (b) for n = 27–80, they found fullerene-like structures which consist of two regions: the exterior fullerene-like region and the interior region consisting of core Si atoms located at random positions. Overall, these studies provided a reference for further investigations of Si clusters for battery applications.

Early studies performed on the reductive decomposition of electrolytes on graphite-based anodes [69–71] made it clear that the electrochemical reduction process is convoluted, and that the corresponding decomposition mechanism may vary with the electrode's surface structure, and composition and nature of the electrolyte components among other factors. The electrochemical reduction mechanism on Si is not an exception. In fact, the electrochemical reduction process that takes place on Si is amplified due to the volume changes that Si undergoes upon cycling [72]. To elucidate the phenomena and elaborate guidelines to exert some control over Si anodes, nanostructured Si and Si composite anode materials have been investigated in experimental and computational works [73–75]. It is important to note that besides battery applications, due to silicon's dominance in the microelectronic industry, the structure and dynamics of Si atomic clusters have attracted attention for their applications in optoelectronics, solar cells, biosensors, and so forth [76].

Since Si anodes in real-life batteries are composed of clusters of particles exhibiting different shapes sometimes in confined spaces, an important question emerges: How reduction proceeds in the limit of ultra-low lithiation? To address this question, researchers implemented several testing models [77] using DFT calculations, emulating the interactions between EC and fluoroethylene carbonate (FEC) on Li^+ adsorbed Si clusters with compositions $LiSi_{15}$ and Li_2Si_{15} which could constitute the very initial stages of Si lithiation. The simulations showed that the one-electron and two-electron electrochemical reduction processes depend on the local concentration of EC and Li-ions. Thus, the one-electron electrochemical reduction process is favored at a high concentration of EC, while the two-electron process is favorable at high Li-ion content. In addition, the configurations of the adsorbed solvent were shown to begin with a Li-O interaction, in contrast to adsorptions observed on surfaces with higher contents of Li where a Si–C bond forms first, thus leading to a different reduction mechanism. Regarding the fate of FEC, the two-electron mechanism was found to be the most favorable reduction path for FEC starts with the formation of a Li-F pair. AIMD simulations were also employed to study the reactions that take place in a semi-lithiated Si cluster [78]. The simulations showed that both the lithium hexafluorophosphate salt ($LiPF_6$) and the EC solvent participate in the SEI layer formation on the Si anode, where the addition of the salt dramatically increases the stability of the solvent. This was a demonstration that by tuning the electrolyte, it is possible to regulate the composition of the SEI, which would be the first step in controlling the issues at the anode (and cathode) sides. Perhaps more important, it has been reported [52] that sub-nanosized Si clusters would potentially work as active anode materials for LiBs in terms of the insertion potential. However, such sub-nanosized size has its disadvantages, as shown by the simulation results that highlight the fact that at these Si clusters, Li may be found irreversibly trapped in the SEI layer rather than intercalated in the Si cluster. Despite the excellent insights provided by the cluster models at ultra-low lithiation degrees, the voltage drops experienced in the cell during the battery operation cycling the Li/Li_xSi_y cell correspond to higher lithiation. For example, potential plateaus are observed at 0.39, 0.37, 0.26, and 0. 1 V, indicating the formation of LiSi, $Li_{12}Si_7$, Li_7Si_3, and $Li_{13}Si_4$, respectively [79, 80]. Section 6.3.3 discusses the effects on the SEI layer stemming from these higher lithiation degrees in the Si anode.

6.3.3 Effect of higher lithiation and surface facet structure on the reductive decomposition of solvents, additives, and salts of electrolyte solutions

DFT calculations provided new insights into the structure and dynamics of electrolyte solutions on lithiated Si slabs, modeling the surface of the anode. Using AIMD methodologies, simulations [81] revealed a dependence of the reduction mechanism on the lithiation degree of the Si slab. Sequential two-electron transfers to the EC molecule were observed for low lithiated Si surfaces such as $LiSi_4$, $LiSi_2$, and LiSi. Meanwhile, a four-electron transfer mechanism is found in highly lithiated Si surfaces such as $Li_{13}Si_4$. Differences in reactivity due to surface functionality and surface structure among the six LiSi surfaces studied (LiSi-100-H, LiSi-100-O, LiSi-100-OH, LiSi-101-H, LiSi-101-O, and LiSi-101-OH) were also reported. The low-coordinated surface Si atoms were saturated with –H, –O, or –OH functional groups. However, the study showed that the reactivity trend was mainly driven by the geometrical differences among LiSi slabs with the various exposed facets. Figure 6.4 illustrates the main features observed in this study regarding how adsorption from solution onto the surface takes place in a very similar way, via the formation of a C–Si bond, independently of the surface functionality. As in the low-lithiation cases, an O–Si bond is also formed. This is because the presence of Li is what triggers the reaction. Therefore, where and how such Li atom is bonded to Si is what matters most because that structural arrangement determines the structure of the electrochemical double layer and the orientation and concentration of

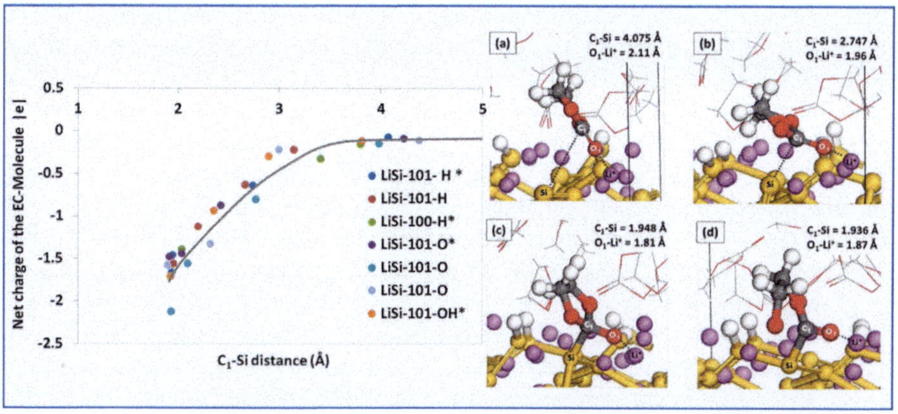

Figure 6.4 *Ab initio MD simulations of electrolyte solutions over LiSi (101) surfaces. Left: Net charges of EC molecules on LiSi surfaces terminated on functional groups (H, O, OH), as a function of their C1−Si distance to the surface. The asterisks denote cases in which the molecule is adsorbed on two-fold coordinated silicon atoms. Other cases refer to molecules adsorbing on three-fold coordinated silicon. In both cases, the Si atom is also coordinated to Li. Right: Time evolution of reduction of adsorbed EC molecule on a LiSi (101) surface terminated in H atoms. The adsorption starts with O1−Li+ interactions (panel a), once the molecule gets closer to the surface (C1−Si <3.5 Å, panel b) adsorption of the molecule takes place (panel c) and a C1−O2 bond breaks (panel d). The net Bader charge of the EC molecule is* $-0.07|e|$, $-0.64|e|$, $-1.46|e|$, *and* $-1.66|e|$ *in panels a, b, c, and d, respectively. Note the ring opening in the last image. Color code: O—red; Li—purple; C—gray; H—white; and Si—yellow. Adapted with permission from [80]. Copyright (2013) American Chemical Society.*

EC molecules approaching the surface. This is in sharp contrast to what is found on graphite anodes where reactivity is dominated by the electrochemical activity of functional groups at the edges of the graphite slab. The strategy followed here represented an important contribution to developing a better understanding of the interfacial reactions at the electrode|electrolyte interface in Si anodes during the initial cycling. At the same time, it served as a boost to further develop computational materials science-based models encompassing a multipronged understanding of Si anodes, electrolyte decomposition, SEI layer composition, and many more aspects such as the chemo-mechanical effects discussed later in this chapter.

As mentioned in Section 6.3.2, researchers have employed nano-structured Si particles to investigate the chances of controlling the effects of volume expansion. Despite the implantation of this strategy, the additional consumption of electrolytes to generate new SEI detected during each cycle demands a highly stable SEI. In this respect, the incorporation of electrolyte additives is one method to generate a stable SEI that results in improved cycle life of Si electrodes. Leung and coworkers [82, 83] used DFT and AIMD in slabs and small clusters to elucidate the most favorable electrochemical reduction processes of vinylene carbonate (VC) and FEC on Si anodes. It was shown that the use of a VC additive leads to an improved polymeric nature due to the electrochemical reduction mechanism initiated by an adsorbed CO_2^- anion and the $OC_2H_2O_2^-$ radical anion. Meanwhile, FEC reduced in a two-electron reduction process resulting in the formation of Li fluoride and CO. As shown in graphite anodes, VC is able to induce the formation of a stable, more compact polymer film. Shkrob *et al.* [84] followed the insights from these atomistic simulations to demonstrate a complex scenario of the redox chemistry of FEC. Moreover, other comprehensive work [85] included several electrolyte compositions based on organic carbonates and $LiPF_6$ aiming to characterize the initial stages of the SEI. The simulations predicted that FEC selectively decomposes over diethyl carbonate (DEC) and EC by a comparatively high rate. The electrochemical reduction process follows a one-electron reduction of the FEC solvent molecule to form a LiF as the main SEI component. As also described in the review papers [86, 87], 50 percent of the EC molecules decomposed with the following reaction:

$$EC + 2e^- \rightarrow O(C_2H_4)O^{2-} + CO_{(ads)} \tag{6.1}$$

and the other 50 percent EC molecules decomposed according to

$$EC + 2e^- \rightarrow O(C_2H_4)OCO^{2-} \tag{6.2}$$

Also, FEC molecules were found to reduce according to the following mechanism:

$$FEC + 2e^- \rightarrow OCO(C_2H_3)FO^{2-}{}_{(ads)} \tag{6.3}$$

$$OCO(C_2H_3)FO^2{}_{(ads)} \rightarrow F^-{}_{(ads)} + OCO(C_2H_3)O^-{}_{(ads)} \tag{6.4}$$

$$OCO(C_1H_3)O^-{}_{(ads)} + 2e^- \rightarrow CO_2{}^{2-}{}_{(ads)} + C_2H_3O^- \tag{6.5}$$

Regarding the salt dissociation and decomposition, three of the F atoms leave the PF_6^- group, almost simultaneously, after about 11 ps of simulation time. The neutral PF_3 group remains in the liquid phase for about 1.3 ps and then is adsorbed onto a Si atom of the surface, resulting in the subsequent total decomposition of the salt. Overall, this work offers a new perspective on the effects of electrolyte composition on SEI reactions. It is noteworthy that the nuclear magnetic resonance (NMR) work by Grey *et al.* [88] confirmed the organic composition of the SEI predicted in these simulations.

After the initial reaction mechanisms that lead to the nucleation stages of the SEI layer takes place; the SEI layer continues to grow beyond the electron tunneling regime [89]. Regarding the SEI growth, AIMD simulations [90] on models containing (LiEDC) and (LiVDC) that mimicked SEI layer nuclei on top of a lithiated Si anode ($Li_{13}Si_4$/(LiF, Li_2O, or Li_2CO_3)/(LiEDC or LiVDC) allowed to investigate how does the SEI layer grows after its thickness goes

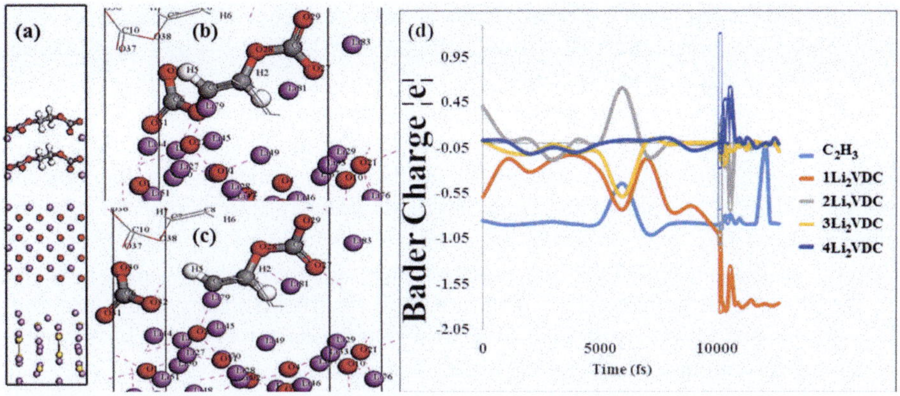

Figure 6.5 *(a) Initial configuration of a model anode/SEI:* $Li_{13}Si_4/Li_2O/Li_2VDC$*. A* C_2H_3 *radical was added near the* Li_2O *surface in the* Li_2VDC *region. (b) After 10,104 fs, one of the* Li_2VDC *molecules near the* Li_2O *surface starts to be reduced and the C (vinyl)–O (carbonate) bond elongated. (c) At 10,115 fs, the C (vinyl)–O (carbonate) bond is broken forming* CO_3 *and* $C_2H_2CO_3$ *radical anions. (d) Bader charges evolution for the decomposing molecule (orange line) and the surrounding* Li_2VDC *molecules (2, 3, and 4) as well as that of* C_2H_3*. Color code as in Figure 6.4. Reproduced with permission from [89]. Copyright (2015) American Chemical Society.*

beyond electron tunneling lengths. It is important to note that in the SEI model proposed by Peled *et al.*, the outer layer of the SEI is mainly formed by a porous organic layer (e.g., LiEDC, LiVDC). However, this organic layer can easily decompose in the presence of Li atoms [91] or radicals (e.g., OH, C_2H_3) formed during the multiple electrolyte decomposition reactions. Based on the fact that LiEDC and LiVDC play a critical role as an outer layer of the SEI, the simulations discovered that these molecules would decompose due to electron transfer from radicals or radical anions present in the SEI environment (generated by the reductive decomposition of an EC-based electrolyte). Specifically, the decomposition of Li_2EDC and LiVDC occurs through the detachment of CO_2 and CO_3 (Figure 6.5). Thus, these simulations discovered that radical species, produced at the initial stages of SEI nucleation, are capable of propagating electrons through the organic layers of the battery and generate new radicals after decomposing electrolytes, thus explaining the extended growth. The analysis in this work identified these radical species as the primary causative agent behind the electron transfer processes that allow the outer SEI layer to grow beyond the electron tunneling regime. They also determined that the stability of the decomposed products is another important factor because it determines the controlled or uncontrolled growth of the SEI layer. This concept was an entirely new finding in the field, making these computational analyses especially significant. Very recently, Chen *et al.* [92] confirmed the proposed SEI growth mechanism by mass-sensitive scanning transmission electron microscopy.

Further work [93] used AIMD simulations to study the stability of SEI components (Li_2CO_3 and LiEDC) on top of the surface of the Li (100) facet. The simulations show that both SEI components are thermodynamically unstable near the equilibrium Li^+/Li (s) potential. The authors surmise that due to its superior stability, the innermost SEI layer on a lithium metal anode should be a thin layer of Li_2O. Zhuo *et al.* [94] presented the first experimental evidence on the dynamic behavior of the SEI layer regarding copper electrodes and electrochemical cycling. The findings in the simulation work were used as a foundation for the experimental test, justifying the authors' results on carbonate decomposition.

To determine the possibility of organic films attaching directly to the Si anode surfaces, a hybrid computational approach combining classical MC and DFT-based geometry optimization and MD [95] proved useful to test the interaction between oligomers (LiEDC and LiVDC) and a highly lithiated Si slab ($Li_{13}Si_4$). The authors reported that the first oligomers attach strongly to the slab surface with calculated adsorption energies closer to −4 eV for both oligomers. Moreover, the results indicate that the mode of adsorption of the first oligomer is sensitive to the chemical composition of the surface. The stability of the interactions at 450 K was explored with AIMD simulations, and the system evolution showed that the proximity of the carbonate groups induces decomposition processes within oligomers. The global behavior is as follows: the oligomers adsorb firmly to the surface and then dissociate into a $CO_2 + OC_2H_4CO_3$ fragment. Interestingly, it was found that the surface-oligomer conformation between the two oligomers is different. The implication of this finding is that the difference in conformation may further affect the compactness of the SEI layer. Moreover, considering that other reactions such formation of LiF, Li_2CO_3 and Li_2O are also favorable simultaneous events; it is important to note that the lack of stability of the organic film detected by the simulations explains why the inorganic films are the ones that will at the end of the process remain attached to the anode surface, whereas the organics are expected to slowly grow in external regions of the SEI mainly due to their instability and porosity that would allow the motion of radicals and electrolyte out, thus favoring continuous SEI growth. Furthermore, this scenario is more favorable due to the expansion of the anode that would create a new surface where these events would continue to take place.

6.3.4 *Effects of Si crystallinity and oxidation degrees on the SEI reactions*

Investigations of the interfacial phenomena (e.g., lithiation, solvent orientation, electrolyte decomposition) that may take place at the interphase region of amorphous Si slabs and amorphous Si dioxide films are also relevant for the formation of the SEI layer. It is noteworthy that oxide-terminated Si films arise due to the exposure of Si surfaces to atmospheric environments. Despite the presence of this natural oxide film, findings through computational materials science studies show that the oxide

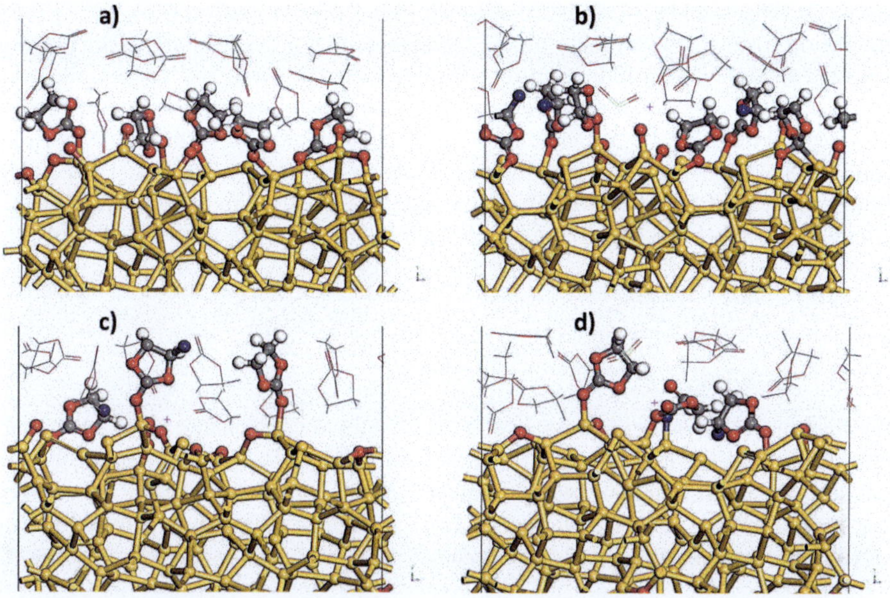

Figure 6.6 *Snapshots at approximately 50 ps of simulation time for the models with (a) 20 wt% FEC, (b) 40 wt% FEC, (c) 50 wt% FEC, and (d) pure FEC. The color code representing the various atoms is as follows: yellow—Si, red—O, gray—C, white—H, and dark blue spheres represent F atoms. For clarity, some solvent molecules were removed from the snapshots. Solvent molecules and salt not interacting with the Si surface are shown in a line display style. Reproduced with permission from [95]. Copyright (2018) American Chemical Society.*

film does not hinder the electrochemical processes occurring in a Si dioxide film. The amorphous Si phase contains under-coordinated sites and over-coordinated sites that can control the electronic and macroscopic properties of the material. It is not possible to characterize with most of the advanced experimental approaches the bulk atomic structure in amorphous Si. Therefore, computational tools can help to elucidate these complex structures.

In a novel study combining experimental and computational techniques, Horowitz *et al.* [96] probed the electrolyte interface structure by sum-frequency vibrational spectroscopy and AIMD simulations and discovered that the electrolyte mixture and additives (e.g., EC combined with FEC) used in the design of an SEI layer on top of an amorphous Si slab are capable of manipulating the orientation of the layer such that ordered, thinner SEI layers are formed, which greatly facilitate the passage of Li-ions through the layer (Figure 6.6). They also showed that a native oxide layer does not affect the overall orientation of EC solvent molecules found closer to the Si anode.

Regarding the reactivity of Si dioxide films, recent work revealed [97] that the insertion of Li atoms may promote the cleavage of the Si–O bonds. During the cleavage of the Si–O bonds, Si atoms are partially reduced, and the simulations indicated that surface Si atoms are more prone to this reduction due to the surface reconstruction of the hydroxylated atoms. The calculated formation energies for lithiation showed that the system would saturate at a Li/Si ratio of 3.48. Thus, even though an oxide film forms on top of a Si slab, this oxide film does not hinder the electrochemical phenomena taking place at the interface [98] and bulk [97] of the anode.

6.3.5 *Allotropes of silicon and implications for the SEI layer formation*

The high initial capacity in Si drastically drops at the second charge cycle due to the irreversible loss of crystalline order. Therefore, finding novel crystalline Si phases that are capable of intercalating alkali metals is key for the development of Si battery technology. In this regard, the thermal "degassing" process by Strobel *et al.* [99] yielded excellent results for the synthesis of Si allotropes capable of hosting Na-ions while showing high metallic conductivity. Si allotropes are particularly striking, due to their open-cage structure that, a priori, could easily accommodate alkali-ions inside the porous cages. It is noteworthy that due to the limited availability of lithium ores coupled with the projected demand of small-scale and large-scale electrochemical energy storage devices, the potential use of Na-ion based batteries is very promising [100].

Carrasco *et al.* [101] employed the VASP code [102–105] together with the climbing-image nudged elastic band (NEB) method [106] to show that Na_xSi_{24} forms a solid solution with minimal volume changes. However, sodium diffusion is predicted to be insufficiently fast for facile kinetics of Na-ion intake. Meanwhile, He *et al.* [107] confirmed the difficulty of sodium diffusion by reporting that Li^+ and Na^+ can diffuse along the channels inside the Si_{24} structure with barrier energies of 0.14 and 0.80 eV, respectively, and the ionic conductivity of $Li_{2.66}Si_{24}$ was calculated to be as high as 1.03×10^{-1} S/cm at 300 K. Marzouk *et al.* [108] investigated the crystalline Si_{24} structure, a type of Si allotrope that, in principle, can intercalate more than four alkali-ions namely Na and Li. Despite its potential use as an anode material, there is very little knowledge surrounding the sodiated/lithiated structure of this Si allotrope, and this information is important in order to analyze quantitatively the electrochemical process. In related work, Marzouk and coworkers [109] used computational models to determine the dynamic behavior of slabs of the sodiated/lithiated Si allotropes, the migration barriers of Na and Li-ions and the solvent decomposition at the interface of the sodiated/lithiated Si allotrope's surface. The AIMD simulations showed that the variation of the decomposed products with the underlying anode might induce differences in the morphology and composition of the SEI layer. Moreover, the simulations also suggested that the open cage Si allotropes are suitable candidates to replace graphite in terms of suitable voltage, cost, capacity, and cycling. These findings regarding anode materials based on Si allotropes provide a promising route for further explorations of these and alternative related materials.

6.4 Consequences of volume expansion and potential solutions

6.4.1 Mechanical stresses and particle cracking

One of the main challenges of the Si anode is the mechanical stability issues induced by volume expansion during lithiation and delithiation cycles since the Si volume can increase three times during charge [29, 55]. The volume change (~300 percent) upon full lithiation and the resultant expansion/shrinkage stress during lithiation/delithiation induces severe cracking and structural pulverization of Si, leading to a rapid capacity fading, resulting in the formation of an unstable solid electrolyte interphase on the Si surface, and causes lithium trapping in active Si material, consequently leading to irreversible fast capacity loss and low initial and subsequent CE.

Capacity can be maintained over many cycles for Si anodes of small feature sizes. The high-volume expansion that a Si electrode suffers during lithiation can be mitigated, accommodating the lithiation-induced strain by plastic deformation using anodes of small feature size. Active materials with sufficiently small sizes can avert lithiation-induced stress accumulation and fracture, and to allow fast charging and discharging. Also, several surface coating materials have been proposed to mitigate the mechanical effects of Si expansion and surface cracking during lithiation.

The volume expansion and cracking phenomena that may occur in nano- or micro-size particle are important issues that require further understanding. Galvez-Aranda *et al.* used classical MD simulations to investigate the origin and effects of cracking and stresses created in Si nanoparticles (SiNPs) during charge. The SEI coating of the Si active NP is modeled entirely as a pure LiF layer of 1.8 nm thickness. The mechanical properties of the SiNP and the SEI layer are strongly linked to the charge-discharge cycling phenomena, since a cracking on the coating will permit a direct contact of the lithiated SiNP and the electrolyte, leading to undesired reactions to happen. A SiNP model of 4.4 nm radius was pre-lithiated to $Li_{3.5}Si$. In total, with the coating LiF layer, the whole spherical SiNP has a radius of 6.2 nm. The size of the NP is similar to the one used in Jaumann's experimental work [110]. The SiNP model was equilibrated for 100 ps, and afterward a time-dependent expanding potential (TDEP) is implemented to emulate different charging currents. The TDEP increases the SiNP volume linearly, emulating constant current charging. For various charging currents, 2.66, 1.33, 0.67, 0.33, 0.22, 0.13, and 0.07 mA, the engineering stress and the volume are analyzed. It was concluded that for expansions of 300 percent or smaller, the behavior of the LiF-SEI depends on the expansion rate of the TDEP, which may be assigned to a pseudo charging current: (a) for large currents (2.66 and 1.33 mA) cracking is limited to sporadic bond breakings; (b) for medium-sized currents (0.67, 0.33, and 0.22 mA) cracking is observed (holes), and (c) for the smallest currents in this study (0.13 and 0.07 mA), the LiF SEI reconstructs and few, small, maxima, and minima on the stress curve are observed due to atoms of the SEI layer rearranging in their positions; however, once the reconstruction cycle ends, the SEI layer ends up divided into few pieces. The kinetic effects of the rapid expansion are observed to reflect on the mechanical properties, whereas thermal effects of high rates are eliminated by not letting the actual ions go through the NP. The corresponding C-rates of the nanobattery when extrapolated to macroscopic batteries yield values consistent with experimental findings (Figure 6.7).

Interestingly, similar chemo-mechanical phenomena manifest not only at the nanoscopic level, but also at larger scales. This hypothesis was verified by a comparative analysis of the results against those of a mesoscopic model [35]. In the mesoscopic model, a lattice-spring formulation accounts for large deformations and is coupled to an analysis based on transport and mechanical interactions defined in a spherical grid using a finite volume framework [34]. The concentration profile inside the Si particle was solved using two diffusion methodologies. For amorphous Si, single-phase diffusion was studied via Fick's law formulation, whereas for crystalline Si, two-phase diffusion was investigated to determine the impact of nature of the diffusion process on fracture characteristics using the Cahn-Hilliard model. On the other hand, the mesoscopic model of Mukherjee and collaborators [34] accounted for a brittle surface film of varying Young's modulus on top of the active material particle. Indicator functions were used to identify the film or active material regions and assignment of different axial and shear spring stiffness constants to each region in accordance with their Young's modulus and Poisson's ratio. Also, the model employed small deformation characteristics for the surface film layer.

Very distinct fracture behavior is observed for amorphous and crystalline Si because of the nature of the underlying intercalated lithium diffusion process, bond breaking, and subsequent different responses to mechanical forces. In fact, Figure 6.8 illustrates that the computed damage density is much higher in crystalline than in amorphous systems. It is interesting that similar differences between amorphous and crystalline fracture behavior such as changes in the damage density and on the tensile strength curves, as shown in Figure 6.7 in relation to the film fracture, were observed using the atomistic and mesoscopic models.

6.4.2 Artificial SEI layers and other anode protection strategies

Artificial SEI layers are potential solutions to control the reactivity of both anodes and cathodes. In the particular case of Si anodes, the expectation is that these layers act as barriers for electron transfer, and perhaps more importantly, as a vehicle for controlling unwanted volume expansion. In this sense, flexible films with self-healing properties would be a significant solution. Here we discuss the modeling of these artificial layers and the insights obtained from such modeling analysis. Alucone (aluminum alkoxide) systems generated by molecular layer deposition (MLD) methods were introduced and tested as coating in Si anodes several years ago [111]. First-principles modeling of these coatings is quite a challenge [112] because the chemical structure is not totally known as they result from a complex layer-by-layer chemical vapor deposition process. A DFT analysis of alucone formation on silicon oxide surfaces. In the first step of the process, trimethyl aluminum, a metal precursor for the coating process, reacts with the surface silanol groups of the native

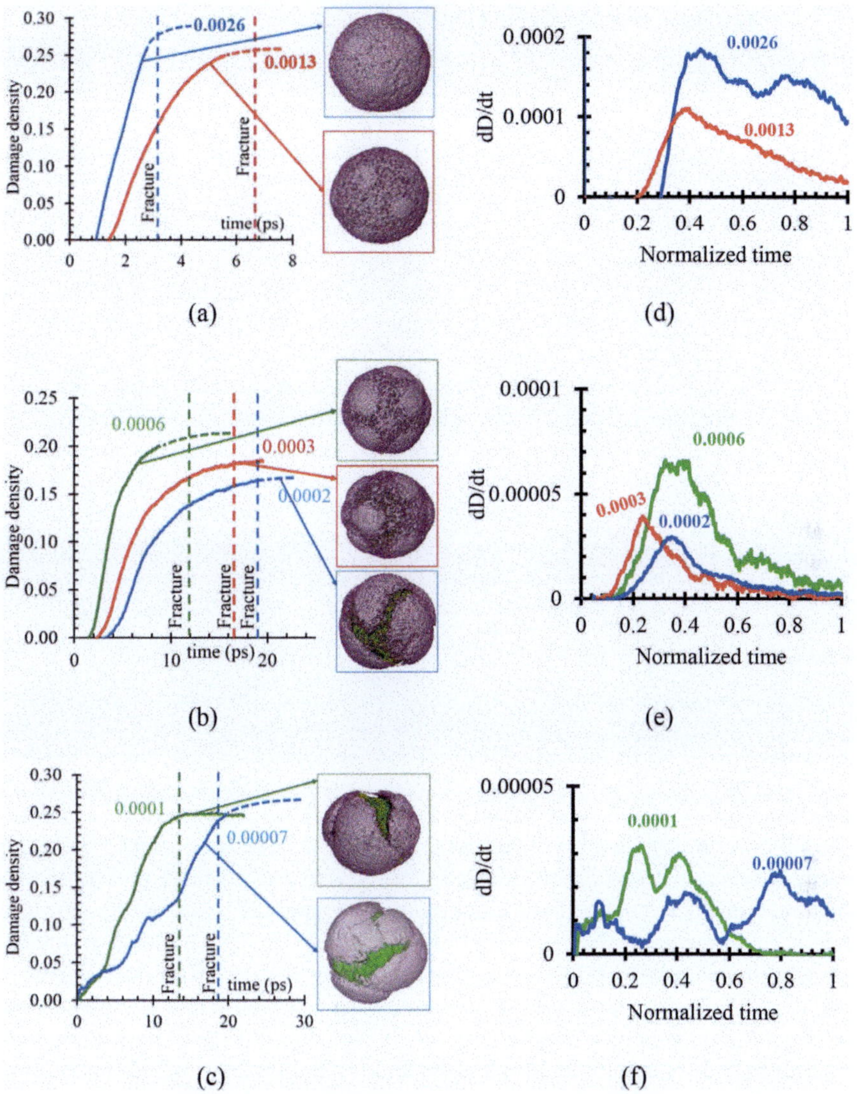

Figure 6.7 *Damage density of a nanoparticle of 6.2 nm radius obtained from atomistic classical MD simulations at (a) high C-rates 0.0026 and 0.0013, (b) medium C-rates 0.0006, 0.0003, and 0.0002, (c) low C-rates 0.0001 and 0.00007. Rate of damage density dD/dt at (d) high C-rates 0.0026 and 0.0013; (e) medium C-rates 0.0006, 0.0003, and 0.0002; (f) low C-rates 0.0001 and 0.00007. Here the damage density represents the total damage density only in the surface film. Cracking behavior of V% versus C-rate of the nanobattery (full dots) extrapolated to real battery C-rates (dotted lines). Adapted from Reference [35]. Copyright © Elsevier, with permission.*

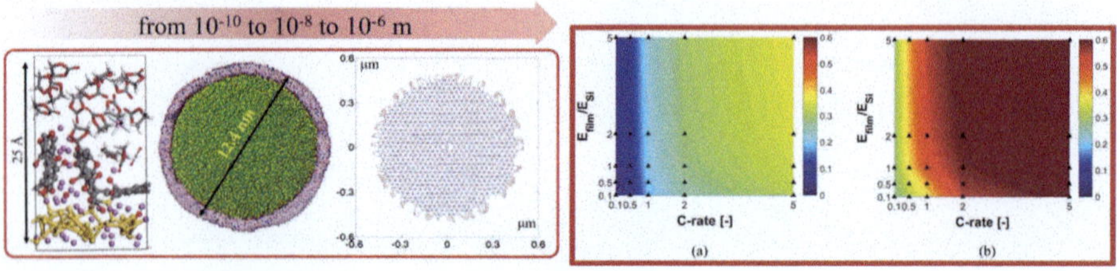

Figure 6.8 Left: Multiscale modeling approach (atomistic, nanoscale, and mesoscopic levels) used to evaluate chemo-mechanical interactions at the electrolyte/lithiated silicon surface. Right: Contour plots for total damage density as a function of C-rate and E_{film}/E_{Si} (SEI film to Si Young's moduli ratio) for (a) amorphous Si with film and (b) crystalline Si with film. The E_{film}/E_{Si} ratio is varied by changing the film Young's modulus while keeping the Si modulus constant. Crystalline Si with film exhibits higher damage density (shown by color scale) as compared to amorphous Si with film. High C-rate and high film's Young modulus contribute to enhanced damage in Si and film composite particle. The effect of C-rate is more pronounced compared to that of the Young's moduli ratio. The simulated Si particle diameter is 500 nm with 50 nm SEI thickness. Here damage density corresponds to the total damage density in the silicon particle and surface film agglomerate. Adapted from [35]. Copyright Elsevier, with permission.

Si oxides. The reactions were found to be thermodynamically favorable. Thus, the alucone coating is covalently bound to the Si surface. The adsorbed CH_3–Al group is able to react with glycols, such as ethylene glycol and glycerol, and other subsequent reactions lead to the final formation of amorphous 3D network structures with chemical composition $-Al_n(-OCH_2-CHO-CH_2O-)_m$ (n:m = 1). This is followed by film growth and partial restructuring of the Si oxide surface. MD modeling was then used to determine the potential polymer structure that would match the experimental density. This structure was built on top of Si slabs and DFT characterization of Li adsorption was employed to evaluate the thermodynamics of lithiation in such systems. It was concluded that the most favorable sites for Li were in the alucone film. Therefore, the film would lithiate first, forming the artificial SEI, followed by Si lithiation. This is because the film is composed of Al–O complexes with 3-O or 4-O coordination (Figure 6.9).

The DFT study revealed that AlO_3 sites are able to capture Li-ions quite strongly, with Li forming three bonds to O. Also, the most favorable locations for Li binding were found to be in the bulk of the film, whereas locations near the interface with the Si surface were less energetically favorable but still a little bit more favorable than inside the Si surface. With respect to Li transport through the film, the DFT analysis of binding geometries and energies along with the characterization of electron-rich regions suggested relatively fast Li transport probably via knock-off mechanism

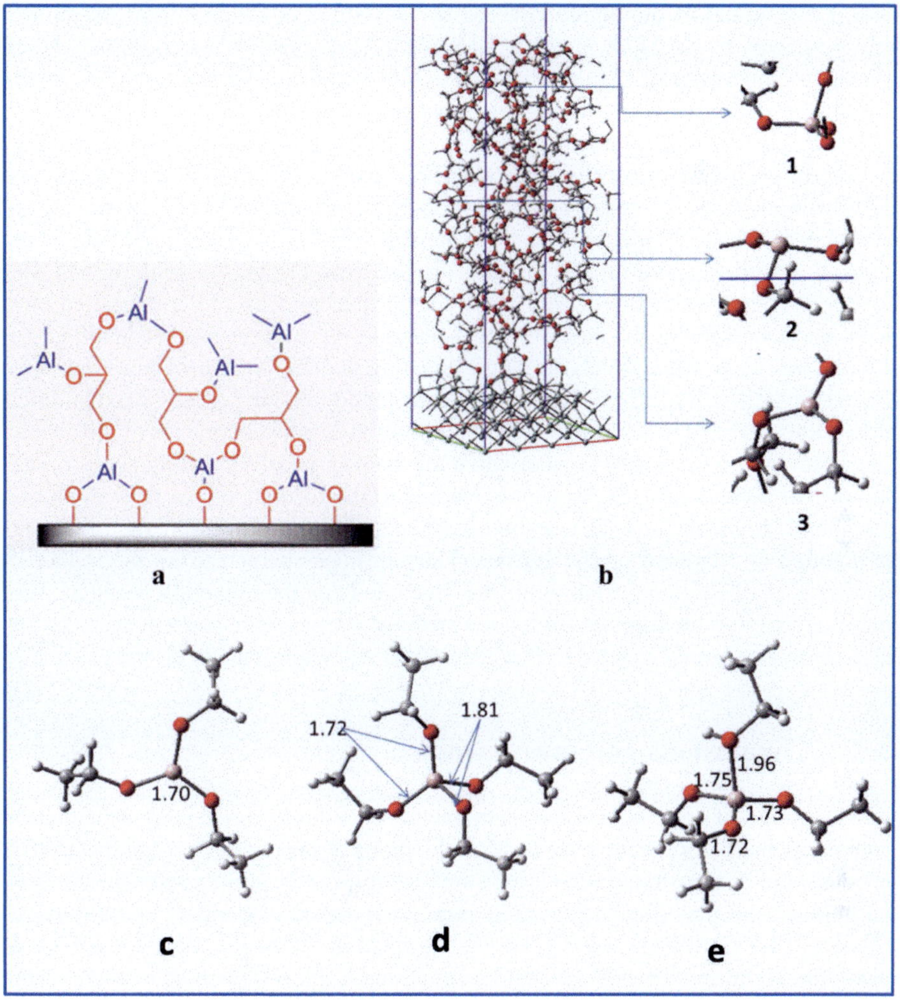

Figure 6.9 *(a) Possible structure of the alucone film deduced from the modeling study with experimental input. (b) DFT optimized structure of the film modeled according to the experimental density; 1, 2, and 3 are modes of coordination of Al atoms in the film; (c) 3-O coordination; (d) 4-O coordination; (e) 4-O coordination including an OH group. Adapted with permission from [111]. Copyright (2015) American Chemical Society.*

through the alucone film. Another important question is related to the reactivity of the film, which is also determined by its electronic conductivity. To address this question, a DFT-Green function analysis was done considering various states of lithiation of the film and Si electrode [112]. The conductivity study was performed on a $LiSi_2$ model of the anode covered by two different films: a "saturated" film represented by $C_{14}Li_5Al_5O_{15}H_{26}$ were all the five Al–Ox groups are coordinated with Li, and a "supersaturated" film where six extra Li atoms added to random locations in the film. In agreement with experimental measurements, it was found that both films were electronically conductive, and the electronic conductivity increased as the extent of lithiation goes beyond saturation of the film. Such result suggests that SEI reactions could be easily induced, even in the presence of the film. Additional modeling work was conducted to elucidate this point [113]. Using AIMD simulations, this work explored the interactions between an electrolyte solution of $LiPF_6$ in EC in contact with the previously modeled alucone film [112] deposited on a $LiSi_2$ surface. For the EC molecule, two reaction mechanisms were identified: (a) a two-electron transfer involving bond breaking and adsorption of the fragments. Interestingly, the second electron transfer was found to be triggered by oxidation of the Al ions in the film, which leads to the formation of new complexes of the opened EC anion with Al ions; (b) an alternative mechanism involves simultaneous breaking of two bonds of the EC molecule leading to the formation of a CO_3 anion that also bonds to Al cations. The second mechanism was observed on EC molecules in solution or adsorbed on the surface. Thus, besides characterizing the reduction mechanisms of EC and the anion, it was concluded that the reduction products were incorporated into the film modifying the artificial coating. The consequences of this new chemistry would require further theoretical and experimental efforts.

6.5 Conclusions and future work

A wide range of electrochemical phenomena involving Si anodes investigated by computational theory were reviewed. Applications of the theoretical treatment of Si anodes were demonstrated via various approaches: (a) ab initio approaches relying on finite cluster methods, (b) ab-initio employing periodic boundary conditions, (c) classical MD simulations, and (e) coarse-grained models. The main advantage of using the ab initio cluster model approach is that it allows quick screenings of the lithiation mechanism, volume expansion, and Si-Si dynamics without imposing a costly computational burden. On the other hand, the periodic boundary approach is crucial to model an infinite Si slab that is useful in order to model the anode electrode component of the battery. Moreover, the periodic ab initio approach also allows studying interfacial reactions, including the effect of the lithiation degree on key aspects such as electrolyte reduction and molecular adsorption at the anode surface. In addition, we have also illustrated the usefulness of classical MD simulations including the applications of electric fields that allow emulation of charge and discharge cycles as well as cracking and fracturing phenomena that is significantly important for Si anodes. These aspects included computational studies of mechanical properties and the coupling between chemical and mechanical phenomena on Si-based materials as anodes, including their lithiation, interactions with the electrolyte solution, and swelling effects and its consequences. In this regard, we also included the perspective from coarse-grained modeling.

The discussion of these topics was centered on issues deemed relevant to understand the initial formation of the SEI layer and its extension into Si-based LiBs as well as the challenges faced from a computational standpoint. A common feature of the Si electrode|electrolyte interaction is the initial reduction of the molecules. Subsequently, molecules such as solvents and anions may undergo reductive decomposition. The probability of these processes to occur varies with the nature of the exposed facet, functionalization groups, and the nature of the electrolyte solution, as well as the lithiation degree. Detailed studies have been devoted to unraveling the complex electrochemistry of the electrode|electrolyte interface. A key finding resulting from these efforts is that not only the stability of the decomposed products influences the growth of the SEI layer, but that the radical species generated as intermediate products are a primary causative agent behind the electron transfer processes that allow the outer SEI layer to grow beyond the electron tunneling regime.

The exploitation of the full benefits of the SEI layer is still limited by the inability to predict and control its multifunctionality as an electron blocker and an excellent ionic conductor. The SEI layer is typically described as a bilayer containing an inner inorganic region and an outer porous region, the so-called mosaic model, leading to the following scenarios (1) if one follows the mosaic model, then the grain boundaries that exist between each SEI block may impact the overall Li^+ transport; (2) if the SEI is considered as a homogeneous phase, then this phase, at the molecular level, must be composed of amorphous SEI blocks that allow Li^+ transport but also allows electron flow. Both of these scenarios call for alternative computation methodologies to describe ionic transport in such complex environments. Also, the reaction pathways leading to the SEI layer described in both scenarios remain uncertain or a matter of debate. As discussed in this chapter, progress in the direction of scenario (1) has been made by several studies that used quantum-based and atomistic simulations to quantitatively assess the electrolyte decomposition, clustering of SEI components and other nano-structural details of the electrolyte decomposition products. Efforts to investigate scenario (2) are still not sufficient; in this respect, the use of kinetic, thermodynamic, and multiscale microstructural models of SEI growth during Li intercalation/deintercalation should be encouraged. This is particularly important for the SEI growth on a Si anode since a model for the SEI growth has been proposed but no experimental observations have been conducted to prove it. Finally, the implementation of data science approaches such as machine learning should be done with the aim of providing answers to questions such as: What are the underlying mechanisms inducing the macroscopic properties (e.g., mechanical strength) when different electrolyte solutions are used? How do the rates of decomposition product change when the salt concentration is varied? This machine learning-based study would represent a starting point for the definition of a chemical kinetics descriptor inherent to any electrolyte solution.

References

[1] Ashuri M., He Q., Shaw L.L. 'Silicon as a potential anode material for Li-ion batteries: where size, geometry and structure matter'. *Nanoscale*. 2016;**8**(1):74–103.

[2] Zhang Y., Du N., Yang D. 'Designing superior solid electrolyte interfaces on silicon anodes for high-performance lithium-ion batteries'. *Nanoscale*. 2019;**11**(41):19086–104.

[3] Peled E. 'The electrochemical behavior of alkali and alkaline earth metals in nonaqueous battery Systems—The solid electrolyte interphase model'. *Journal of the Electrochemical Society*. 2047;**1979**:126.

[4] Peled E., Menkin S. 'Review—SEI: past, present and future'. *Journal of the Electrochemical Society*. 2017;**164**(7):A1703–19.

[5] An S.J., Li J., Daniel C., Mohanty D., Nagpure S., Wood D.L. 'The state of understanding of the lithium-ion-battery graphite solid electrolyte interphase (SEI) and its relationship to formation cycling'. *Carbon*. 2016;**105**:52–76.

[6] Peled E., Golodnitsky D., Ardel G. 'Advanced model for solid electrolyte interphase electrodes in liquid and polymer electrolytes'. *Journal of the Electrochemical Society*. 1997;**144**(8):L208–10.

[7] Veith G.M., Doucet M., Sacci R.L., Vacaliuc B., Baldwin J.K., Browning J.F. 'Determination of the solid electrolyte interphase structure grown on a silicon electrode using a Fluoroethylene carbonate additive'. *Scientific Reports*. 2017;**7**(1):6326.

[8] Meunier M. *Industrial Applications of Molecular Simulations*. Boca Raton: CRC Press; 2012.

[9] LeSar R. *Introduction to Computational Materials Science: Fundamentals to Applications*. New York City: Cambridge University Press; 2013.

[10] Rai B. *Molecular Modeling for the Design of Novel Performance Chemicals and Materials*. Boca Raton: CRC Press; 2012.

[11] Vanderbilt D. 'Soft self-consistent pseudopotentials in a generalized eigenvalue formalism'. *Physical Review B*. 1990;**41**(11):7892–5.

[12] Perdew J.P., Burke K., Ernzerhof M. 'Generalized gradient approximation made simple'. *Physical Review Letters*. 1996;**77**(18):3865–8.

[13] Tran F., Laskowski R., Blaha P., Schwarz K. 'Performance on molecules, surfaces, and solids of the Wu-Cohen GGA exchange-correlation energy functional'. *Physical Review B*. 2007;**75**(11):115131.

[14] Paier J., Hirschl R., Marsman M., Kresse G. 'The Perdew-Burke-Ernzerhof exchange-correlation functional applied to the G2-1 test set using a plane-wave basis set'. *The Journal of Chemical Physics*. 2005;**122**(23):234102–13.

[15] Arias T.A., Payne M.C., Joannopoulos J.D. '*Ab initio* molecular dynamics: Analytically continued energy functionals and insights into iterative solutions'. *Physical Review Letters*. 1992;**69**(7):1077–80.

[16] Alfè D. 'Ab initio molecular dynamics, a simple algorithm for charge extrapolation'. *Computer Physics Communications*. 1999;**118**(1):31–3.

[17] Nosé S. 'A molecular dynamics method for simulations in the canonical ensemble'. *Molecular Physics*. 1984;**52**(2):255–68.

[18] Allen M.P., Tildesley D.J. *Computer Simulation of Liquids*. Oxford: Oxford University Press; 1990.

[19] Alvarado J., Schroeder M.A., Pollard T.P., *et al*. 'Bisalt ether electrolytes: a pathway towards lithium metal batteries with Ni-rich cathodes'. *Energy & Environmental Science*. 2019;**12**(2):780–94.

[20] Borodin O., Smith G.D. 'Quantum chemistry and molecular dynamics simulation study of dimethyl carbonate: ethylene carbonate electrolytes doped with LiPF6'. *The Journal of Physical Chemistry B*. 2009;**113**(6):1763–76.

[21] Borodin O., Smith G.D. 'Development of many-body polarizable force fields for Li-battery applications: 2. LiTFSI-doped oligoether, polyether, and carbonate-based electrolytes'. *The Journal of Physical Chemistry B*. 2006;**110**(12):6293–9.

[22] Bauschlicher C.W., Haskins J.B., Bucholz E.W., Lawson J.W., Borodin O. 'Structure and energetics of Li+-(BF4-)(n ') Li+-(FSI-)(n '), and Li+-(TFSI-)(n): ab initio and polarizable force field approaches'. *The Journal of Physical Chemistry. B*. 2014;**118**:10785–94.

[23] Borodin O., Bedrov D. 'Interfacial structure and dynamics of the lithium alkyl dicarbonate SEI components in contact with the lithium battery electrolyte'. *The Journal of Physical Chemistry C*. 2014;**118**(32):18362–71.

[24] Li T., Balbuena P.B. 'Theoretical studies of lithium perchlorate in ethylene carbonate, propylene carbonate, and their mixtures'. *Journal of the Electrochemical Society*. 1999;**146**(10):3613–22.

[25] Kovačević G., Pivac B. 'Modeling the interface between crystalline silicon and silicon oxide polymorphs'. *Physica Status Solidi*. 2013;**210**(4):717–22.

[26] Kim S.-P., Duin A.C.Tvan., Shenoy V.B. 'Effect of electrolytes on the structure and evolution of the solid electrolyte interphase (SEI) in Li-ion batteries: a molecular dynamics study'. *Journal of Power Sources*. 2011;**196**(20):8590–7.

[27] Bedrov D., Smith G.D., van Duin A.C.T. 'Reactions of singly-reduced ethylene carbonate in lithium battery electrolytes: a molecular dynamics simulation study using the ReaxFF'. *The Journal of Physical Chemistry A*. 2012;**116**(11):2978–85.

[28] Ponce V., Galvez-Aranda D.E., Seminario J.M. 'Analysis of a Li-ion nanobattery with graphite anode using molecular dynamics simulations'. *The Journal of Physical Chemistry C*. 2017;**121**(23):12959–71.

[29] Galvez-Aranda D.E., Ponce V., Seminario J.M. 'Molecular dynamics simulations of the first charge of a Li-ion—Si-anode nanobattery'. *Journal of Molecular Modeling*. 2017;**23**(4):120.

[30] Selis L.A., Seminario J.M. 'Dendrite formation in Li-metal anodes: an atomistic molecular dynamics study'. *RSC Advances*. 2019;**9**(48):27835–48.

[31] Selis L.A., Seminario J. M. 'Dendrite formation in silicon anodes of lithium-ion batteries'. *RSC Advances*. 2018;**8**(10):5255–67.

[32] van Duin A.C.T., Dasgupta S., Lorant F., Goddard W.A. 'ReaxFF: a reactive force field for hydrocarbons'. *The Journal of Physical Chemistry A*. 2001;**105**(41):9396–409.

[33] Bertolini S., Balbuena P.B. 'Buildup of the solid electrolyte interphase on lithium-metal anodes: reactive molecular dynamics study'. *The Journal of Physical Chemistry C*. 2018;**122**(20):10783–91.

[34] Verma A., Mukherjee P.P. 'Mechanistic analysis of mechano-electrochemical interaction in silicon electrodes with surface film'. *Journal of the Electrochemical Society*. 2017;**164**(14):A3570–81.

[35] Galvez-Aranda D.E., Verma A., Hankins K., Seminario J.M., Mukherjee P.P., Balbuena P.B. 'Chemical and mechanical degradation and mitigation strategies for Si anodes'. *Journal of Power Sources*. 2019;**419**:208–18.

[36] Methekar R.N., Northrop P.W.C., Chen K., Braatz R.D., Subramanian V.R. 'Kinetic monte carlo simulation of surface heterogeneity in graphite anodes for lithium-ion batteries: passive layer formation'. *Journal of the Electrochemical Society*. 2011;**158**(4):A363–70.

[37] Zhang X.-Q., Jansen A.P.J. 'Kinetic Monte Carlo method for simulating reactions in solutions'. *Physical Review E*. 2010;**82**(4).

[38] Li H.Y., Chavez A.D., Li H.F., Li H., Dichtel W.R., Bredas J.L. 'Nucleation and growth of covalent organic frameworks from solution: the example of COF-5'. *Journal of the American Chemical Society*. 2017;**139**:16310–8.

[39] Liu Z., Wood D.L., Mukherjee P.P. 'Evaporation induced nanoparticle – binder interaction in electrode film formation'. *Physical Chemistry Chemical Physics*. 2017;**19**(15):10051–61.

[40] Callejas-Tovar R., Diaz C.A., de la Hoz J.M.M., Balbuena P.B. 'Dealloying of platinum-based alloy catalysts: kinetic Monte Carlo simulations'. *Electrochimica Acta*. 2013;**101**:326–33.

[41] Methekar R.N., Northrop P.W.C., Chen K., Braatz R.D., Subramanian V.R. 'Kinetic monte carlo simulation of surface heterogeneity in graphite anodes for lithium-ion batteries: passive layer formation'. *Journal of the Electrochemical Society*. 2011;**158**(4):A363–70.

[42] Taiwo O.O., Paz-García J.M., Hall S.A., *et al*. 'Microstructural degradation of silicon electrodes during lithiation observed via operando X-ray tomographic imaging'. *Journal of Power Sources*. 2017;**342**:904–12.

[43] Chew H.B., Hou B., Wang X., Xia S. 'Cracking mechanisms in lithiated silicon thin film electrodes'. *International Journal of Solids and Structures*. 2014;**51**(23-24):4176–87.

[44] Lyalin A., Kuznetsov V.G., Nakayama A., *et al*. 'Lithiation products of a silicon anode based on soft X-ray emission spectroscopy: a theoretical study'. *The Journal of Physical Chemistry C*. 2018;**122**(20):11096–108.

[45] Morris A.J., Grey C.P., Pickard C.J. 'Thermodynamically stable lithium silicides and germanides from density functional theory calculations'. *Physical Review B*. 2014;**90**(5):054111.

[46] Johari P., Qi Y., Shenoy V.B. 'The mixing mechanism during lithiation of Si negative electrode in Li-ion batteries: an ab initio molecular dynamics study'. *Nano Letters*. 2011;**11**(12):5494–500.

[47] Valencia-Jaime I., Sarmiento-Pérez R., Botti S., *et al*. 'Novel crystal structures for lithium–silicon alloy predicted by minima hopping method'. *Journal of Alloys and Compounds*. 2016;**655**:147–54.

[48] Goedecker S. 'Minima hopping: an efficient search method for the global minimum of the potential energy surface of complex molecular systems'. *The Journal of Chemical Physics*. 2004;**120**(21):9911–7.

[49] Chan K.S., Liang W.-W., Chan C.K. 'First-principles studies of the lithiation and delithiation paths in Si anodes in Li-ion batteries'. *The Journal of Physical Chemistry C*. 2019;**123**(37):22775–86.

[50] Pedersen A., Bieri M., Luisier M., Pizzagalli L. 'Lithiation of silicon nanoclusters'. *Physical Review Applied*. 2017;**7**(5):054012.

[51] Bunjaku T., Bauer D., Luisier M. 'Structural and electronic properties of lithiated Si nanowires: an *ab initio* study'. *Physical Review Materials*. 2019;**3**(10):105402.

[52] Zhang Q., Cui Y., Wang E. 'First-principles approaches to simulate lithiation in silicon electrodes'. *Modelling and Simulation in Materials Science and Engineering*. 2013;**21**(7):074001.

[53] Jerliu B., Hüger E., Dörrer L., *et al*. 'Volume expansion during lithiation of amorphous silicon thin film electrodes studied by in-operando neutron reflectometry'. *The Journal of Physical Chemistry C*. 2014;**118**(18):9395–9.

[54] Jung H., Lee M., Yeo B.C., Lee K.-R., Han S.S. 'Atomistic observation of the lithiation and delithiation behaviors of silicon nanowires using reactive molecular dynamics simulations'. *The Journal of Physical Chemistry C*. 2015;**119**(7):3447–55.

[55] Galvez-Aranda D.E., Seminario J.M. 'Simulations of a LIF solid electrolyte interphase cracking on silicon anodes using molecular dynamics'. *Journal of the Electrochemical Society*. 2018;**165**(3):A717–30.

[56] Sporea C., Rabilloud F., Aubert-Frécon M. 'Charge transfers in mixed silicon–alkali clusters and dipole moments'. *Journal of Molecular Structure: THEOCHEM*. 2007;**802**(1-3):85–90.

[57] Frisch M.J., Trucks G.W., Schlegel H.B., *et al*. *Gaussian 16 Rev. C.01*. Wallingford, CT; 2016.

[58] Sporea C., Rabilloud F. 'Stability of alkali-encapsulating silicon cage clusters'. *The Journal of Chemical Physics*. 2007;**127**(16):164306.

[59] Hao D., Liu J., Yang J. 'A Gaussian-3 Theoretical Study of Small Silicon-Lithium Clusters: Electronic Structures and Electro$_n$ Affinities of Si$_n$ Li$^-$ (n=2-8)'. *The Journal of Physical Chemistry A*. 2008;**112**(41):10113–9.

[60] Santos J.C., Contreras M., Merino G. 'Structure and stability of Si_6Li_6: aromaticity vs polarizability'. *Chemical Physics Letters*. 2010;**496**(1–3):172–4.

[61] Kim K., Jordan K.D. 'Comparison of density functional and MP2 calculations on the water monomer and dimer'. *The Journal of Physical Chemistry*. 1994;**98**(40):10089–94.

[62] Stephens P.J., Devlin F.J., Chabalowski C.F., Frisch M.J. 'Ab initio calculation of vibrational absorption and circular dichroism spectra using density functional force fields'. *The Journal of Physical Chemistry*. 1994;**98**(45):11623–7.

[63] Bae G-T. 'Density functional theory calculations of geometrical and electronic properties of neutral and charged silicon oxide clusters'. *Bulletin of the Korean Chemical Society*. 2019;**40**(8):780–6.

[64] Ding L.P., Zhang F.H., Zhu Y.S., *et al*. 'Understanding the structural transformation, stability of medium-sized neutral and charged silicon clusters'. *Scientific Reports*. 2015;**5**(1):15951.

[65] Yoo S., Zeng X.C. 'Motif transition in growth patterns of small to medium-sized silicon clusters'. *Angewandte Chemie International Edition*. 2005;**44**(10):1491–4.

[66] Yoo S., Zeng X.C. 'Structures and stability of medium-sized silicon clusters III. Reexamination of motif transition in growth pattern from Si_{15} to Si_{20}'. *The Journal of Chemical Physics*. 2005;**123**:164303.

[67] Heydariyan S., Nouri M.R., Alaei M., Allahyari Z., Niehaus T.A. 'New candidates for the global minimum of medium-sized $silico_n$ clusters: a hybrid DFTB/DFT genetic algorithm applied to Si_n, n = 8-80'. *The Journal of Chemical Physics*. 2018;**149**(7):074313.

[68] Oganov A.R., Lyakhov A.O., Valle M. 'How evolutionary crystal structure prediction works—and why'. *Accounts of Chemical Research*. 2011;**44**(3):227–37.

[69] Dey A.N., Sullivan B.P. 'The electrochemical decomposition of propylene carbonate on graphite'. *Journal of the Electrochemical Society*. 1970;**117**(2):222–4.

[70] Aurbach D., Ein-Ely Y., Zaban A. 'The surface chemistry of lithium electrodes in alkyl carbonate solutions'. *Journal of the Electrochemical Society*. 1994;**141**(1):L1–3.

[71] Naji A., Ghanbaja J., Humbert B., Willmann P., Billaud D. 'Electroreduction of graphite in $LiClO_4$-ethylene carbonate electrolyte: characterization of the passivating layer by transmission electron microscopy and Fourier-transform infrared spectroscopy'. *Journal of Power Sources*. 1996;**63**(1):33–9.

[72] Ko M., Chae S., Cho J. 'Challenges in accommodating volume change of Si anodes for Li-ion batteries'. *ChemElectroChem*. 2015;**2**(11):1645–51.

[73] Xiao Z., Xia N., Song L., Cao Z., Zhu H. 'Mesoporous Si/C composite anode material: experiments and first-principles calculations'. *Ionics*. 2019.

[74] Wang J., Xu T., Huang X., Li H., Ma T. 'Recent progress of silicon composites as anode materials for secondary batteries'. *RSC Advances*. 2016;**6**(90):87778–90.

[75] Wang A., Kadam S., Li H., Shi S., Qi Y. 'Review on modeling of the anode solid electrolyte interphase (SEI) for lithium-ion batteries'. *npj Computational Materials*. 2018;**4**(1):15.

[76] Khriachtchev L., Ossicini S., Iacona F., Gourbilleau F. 'Photonic properties of silicon-based materials'. *International Journal of Photoenergy*. 2012;**2012**:2–2.

[77] Ma Y., Balbuena P.B. 'Dft study of reduction mechanisms of ethylene carbonate and fluoroethylene carbonate on Li^+-adsorbed Si clusters'. *Journal of the Electrochemical Society*. 2014;**161**(8):E3097–109.

[78] Hankins K., Soto F.A., Balbuena P.B. 'Insights into the Li intercalation and SEI formation on LiSi nanoclusters'. *Journal of the Electrochemical Society*. 2017;**164**(11):E3457–64.

[79] Hautier G., Jain A., Ong S.P., *et al.* 'Phosphates as lithium-ion battery cathodes: an evaluation based on high-throughput *ab Initio* calculations'. *Chemistry of Materials*. 2011;**23**(15):3495–508.

[80] Jain A., Hautier G., Moore C.J., *et al.* 'A high-throughput infrastructure for density functional theory calculations'. *Computational Materials Science*. 2011;**50**(8):2295–310.

[81] Martinez de la Hoz J.M., Leung K., Balbuena P.B. 'Reduction mechanisms of ethylene carbonate on Si anodes of lithium-ion batteries: effects of degree of lithiation and nature of exposed surface'. *ACS Applied Materials & Interfaces*. 2013;**5**(24):13457–65.

[82] Martínez de la Hoz J.M., Balbuena P.B. 'Reduction mechanisms of additives on Si anodes of Li-ion batteries'. *Physical Chemistry Chemical Physics*. 2014;**16**(32):17091–8.

[83] Leung K., Rempe S.B., Foster M.E., *et al.* 'Modeling electrochemical decomposition of fluoroethylene carbonate on silicon anode surfaces in lithium ion batteries'. *Journal of the Electrochemical Society*. 2014;**161**(3):A213–21.

[84] Shkrob I.A., Wishart J.F., Abraham D.P. 'What makes fluoroethylene carbonate different?' *The Journal of Physical Chemistry C*. 2015;**119**(27):14954–64.

[85] Martinez de la Hoz J.M., Soto F.A., Balbuena P.B. 'Effect of the electrolyte composition on sei reactions at Si anodes of Li-ion batteries'. *The Journal of Physical Chemistry C*. 2015;**119**(13):7060–8.

[86] Soto F.A., Martinez de la Hoz J.M., Seminario J.M., Balbuena P.B. 'Modeling solid-electrolyte interfacial phenomena in silicon anodes'. *Current Opinion in Chemical Engineering*. 2016;**13**:179–85.

[87] Ramos-Sanchez G., Soto F.A., Martinez de la Hoz J.M., *et al.* 'Computational studies of interfacial reactions at anode materials: initial stages of the solid-electrolyte-interphase layer formation'. *Journal of Electrochemical Energy Conversion and Storage*. 2016;**13**(3).

[88] Michan A.L., Divitini G., Pell A.J., Leskes M., Ducati C., Grey C.P. 'Solid electrolyte interphase growth and capacity loss in silicon electrodes'. *Journal of the American Chemical Society*. 2016;**138**(25):7918–31.

[89] Benitez L., Seminario J.M. 'Electron transport and electrolyte reduction in the solid-electrolyte interphase of rechargeable lithium ion batteries with silicon anodes'. *The Journal of Physical Chemistry C*. 2016;**120**(32):17978–88.

[90] Soto F.A., Ma Y., Martinez de la Hoz J.M., Seminario J.M., Balbuena P.B. 'Formation and growth mechanisms of Solid-Electrolyte interphase layers in Rechargeable batteries'. *Chemistry of Materials*. 2015;**27**(23):7990–8000.

[91] Kim S.-P., Duin A.C.Tvan., Shenoy V.B. 'Effect of electrolytes on the structure and evolution of the solid electrolyte interphase (SEI) in Li-ion batteries: a molecular dynamics study'. *Journal of Power Sources*. 2011;**196**(20):8590–7.

[92] Hou C., Han J., Liu P., *et al*. 'Operando observations of sei film evolution by mass-sensitive scanning transmission electron microscopy'. *Advanced Energy Materials*. 2019;**9**(45):1902675.

[93] Leung K., Soto F., Hankins K., Balbuena P.B., Harrison K.L. 'Stability of solid electrolyte interphase components on lithium metal and reactive anode material surfaces'. *The Journal of Physical Chemistry C*. 2016;**120**(12):6302–13.

[94] Zhuo Z., Lu P., Delacourt C., *et al*. 'Breathing and oscillating growth of solid-electrolyte-interphase upon electrochemical cycling'. *Chemical Communications*. 2018;**54**(7):814–7.

[95] Soto F.A., Balbuena P.B. 'Elucidating oligomer-surface and oligomer-oligomer interactions at a lithiated silicon surface'. *Electrochimica Acta*. 2016;**220**:312–21.

[96] Horowitz Y., Han H.-L., Soto F.A., Ralston W.T., Balbuena P.B., Somorjai G.A. 'Fluoroethylene carbonate as a directing agent in amorphous silicon Anodes: electrolyte interface structure probed by sum frequency vibrational spectroscopy and ab initio molecular dynamics'. *Nano Letters*. 2018;**18**(2):1145–51.

[97] Perez-Beltran S., Ramírez-Caballero G.E., Balbuena P.B. 'First-principles calculations of lithiation of a hydroxylated surface of amorphous silicon dioxide'. *The Journal of Physical Chemistry C*. 2015;**119**(29):16424–31.

[98] Perez-Beltran S., Ramírez-Caballero G.E., Balbuena P.B. 'Ethylene carbonate reduction on lithiated surfaces of hydroxylated amorphous silicon dioxide'. *Journal of the Electrochemical Society*. 2016;**163**(10):A2197–202.

[99] Kim D.Y., Stefanoski S., Kurakevych O.O., Strobel T.A. 'Synthesis of an open-framework allotrope of silicon'. *Nature Materials*. 2015;**14**(2):169–73.

[100] Hwang J.-Y., Myung S.-T., Sun Y.-K. 'Sodium-ion batteries: present and future'. *Chemical Society Reviews*. 2017;**46**(12):3529–614.

[101] Arrieta U., Katcho N.A., Arcelus O., Carrasco J. 'First-principles study of sodium intercalation in crystalline NaxSi24 ($0 \leq x \leq 4$) as anode material for Na-ion batteries'. *Scientific Reports*. 2017;**7**(1):5350–5350.

[102] Kresse G., Furthmüller J. 'Efficiency of ab-initio total energy calculations for metals and semiconductors using a plane-wave basis set'. *Computational Materials Science*. 1996;**6**(1):15–50.

[103] Kresse G., Joubert D. 'From ultrasoft pseudopotentials to the projector augmented-wave method'. *Physical Review B*. 1999;**59**(3):1758–75.

[104] Kresse G., Hafner J. '*Ab initio* molecular-dynamics simulation of the liquid-metal–amorphous-semiconductor transition in germanium'. *Physical Review B*. 1994;**49**(20):14251–69.

[105] Kresse G., Hafner J. '*Ab initio* molecular dynamics for liquid metals'. *Physical Review B*. 1993;**47**(1):558–61.

[106] Henkelman G., Uberuaga B.P., Jónsson H. 'A climbing image nudged elastic band method for finding saddle points and minimum energy paths'. *The Journal of Chemical Physics*. 2000;**113**(22):9901–4.

[107] He Y., Lu X., Kim D.Y. 'A first-principles study on Si_{24} as an anode material for rechargeable batteries'. *RSC Advances*. 2018;**8**(36):20228–33.

[108] Marzouk A., Balbuena P.B., El-Mellouhi F. 'Open framework allotropes of silicon: potential anode materials for Na and Li-ion batteries'. *Electrochimica Acta*. 2016;**207**(2):301–7.

[109] Marzouk A., Soto F.A., Burgos J.C., Balbuena P.B., El-Mellouhi F. 'Dynamics of the lithiation and sodiation of silicon allotropes: from the bulk to the surface'. *Journal of the Electrochemical Society*. 2017;**164**(7):A1644–50.

[110] Jaumann T., Balach J., Klose M., *et al.* 'SEI-component formation on sub 5 nm sized silicon nanoparticles in Li-ion batteries: the role of electrode preparation, FEC addition and binders'. *Physical Chemistry Chemical Physics*. 2015;**17**(38):24956–67.

[111] Luo L., Yang H., Yan P., *et al.* 'Surface-coating regulated lithiation kinetics and degradation in silicon nanowires for lithium ion battery'. *ACS Nano*. 2015;**9**(5):5559–66.

[112] Ma Y., Martinez de la Hoz J.M., Angarita I., *et al.* 'Structure and reactivity of alucone-coated films on Si and LixSiy surfaces'. *ACS Applied Materials & Interfaces*. 2015;**7**(22):11948–55.

[113] Gomez-Ballesteros J.L., Balbuena P.B. 'Reduction of electrolyte components on a coated Si anode of lithium-ion batteries'. *The Journal of Physical Chemistry Letters*. 2017;**8**(14):3404–8.

Chapter 7

Nanostructure silicon for Li-ion batteries

Jiangyan Wang[1] and Yi Cui[1,2]

7.1 Introduction

Rechargeable lithium-ion batteries (LIBs) have been widely used for portable electronics and electric vehicles, and they have shown great promise for large-scale stationary energy storage [1, 2]. The advancement of next-generation LIBs with lower cost and higher energy density greatly depends on the development of low-cost, energy-dense electrode materials [3–5]. Intense research has been devoted to developing higher-capacity cathode and anode materials. Sulfur- and oxygen-based materials have been intensely explored for the cathode side [6–9], while silicon (Si), tin, and Li are mainly studied for the anode side [10–12]. Among these materials, Si, as the second most abundant element in the earth, is cheap, nontoxic and most importantly possesses a theoretical capacity that is an order magnitude higher than the conventional graphite anode [13, 14]. Successful implementation of a Si anode will fundamentally enable high-energy Li battery and reduce the cell cost. Despite its high capacity, Si suffers from short cycling life, which is caused by its large volume change during lithiation/delithiation processes and the serious issues stemming from this volume change, such as unstable solid electrolyte interphase (SEI), particle fracture, and cracking of the electrode structure [15–17].

During the past decade, nanotechnology has been providing unparallel solutions to solve these challenges. First, by reducing the size of Si to the nanoscale in at least one dimension, the stress induced upon lithiation/delithiation can be effectively alleviated, which can even prevent particle fracture. In addition, nanostructures offer a facilitated ion/electron transport and thus enhance the rate capability. Following the initial insight of Si nanowires (NWs) [10], various Si nanostructures including nanoparticles (NPs) [18], nanorods (NRs) [19, 20], NWs [10, 21, 22], nanotubes (NTs) [23], and nanosheets (NSs) [24, 25] have been developed and demonstrated an improved cycling life compared to their bulk counterparts. Second, a new effective concept combining pre-reserved void space to buffer volume expansion with a

[1]Department of Materials Science and Engineering, Stanford University, Stanford, CA 94305, USA
[2]Stanford Institute for Materials and Energy Sciences, SLAC National Accelerator Laboratory, Menlo Park, CA, USA

stable shell to stabilize SEI layer was pioneered by a Si–SiO_x double-walled NTs [26] and further confirmed by Si core–C shell composites [27, 28]. Apart from that, nanostructured Si is embedded in conductive matrixes including graphene [29], carbon NTs [30], or conductive polymer [31] to effectively enhance the electrical conductivity, limit the volume change of Si, and stabilize the SEI to some extent.

This chapter aims to review the synthesis of Si-based nanostructures and highlight the crucial effects of nanotechnology on the electrochemical properties of Si-based anode materials. Limited by space, representative examples of rational nanostructure design for Si-based anode are selectively discussed. Nanostructured Si materials with different dimensions and morphologies are discussed to reflect the power of nanotechnology in tackling the particle pulverization problem and enhancing rate capability. Subsequently, we summarize the efforts of constructing nanostructured Si-based composite to improve electronic conductivity, optimize electrode-electrolyte contact area, and stabilize SEI formation. In a commercialized battery, Si-based anodes are coupled with specific cathodes, wherein optimization terms such as the capacity ratio are taken into intimate consideration. Thus, we then discuss recent progress achieved in full cells using Si-based materials as anode. Finally, we provide some guidelines for the future design of nanostructured Si anodes to further improve electrochemical performance and better satisfy the practical requirements of the battery application.

7.2 Opportunities and challenges for Si anodes

Si material has been considered as one of the most promising candidates for next-generation high-energy anodes. Table 7.1 compares the properties of different anode materials. Li metal anode has a high specific capacity. However, its high chemical and electrochemical activity induces severe SEI formation, moreover, dendritic Li growth could cause an internal short circuit, which results in serious safety concern.

Table 7.1 Physical and electrochemical properties of various anode materials

Materials	**Li**	**C**	**P**	**$Li_4Ti_5O_{12}$**	**Si**
Density (g/cm^3)	0.53	2.25	1.82	3.5	2.3
Lithiated phase	Li	LiC_6	Li_3P	$Li_7Ti_5O_{12}$	$Li_{15}Si_4$
Theoretical gravimetric capacity (mAh/g)	3 862	372	2 596	175	3 579
Theoretical volumetric capacity (mAh/cm^3)	2 047	810	3 790	633	4 330
Volume change (%)	100	12	300	1	280
Potential versus Li (V)	0	0.05	0.7	1.6	0.4
Conductivity (S/m)	1.1×10^7	1×10^5	1×10^7	1×10^{-11}	1×10^3

Note: Theoretical volumetric capacity is calculated by (theoretical gravimetric capacity) * (density of lithiated phase).

Although the carbon material is the dominant anode for the state-of-the-art LIB, its low specific capacity suspends it from next-generation LIBs. Si has stood up as the best replacement for carbon-based anodes because it provides a theoretical capacity (4200 mAh/g, $Li_{22}Si_5$ at 415 °C; 3579 mAh/g, $Li_{15}Si_4$ at room temperature) 10 times higher than that of carbon anodes. In addition to its natural abundance and environmental-kindness, Si has an average operating potential at 0.4 V (vs. Li^+/Li), which not only avoids the Li metal plating at a high current density diffusion-induced stress. The volum but also offers a relatively high energy density for full cells [32].

Accompanying with the high capacity, Si suffers from the huge volume expansion upon lithiation- and diffusion-induced stress. The volume expansion is about 420 percent when Si is fully lithiated to $Li_{22}Si_5$, while the stress reaches ~GPa level [33]. This undesirable volume change causes unstable and accumulated SEI, particle fracture, and electrode cracking, which results in a poor capacity retention and limits the practical commercialization of Si anodes [34]. Based on previous studies, four main challenges are outlined here.

(1) Extreme volume expansion and particle fracture

The huge volume expansion/contraction upon lithiation/delithiation induces large mechanical stress, which leads to the cracking and pulverization of the Si particles. This phenomenon causes the loss of electrical contact between Si fragments and electrical isolation of Si fragments from the current collector. As a result, the capacity decays dramatically over a number of cycles, especially for bulk Si, Si films, and large Si particles (Figure 7.1a).

(2) Electrode-level fracture and volume change

The repeated huge volume change also causes the pulverization of the whole electrode structure. Si particles expand and impinge on each other upon lithiation, while contract during delithiation, which can lead to their detachment from the surrounding electrical contacts. Besides, the expansion of Si active materials induces a large increase in electrode thickness, which causes big challenge for the cell design to retain structural integrity and good electrical contact under big change. In addition, the drastic expansion and contraction of the whole Si electrode could also lead

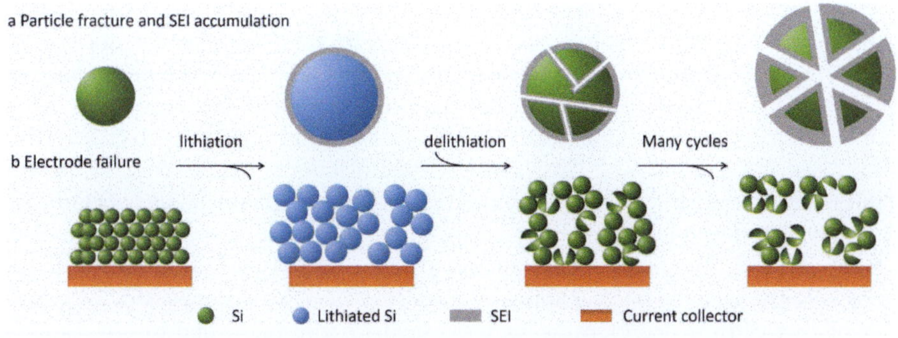

Figure 7.1 *Si anode failure mechanism. (a) Particle-level fracture and SEI accumulation. (b) Electrode-level cracking. Reproduced from [34].*

to the electrode peel-off and failure, which further contributes to the capacity decay (Figure 7.1b).

(3) Unstable SEI

Organic electrolyte spontaneously starts to decompose at the electrode surface when the voltage of the anode goes below ~1 V (vs. Li^+/Li). The decomposition products form a SEI layer at the electrolyte-electrode interface. The SEI film on Si surface is mainly composed of Li_xSiO_y, Li_2O, LiF, Li_2CO_3, organic carbonates, and nonconductive polymers [35–37]. The SEI layer should be ionically conducting to allow the penetration of Li ions and also electronically insulting to prevent further side reactions, thus SEI layer stops growing at a certain thickness. However, the repeated volume expansion and contraction induces a migrating electrode-electrolyte interface as well as breaks the SEI layer. Moreover, the fracture of Si particle creates new active surface for SEI formation, which leads to the continuous SEI formation (Figure 7.1a). The accumulation of thick SEI layer leads to the drastic capacity fading, which is due to the following facts: first, the continuous SEI formation indicates the ever-increasing consumption of Li ions and electrolyte, which causes increasing resistance; second, the passivation effect of SEI layer could also increase electrode impedance/polarization; third, the thick SEI layer prolongs the Li-ion diffusion path and could even electrically isolate Si material from the current collector.

(4) Low intrinsic electronic and ionic conductivity

Besides, compared to Li or carbon anode, Si has a relatively low intrinsic electronic conductivity (Table 7.1), which could increase orders magnitude higher after lithiation. In addition, the Li diffusion coefficient in Si is also low, which is between 10^{-14} and 10^{-10} cm^2/s, with an average of 10^{-12} cm^2/s [38]. These intrinsic properties also hinder the realization of full capacity and limit the rate capability of Si anodes.

7.3 Synthesis and nanoeffect of nanostructured Si anodes

Nanostructured Si materials, with the size within nanoscale in at least one dimension, have the potential to tackle the challenges of Si anodes and create new opportunities for LIBs. The main advantages of nanostructured Si anode include: (1) Nanoscale dimensions allow quick relaxation of stress, making nanostructures more resistant to fracture than bulk counterparts. As revealed by scanning electron microscopy (SEM) and in situ transmission electron microscopy (TEM) observations combined with simulation, crystal Si particle with a diameter less than 200 nm [39], wires or pillars with a diameter smaller than 300 nm, and films with a thickness below 100 nm can resist fracture [40–42]; (2) Nanostructured electrodes can be engineered with hollow or porous structures to provide free space to accommodate expansion during alloying, thus minimizing material stress and electrode pulverization to achieve better reversibility and cycling stability; (3) The use of nanoscale morphologies may also enhance the rate capability and specific capacity. As expected, the capacity of Si electrode decreases at high charge/discharge rates, due to low Li-ionic conductivity in Si and sluggish mass transfer at the electrode interface. By reducing electrode dimensionality to increase the surface-to-volume ratio and provide shorter Li diffusion distances, electrode polarization may be reduced, allowing high capacities to be realized. Nanoscaled Si structures with diverse morphologies have been explored as anodes for LIBs, which can be morphologically classified to four groups: zero-dimensional (0D), one-dimensional (1D), two-dimensional (2D) and three-dimensional (3D).

7.3.1 0D Si nanostructure

Bare micron-sized Si particle anodes could lose over 90 percent of the initial capacity within 5 cycles [15]. Nanosized Si particle significantly decreases the electron and Li-ion diffusion path, more importantly is able to undergo large volumetric strains without mechanical fracture, thus enabling reversible cycling with moderate high capacity. Kim *et al.* produced well-dispersed Si NPs using reverse micelle that consisted of $SiCl_4$ source and trimethyloctadecylammonium bromide surfactant, at high pressure and temperature in a bomb [43]. The Si NPs with an average size of 10 nm delivered a high initial delithiation capacity of 3 380 mAh/g and a high-capacity retention of 81 percent after 40 cycles. Liu *et al.* discovered a strong size dependence of fracture for crystal Si particles through in situ transmission electron microscopy: the critical size for fracture is estimated to be about 150 nm, below which the particles neither cracked nor fractured upon first lithiation, and above which the particles fractured due to lithiation-induced swelling [39]. Based on these understanding, stable reversible capacity was achieved in Si-based anodes using Si NPs with diameter below 150 nm, despite different shapes.

In addition to solid Si NPs, porous NPs have also been reported to extend the cycling life. Ge *et al.* report the synthesis of porous boron-doped Si NPs through an electroless etching method [18]. Solid Si NPs less than 200 nm were first doped with boron, then etched in an etchant solution with silver nitrate and hydrofluoric acid to produce porous Si NPs. Interior nanopores can provide additional room for volume change, but the effective Li-ion pathway was not significantly reduced because of its smaller initial size. When coupled with reduced graphene oxide (rGO), capacities around 1 400 mAh/g achieved at a current rate of 1 A/g, and 1 000 mAh/g achieved at 2 A/g, and stable operation tested over up to 200 cycles.

Moreover, Si hollow spheres (HSs)-based anodes have also demonstrated good LIB performances. The inner cavity can well accommodate the volume expansion and relieve the diffusion-induced stress during cycling, additionally, the thin shell of Si HSs can significantly shorten the Li^+-ion/electron diffusion distance. Yao *et al.* reported the synthesis of interconnected Si HSs through a chemical vapor deposition (CVD) method using SiH_4 as Si-source and SiO_2 spheres as hard templates [44]. The Si HSs with a thin shell thickness of about 25 nm achieved an initial reversible capacity of 2 725 mAh/g at a rate of 0.1 C and a discharge capacity of 1420 mAh/g after 700 cycles at 0.5 C. To avoid the toxic SiH_4 and make it more feasible for mass production, Chen *et al.* reported a scalable method for the preparation of mesoporous Si HSs through the magnesiothermic reduction of mesoporous silica HSs [45]. Both the interior void and the porosity in the shell can effectively cushion the volume expansion during alloying and enable the Si host a better Li accessibility.

With an average diameter and shell thickness of approximately 120 and 20 nm, respectively, these Si HSs displayed a much better electrochemical performance than commercial Si NPs.

Despite the significantly enhanced cycling stability compared to the bulk Si materials, the cycling lifetime of 0D Si nanostructure-based anode is still limited, which may be attributed to a combination of delamination of the particulate electrode films upon repeated expansion/contraction, severe irreversible side reactions consuming the electrolyte, and poor electrical connection between 0D Si nanostructure and the current collector [46]. The large surface area of 0D Si nanostructure resulting from the reduced dimensions, induces a severe irreversible capacity loss at the initial cycle, thus leading to a low initial coulombic efficiency (ICE). Moreover, the repeated expansion/contraction upon alloying/dealloying leads to the unstable SEI layers; continuous electrolyte decomposition occurs at the cracked sites, consequently, continuous growth of SEI layers which accumulate at the surface forming an electrically insulating layer that prohibits electron transport.

To solve the above problems, diverse engineering methods including secondary architecture assembly, surface coating and electrically conductive glue connection, have been investigated. Liu *et al.* reported a pomegranate micro-sized secondary architecture assembled by primary Si NPs [27]. The specific surface area of the secondary architecture is significantly decreased compared to the primary Si NPs, 90 m^2/g for the NP while only 15 m^2/g for 1 μm secondary architecture and 1.5 m^2/g for 10 μm secondary architecture. Therefore, the SEI formation is greatly decreased benefited from the reduced surface area of the secondary architecture, thus effectively improving the ICE. Besides, surface coating with carbon, oxide, or conductive polymer has also been proved to effectively improve the ICE and stabilize the SEI layer. Liu *et al.* designed a yolk–shell structure with 100 nm Si NP as the yolk and 5–10 nm carbon as the shell [47]. The carbon shell is both electronically and ionic conductive, with few pinholes to allow electrolyte penetration, thus limiting the SEI formation to the outside of the carbon shell. The void space between Si NP yolk and carbon shell allows volume expansion upon lithiation. The robust and self-supporting carbon framework guarantees a static stable SEI and prevents rupturing and reforming of the SEI, resulting in a high CE reaching 99.84 percent.

In addition, great efforts have also been devoted to developing novel methods for electrically connecting Si NPs together and with the current collector. Cui *et al.* deposited amorphous Si onto Si NP electrode films through a CVD method [48]. The amorphous Si acts as an inorganic glue to fuse the NPs together as well as bind them onto the current collector (Figure 7.2a and b), thus effectively solving the issue of the poor electrical contact in Si NP anodes prepared through conventional slurry methods. With a limited charge capacity of 1 200 mAh/g, the prepared 200 nm Si NP anode showed stable cycling up to 130 cycles, as shown in Figure 7.2c. In another study, Hu *et al.* used the CVD-deposited inorganic glue to physically and electrically anchor Si NPs to conductive Si NWs, significantly improving the areal capacity from 0.6 to 2.5 mAh/cm^2 without using any binder (Figure 7.2d–f) [49].

Another way to improve the electrochemical cycling performance of Si NP electrodes is to explore new polymer binder materials. Song *et al.* found that compared

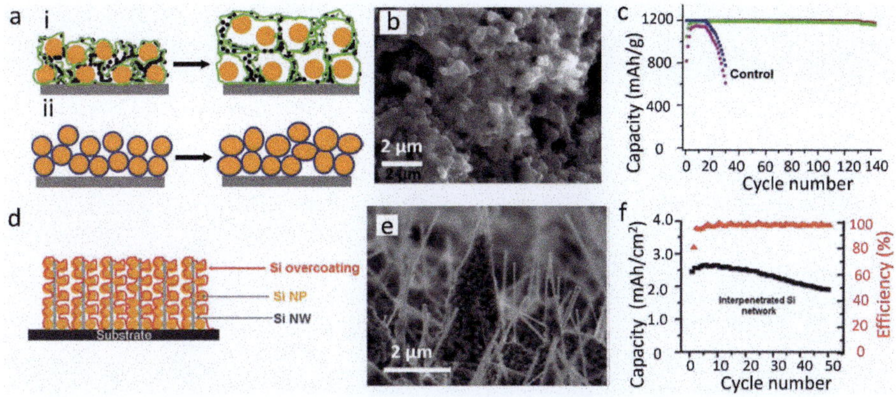

Figure 7.2 Amorphous Si glue fuses the 0D NPs together and binds them with the current collector. (a) Illustration showing the morphology change of Si NP electrode films before and after cycling test: (i) Si NP electrode film prepared by conventional slurry method, (ii) Si NP fused together and bond onto the current collector by amorphous Si glue (the blue rings). (b) SEM images of Si NP electrode films prepared by glue method. (c) The cycling life of Si NP electrode prepared by amorphous Si glue method. (d) Scheme showing the network of Si NPs-decorated Si NWs. (e) Cross-section SEM image, and (f) Cycling performance of Si NPs-decorated Si NWs. (a–c) reproduced from [48]. (d–f) reproduced from [49].

to other binders such as polyvinylidene fluoride (PVDF) and sodium carboxymethyl cellulose (NaCMC), polyacrylic acid-polyvinyl alcohol (PAA-PVA) is more suitable as an interpenetrated gel polymer matrix for commercial Si NPs [50]. This is due to that PAA-PVA binder provided a better electrode integrity and sustainable electric network, achieving a high CE even at a high rate of 4 A/g and a high areal capacity of ~4.3 mAh/cm^2 with good cycling stability. In another study, Yushin *et al.* showed that mixing Si NPs with alginate, a natural polysaccharide extracted from brown algae, yields stable Si NP anodes [51]. Liu *et al.* developed a new poly (9, 9-dioctylfluorene-co-fluorenoneco-methylbenzoic acid) (PFFOMB) binder, which played dual functionalities providing both improved electronic conductivity and robust mechanical binding forces, thus simultaneously maintaining electrical connectivity and accommodating the Si volume change [52]. Si NP electrodes based on p-type polyaniline (PANi) and PVDF binders showed low capacity and poor capacity retention due to the insulating nature of the polymer binder, while the Si/AB/PVDF, and Si/PFFO suffered from fast capacity fading due to mechanical failure during cycling. On the contrary, without adding any conductive additive, electrodes based on Si NPs and the PFFOMB binder maintained a high capacity of 2 100 mAh/g for Si at a current density of 4.2 A/g after 650 cycles. These applications

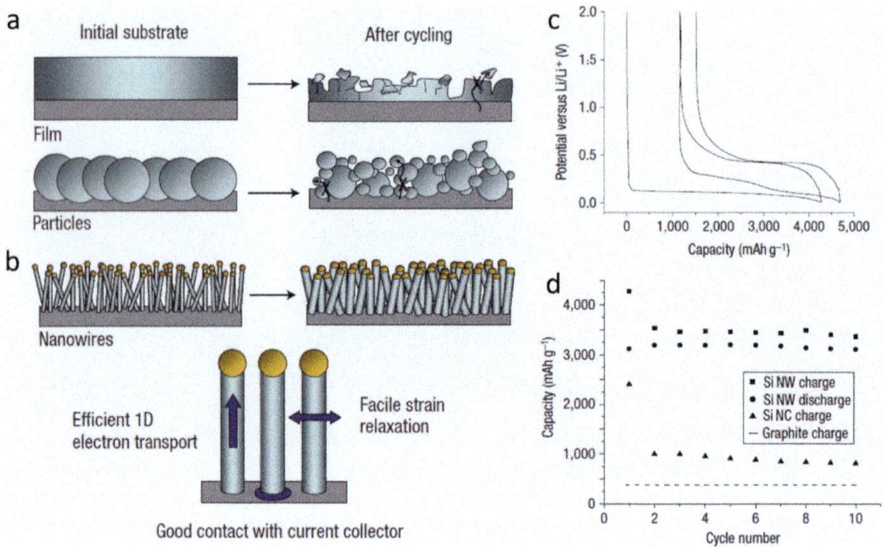

Figure 7.3 *1D Si NWs as anode materials for LIBs. Scheme illustrating that (a) the Si films and large particles pulverize during cycling, while (b) Si NWs grown directly on the current collector do not break during cycling. (c) The voltage profiles for the first and second cycles of the Si NW-based electrode. (d) Cycling stability of the Si NWs compared with those of the Si nanocrystals and graphite. Reproduced from [10].*

of new high-performance binders are low-cost and compatible with current slurry manufacturing and as such are very attractive.

7.3.2 1D Si nanostructure

1D Si nanostructures including NW, NT, NR, etc. possess many desired electrochemical features. As a pioneering work, Chan *et al.* reported the growth of Si NWs array on stainless steel through a vapor–liquid–solid (VLS) method at the end of 2007 [10]. These Si NWs-based anodes for LIBs exhibited much improved performance than the thick film and large particles, achieving nearly the theoretical lithiation capacity of Si and maintaining a delithiation capacity close to 75 percent of this maximum, with little fading over 10 cycles. The superior electrochemical performance is benefited from the following attractive features of Si NWs anodes as shown in Figure 7.3: (1) Each Si NW is bonded to the metal substrate, that's electrically connected to the current collector, thus enabling robust electrical contact, making conductive additives and binders free, and allowing for all NWs to contribute to the capacity; (2) The NW feature provides continuous 1D electronic pathways, which can enable much more efficient charge transport than that through hopping of electrons between particles in traditional slurry-coated battery electrodes; (3) The NW array provides sufficient empty space between adjacent NWs to accommodate the volume change associated with lithiation/delithiation process; (4) The Si NWs with an average diameter of only 89 nm are resistant to fracture. For small nanostructures, the stress-relief volume accompanying crack growth may be not significant enough to overcome the surface energy penalty associated with the crack growth. As a result, despite the large increase of diameter to 141 nm, the Si NWs remained intact and also remain in contact with the current collector, suggesting minimal capacity fade due to electrically disconnected material during cycling. Following this first demonstration of rational Si NW design, there were many other related studies. For example, Chan *et al.* reported a scalable supercritical fluid–liquid–solid method to prepare Si NWs. These solution-based Si NWs can be coated with carbon and demonstrated a reversible capacity of 1 500 mAh/g after 30 cycles [53]. Schulz *et al.* demonstrated that using Si_6H_{12} as the Si precursor in a polymer carrier, amorphous Si NWs anode could be synthesized by an electrospinning method followed with a thermal treatment process [54]. Peng *et al.* [55] and Xu *et al.* [56] reported Si NW anodes fabricated by metal-induced chemical etching of Si wafers. This electroless etching method has the potential for precise controlling the crystal orientation, the Si NWs retained the (100) orientation of the parent wafer with (110) sidewalls.

Si, NR as another typical 1D nanostructure, has also drawn extensive attention as anode materials for Li-ion batteries. Si NRs have been prepared through modified metal-assisted chemical etching technique of Si film [57] or oblique angle deposition [58] on copper foils. These Si NR-based anodes demonstrated a high reversible capacity and good capacity retention, which is due to the pre-reserved free vacancies between NRs to buffer the volume variation and relieve the stress during lithiation process. However, these synthesis methods involve harsh reaction conditions and tedious procedures. In 2014, Zhou *et al.* reported the synthesis of mesoporous Si NRs assembled by small nanocrystals by magnesiothermically reducing SiO_2 deposited on multiwalled carbon nanotubes (MWCNTs) templates, as shown in Figure 7.4a [59]. This synthesis method is comparatively eco-friendly and facile. Besides, due to the mesoporous feature and nanoscale size that enable a fast Li penetration and cushion the volume expansion, these Si NRs exhibited a much better electrochemical performance than the bulk Si and porous Si networks, a reversible capacity of approximately 1 038 mAh/g was still maintained after 170 cycles at a current density of 0.2 A/g (Figure 7.4c).

Compared to solid NW or NR, Si NTs as anode materials for LIBs possess other benefits, first, it provides additional space for volume expansion during lithiation, thus improving cycling stability; second, it enables both inner and outer walls of Si accessible to the electrolyte, thus shortening the Li diffusion length and enhance rate capability (Figure 7.4d). Park *et al.* reported the fabrication of Si NTs by reductively decomposing $SiCl_4$ precursor in an aluminum template [23]. These NTs showed a high reversible charge capacity of approximately 3 247 mAh/g with a high CE of 89 percent at a rate of 1 C, and a superior capacity retention even at a high rate of 5 C (15 A/g). Yoo *et al.* prepared Si NTs using electrospun polyacrylonitrile NWs as templates [60]. These Si NTs also displayed a significantly improved electrochemical performance compared to the commercial Si NPs and Si NWs, at a current

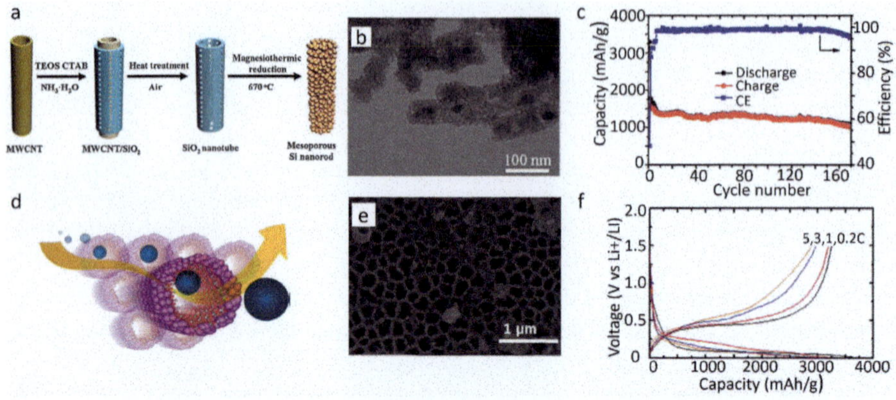

Figure 7.4 *1D Si NRs and NTs as anode materials for LIBs. (a) Scheme illustration of the synthesis process, (b) TEM image and (c) cycling life of mesoporous Si NRs. (d) Scheme showing the lithium-ion transport in Si NTs. (e) SEM image of Si NTs array. (f) Voltage profiles of Si NT-based anode at different rates from 0.2 to 5 C (15 A/g). (a–c) reproduced from [59]. (d–f) reproduced from [23].*

density of 0.4 mA/g, an initial discharge capacity of approximately 1 900 mAh/g was achieved and good capacity retention was also demonstrated.

To better maintain a good electrical connection and structural integrity, Cui's group demonstrated the first core–shell NW concept, where the core material is mechanically stable and the shell material is active Si for Li-storage [21]. Through a simple CVD growth using SiH_4 as Si source, crystalline–amorphous core–shell Si NWs grew directly on stainless steel current collectors (Figure 7.5). The transition metals in stainless steel, such as Ni, Fe, or Cr, are believed to catalyze the Si NW growth by a direct vapor–solid process. The thickness of amorphous shell increases linearly with growth time while the core radius remains unchanged, which indicates that crystalline core grows first then amorphous shell is coated onto the cores. By controlling the charging potential to remain above the amorphization potential of crystalline Si (0.12 V vs. Li^+/Li), only the amorphous Si shell is lithiated (at about 0.4 V), thus preserving the crystalline core as a mechanical support and an efficient electronic transport pathway. As a result, a high capacity of 1 060 mAh/g is achieved, with a 90 percent capacity retention over 100 cycles. In addition, this core–shell concept has been realized by using other core materials, such as carbon nanofibers (NFs), metal silicides, nitride, and carbon NTs, and so forth.

Although the pulverization issue can be effectively addressed by introducing the nanotubular structures, the SEI layer is still unstable due to the repeated volume expansion/extraction along with lithiation/delithiation process, as shown in Figure 7.6a. A robust coating/deposition at the active Si surface have shown to stabilize the SEI layer. Wu *et al.* proposed a novel double-walled Si–SiO_x NT (DWSiNT) structure to stabilize SEI layer [26]. An electrospun NF templating method combined

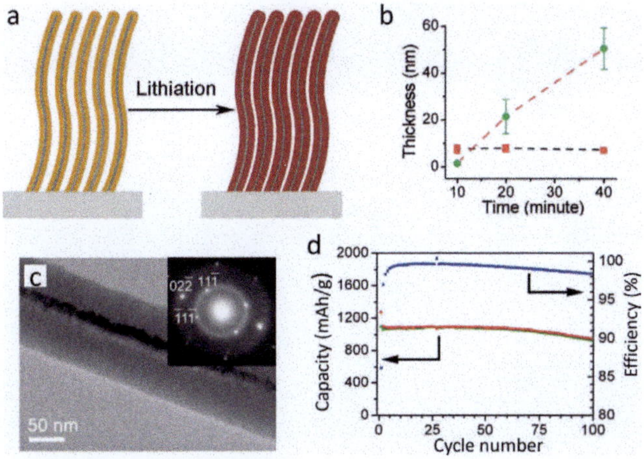

Figure 7.5 *Crystalline–amorphous core–shell Si NWs as anode materials for LIBs. (a) Scheme illustration of the lithiation of the core–shell Si NWs grown on a stainless-steel substrate. (b) Statistics of the core radius (red) and shell thickness (green) versus growth time. (c) TEM image of one Si core–shell NW inserted is the corresponding SAED image. (d) Cycling performance at a cutoff voltage of 0.15 V. Reproduced from [21].*

with a subsequent thermal oxidation treatment was adopted to realize the DWSiNT structures. These DWSiNT-based anode for Li batteries showed much better electrochemical performance than Si NTs without oxide and Si NWs, achieving high specific charge capacities, stable cycling up to 6 000 cycles with an 85 percent capacity retention, and high rate capability (up to 20 C). The highly improved electrochemical performance is attributed to the outside thin SiO_x coating layer: first, it is mechanically rigid and can limit Si to expand only inward while still allowing Li^+ to pass through; second, the electrolyte only wet the outside of the NTs due to the high aspect ratio of the continuous NTs, along with the rigid SiO_x layer, the electrolyte/electrode interface is stabilized, resulting in a stable SEI on the outer surface, which was proved by the SEM image of NTs after 2 000 cycles and impedance measurements for NTs after different number of cycles. In addition to SiO_x coating, other materials such as carbon, metal, polymer are also utilized to coat at the surface of 1D Si nanostructure to stabilize the SEI layer.

7.3.3 2D Si nanostructure

In early studies, 2D Si nanostructures were equated with thin film electrodes. The performance of thin film improves as film thickness decreases, which is consistent with the Griffith–Irwin relation [61] stating that the critical fracture stress increases as film thickness decreases:

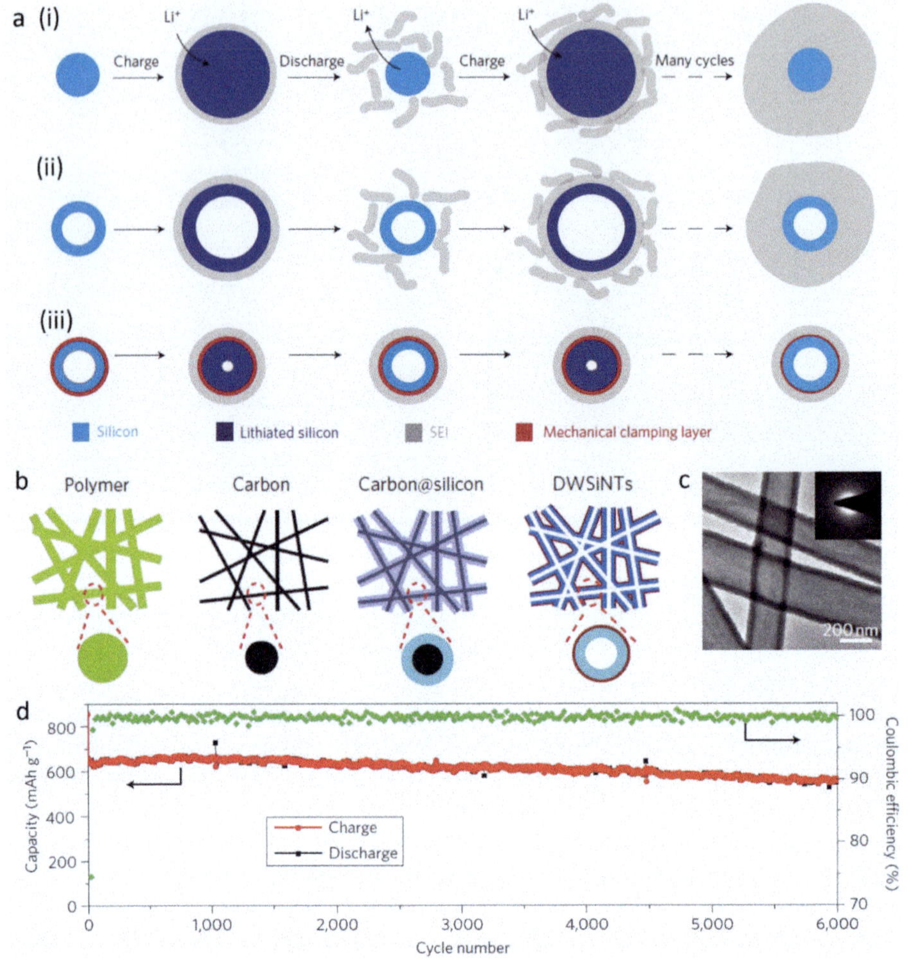

Figure 7.6 *DWSiNTs as anode materials for LIBs. (a) Scheme illustration of the SEI formation on silicon surfaces with different morphologies: (i) SEI on a solid Si NW can break down upon Si NW expansion; (ii) thick SEI formed at the outside of bare Si NT is also unstable; (iii) a thin and stable SEI formed on a double-walled Si NT with a mechanical constraining layer preventing Si from expanding outside, (b) Scheme showing the synthesis process, (c) TEM image, and (d) long-time cycling performance of DWSiNTs. Reproduced from [26].*

$$\sigma_{fracture} = \frac{K}{\sqrt{\pi d}} \tag{7.1}$$

where $\sigma_{fracture}$ is the critical fracture stress, K is the fracture toughness of the material, and d is the film thickness.

Although high first cycle delithiation capacities can be achieved by moderately thick Si films, the reversible capacity and cycling stability decrease as film thickness increases. Ohara *et al.* observed that Si films deposited on nickel foil through evaporation in vacuum exhibited a good reversibility and high reversible capacity when the film thickness was 50 nm or less, achieving a reversible capacity over 3 500 mAh/g maintained for 200 cycles under a rate of 2 C [62]. Comparatively, Si film with a thickness of 150 nm delivered a much lower capacity of 2 200 mAh/g even under a lower rate of 1 C. Maranchi *et al.* discovered that amorphous Si films deposited on copper foil through radio-frequency magnetron sputtering demonstrated a reversible capacity as high as 3 800 mAh/g over 29 cycles when the thickness is 250 nm [63]. In contrast, for a 1 mm thick Si film, the reversible capacity maintained at 3 000 mAh/g for only 13 cycles, followed by a quick decline. Despite the small film thickness, the capacity still decreases along with longer cycling, which may be due to the cracking of Si films. The volumetric expansion during lithiation of thin films occurs anisotropically. The expansion occurs perpendicular to the film surface, while contraction during delithiation occurs both perpendicular to and in plane with the substrate [64]. As a result, the film cracks after the lithiation/delithiation cycling, forming 1–3 μm diam islands. Despite the fracture, the active material can still be cycled since the electrical contact between Si active material and the current collector is retained.

Similar to carbon-based electrodes, doping Si with boron or phosphorus can enhance the reversibility and capacity; however, the enhancing mechanisms are different. As for carbon electrodes, the interlayer spacing is increased with dopant substitution, which facilitates the Li insertion and extraction and leads to a better cycling life. As for Si, since the poor reversibility of Si is ascribed from the large volume expansion that could not be altered by the low-concentration dopants, the enhanced performance in doped Si anode deserves an alternative mechanism. One explanation is that the ohmic polarizations reduce and the alloying of Si becomes more uniform after doping, thus resulting in a higher capacity in n-type Si films than that in the intrinsic Si film [46]. The intrinsic Si possesses a low electrical conductivity, while the Li-Si alloys have a much higher conductivity; therefore, it could be more favorable to further lithiate the lithiated regions than the fresh Si region. As a result, huge stress gradients could generate due to the large expansion gradients, which leads to the low capacity and fracture. As for the doped Si, both the Si before and after lithiation possess a good electrical conductivity, which favors the uniform lithiation and reduces stress gradients, thus achieving a higher capacity and better reversibility.

In addition to doping, coating the surface of Si film with polymerized fullerenes has also been proven to improve the electrochemical performance [65]. Moreover, the stabilizing effect of the fullerene layer increases along with the increase of polymerization degree, which could be controlled via adjusting the plasma power during the evaporation of the coating layer. The fullerene coating layer worked as an artificial SEI layer, stabilized the electrode-electrolyte interfacial contact, and prevented the Si film from pulverization. As revealed by SEM characterization, a severely fragmented surface was observed for the uncoated 300 nm thick Si film

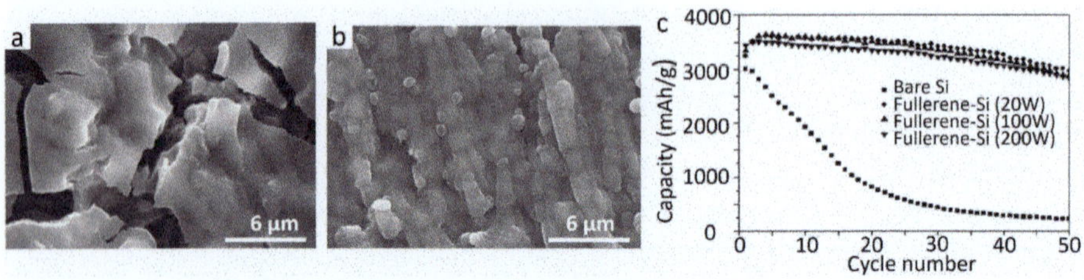

Figure 7.7 Polymerized fullerene coating enhances the electrochemical performance of 2D Si film. SEM images of the surfaces of (a) bare Si film, (b) fullerene-coated Si film, and (c) cycling performance of bared Si and fullerene-coated Si films under different plasma powers. Reproduced from [65].

after charge/discharge cycling, while no cracks were observed for the fullerene-coated Si film electrode with the same thickness (Figure 7.7a and b). Consequently, Si film anode without coating shows a rapid decrease of capacity. In contrast, the capacities of the fullerene-coated films decrease much more slowly, retaining capacities between 2 500 and 3 000 mAh/g after 50 cycles (Figure 7.7c).

Nevertheless, the thin 2D Si films suffer from some intrinsic limitations, mainly including low energy density ascribed from the low mass loading of active material and the other is the difficult scale-up fabrication. Mostly, the thin 2D Si films are produced through sputtering or CVD method in a space-limited chamber, which hinders the scale-up fabrication. In addition, to insurance a good cycling stability, the deposition thickness is only hundreds of nanometers, inevitably inducing a low areal energy density. In this regard, thin 2D Si films are prohibited from commercialization. Instead, other 2D-structrued Si materials, such as NSs and nanoflakes (NFs), which provide fast kinetics for charge carriers and Li-ion diffusion as well as effectively buffer the stress generated during the lithiation/delithiation processes, have been intensively explored as anodes for LIBs.

For instance, Ryu *et al.* prepared Si NSs from low-cost natural clay source via a one-step simultaneous molten-salt-induced exfoliation and reduction process [24]. The obtained ultrathin mesoporous Si NSs have an average dimension of ~2 μm and a thickness of ~5 nm. As a control, two-step process of pre-exfoliating and subsequent chemical reduction could not sustain original 2D structure and produced NPs, as shown in Figure 7.8. After coating with carbon, these Si NSs delivered a high reversible capacity of 865 mAh/g at a current density of 1 A/g, a good capacity retention of 92.3 percent after 500 successive cycles. Besides, the fast kinetics facilitated by the ultra-small thickness enabled an outstanding rate capability with 60 percent capacity retained at 20 A/g relative to that at 2 A/g. Another porous Si flakes were obtained in micrometer-sized particles through a similar approach using negative catalytic metal oxides [66]. The distinctive morphology of the porous Si flakes enables an impressive ICE of 92 percent and high later-cycle CEs of over 99.5 percent. Kim *et al.* also reported the synthesis of Si NSs from the most abundant

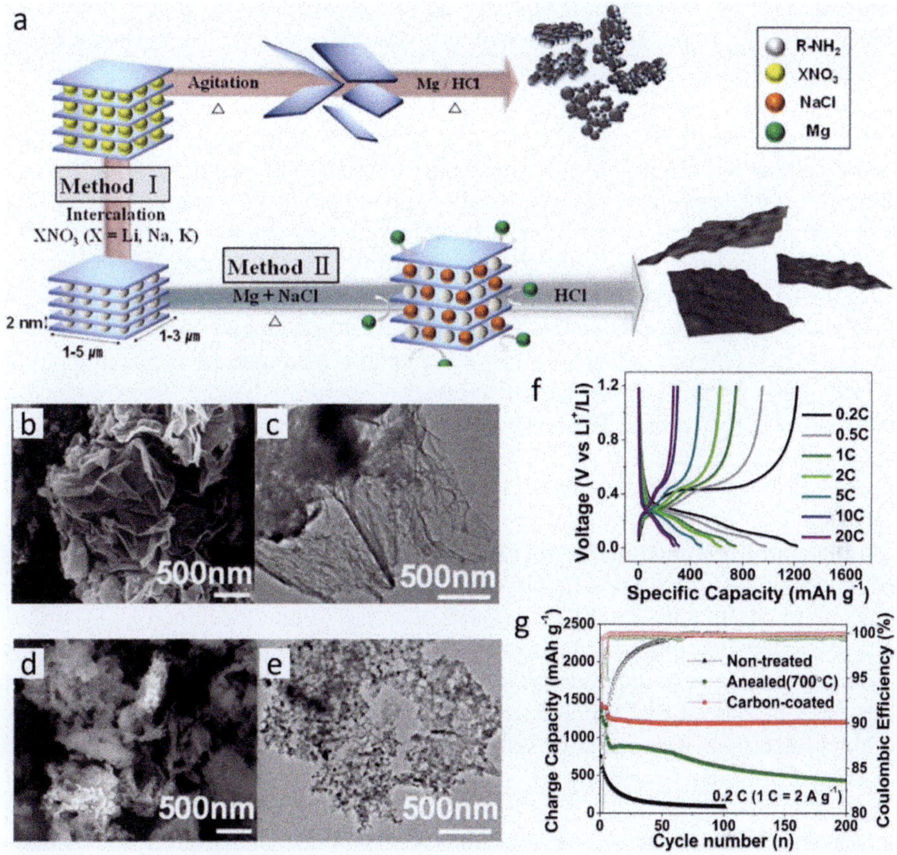

Figure 7.8 *Ultrathin Si NSs as anode materials for LIBs. (a) Schematic illustration of the synthesis of Si NSs through two methods, method I includes two steps, while method II is a one-step process. (b) SEM and (c) TEM images of Si NSs produced through method II. (d) SEM and (e) TEM images of Si NPs produced through method I. (f) Voltage curves at different rates of carbon-coated Si NSs anode. (g) Cycling performance of Si NSs anode. Reproduced from [24].*

Si source of sand via an easily scale-up synthetic method of reducing silica with Mg [67], which is promising for commercial manufacturing.

7.3.4 3D Si nanostructure

3D Si nanostructures mainly indicate the micrometer-sized secondary particles assembled with low-dimensional subunits. Due to the abundant variations in 3D structures, their Li-storage performance can be affected by their size, geometry, surface structure, and dimension, which has been extensively investigated in the past decade. Despite the polycrystalline nature of 3D-structured materials, anisotropic expansion upon Li-insertion can be observed in a single crystal Si subunit. Compared to their bulk or low-dimensional counterparts, 3D-structured materials possess many advantages as anode for Li-ion batteries: first, their larger specific surface area and higher surface to volume ratio than bulk particles, help to remedy large volume expansions, relieve particle pulverization, and avoid consequent delamination of electrode materials from the current collectors; second, 3D nanostructures provide open pores to enable the electrolytes easily diffuse into the interior part of Si anode, thus improve Li-ion transport and reduce the impedance originating from polarization of Li-ion concentration at the interface; third, compared to low-dimensional (0D, 1D, or 2D) materials, 3D-structured materials have a relatively better structural integrity and higher packing density thus able to realize a higher volumetric capacity and higher energy density, which are desirable for practical applications. It's worth noting that maintaining structural integrity and providing an effective electric pathway in 3D-structured Si anodes over long-time cycle life are quite challenging. Below, the fabrication of 3D Si nanostructures through various methods and the effect of structural characteristics on their Li-storage behaviors are discussed in detail.

Generally, there are three main challenges for nanomaterials, that's low ICE due to the severe SEI formation resulting from the large specific surface area, unstable SEI due to continuous change of electrode/electrolyte interface upon Li-insertion/desertion, and low volumetric capacity due to the low tap density. To tackle all these problems, Liu *et al.* proposed a 3D micro-sized secondary particle assembled with carbon-coated Si NP subunits and then further encapsulated by a carbon layer in micrometer-sized pouches to stabilize the electrode/electrolyte interface [27]. As a result, a high later-cycle CE of 99.87 percent was obtained benefited from the reduced electrode–electrolyte contact area. Although these nanostructured Si secondary clusters effectively reduce side reactions and increase the tap density, yet the scalability and tap density could still be further improved. Lin *et al.* proposed a mechanical approach for secondary microparticle fabrication, Si bulks were first made from NPs through high-pressure dry press and subsequently broken into micrometer-sized powder through high-energy ball milling [68]. This mechanical approach demonstrated an over 20 g/batch production of secondary microparticles with a high yield of 95 percent. However, the first cycle CE is still limited due to the large electrode/electrolyte contact area, with close to 80 percent obtained. Wang *et al.* proposed a surface-coating method to simultaneously improve the ICE and mechanical property of 3D secondary particles, a dense Si shell is deposited on secondary micrometer particles, each consisting of many Si NPs (Figure 7.9) [16]. The Si skin layer significantly improves mechanical stability, while the inner porous structure efficiently accommodates the volume expansion. Such a structure can resist a high pressure of over 100 MPa and is well-maintained after the calendering process, demonstrating a high volumetric capacity of 2 041 mAh/cm^3. In addition, the dense Si shell decreases the surface area and significantly increases the ICE to 90.4 percent.

In addition to bottom-up method to fabricate 3D Si nanostructure, electroless etching and electrochemical etching are also effective methods to prepare 3D

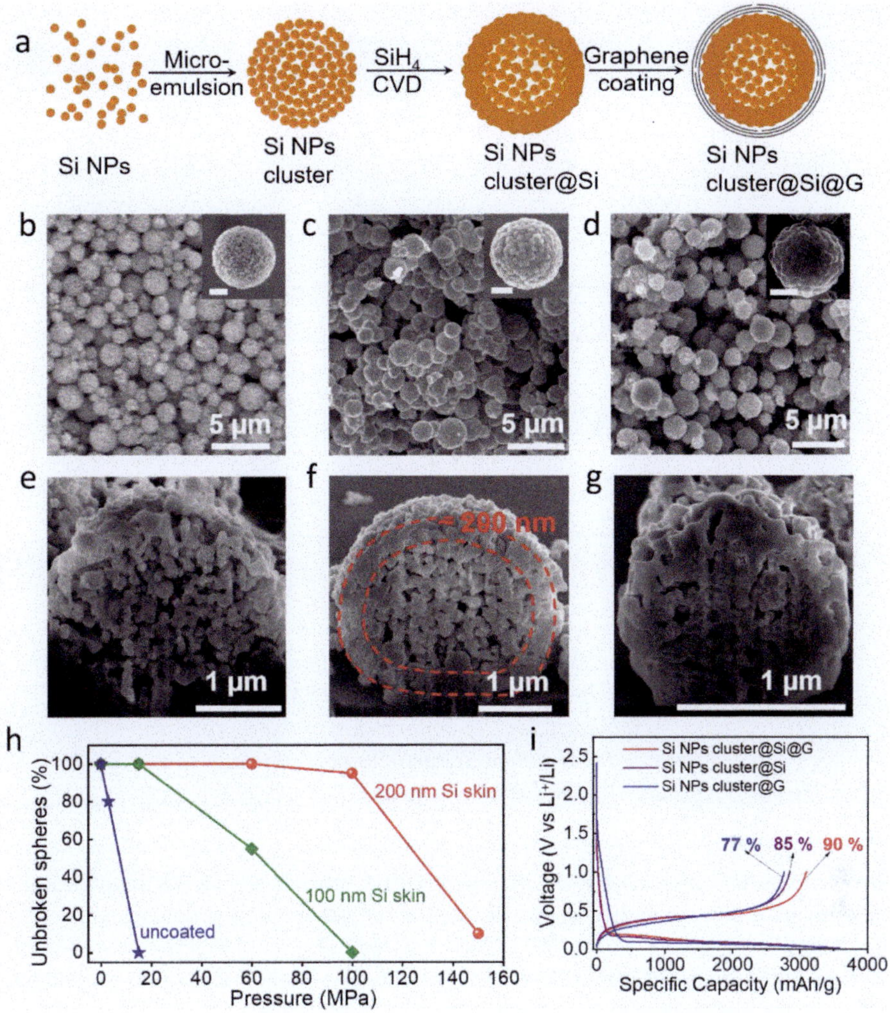

Figure 7.9 *Si skin-coated 3D Si secondary microparticles as high ICE and pressure-resistant anode materials for LIBs. (a) Schematic illustration of the synthesis of Si NPs cluster@Si@G. (b–d) SEM and (e–g) FIB-SEM images of (b,e) Si NPs cluster, (c,f) Si NPs cluster@Si, (d,g) Si NPs cluster@Si@G. Inset in images b–d is the corresponding SEM image of one single particle, scale bar is 1 μm. (h) Diagram of the fraction of unbroken spheres after being pressured as a function of applied pressure. (i) First-cycle voltage curves of individual cells with corresponding ICEs. Reproduced from [16].*

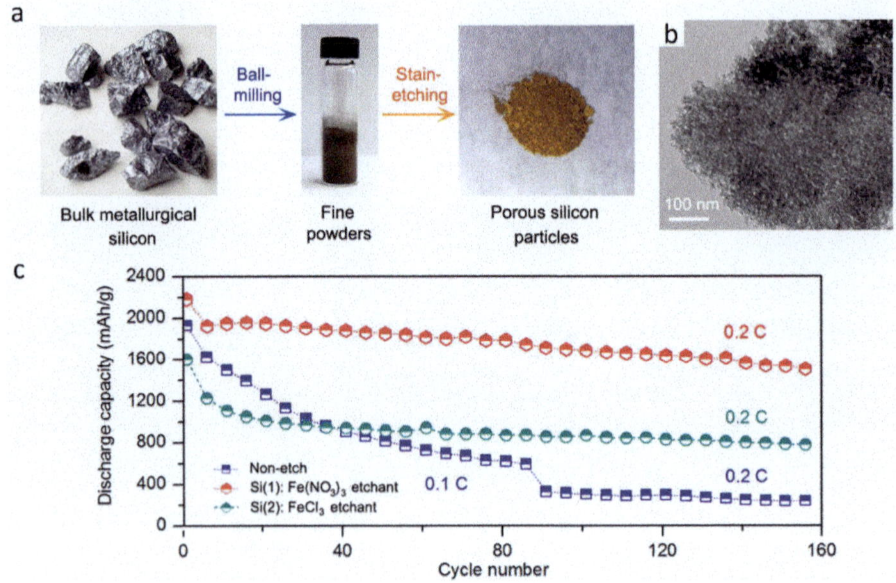

Figure 7.10 *3D micro-sized porous Si particles. (a) Si micro-sized particles prepared through combining ball milling and iron-assisted chemical etching bulk Si powder, (b) TEM image of porous Si micro-sized particles, (c) comparison of the cycling stability of unetched and etched Si particles at a current rate of 0.2 C. Reproduced from [70].*

porous microparticles. Bang *et al.* reported the fabrication of 3D macroporous bulk Si materials through a silver-assisted chemical etching process using commercial bulk Si powder source [69]. The prepared 3D porous Si structures have an average particle size of about 7 μm with hundreds-of-nanometer pores embedded inside. After further coated with carbon layer to increase the electrical conductivity, the carbon-coated 3D microporous bulk Si exhibited a high storage capacity of 2 050 mAh/g at a rate of 0.4 A/g with a remarkable CE of 94.4 percent. Ge *et al.* synthesized porous micro-sized Si particles from low-cost bulk metallurgical Si by combining ball milling and iron-assisted chemical etching (Figure 7.10) [70]. Abundant pores of around 3 nm were observed in the produced micro-sized Si with a particle size of around 1 μm. These porous Si-based LIB anode delivered a much enhanced performance than that of the dense bulk Si electrode, a reversible capacity of 2 900 mAh/g was retained at a rate of 0.4 A/g, and a stable capacity above 1 100 mAh/g was retained for extended 600 cycles at 2 A/g. Besides the electroless chemical etching method, Li *et al.* reported the preparation of porous Si through anodic electrochemical etching of crystal Si, the pore size can be controlled to be between 2 and several μm and the porosity can be controlled from 5 to 95 percent [71]. The as-prepared porous sponge-like Si shows irregular bricks or rocks with a dimension of over 20 μm, in which large pores of 50 nm are surrounded by thin walls with thickness of 10 nm. Benefited from the mesoporous sponge-like structure, the particle pulverization is effectively inhibited, realizing a capacity of up to 750 mAh/g based on the total electrode weight with 80 percent capacity retention over 1 000 cycles.

Besides bulk Si materials, other natural diatomaceous earth materials or well-structured SiO, SiO_2 have been adopted as Si source and/or template to fabricate 3D Si nanostructures. For example, Liu *et al.* using low-cost rice husks as raw material source and convert it to Si NPS through a first thermal calcination process followed with a magnesiothermic reduction process. Due to their small size (10–40 nm) and porous feature, these recovered Si porous structure exhibited a high battery performance, delivered a high reversible capacity of 2 790 mAh/g, corresponding to 86 percent capacity retention over 300 cycles [72]. In another example, 3D porous Si was synthesized by magnesiothermically reducing the diatomaceous earth, which works as both the Si source and the template [73]. In addition, Wang *et al.* fabricated mesoporous Si microparticles by thermal disproportionation of low-cost SiO microparticles and subsequent removal of the SiO_2 by-product [74]. Jia *et al.* reported a lotus-root-like 3D mesoporous Si by a magnesiothermically reducing SBA-15 silica, which serves as both template and Si source [75]. The obtained material demonstrates uniform mesopores within the fine particles composed of 30 nm-sized Si crystallites. After a thin carbon layer coating on the mesoporous Si to enhance the surface electric conductivity, the mesoporous Si@C composite displayed a stable capacity of 1 500 mAh/g for 100 cycles at 1 C and high-rate capability up to 15 C. In another case, Esmanski *et al.* fabricated 3D porous Si structure through a CVD method using SiO_2 hard templates. First, SiO_2 opal structure is prepared through evaporation-induced self-assembly of monodispersed SiO_2 spheres onto a substrate. Subsequently, hydrogenated amorphous Si was deposited in the interstices of SiO_2 spheres by dynamic cold-wall low-pressure CVD using Si_2H_6 as a Si source. Finally, porous Si was obtained by removing SiO_2 in hydrofluoric acid solution [76].

Above all, nanostructured Si with different dimensions and morphologies has their inherent characteristics. For instance, 0D materials have a relatively low ICE because the severe SEI formation as a result of the large surface area. As for the 1D structure, it shows promising electrochemical performances in many cases. However, most of them are obtained through elaborately designed techniques that are inapplicable for scale-up fabrication. As for the 2D structure, despite their superior electrochemical results, the ICE and swelling suppression still need to be further developed using fundamental studies. 3D structured Si could deliver a relatively higher ICE benefited from its micro-scaled particle size, thus it has been considered the most promising candidate for practical application. However, the cycling life and rate capability need to be further improved to satisfy the practical requirements. In summary, tremendous efforts should be continuously devoted to optimizing Si materials through nanostructuring techniques and surface modifications.

7.4 Synthesis and nanoeffect of nanostructured Si-based composite anodes

Apart from the quests for different dimensions, structures, and morphologies of Si materials, Si-based composites have also been intensively studied to overcome the problems of Si anodes resulting from the intrinsic properties of Si. As discussed above, Si anodes suffer from the significant volume change, which induces electrode structure cracking, loss of electrical contact, and unstable SEI formation; these issues cannot be all tackled by solely engineering Si dimensions, structures, and morphologies. Different conductive additives have proven to not only enhance the electrical conductivity of Si electrodes but also buffer the volume change of Si and stabilize the SEI layer to some extent. However, simply adding or mixing conductive agents with Si active materials cannot provide an intact and robust conductive matrix to buffer volume expansion and stabilize SEI formation, thus leading to a limited performance improvement. Well-engineered Si-conductive agent composites have proven to be able to provide a more efficient, conductive, and protective network; thus tremendous efforts have been devoted to the design and fabrication of Si-based composites. During the past decades, a variety of Si composites have been studied, including Si/C composites, Si/graphene, Si/CNT (carbon NT) composites, Si/conductive polymers, Si alloys, and Si/metal oxide composites.

7.4.1 Si/amorphous carbon composites

The carbon coating improves the electric conductivity and restricts the volume change of Si upon cycling. In addition, such carbon coating offers a rigid stable surface for SEI formation, thus stabilizing SEI layer and inhibiting continuous SEI growth upon repeated discharge/charge cycling. Various methods such as CVD by using toluene or propylene gas and pyrolysis of carbonaceous precursors have been employed to fabricate carbon coating on the surface of Si particles.

Commonly, Si is mixed with carbon-precursor through high-energy ball milling or through solution-based mixing, then pyrolysis is followed to carbonize the carbon-precursor. Si *et al.* studied the effects of both the size of Si particles and the carbon-precursor on the battery performance of their Si/C composites [77]. It was found that the composites with nanosized Si outperformed those with larger-sized Si particles. Besides, the chlorine content in the carbon-precursor significantly affects the yield, structure, and properties of carbon coating, thus resulting in varied electrode performance. Magasinski *et al.* prepared Si NPs/carbon composite granules by using a CVD method to drive hierarchical bottom-up assembly [78]. Carbon black NPs branched into short chains during high-temperature annealing, subsequently Si NPs were deposited onto the carbon black chains through a CVD method using SiH_4 gas as Si precursor. Finally, these multibranched Si-C composite was self-assembled into large porous spherical granules during another CVD process for carbon deposition. The sizes of the branches, Si NPs, and pores can be controlled during the CVD process for Si deposition, while the diameter of the composite can be controlled during the following CVD process for carbon coating. The size of the produced composite ranges from 15 to 35 μm and pore sizes ranges from 30 to 100 nm. The Si NPs/C composite particle is rather robust, remains a barely changed morphology after mixing, calendering, and cycling test. Benefited from the conductive carbon network, pore structure, and its exceptional robustness, the Si NPs/C composite delivered a high-reversible capacity of 1 950 mAh/g at 0.05 C and a capacity retention of 82 percent is maintained at 1 C after 100 cycles.

Beyond Si-C core–shell structure, Si/C composites have been advanced by introducing void space between Si and carbon, forming yolk–shell structures. Various Si dimensions and morphologies have been adopted in yolk–shell Si/C composites, combining the advantage of both versatile Si dimensions and the yolk–shell structures. Liu *et al.* designed a hierarchical pomegranate-like Si NPs/C composite through a bottom-up emulsion approach and a following polymer-pyrolysis method [27]. As shown in Figure 7.11, commercial Si NPs with a size around 80 nm were first coated with a SiO_2 layer then self-assembled to form secondary microparticle with a size ranging from 500 nm to 10 μm through an emulsion method. Subsequently, a resorcinol-formaldehyde resin layer was coated on the surface of the Si@SiO_2 NP clusters through a wet-chemical method, and then carbonized after annealing under inert gas. Finally, pomegranate-like Si NPs/C composites with controllable void space between the Si NPs and carbon coating was obtained after removing SiO_2 with hydrofluoric acid solution. Considering that a sufficient and well-defined void space is able to guarantee the structural integrity during repeated lithiation/delithiation, the gap size is accurately adjusted by controlling the thickness of SiO_2 sacrificial layer. Taking into account the size of the Si NPs and the expansion ratio of Si NPs after lithiation, the critical gap size is calculated and controlled to be 30–40 nm. Benefited from the robust carbon coating that can not only provide a conductive network but also limit the SEI mainly formed outside, as well as the well-defined void space that allows the Si NPs to expand without breaking the structural integrity of the secondary microparticles, Si NPs/C composite anode achieved a high later-cycle CE of 99.87 percent, a reversible capacity of 2 350 mAh/g at a rate of 0.05C, and maintained a capacity retention higher than 97 percent from the 2nd to 1 000th cycle at a rate of 0.05C. Similarly, Tao *et al.* also reported the fabrication of yolk–shell Si@void@C nanocomposites by first coating Si NPs with SiO_2, then further coating with carbon through CVD method, and finally removing SiO_2 with hydrofluoric acid solution [79]. Su *et al.* synthesized similar yolk–shell Si NPs/C nanocomposites via a similar resorcinol-formaldehyde resin coating and subsequent lithium hydroxide solution etching to generate void space [80]. All these yolk–shell Si@void@C composites showed much better electrochemical performance than raw Si particle or core–shell Si-C composites.

In addition, pores are generated in the carbon shell to enable a better access of Li ions to the interior Si cores. Yang *et al.* designed a yolk–shell Si–mesoporous carbon composite, in which both gap size and pores in the carbon shell are carefully controlled [81]. As shown in Figure 7.12, Si NPs were first encapsulated by a solid SiO_2 layer, followed by the coating of another mesoporous SiO_2 layer through a surfactant-templating sol-gel method. This mesoporous SiO_2 layer then works as template for mesoporous carbon layer growth via the pyrolysis of phenolic resin precursor at

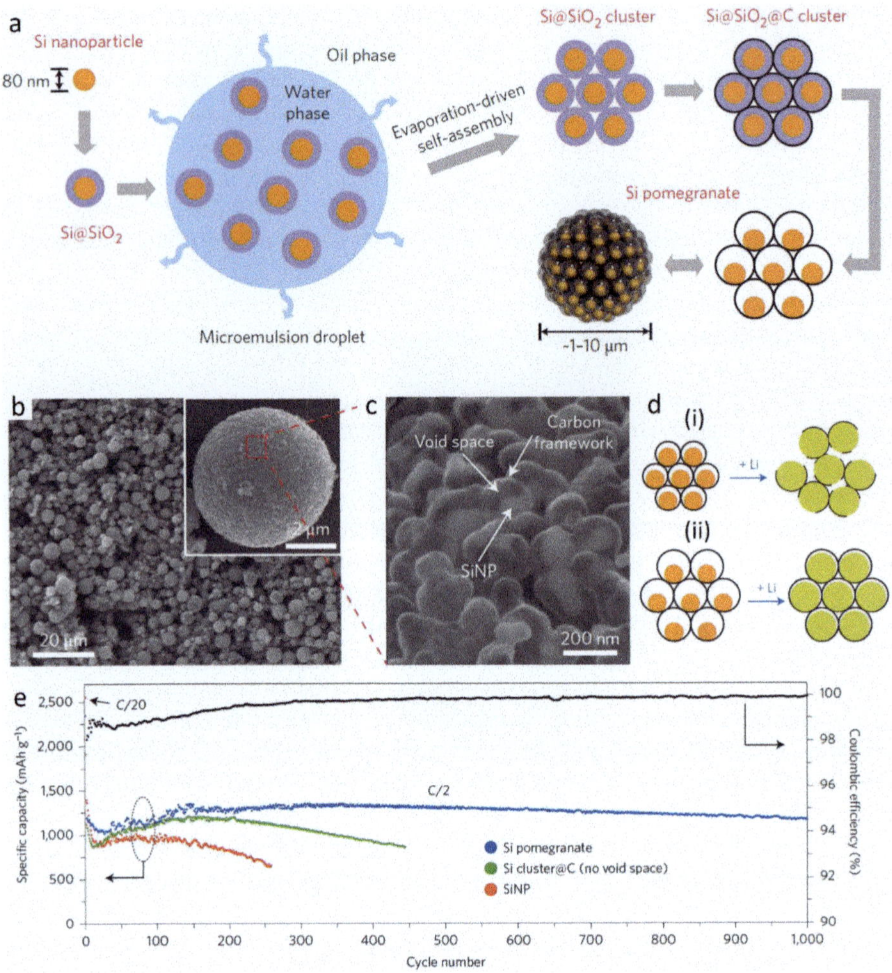

Figure 7.11 *Si NPs/C pomegranate composites as anode materials for LIBs. (a) Schematic illustration of the synthesis process and (b) SEM and (c) high-magnification SEM images of Si NPs/C pomegranate composite. (d) Schematic of the lithiation of Si pomegranates with (i) insufficient and (ii) sufficient void space between Si yolk and C shell. (e) Comparison of the cycling performance of Si-based anode with different structures. Reproduced from [27].*

high temperature. Void space between Si core and carbon shell and pores in carbon shell are generated after removing both the SiO_2 layers through hydrofluoric acid etching. The as-prepared structure possesses several merits as anode for LIBs: the sufficient void space allows Si expansion while maintaining the structural stability; the carbon-coating layer provides a conductive matrix for electrons while

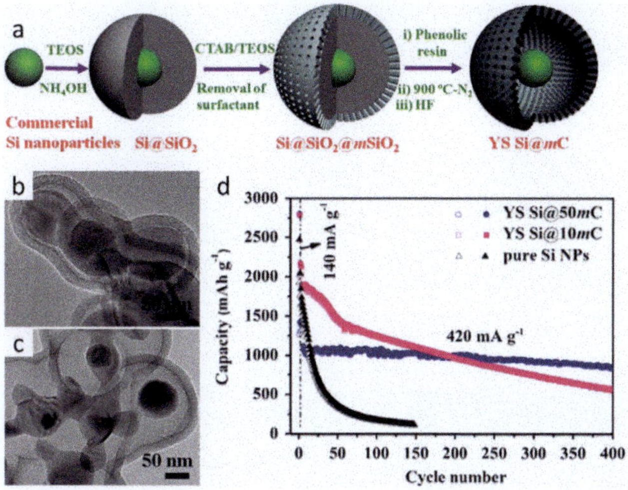

Figure 7.12 *Yolk–shell Si–mesoporous carbon composites as anode materials for LIBs. (a) Schematic illustration of the synthesis process and (b,c) TEM images of Si–mesoporous carbon composite with void space of (b) 10 nm (YS Si@10 mC) and (c) 50 nm (YS Si@50 mC). (d) Comparison of the cycling performance Si-based anode with different structures. Reproduced from [81].*

the inside pores work as Li-ion diffusion pathway. As a result, the yolk–shell Si–mesoporous carbon composite exhibited a long cycling life and a good rate capability, achieving a high reversible capacity of about 1 000 mAh/g after 400 cycles, and 62.3 percent capacity retention at a high current density of 8.4 A/g.

7.4.2 Si/graphitic carbon composites

In addition to amorphous carbon, graphitic carbon including carbon NF (CNF), CNT, graphene, etc., not only possesses the intrinsic advantages of carbon materials but also has a relatively better electrical conductivity and less irreversible trap of Li, thus has been considered as more attractive coating material to improve the electrochemical performance of Si anodes.

Carbon nanofibers. Encapsulation Si into CNFs has been proven to effectively improve the batter performance of Si anodes. Electrospinning is the most common approach for preparing Si/CNF composites. Li *et al.* fabricated Si/CNFs from electrospun Si/polyacrylonitrile precursor and subsequent carbonization treatment under argon gas [82]. Two different surfactants that's sodium dodecanoate (SD) and cetyl trimethyl ammonium bromide (CTAB) were investigated, which could improve the dispersion of Si NPs in polyacrylonitrile, thus enhancing the electrical contact between Si and carbon as well as helping to preserve the structural integrity. It was found that the discharge capacity of Si/CNFs after the addition of CTAB surfactant is similar to that without surfactant, while that of the Si/CNFs employing SD

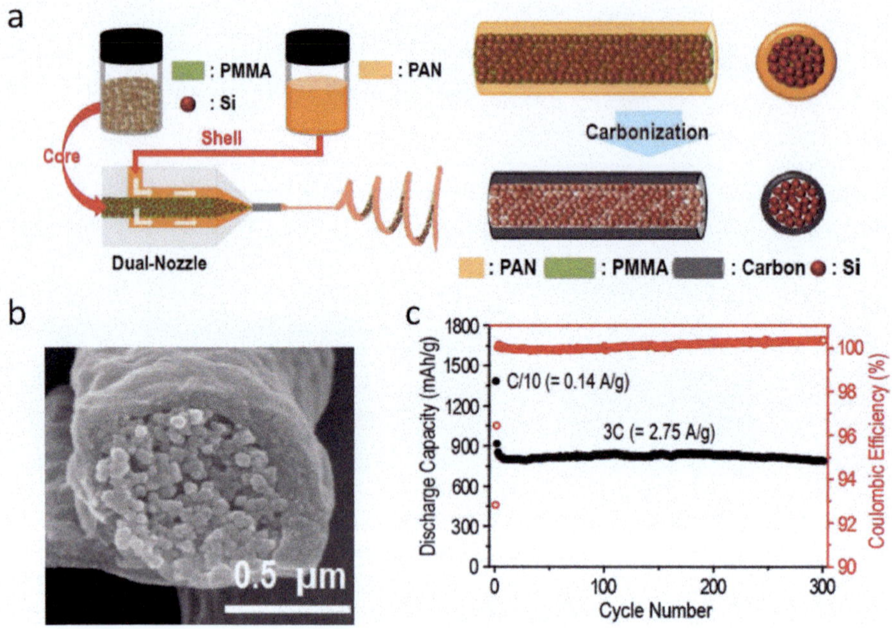

Figure 7.13 Si–carbon NF composites as anode materials for LIBs. (a) Schematic illustration of the synthesis process, (b) SEM image, and (c) cycling performance of core–shell Si NPs/CNFs composite through an electrospinning process using a dual nozzle. Reproduced from [84].

surfactant is 20 percent higher ascribed to the better dispersion of Si NPs. Zhou *et al.* confined Si NPs in porous CNFs (Si@PCNF) utilizing a single-nozzle electrospinning technique without adding any additive or surfactant and following annealing and etching [83]. Si NPs were first coated with SiO_x, the SiO_x-coating layer not only helps the homogenous dispersion in the solution for electrospinning but also functions as sacrificial template for producing void space between Si core and carbon shell. The highly conductive porous CNF sheath promotes the electron transport while the void space cushion the volume expansion of Si NPs during lithiation. As a consequence, this Si@PCNF hybrid delivered a good cycling life with 1 104 mAh/g maintained after 100 cycles at 0.5 A/g, and an excellent rate capability with 485 mAh/g still obtained at a high rate of 10 A/g. Hwang *et al.* developed a core–shell Si NPs/CNFs composite also through an electrospinning process using a dual nozzle (Figure 7.13) [84]. Si NPs and poly(methyl methacrylate) were dispersed in the cosolvent of N,N-dimethylformamide and acetone, and then injected into the core channel of the nozzle, simultaneously polyacrylonitrile/N,N-dimethylformamide solution was injected into the shell channel of the nozzle. After a subsequent carbonization treatment, the Si NPs/CNFs composite with a diameter of about 1 μm was obtained. The Si NPs were completely wrapped by a carbon shell that provides a continuous conductive network, protects the Si NPs from directly contacting the electrolyte thus reducing and stabilizing SEI formation, and alleviates the volume change during lithiation/delithiation processes. This Si NPs/CNFs composite delivered a high capacity of 1 384 mAh/g at 0.12 A/g based on the total weight of Si and C, a good cycling life with only 1 percent loss after 300 cycles at 2.75 A/g, and an impressive rate capability with 52.2 percent of the original capacity was still maintained when the rate increased from 0.12 to 9.75 A/g.

Carbon nanotubes. CNT is another emerging star. Due to their high-electrical conductivity, good mechanical property, and chemical stability, CNTs have been widely explored in various areas. The study of Si/CNT composites for LIB anodes started a decade ago with the boosting of the fabrication of Si/CNT composites. Physical mixing Si with CNTs is a simple and effective way to improve the conductivity of Si anode; however, it suffers from low homogeneity and poor consistency in cycling performance. Chemical bonding enables a much better connectivity between Si and CNTs; however, the C–Si bonds generate defects in the CNTs, which decreases the conductivity. During the past decade, both the deposition of Si on CNTs and CNTs on Si have been intensively studied.

Wang *et al.* prepared Si NPs/CNTs hybrid by using a template-free two-step CVD method [85]. First, the MWCNTs with a diameter of 40–50 nm grew on quartz microscope slides through a liquid injection-based CVD reactor using xylene as the hydrocarbon source and iron as the catalyst, which comes from the decomposition of ferrocene. Subsequently, Si NPs with a diameter around 40 nm grow on CNTs through a following CVD process using silane as the Si source. Finally, Si NPs/CNTs hybrid were scraped off from the quartz slides for the electrode preparation. The Si NPs/CNTs composite anode delivered a high reversible capacity and a good rate capability with 1 000 mAh/g still maintained at a high current density of 2.5 C. These good electrochemical performances are ascribed to the CNTs support, which provide a superior mechanical strength and great kinetic transport for both electrons and Li ions, besides, their tubular morphology and attractive tortuosity are able to maintain the Si NPs in contact despite the phase-related pulverization. Fan *et al.* reported another Si growth on CNTs core–shell composite [86]. As shown in Figure 7.14, vertically aligned CNTs core grow on Ni catalyst deposited stainless steel current collector via a plasma-enhanced CVD process. Then the cone-shaped Si shell with larger diameter on the top grows on CNT by using a Radio Frequency (RF) magnetron sputtering system. The length of the CNTs varies a little, while the in-between vacancy is uniform. The cone structure of Si shell is more beneficial than the uniform Si shell given the following two considerations: first, the Si shell on the top has a better contact with electrolyte and provides faster Li-ion kinetics, which enable better lithiation than that in the root; second, the volume expansion of Si on the top would avoid excessive strain accumulation at the root, thus improving the structural stability. As a result, the CNT/Si core–shell composite demonstrates a good cycling life with more than 90 percent remained over 100 cycles, and a good rate capability with 1 150 and 700 mAh/g retained at a high rate of 8 and 15 C, respectively. Another Si/MWCNTs nanocomposites are obtained through a grafting technique using the diazonium chemistry

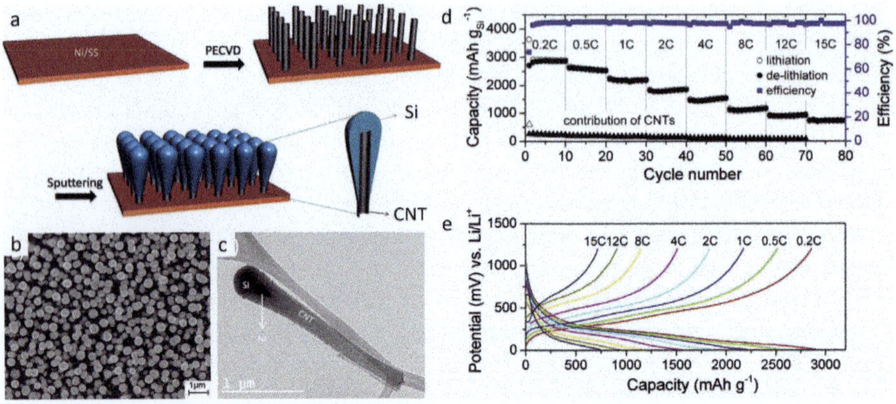

Figure 7.14 Si-CNT composites as anode materials for LIBs. (a) Schematic illustration of the synthesis process, (b) SEM images, (c) TEM images, (d) rate capability, and (e) voltage profiles at different rates of Si/CNT composite anode. Reproduced from [86].

[87]. The surface of MWCNT is first decorated with aminophenyl groups, then Si NPs are linked to the MWCNTs via the phenyl bridge through the in situ diazotation of the aminophynyl-modified MWCNTs with tert-butylnitrite. A capacity of 1 000 mAh/g was delivered by this Si NP/MWCNT composite after 35 cycles at a rate of 500 mA/g.

In addition to the deposition of Si on CNTs, study has also been explored on the deposition of CNTs on Si. Shu *et al.* reported the in situ growth of coiled CNTs on 1–2 μm Si particles through a CVD method [88]. The CNTs enwrap the Si particles tightly forming a cage-like Si/CNT composites, which achieved a much better battery performance than the mechanically mixed Si/CNTs composites. Besides, it was found that the cage-like Si/CNTs with longer CNTs exhibited a better cycling stability than the one with shorter CNTs. The tight wrapping of CNTs on the Si particles serves as highly conductive network and helps to stabilize the SEI layer. As a result, the first cycle CE is about 80 percent and the capacity retention after 20 cycles is 80 percent with 940 mAh/g remained.

Graphene nanosheets. Since 2004, graphene NSs have drawn tremendous attention and research interest due to their spectacular surface, mechanical, and electronic properties. Early as in 2010, Lee *et al.* first reported the adopting of graphene in Si NPs/graphene composites as anodes for Li-ion batteries [89], in which graphene NS was prepared by reducing GO that was first prepared through oxidation and exfoliation of graphite. Si NPs and GO were mixed together and dispersed uniformly in aqueous solution via sonication. Then paper-like Si NP/GO composite was obtained by filtering the above suspension with vacuum suction. The free-standing paper was used as electrode directly after thermally reducing GO with hydrogen gas. The paper-like composite achieves a good mechanical robustness and an enhanced electrical conductivity in both horizontally and vertically directions, due to the intimate wrapping of Si NPs by graphene and restacking of graphene sheets. As the pioneering work of Si/graphene composite, it delivered a much better electrochemical performance than bare Si NPs, showing a capacity retention of 50 percent after 300 cycles. Chabot *et al.* prepared Si NPs/graphene via a first freeze-drying treatment of the Si NPs and GO aqueous suspension and a following thermal reduction of GO process [90]. Zhou *et al.* reported the uniform dispersion and effective encapsulation of Si NPs by graphene via electrostatic attraction [29]. Si NPs are negatively charged due to the native SiO_x layer on the surface. To realize the self-assembly of Si NPs and GO, Si NPs are first anchored with polyelectrolyte poly(diallydimethylammonium chloride) (PDDA) to make surface positively charged, then assembled with negatively charged GO through electrostatic attraction. Subsequently, the Si NPs-graphene suspension was freeze-dried, thermally reduced, and followed by hydrofluoric acid treatment to remove native SiO_x layer. The as-prepared Si/graphene nanocomposite delivered a good cycling stability with 1205 mAh/g retained after 150 cycles at 100 mA/g, and a good rate capability showing a capacity of 990 mAh/g at a higher rate of 1600 mA/g. Apart from Si NPs, Si NWs was also anchored into graphene support to enhance the electrochemical property [91]. Hassan *et al.* reported one-pot growth of (111)-orientated Si NWs on nickel NP catalyst generated from nickel nitrate precursor. Nickel NP catalyst is anchored to graphene, which is simultaneously formed from reduced GO during the annealing process. The Si NWs-graphene composite delivered a high capacity of 550 mAh/g after 100 cycles at a high current rate of 6.8 A/g, creating a promising avenue for high rate LIBs.

In addition to the thermal reduction of Si/GO synthetic approach, Si/graphene composites have also been prepared via the combination of wet chemical polymer coating and a following metal-catalyzed graphitization process. Si NPs were dispersed in the mixture of sulfuric acid and ammonium fluoride aqueous solution to remove the native oxide layer and form hydrogen-terminated surface (Figure 7.15a–c) [92]. Then silver sacrificial layer was deposited onto the surface of Si NPs via the reaction effect of the surface hydride. Subsequently, the obtained Si@Ag nanocomposites were mixed with Iron(II) sulfate ($FeSO_4$) and glucose uniformly. Finally, Si@void@graphene composites were prepared by thermal graphitization of glucose and following removal of Ag layer. The obtained Si NP@graphene composite delivered a high capacity of 1410 mAh/g even after 600 cycles. Besides Si NPs, graphene cage was encapsulated on Si microparticles. Li *et al.* reported the highly conformal encapsulation of graphene on Si microparticles through a simple approach using a dual-purpose nickel template that works as both catalysis for low-temperature graphene growth and sacrificial layer for generating void space [15]. Si microparticles were first conformally coated with nickel (Ni) through an electroless deposition method. Then a benign carburization process on SiMP@Ni composite was followed by graphene growth at 450 °C through a dissolution-precipitation mechanism. Finally, SiMP@void@graphene yolk–shell composite was obtained after etching away Ni template. The graphene cage provides multiple advantages: first, regardless of the fracture of the SiMP during cycling, the mechanically strong and flexible graphene cage confines all broken Si pieces within and the integrity of the whole

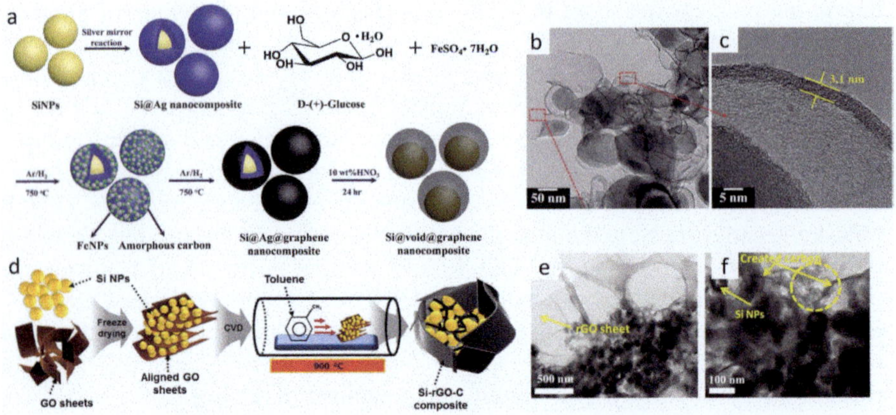

Figure 7.15 *Si-graphene composite materials prepared through different methods. (a, d) Schematic illustration of the synthesis process, (b, e, f) TEM images, and (c) HRTEM image of Si/graphene composite prepared via (a–c) combination of wet chemical polymer coating and a following metal-catalyzed graphitization process and (d–f) carbon network engineered Si/graphene composite. (a–c) Reproduced from [92]. (d–f) reproduced from [93].*

composite is well-preserved; second, the good electronic conductivity and ionic permeability through defects of graphene cage enable SiMPs high electrochemical activity; last, the SEI is mainly limited to the surface of graphene cage, inducing a thin and stabilized SEI layer. Benefited from these merits, the SiMP@graphene composite achieved a high first cycle CE of 93.2 percent, high later-cycle CEs as high as 99.9 percent, and good cycling stability in full cells.

Moreover, auxiliary agents have been integrated with Si/graphene composite to further improve the battery performance of Si anodes. Feng *et al.* engineered carbon network in Si/graphene composite by coating on Si NPs with carbon and interconnecting these NPs with graphene flakes through carbon rods (Figure 7.15d–f) [93]. Commercial Si NPs and GO were homogeneously dispersed in water and then freeze-dried, then the GO is reduced to graphene with simultaneously formed carbon coating and carbon branches using toluene precursor during the following CVD process. The additional carbon-coating layer provides extra protection from fracture and helps to stabilize SEI layer. Besides, the carbon rods create a macro-scale conductive network between Si NPs and graphene flakes. As a result, the as-produced Si NP/graphene/carbon composite achieved a better electrochemical performance than that of the regular Si NP/graphene composite, reaching a capacity of 894 mAh/g based on the total mass of the entire composite at 100 mA/g, and retaining a capacity retention of 94 percent after 300 cycles. Ji *et al.* reported another composite Si-C electrode by wrapping Si NPs with GO and then drop casting on to graphite foam [94]. After thermal reduction of GO, the Si/graphene-graphite composite electrode is used directly as anode for LIBs, which is able to deliver 983 mAh/g based on the total mass of the whole electrode.

7.4.3 Si/conductive polymer composites

Conductive polymers, such as polypyrrole (PPy) and polyaniline, have gained wide popularity for enhancing the electrochemical performance of Si anode. They possess many attractive features including good electrical conductivity, superior chemical stability, highly tunable morphologies, and high structural flexibility.

In 2005, Guo *et al.* reported the synthesis of Si/PPy composites through a simple high-energy ball milling method [95]. The obtained Si/PPy composites showed an enhanced electrochemical performance compared to the bare Si NPs. However, the PPy coating on Si NPs is not uniform through mechanical milling, which impacted the establishment of a continuous electron transport pathway and limited the improvement of battery performance. In 2013, through an in situ polymerization method, a well-connected 3D network structure composed with Si NPs conformally coated by conducting polyaniline was fabricated by Wu *et al.* (Figure 7.16a and b) [31]. Such a hierarchical structure has abundant merits: first, the porous hydrogel matrix provides void space for the volume expansion of the Si NPs during lithiation process; second, the PANi framework enables a continuous and conductive pathway for electrons, prevents the aggregation of Si NPs, and maintains the structural integrity during repeated charge/discharge cycles; third, each Si NP was conformally coated by PANi, which guarantees a good electric contact between Si NPs and the current collector, reduces the Si-electrolyte contact area, stabilizes the SEI. In addition, the Si NPs/PANi gel could be directly bladed onto a current collector without binder or conducting additives (Figure 7.16b–f). As a result, this Si NPs/PANi composite achieved a much better cycling stability than PVDF-bound Si or Si NPs-PANi mixture, maintaining a capacity of 1 600 mAh/g at a high current density of 1.0 A/g for 1 000 cycles, and delivering an impressive rate capability with a capacity varied from 2 500 to 1 100 mAh/g as current rate changed from 0.3 to 3.0 A/g (Figure 7.16g and h).

To further enhance the performance of Si anode, different nanostructures in addition to solid structure, including hollow structure and porous structure have been adopted in Si-conductive polymer composite. Du *et al.* fabricated porous Si hollow nanospheres through the magnesiothermic reduction of mesoporous SiO_2 hollow nanospheres, then coated with PPy [96]. The porous hollow Si structure possesses sufficient void space for volume expansion and maintains well after 250 cycles, while the PPy coating improves the electrical conductivity and limits the Si to expand mainly toward inside. Benefited from the synergistical improvement effect, the as-prepared hollow Si/PPy composite achieved a high initial lithiation capacity of 2 603 mAh/g and a good capacity retention of 88 percent against that at the second cycle after 250 cycles. In addition, different coating materials are combined with conductive polymer to further improve the battery performance. For instance, Zhang *et al.* prepared SiO_x microparticles with homogeneously embedded Si nanocrystals and nanopores through magnesiothermic reduction of SiO_2, then these SiO_x/Si

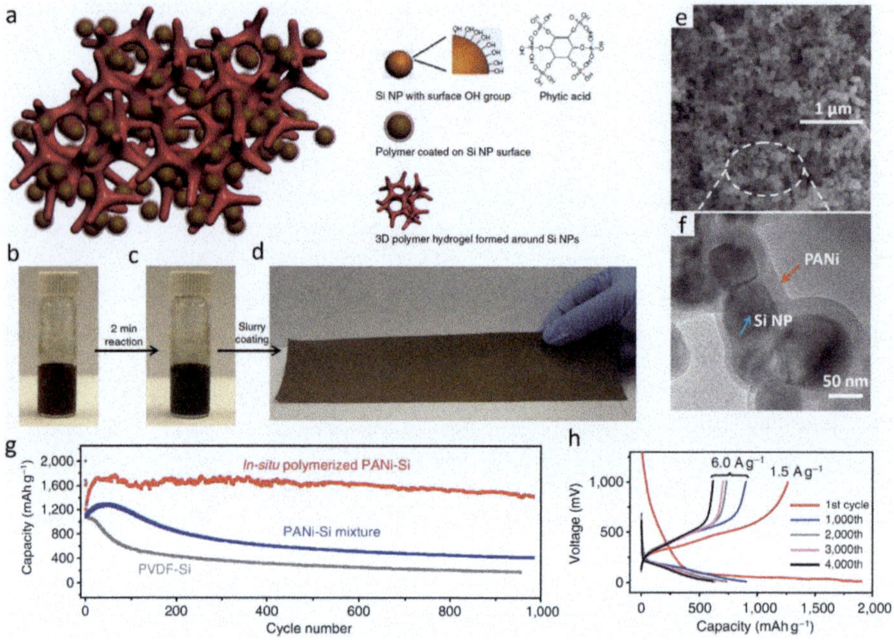

Figure 7.16 *Si NPs-PANi composites as anode materials for LIBs. (a) Schematic illustration of the synthesis process. (b–d) optical images showing the key steps of the electrode fabrication process, (b) Si NPs were dispersed in PANi, (c) chemical reaction induces the crosslinking between Si and PANi, (d) the bladed electrode. (e) TEM image and (f) HRTEM image of Si NPs-PANi composites. (g) Cycling stability and (h) voltage curves of Si NPs-PANi composite anode. Reproduced from [31].*

microparticles were further coated with PANi–Ag shell via in situ polymerization [97]. Profited from the porous structure that provides void space for volume expansion, and the highly electrical conductivity and flexibility of the PANi–Ag layers, the as-generated composite exhibited a high reversible capacity of 1 149 mAh/g after 100 cycles at a rate of 100 mA/g.

7.4.4 Si/metal or oxide composites

In addition to the incorporation of carbon and conducting polymer for Si anode, Si/metal and Si/oxide composites have also been exhaustively studied for LIB anodes. By surface modification with metal NPs, the electrochemical performance is effectively improved, which is profited from the following merits: first, the metal NPs offer electron percolation paths from the current collector to each individual Si NP, which effectively improves the rate capability; second, these coating layers can prevent the aggregation of Si NPs and reduce the Si-electrolyte contact area, thus

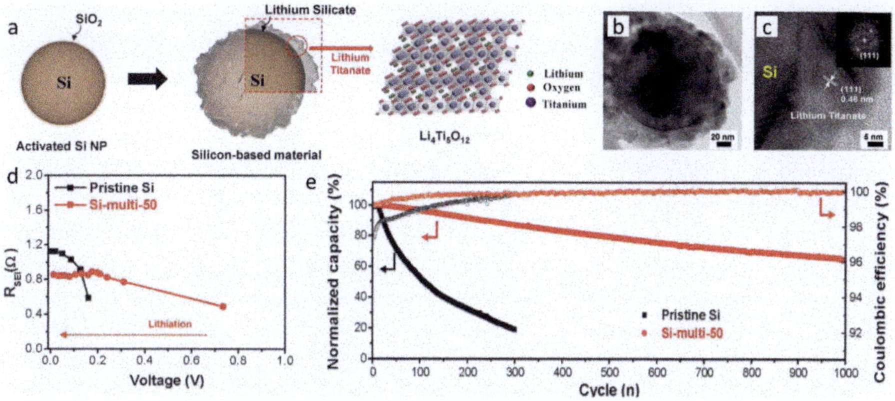

Figure 7.17 Si-lithium silicate-lithium titanate composites as anode materials for LIBs. (a) Schematic illustration of the synthesis process. (b) TEM image of one composite particle and (c) its magnification showing the crystalline phase. Comparison of (d) the resistance value and (e) the long-term cycling stability of the pristine Si and the composite. Reproduced from [99].

reducing SEI formation as well as stabilizing the SEI layer. For instance, Yoo *et al.* coated Si NPs with Ag NPs via a chemical reduction process using silver nitrate ($AgNO_3$) as an Ag source and butylamine as a reducing agent [98]. Due to the highly dispersive and stable Ag NP anchoring, the Si NPs/Ag composite achieved a much improved performance than that of the bare Si NPs. The first cycle CE increased from 76.9 to 80 percent after Ag NP coating, a reversible capacity of 1 320 mAh/g was still maintained for 100 cycles at a rate of 0.2 C, corresponding to nearly 100 percent capacity retention against that the second cycle, besides, a high-rate capability was achieved showing a capacity retention of 46.4 percent at a rate as high as 5 C.

Besides metal, oxide can also coat on the surface of Si and prevent the direct contact of Si with the electrolyte and stabilize the SEI layer. As we described in Section 7.3.1, SiO_x coating layer on Si NT can effectively improve the electrochemical performance by preventing the Si NT expand toward outside and allow for the development of a stable SEI layer. As another typical example, Lee *et al.* have coated Si NPs with a multifunctional layer that consists of lithium silicate (Li_2SiO_3 and $Li_2Si_2O_5$) and lithium titanate ($Li_4Ti_5O_{12}$) and demonstrated a significant improvement in performance (Figure 7.17) [99]. The Li-ion diffusivity in lithium titanate (8.36×10^{-11} cm^2/s) is a few times higher than that in conventional SEI layer formed on pristine Si NPs (1.56×10^{-11} cm^2/s), thus offering a more efficient pathway for Li diffusion. In addition, $Li_4Ti_5O_{12}$ is converted to the spinel $Li_7Ti_5O_{12}$ during lithiation process, while $Li_7Ti_5O_{12}$ has an electrical conductivity of 10^{-2} S/cm, thus significantly increasing the electrical conductivity. As for the lithium silicate coating layer, it is electrochemically stable in a wide voltage range, which not only maintains the structural integrity but also facilitates to form a stable SEI layer. As shown in Figure 7.17d, the resistance of the composite maintained a constant value, while that of the pristine Si significantly increased due to the unstable SEI layer. Benefited from the synergistic coupling of both multifunctional coating layers, the composite delivered a high capacity of 1 000 mAh/g even after 1 000 cycles at 1 C and also achieved an excellent rate capability with a capacity of 800 mAh/g still maintained at a high rate of 10 C, whereas the capacity of uncoated Si NPs severely declined at only 3 C.

Above all, it is worth mentioning that to fairly judge the promise of a material, many other factors should also be taken into consideration, especially on the practical aspects of new materials, which includes the cost, electrode and fabrication details, full-cell LIB data, etc.

7.5 Si-based nanostructure anodes for high-energy full cells

As discussed above, by optimizing the dimension, morphology, and structure of Si and incorporating with carbon, conductive polymer, metal, and oxide, Si-based anodes have delivered very promising performance in half-cells paired with Li foil reference electrode. Nevertheless, in a real commercialized battery, Si-based anodes are coupled with specific cathodes, typically Li metal oxide materials. To evaluate a full cell, many more optimization terms should be taken into intimate considerations, such as the capacity ratio of the anode to the cathode with the mass balance, the electrolyte amount, the exteriors, and some other process parameters, etc. Recently, impressive progress has been achieved in full cells using Si-based materials as anode. Several typical examples are discussed below.

7.5.1 Si-based anode paired with LMO cathode

According to the equation $E = \frac{1}{2}CV^2$, the energy density (E) is proportional to the cell capacity (C) and operating voltage (V). Therefore, by building a full cell with a large potential difference between the anodes and cathodes and/or utilizing high-capacity electrode materials can effectively improve the energy density of a full cell. Among various anode materials, Si has the lowest voltage after graphite-based anode (0.4 vs. 0.1 V) and an around 10 times higher capacity than that of graphite. However, the capacity of Li metal oxide (LMO) cathodes is far behind the requirements for extended applications. This background will ultimately drive the research on Si-based full cells coupled with appropriate cathodes to achieve a high energy density as well as other parameters.

Ko *et al.* demonstrated amorphous Si-nanolayer-embedded graphite/carbon (SGC) hybrids through an easily scaled-up CVD method [100]. The nanosized Si layer was deposited on the inner pores and outer shell of commercial graphite microparticles using silane as Si source through a CVD process, subsequently amorphous carbon is further coated at the surface of Si (Figure 7.18). The Si layer can significantly improve the total energy density of the hybrid anode. Its nanosized thickness not only allows the relief of induced stress but also facilitates Li-ion transport and reactivity, while inner pores of graphite particles can effectively buffer

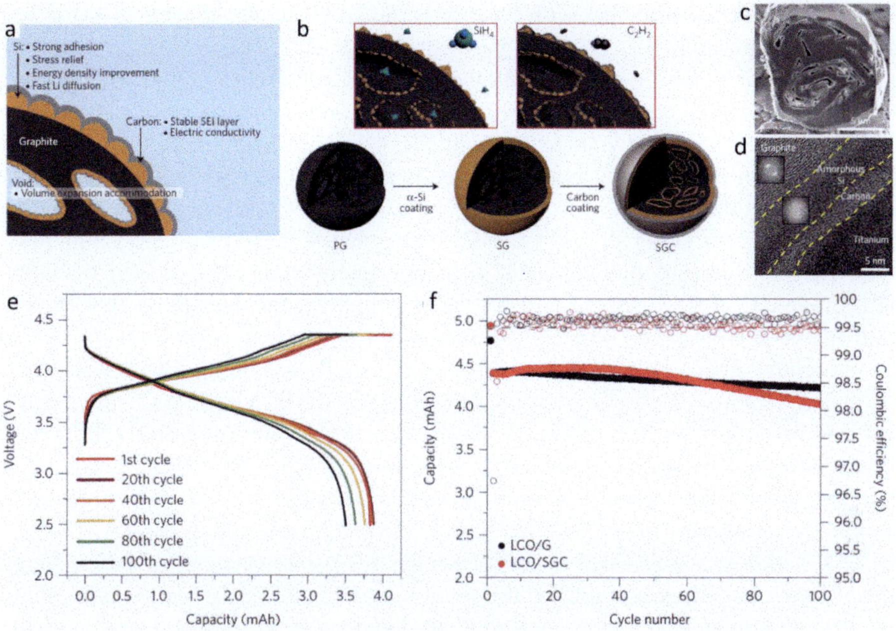

Figure 7.18 *SGC hybrids as anode materials for full cells paired with LCO cathode. (a) Cross-sectional schematic view showing the detailed structural characteristics of one SGC hybrid particle. (b) Schematic of the synthesis process. (c) SEM image of cross-sectional SGC hybrid particle. (d) HRTEM image at the interfacial region of the SGC hybrid with fast Fourier transform inset images. (e) Voltage curves of the LCO/SGC full cell. (f) Cycling stability up to 100 cycles for LCO/SGC and LCO/graphite full cells. Reproduced from [100].*

the volume expansion during lithiation. The carbon-coating layer enhances electric and ionic conductivity and stabilizes the SEI formation. Consequently, these SGC achieved a much better performance than that of graphite in full cells paired with lithium cobalt oxide (LCO) cathodes. The SGC/LCO full cell delivered an average voltage of ~3.77 V, offering a high energy density of 1 043 Wh/l with 92 percent maintained after 100 cycles, which is higher than 900 Wh/l for LCO/graphite.

Chae *et al.* prepared another Si/C composite internally wired with graphene sheets (Si/C-IWGS) and coupled with a novel Ni-rich cathode ($Li[Ni_{0.75}Co_{0.1}MnO_{0.15}O]_2$) to build a full cell [101]. The Si/C-IWGS composite is prepared through a simple and scalable sol-gel process, with graphene sheets highly dispersed in Si/C particle aggregates. The embedded graphene sheets generate enough void space for volume expansion and enhance the electrical conductivity, which results in a good cycling stability up to 200 cycles and a high capacity of 880 mAh/g at 100 mA/g for the Si/C-IWGS composite anode. The $Li[Ni_{0.75}Co_{0.1}MnO_{0.15}O]_2$ cathode is also structurally optimized with Ni-rich core and Ni-depleted shell showing an inside-out continuous concentration gradient, which also obtained a high capacity and good cycling stability. As a result, the Si/C-IWGS/$Li[Ni_{0.75}Co_{0.1}MnO_{0.15}O]_2$ full cell achieved a high energy density of 240 Wh/kg and a good cycling stability up to 750 cycles at 1 C in a voltage range of 2.7–4.2 V, corresponding to a capacity fading of 0.016 percent per cycle. Lee *et al.* prepared another Ni-rich cathode with even higher Ni ratio ($Li[Ni_{0.85}Co_{0.05}Mn_{0.10}]O_2$) and coupled with Si-embedded CNTs anode prepared through ball milling [102]. The pouch full cell exhibited a high first cycle CE of 94 percent, maintained 81 percent capacity after 500 cycles at 1 C with an average CE of 99.8 percent and delivered a high energy density of 348 Wh/kg with an average working voltage of 3.63 V.

Recently, Zhang *et al.* reported a new yolk–shell Si/C composite with a high tap density; the inside core is composed of multiple Si NPs and CNTs with embedded iron oxide (Fe_2O_3) NPs, while the shell is a dual layer of C and SiO_2 [103]. The inner void space between Si NPs, Fe_2O_3 NPs, and CNTs can effectively accommodate the volume expansion during lithiation, while the CNTs work as a "highway" for electron transport. When coupled with homemade triple-shelled hollow Li-rich $Li_2V_2O_5$ cathode, the full cell delivered a high reversible capacity of 260 mAh/g after 300 cycles. An *et al.* reported the synthesis of an ant-nest-like porous Si from magnesium-silicon alloy using a scalable top-down method [104]. The ant-nest-like porous Si consisted of interconnected Si nanoligaments and nanopores to accommodate volume expansion and could expand/shrink repeatedly without pulverization. When coupled with the $Li(Ni_{1/3}Co_{1/3}Mn_{1/3})O_2$ cathode, the full cell delivered a high energy density of 502 Wh/kg and a good capacity retention of 84 percent after 400 cycles.

7.5.2 Si-based anode paired with sulfur-based cathode

Sulfur or lithium sulfide cathode has a much higher capacity and is able to provide a much higher energy density than that of the Li metal oxide cathode. Compared to Li metal anodes that suffer from dendritic growth, high reactivity, and infinite volume change, Si-based anodes have shown a brighter future for commercial full cell constructed with sulfur or lithium sulfide as cathodes.

In 2010, Yang *et al.* constructed a full cell by coupling Si NW anodes with Li_2S/mesoporous carbon composite, this Li_2S/Si full cell yields a high theoretical energy density of 1 550 Wh/kg, which is about four times that of the conventional $LiCoO_2$/graphite battery (~410 Wh/kg) [105]. The thin Si NW anode could undergo repeated lithiation/delithiation processes without pulverization. As for the cathode, Li_2S was embedded in mesoporous carbon that provided electronic access to insulating Li_2S. The full cell delivered an initial discharge capacity of 482 mAh/g and an average discharge voltage of 1.7 V, corresponding to an energy density of 630 Wh/kg. Although the preliminary results of the full cell were not satisfactory, this battery system inspired the development of Li metal-free S/Si batteries. Later in 2004, Agostini *et al.* constructed another Li metal-free full cell that consisted of Li_2S-MCMB (mesophase carbon micro beads) composite cathode and electrode posited Si-based anode [106]. This full cell maintained a stable specific capacity of 280 mAh/g (with respect to Li_2S) and evolved with an average voltage of 1.4 V, resulting in a theoretical energy density of 400 Wh/kg. Balach *et al.* also coupled nanosized Li_2S-based cathode with Si-based anode for constructing full cells [107]. Nanosized Li_2S was produced through in-cell electrochemical full lithiation and irreversible decomposition of MoS_2 NPs. When coupled with nanostructured silicon/carbon composite anode, the Li_2S/Si full cell obtained a high initial capacity of about 780 mAh/g based on the sulfur mass.

In addition to Li_2S/Si full cell, Si anode is prelithiated to form Li_xSi alloy and then paired up with Li-free sulfur cathode. Pu *et al.* reported a full cell constructed with lithiated Si anode and sulfur cathode [108]. Mesoporous Si was prepared through magnesiothermical reduction of mesoporous silica and subsequently coated with carbon. Before utilized as anode for the full cell, the mesoporous Si/carbon composite was completely prelithiated. Sulfur was embedded in mesoporous carbon matrix that worked as electrically conducting reservoir for both sulfur and polysulfides. By engineering mesopores and conductive carbon for both Si-based anode and sulfur cathode, the full cell maintained a capacity retention of 80 percent after 100 cycles at a rate of 0.5 C and delivered a discharge capacity of 465 mAh/g at a high rate of 3 C. Bruckner *et al.* sputter-deposited an amorphous Si layer onto a 3D flexible carbon non-woven substrate as the anode, which showed a stable cycling up to 300 cycles when cycled at a capacity limit of 1 000 mAh/g [109]. The sulfur/carbon composite cathode was prepared by infiltrating sulfur into hollow carbon spheres, which delivered around 800 mAh/g after 135 cycles. To pair with the Li-free sulfur cathode, the silicon/carbon anode is prelithiated via short circuiting versus Li foil. The S/Li_xSi full cell obtained a high discharge capacity of 765 mAh/g for the fifth cycle at a current density of 836 mA/g based on sulfur and stably cycled for 1 400 cycles with a slow capacity fading of about 0.08 percent. Zhao *et al.* reported the fabrication of large-scale and free-standing Li_xSi/graphene foils with densely- packed Li_xSi NPs encapsulated by large graphene sheets (Figure 7.19) [110]. The fully expanded Li_xSi was confined in the graphene matrix, which avoided the volume expansion and dendrite Li metal growth. As a result, in half cells paired with Li foil reference electrode, the Li_xSi/graphene foils delivered a volumetric capacity of 1 800–2 000 mAh/cm^3, stably cycled for 400 cycles with 98 percent capacity retention at an areal capacity of ~ 2.4 mAh/cm^2. The Li_xSi/graphene foil was coupled with Li-free sulfur cathode to build full cells. The foil-S full cell maintained a high capacity of 858 mAh/g after 110 cycles with a CE of 99.5 percent at a rate of 0.5 C (1 C = 1 672 mA/g), which is higher than that of the conventional Li-S cell.

Besides, Lee *et al.* demonstrated an ultra-stable full cell consisting of lithiated Si–SiO_x nanosphere anode and dual-type (solid sulfur–polysulfide catholyte) sulfur cathode [111]. Hydrogen silsesquioxane nanospheres that were obtained through a sol-gel method decomposed to Si nanocrystals and amorphous SiO_x matrix. Prior to cycling, Si–SiO_x nanospheres were completely lithiated via directly contacting with Li foil in the electrolyte. The lithiated Si-SiO_x/Li half cell exhibited a very low first cycle irreversible capacity loss and a stable cycling up to 100 cycles with reversible capacity of 830 mAh/g maintained at 200 mA/g. The sulfur cathode

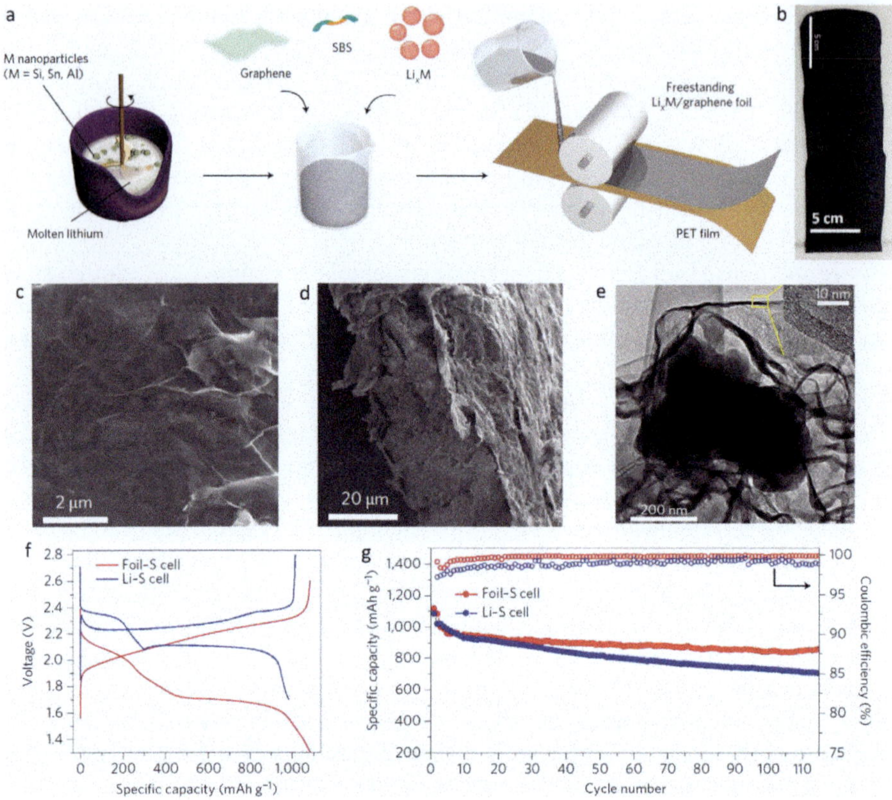

Figure 7.19 *Li_xSi/graphene foil as anode materials for full cells paired with sulfur cathode. (a) Schematic of the synthesis process of Li_xSi/graphene foil. (b) Photograph of a large Li_xSi/graphene foil. (c) Top-view and (d) cross-sectional view SEM images of the Li_xSi/graphene foil. (e) TEM image of the graphitic carbon-encapsulated sulfur composite. (f) The voltage profiles and the cycling performance of Li_xSi/graphene foil-S and Li-S cells. Reproduced from [110].*

consisted of sulfur-carbon composite prepared by infiltrating sulfur into activated carbon particles on the surface of a gas-diffusion-layer electrode in contact with Li_2S_8 catholyte solution. When paired with Li foil reference electrode, the sulfur cathode showed a high reversible capacity of 1 000 mAh/g after 100 cycles at a C/3 rate. As a result, the full cell constructed with the lithiated Si-SiO_x anode and the dual-type S-Li_2S_8 cathode achieved a capacity of 1 100 mAh/g at a rate of 0.1 C, evolving with an average voltage of about 1.8 V. Based on the cell capacity of 750 mAh/g at a 1 C rate and the weight of the electrode materials, the energy density was estimated to be 497 Wh/kg, which is more than double that of the conventional LIBs. Besides, the full cell maintained about 85.5 percent of the initial capacity after 500 cycles, which is comparable to the cycle life of the commercially available LIBs.

7.6 Summary and outlook

In this chapter, we have reviewed the synthesis of Si-based nanostructures and highlighted the crucial effects of nanotechnology on the electrochemical properties of Si-based anode materials. Nanostructured Si materials with different dimensions and morphologies are discussed to reflect the power of nanotechnology in tackling the particle pulverization problem and enhancing rate capability. In addition, the efforts of constructing nanostructured Si-based composite to improve the electronic conductivity, optimize electrode-electrolyte contact area, and stabilize SEI formation have also demonstrated. Moreover, recent progress achieved in full cells using Si-based anodes coupled with specific cathodes are discussed wherein optimization terms such as the capacity ratio are taken into intimate consideration. In all, the future of Si-based materials is bright with nanoscale structuring, more efforts are demanded to optimize the low-cost, large-scale synthesis to further improve the electrochemical performance and better satisfy the practical requirements of battery application.

References

[1] Dunn B., Kamath H., Tarascon J.-M. 'Electrical energy storage for the grid: a battery of choices'. *Science*. 2011;**334**(6058):928–35.

[2] Cao Y., Li M., Lu J., Liu J., Amine K. 'Bridging the academic and industrial metrics for next-generation practical batteries'. *Nature Nanotechnology*. 2019;**14**(3):200–7.

[3] Whittingham M.S. 'Lithium batteries and cathode materials'. *Chemical Reviews*. 2004;**104**(10):4271–302.

[4] Aricò A.S., Bruce P., Scrosati B., Tarascon J.-M., van Schalkwijk W. 'Nanostructured materials for advanced energy conversion and storage devices'. *Nature Materials*. 2005;**4**(5):366–77.

[5] Armand M., Tarascon J.-M. 'Building better batteries'. *Nature*. 2008;**451**(7179):652–7.

[6] Ogasawara T., Débart A., Holzapfel M., Novák P., Bruce P.G. 'Rechargeable Li_2O_2 electrode for lithium batteries'. *Journal of the American Chemical Society*. 2006;**128**(4):1390–3.

[7] Ji X., Lee K.T., Nazar L.F. 'A highly ordered nanostructured carbon–sulphur cathode for lithium–sulphur batteries'. *Nature Materials*. 2009;**8**(6):500–6.

[8] Wei Seh Z., Li W., Cha J.J., *et al.* 'Sulphur-TiO_2 yolk-shell nanoarchitecture with internal void space for long-cycle lithium-sulphur batteries'. *Nature Communications*. 2013;**4**(1):1331.

[9] Bruce P.G., Freunberger S.A., Hardwick L.J., Tarascon J.-M. 'Li–O_2 and Li–S batteries with high energy storage'. *Nature Materials*. 2012;**11**(1):19–29.

[10] Chan C.K., Peng H., Liu G., *et al.* 'High-performance lithium battery anodes using silicon nanowires'. *Nature Nanotechnology*. 2008;**3**(1):31–5.

[11] Yang S., Zavalij P.Y., Whittingham M.S. 'Anodes for lithium batteries: tin revisited'. *Electrochemistry Communications*. 2003;**5**(7):587–90.

[12] Zheng G., Lee S.W., Liang Z., *et al.* 'Interconnected hollow carbon nanospheres for stable lithium metal anodes'. *Nature Nanotechnology*. 2014;**9**(8):618–23.

[13] Wu H., Cui Y. 'Designing nanostructured Si anodes for high energy lithium ion batteries'. *Nano Today*. 2012;**7**(5):414–29.

[14] Wang J., Cui Y., Wang D. 'Design of hollow nanostructures for energy storage, conversion and production'. *Advanced Materials*. 2019;**31**(38):1801993.

[15] Li Y., Yan K., Lee H.-W., Lu Z., Liu N., Cui Y. 'Growth of conformal graphene cages on micrometre-sized silicon particles as stable battery anodes'. *Nature Energy*. 2016;**1**(2):15029.

[16] Wang J., Liao L., Li Y., *et al.* 'Shell-protective secondary silicon nanostructures as pressure-resistant high-volumetric-capacity anodes for lithium-ion batteries'. *Nano Letters*. 2018;**18**(11):7060–5.

[17] Wang J., Liao L., Lee H.R., *et al.* 'Surface-engineered mesoporous silicon microparticles as high-Coulombic-efficiency anodes for lithium-ion batteries'. *Nano Energy*. 2019;**61**:404–10.

[18] Ge M., Rong J., Fang X., Zhang A., Lu Y., Zhou C. 'Scalable preparation of porous silicon nanoparticles and their application for lithium-ion battery anodes'. *Nano Research*. 2013;**6**(3):174–81.

[19] Hieu N.S., Lim J.C., Lee J.K. 'Free-standing silicon nanorods on copper foil as anode for lithium-ion batteries'. *Microelectronic Engineering*. 2012;**89**:138–40.

[20] Soleimani-Amiri S., Safiabadi Tali S.A., Azimi S., Sanaee Z., Mohajerzadeh S. 'Highly featured amorphous silicon nanorod arrays for high-performance lithium-ion batteries'. *Applied Physics Letters*. 2014;**105**(19):193903.

[21] Cui L.-F., Ruffo R., Chan C.K., Peng H., Cui Y. 'Crystalline-amorphous core-shell silicon nanowires for high capacity and high current battery electrodes'. *Nano Letters*. 2009;**9**(1):491–5.

[22] Xiao Y., Hao D., Chen H., Gong Z., Yang Y. 'Economical synthesis and promotion of the electrochemical performance of silicon nanowires as anode material in Li-ion batteries'. *ACS Applied Materials & Interfaces*. 2013;**5**:1681–7.

[23] Park M.-H., Kim M.G., Joo J., *et al.* 'Silicon nanotube battery anodes'. *Nano Letters*. 2009;**9**(11):3844–7.

[24] Ryu J., Hong D., Choi S., Park S. 'Synthesis of ultrathin Si nanosheets from natural clays for lithium-ion battery anodes'. *ACS Nano*. 2016;**10**(2):2843–51.

[25] Kim W.-S., Hwa Y., Shin J.-H., Yang M., Sohn H.-J., Hong S.-H. 'Scalable synthesis of silicon nanosheets from sand as an anode for Li-ion batteries'. *Nanoscale*. 2014;**6**(8):4297–302.

[26] Wu H., Chan G., Choi J.W., *et al.* 'Stable cycling of double-walled silicon nanotube battery anodes through solid–electrolyte interphase control'. *Nature Nanotechnology*. 2012;**7**(5):310–15.

[27] Liu N., Lu Z., Zhao J., *et al.* 'A pomegranate-inspired nanoscale design for large-volume-change lithium battery anodes'. *Nature Nanotechnology*. 2014;**9**(3):187–92.

[28] Lu Z., Liu N., Lee H.-W., *et al.* 'Nonfilling carbon coating of porous silicon micrometer-sized particles for high-performance lithium battery anodes'. *ACS Nano*. 2015;**9**(3):2540–7.

[29] Zhou X., Yin Y.-X., Wan L.-J., Guo Y.-G. 'Self-assembled nanocomposite of silicon nanoparticles encapsulated in graphene through electrostatic attraction for lithium-ion batteries'. *Advanced Energy Materials*. 2012;**2**(9):1086–90.

[30] Gohier A., Laïk B., Kim K.-H., *et al.* 'High-rate capability silicon decorated vertically aligned carbon nanotubes for Li-ion batteries'. *Advanced Materials*. 2012;**24**(19):2592–7.

[31] Wu H., Yu G., Pan L., *et al.* 'Stable Li-ion battery anodes by in-situ polymerization of conducting hydrogel to conformally coat silicon nanoparticles'. *Nature communications*. 2013;**4**:4.

[32] Ryu J., Hong D., Lee H.-W., Park S. 'Practical considerations of Si-based anodes for lithium-ion battery applications'. *Nano Research*. 2017;**10**(12):3970–4002.

[33] McDowell M.T., Lee S.W., Nix W.D., Cui Y. '25th anniversary article: understanding the lithiation of silicon and other alloying anodes for lithium-ion batteries'. *Advanced Materials*. 2013;**25**(36):4966–85.

[34] Sun Y., Liu N., Cui Y. 'Promises and challenges of nanomaterials for lithium-based rechargeable batteries'. *Nature Energy*. 2016;**1**(7):16071.

[35] Huang W., Wang J., Braun M.R., *et al.* 'Dynamic structure and chemistry of the silicon solid-electrolyte interphase visualized by cryogenic electron microscopy'. *Matter*. 2019;**1**(5):1232–45.

[36] Chan C.K., Ruffo R., Hong S.S., Cui Y. 'Surface chemistry and morphology of the solid electrolyte interphase on silicon nanowire lithium-ion battery anodes'. *Journal of Power Sources*. 2009;**189**(2):1132–40.

[37] Verma P., Maire P., Novák P. 'A review of the features and analyses of the solid electrolyte interphase in Li-ion batteries'. *Electrochimica Acta*. 2010;**55**(22):6332–41.

[38] Wang Z., Su Q., Deng H., Fu Y. 'Composition dependence of lithium diffusion in lithium silicide: a density functional theory study'. *ChemElectroChem*. 2015;**2**(9):1292–7.

[39] Liu X.H., Zhong L., Huang S., Mao S.X., Zhu T., Huang J.Y. 'Size-dependent fracture of silicon nanoparticles during lithiation'. *ACS Nano*. 2012;**6**(2):1522–31.

[40] Ryu I., Choi J.W., Cui Y., Nix W.D. 'Size-dependent fracture of Si nanowire battery anodes'. *Journal of the Mechanics and Physics of Solids*. 2011;**59**(9):1717–30.

[41] Lee S.W., McDowell M.T., Berla L.A., Nix W.D., Cui Y. 'Fracture of crystalline silicon nanopillars during electrochemical lithium insertion'. *Proceedings of the National Academy of Sciences*. 2012;**109**(11):4080–5.

[42] Chew H.B., Hou B., Wang X., Xia S. 'Cracking mechanisms in lithiated silicon thin film electrodes'. *International Journal of Solids and Structures*. 2014;**51**(23-24):4176–87.

[43] Kim H., Seo M., Park M.-H., Cho J. 'A critical size of silicon nano-anodes for lithium rechargeable batteries'. *Angewandte Chemie International Edition*. 2010;**49**(12):2146–9.

[44] Yao Y., McDowell M.T., Ryu I., *et al.* 'Interconnected silicon hollow nanospheres for lithium-ion battery anodes with long cycle life'. *Nano Letters*. 2011;**11**(7):2949–54.

[45] Chen D., Mei X., Ji G., *et al.* 'Reversible lithium-ion storage in silver-treated nanoscale hollow porous silicon particles'. *Angewandte Chemie International Edition*. 2012;**51**(10):2409–13.

[46] Szczech J.R., Jin S. 'Nanostructured silicon for high capacity lithium battery anodes'. *Energy & Environmental Science*. 2011;**4**(1):56–72.

[47] Liu N., Wu H., McDowell M.T., Yao Y., Wang C., Cui Y. 'A yolk-shell design for stabilized and scalable Li-ion battery alloy anodes'. *Nano Letters*. 2012;**12**(6):3315–21.

[48] Cui L.-F., Hu L., Wu H., Choi J.W., Cui Y. 'Inorganic glue enabling high performance of silicon particles as lithium ion battery anode'. *Journal of The Electrochemical Society*. 2011;**158**(5):A592–6.

[49] Hu L., Wu H., Hong S.S., *et al.* 'Si nanoparticle-decorated Si nanowire networks for Li-ion battery anodes'. *Chemical Communications*. 2011;**47**(1):367–9.

[50] Song J., Zhou M., Yi R., *et al.* 'Interpenetrated gel polymer binder for high-performance silicon anodes in lithium-ion batteries'. *Advanced Functional Materials*. 2014;**24**(37):5904–10.

[51] Kovalenko I., Zdyrko B., Magasinski A., *et al.* 'A major constituent of brown algae for use in high-capacity Li-ion batteries'. *Science*. 2011;**334**(6052):75–9.

[52] Liu G., Xun S., Vukmirovic N., *et al.* 'Polymers with tailored electronic structure for high capacity lithium battery electrodes'. *Advanced Materials*. 2011;**23**(40):4679–83.

[53] Chan C.K., Patel R.N., O'Connell M.J., Korgel B.A., Cui Y. 'Solution-grown silicon nanowires for lithium-ion battery anodes'. *ACS Nano*. 2010;**4**(3):1443–50.

[54] Schulz D.L., Hoey J., Smith J., *et al.* 'Si_6H_{12}/polymer inks for electrospinning a-Si nanowire lithium-ion battery anodes'. *Electrochemical and Solid-State Letters*. 2010;**13**(10):A143–5.

[55] Peng K., Jie J., Zhang W., Lee S.-T. 'Silicon nanowires for rechargeable lithium-ion battery anodes'. *Applied Physics Letters*. 2008;**93**(3):033105.

[56] Xu W., Vegunta S.S.S., Flake J.C. 'Surface-modified silicon nanowire anodes for lithium-ion batteries'. *Journal of Power Sources*. 2011;**196**(20):8583–9.

[57] Nguyen S.H., Lim J.C., Lee J.K. 'Electrochemical characteristics of bundle-type silicon nanorods as an anode material for lithium ion batteries'. *Electrochimica Acta*. 2012;**74**(30):53–8.

[58] Teki R., Datta M.K., Krishnan R., *et al.* 'Nanostructured silicon anodes for lithium ion rechargeable batteries'. *Small*. 2009;**5**(20):2236–42.

[59] Zhou Y., Jiang X., Chen L., *et al.* 'Novel mesoporous silicon nanorod as an anode material for lithium ion batteries'. *Electrochimica Acta*. 2014;**127**:252–8.

[60] Yoo J.-K., Kim J., Jung Y.S., Kang K. 'Scalable fabrication of silicon nanotubes and their application to energy storage'. *Advanced Materials*. 2012;**24**(40):5452–6.

[61] Aifantis K.E., Hackney S.A. 'Nanoscale engineering for the mechanical integrity of Li-ion electrode materials'. *Nanostructured Materials in Electrochemistry*. 2008:319–47.

[62] Ohara S., Suzuki J., Sekine K., Takamura T. 'A thin film silicon anode for Li-ion batteries having a very large specific capacity and long cycle life'. *Journal of Power Sources*. 2004;**136**(2):303–6.

[63] Maranchi J.P., Hepp A.F., Kumta P.N. 'High capacity, reversible silicon thin-film anodes for lithium-ion batteries'. *Electrochemical and Solid-State Letters*. 2003;**6**(9):A198–201.

[64] Beaulieu L.Y., Eberman K.W., Turner R.L., Krause L.J., Dahn J.R. 'Colossal reversible volume changes in lithium alloys'. *Electrochemical and Solid-State Letters*. 2001;**4**(9):A137–40.

[65] Arie A.A., Song J.O., Lee J.K. 'Materials chemistry and physics'. 2009;**113**:249–54.

[66] Ryu J., Hong D., Shin M., Park S. 'Multiscale hyperporous silicon flake anodes for high initial Coulombic efficiency and cycle stability'. *ACS Nano*. 2016;**10**(11):10589–97.

[67] Kim W.-S., Hwa Y., Shin J.-H., Yang M., Sohn H.-J., Hong S.-H. 'Scalable synthesis of silicon nanosheets from sand as an anode for Li-ion batteries'. *Nanoscale*. 2014;**6**(8):4297–302.

[68] Lin D., Lu Z., Hsu P.-C., *et al.* 'A high TAP density secondary silicon particle anode fabricated by scalable mechanical pressing for lithium-ion batteries'. *Energy & Environmental Science*. 2015;**8**:2371–6.

[69] Bang B.M., Lee J.-I., Kim H., Cho J., Park S. 'High-performance macroporous bulk silicon anodes synthesized by template-free chemical etching'. *Advanced Energy Materials*. 2012;**2**(7):878–83.

[70] Ge M., Lu Y., Ercius P., *et al.* 'Large-scale fabrication, 3D tomography, and lithium-ion battery application of porous silicon'. *Nano Letters*. 2014;**14**(1):261–8.

[71] Li X., Gu M., Hu S., *et al.* 'Mesoporous silicon sponge as an anti-pulverization structure for high-performance lithium-ion battery anodes'. *Nature Communications*. 2014;**5**:4105.

[72] Liu N., Huo K., McDowell M.T., Zhao J., Cui Y. 'Rice husks as a sustainable source of nanostructured silicon for high performance Li-ion battery anodes'. *Scientific Reports*. 2013;**3**(1):1919.

[73] Shen L., Guo X., Fang X., Wang Z., Chen L. 'Magnesiothermically reduced diatomaceous earth as a porous silicon anode material for lithium ion batteries'. *Journal of Power Sources*. 2012;**213**:229–32.

[74] Wang J., Liao L., Lee H.R., *et al*. 'Surface-engineered mesoporous silicon microparticles as high-Coulombic-efficiency anodes for lithium-ion batteries'. *Nano Energy*. 2019;**61**:404–10.

[75] Jia H., Gao P., Yang J., Wang J., Nuli Y., Yang Z. 'Novel three-dimensional mesoporous silicon for high power lithium-ion battery anode material'. *Advanced Energy Materials*. 2011;**1**(6):1036–9.

[76] Esmanski A., Ozin G.A. 'Silicon inverse-opal-based macroporous materials as negative electrodes for lithium ion batteries'. *Advanced Functional Materials*. 2009;**19**(12):1999–2010.

[77] Si Q., Hanai K., Imanishi N., *et al*. 'Highly reversible carbon–nano-silicon composite anodes for lithium rechargeable batteries'. *Journal of Power Sources*. 2009;**189**(1):761–5.

[78] Magasinski A., Dixon P., Hertzberg B., Kvit A., Ayala J., Yushin G. 'High-performance lithium-ion anodes using a hierarchical bottom-up approach'. *Nature Materials*. 2010;**9**(4):353–8.

[79] Tao H., Fan L.-Z., Song W.-L., Wu M., He X., Qu X. 'Hollow core–shell structured Si/C nanocomposites as high-performance anode materials for lithium-ion batteries'. *Nanoscale*. 2014;**6**:3138–42.

[80] Su L., Xie J., Xu Y., Wang L., Wang Y., Ren M. 'Preparation and lithium storage performance of yolk-shell Si@void@C nanocomposites'. *Physical Chemistry Chemical Physics*. 2015;**17**(27):17562–5.

[81] Yang J., Wang Y.-X., Chou S.-L., *et al*. 'Yolk-shell silicon-mesoporous carbon anode with compact solid electrolyte interphase film for superior lithium-ion batteries'. *Nano Energy*. 2015;**18**:133–42.

[82] Li Y., Lin Z., Xu G., *et al*. 'Electrochemical performance of carbon nanofibers containing an enhanced dispersion of silicon nanoparticles for lithium-ion batteries by employing surfactants'. *ECS Electrochemistry Letters*. 2012;**1**(2):A31–3.

[83] Zhou X., Wan L.-J., Guo Y.-G. 'Electrospun silicon nanoparticle/porous carbon hybrid nanofibers for lithium-ion batteries'. *Small*. 2013;**9**(16):2684–8.

[84] Hwang T.H., Lee Y.M., Kong B.-S., Seo J.-S., Choi J.W. 'Electrospun core–shell fibers for robust silicon nanoparticle-based lithium ion battery anodes'. *Nano Letters*. 2012;**12**(2):802–7.

[85] Wang W., Kumta P.N. 'Nanostructured hybrid silicon/carbon nanotube heterostructures: reversible high-capacity lithium-ion anodes'. *ACS Nano*. 2010;**4**:2233–41.

[86] Fan Y., Zhang Q., Xiao Q., Wang X., Huang K. 'High performance lithium ion battery anodes based on carbon nanotube–silicon core–shell nanowires with controlled morphology'. *Carbon*. 2013;**59**(6861):264–9.

[87] Martin C., Crosnier O., Retoux R., Bélanger D., Schleich D.M., Brousse T. 'Chemical coupling of carbon nanotubes and silicon nanoparticles for improved negative electrode performance in lithium-ion batteries'. *Advanced Functional Materials*. 2011;**21**(18):3524–30.

[88] Shu J., Li H., Yang R., Shi Y., Huang X. 'Cage-like carbon nanotubes/Si composite as anode material for lithium ion batteries'. *Electrochemistry Communications*. 2006;**8**(1):51–4.

[89] Lee J.K., Smith K.B., Hayner C.M., Kung H.H. 'Silicon nanoparticles-graphene paper composites for Li ion battery anodes'. *Chemical Communications*. 2010;**46**(12):2025–7.

[90] Zhou X., Yin Ya-Xia., Wan Li-Jun., Guo Yu-Guo. 'Self-assembled nanocomposite of silicon nanoparticles encapsulated in graphene through electrostatic attraction for lithium-ion batteries'. *Electrochimica acta*. 2014;**130**:127–34.

[91] Hassan F.M., Elsayed A.R., Chabot V., Batmaz R., Xiao X., Chen Z. 'Subeutectic growth of single-crystal silicon nanowires grown on and wrapped with graphene nanosheets: high-performance anode material for lithium-ion battery'. *ACS Applied Materials & Interfaces*. 2014;**6**(16):13757–64.

[92] Zhang J., Zhang L., Xue P., *et al.* 'Silicon-nanoparticles isolated by in situ grown polycrystalline graphene hollow spheres for enhanced lithium-ion storage'. *Journal of Materials Chemistry A*. 2015;**3**(15):7810–21.

[93] Feng K., Ahn W., Lui G., *et al.* 'Implementing an in-situ carbon network in Si/reduced graphene oxide for high performance lithium-ion battery anodes'. *Nano Energy*. 2016;**19**:187–97.

[94] Ji J., Ji H., Zhang L.L., *et al.* 'Graphene-encapsulated Si on ultrathin-graphite foam as anode for high capacity lithium-ion batteries'. *Advanced Materials*. 2013;**25**(33):4673–7.

[95] Guo Z.P., Wang J.Z., Liu H.K., Dou S.X. 'Study of silicon/polypyrrole composite as anode materials for Li-ion batteries'. *Journal of Power Sources*. 2005;**146**(1-2):448–51.

[96] Du F.-H., Li B., Fu W., Xiong Y.-J., Wang K.-X., Chen J.-S. 'Surface binding of polypyrrole on porous silicon hollow nanospheres for Li-ion battery anodes with high structure stability'. *Advanced materials*. 2014;**26**(35):6145–50.

[97] Zhang P., Wang L., Xie J., Su L., Ma Chun-an. 'Micro/nano-complex-structure SiOx–PANI–Ag composites with homogeneously-embedded Si nanocrystals and nanopores as high-performance anodes for lithium ion batteries'. *Journal of Materials Chemistry A*. 2014;**2**(11):3776–82.

[98] Yoo S., Lee J.-I., Ko S., Park S. 'Highly dispersive and electrically conductive silver-coated Si anodes synthesized via a simple chemical reduction process'. *Nano Energy*. 2013;**2**(6):1271–8.

[99] Lee J.-I., Ko Y., Shin M., *et al.* 'High-performance silicon-based multicomponent battery anodes produced via synergistic coupling of multifunctional coating layers'. *Energy & Environmental Science*. 2015;**8**(7):2075–84.

[100] Ko M., Chae S., Ma J., *et al.* 'Scalable synthesis of silicon-nanolayer-embedded graphite for high-energy lithium-ion batteries'. *Nature Energy*. 2016;**1**(9):16113.

[101] Chae C., Noh H.-J., Lee J.K., Scrosati B., Sun Y.-K. 'A high-energy Li-ion battery using a silicon-based anode and a nano-structured layered composite cathode'. *Advanced Functional Materials*. 2014;**24**(20):3036–42.

[102] Lee J.H., Yoon C.S., Hwang J.-Y., *et al.* 'High-energy-density lithium-ion battery using a carbon-nanotube-Si composite anode and a compositionally graded Li[$_{Ni0.85Co0.}$05Mn$_{0.10}$]O_2 cathode'. *Energy & Environmental Science*. 2016;**9**:2152–8.

[103] Zhang L., Wang C., Dou Y., *et al.* 'A yolk-shell structured silicon anode with superior conductivity and high TAP density for full lithium-ion batteries'. *Angewandte Chemie International Edition*. 2019;**58**(26):8824–8.

[104] An W., Gao B., Mei S., *et al.* 'Scalable synthesis of ant-nest-like bulk porous silicon for high-performance lithium-ion battery anodes'. *Nature Communications*. 2019;**10**(1):1447.

[105] Yang Y., McDowell M.T., Jackson A., Cha J.J., Hong S.S., Cui Y. 'New nanostructured Li_2S/silicon rechargeable battery with high specific energy'. *Nano Letters*. 2010;**10**(4):1486–91.

[106] Agostini M., Hassoun J., Liu J., *et al.* 'A lithium-ion sulfur battery based on a carbon-coated lithium-sulfide cathode and an electrodeposited silicon-based anode'. *ACS Applied Materials & Interfaces*. 2014;**6**(14):10924–8.

[107] Balach J., Jaumann T., Giebeler L. 'Nanosized Li_2S-based cathodes derived from MoS_2 for high-energy density Li–S cells and Si–Li_2S full cells in carbonate-based electrolyte'. *Energy Storage Materials*. 2017;**8**:209–16.

[108] Pu X., Yang G., Yu C. 'Safe and reliable operation of sulfur batteries with lithiated silicon'. *Nano Energy*. 2014;**9**:318–24.

[109] Brückner J., Thieme S., Böttger-Hiller F., *et al.* 'Carbon-based anodes for lithium sulfur full cells with high cycle stability'. *Advanced Functional Materials*. 2014;**24**(9):1284–9.

[110] Zhao J., Zhou G., Yan K., *et al.* 'Air-stable and freestanding lithium alloy/graphene foil as an alternative to lithium metal anodes'. *Nature Nanotechnology*. 2017;**12**(10):993–9.

[111] Lee S.-K., Oh S.-M., Park E., *et al.* 'Highly cyclable lithium-sulfur batteries with a dual-type sulfur cathode and a lithiated Si/SiOx nanosphere anode'. *Nano Letters*. 2015;**15**(5):2863–8.

Chapter 8

Binder additive for silicon anodes

Gao Liu[1]

8.1 Introduction

Unlike conventional carbonaceous anode materials, Si undergoes over 300% volume change during lithiation and delithiation, exerting a large amount of stress on the polymer adhesive binders used to make up the composite electrode. Mechanical failure is common in high-Si-content electrodes (Figure 8.1). Recent polymeric electrode binder developments are largely stimulated by the promise of high-capacity Si anode materials for Li-ion applications. This chapter begins with a discussion of conventional composite electrodes including the functionality of polymer adhesive binders and their mechanism of adhesion, then lists a few common types of binders and new binder development strategies for Si materials, and finally concludes with future prospects.

Polymer electrode binders are used to adhere battery active material (AM) powders and conductive additive powders to each other in order to form a composite electrode, as well as to adhere this composite to the current collector. This provides mechanical stability to the laminate as well as permanent electrical connections in the composite [1]. Many polymer electrode binders have been discussed in literature, but only a few types have found commercial success in lithium-ion battery applications. This chapter will cover both polymer binders for commercial lithium-ion batteries as well as new types of binders that have been proposed in literature for use in Si electrodes.

Electrode binders are a critical component of lithium-ion batteries, but they have not received much emphasis until recently. Because polymer binders do not provide energy storage capability, less binder content improves energy density. Polymer binder content has been reduced to as little as 1% of the weight of the electrode laminate in commercial high-energy consumer electronic (CE) batteries. However, this reduction of the amount of binder in the electrode places higher demand on its ability to form and maintain the mechanical and electrical integrity of the electrode. Moreover, emerging high-capacity Si materials for advanced lithium-ion chemistry experience large volume changes during charge and discharge, putting even more

[1]Lawrence Berkeley National Laboratory, Berkeley, CA, USA

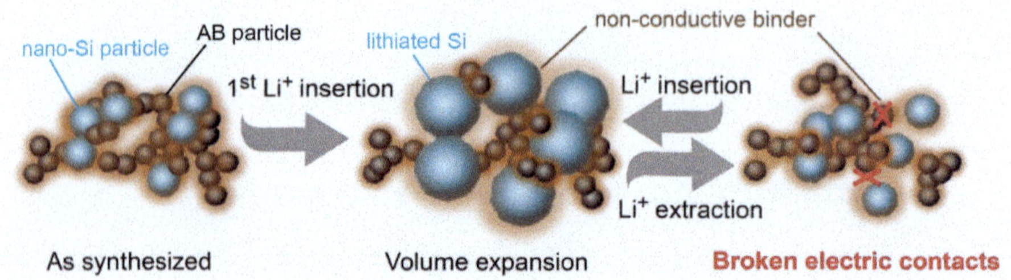

Figure 8.1 The mechanical failure of Si electrode made with nano-silicon particles, acetylene black conductive additive and a polymer binder. The repeat volume change of Si particle during lithiation and delithiation tend to break the electric pathways built through acetylene black conductive additives.

demand on multifunctionality of binders [2]. Functional electrode binders have thus become an enabling component for Si anode chemistry, causing a recent explosive growth of research as reflected in recent literature related to binder development and engineering [3–6].

This chapter first reviews the importance of polymer binder in lithium-ion electrode and cell design. A lithium-ion electrode is composed of micrometer-sized AM particles and nanometer-sized conductive additives bonded together by a polymer binder, then laminated on a thin copper current collector. A porous laminate thickness of less than 100 micrometers allows good lithium-ion transport when filled with organic electrolyte. The electrodes are separated by a thin, porous plastic separator and wound into a jelly roll design, or stacked together into multi-layers to enhance both power and energy density. The mechanical properties of these laminates are important not only to hold particles together for processing, but also to maintain the electrical connections of particles during operation. Lithium-ion cells are operated at 4V nominally, with the next generation aimed at 4.5V; this environment is fairly oxidative within the cathode and very reductive at the anode. An organic electrolyte mixture is also required for operation of the electrodes. Altogether, this is an extremely demanding chemical and physical environment in which polymer binders must operate as an adhesive agent throughout the lifetime of the electrode.

Electrochemical and mechanical stability, as well as intrinsic materials cost and ease of processing, are all key factors for successful commercialization of a polymer binder. The next generation of electrode binders for high-energy-density batteries needs to be stable at high voltage in the cathode, stable at low voltage in the anode, accommodate large volume change at the anode, possibly provide multifunctional capabilities for future batteries, and ideally be processable in an environmentally friendly medium. With such tall demands and stringent criteria, polymer chemistry and binder processing are becoming enabling factors in cell performance, as stated in literature. Therefore, we will review recent notable studies on the synthetic chemistry of polymers and their derivative binders. Based on molecular structure, the chemical and electrochemical properties of these materials will be reviewed

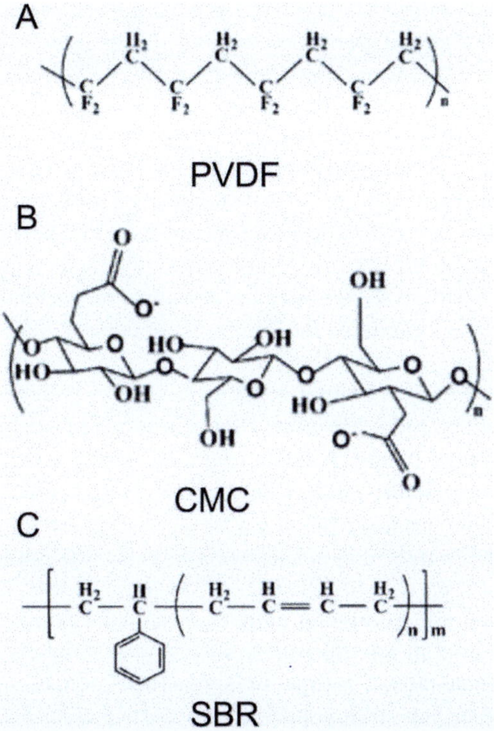

Figure 8.2 *The chemical structures of PVDF binder, and CMC/SBR binder combinations. (A) Chemical structure of PVDF. (B) The MW of CMC is determined by the nature sources. The degree of sodium ion substitution and MW of CMC affects the the binder properties. (C) SBR is a copolymer of styrene and butadiene. The SBR used for battery binders are aqueous-based emulsions. The ratio of n/m affect the adhesion properties of the SBR.*

and discussed. We will also summarize engineering aspects of polymer binders, for example how processing and electrode fabrication can impact binder distribution in the electrode, hence impacting cell performance.

The chemical stability of a polymer binder is a major factor in its selection for electrochemical applications (Figure 8.2). Polyvinylenedifluoride (PVDF) and its co-polymer derivatives have become dominant binder materials in the lithium-ion battery industry due to their superb chemical stabilities and good adhesion properties [7, 8]. There are a variety of PVDF-based polymer binders with different molecular weights and functionalities, used for different electrode materials due to different desired adhesion and swelling characteristics. Copolymers of PVDF are used by researchers and cell manufacturers alike for their desirable adhesion and elastic properties. Both Arkema and Kureha are major suppliers of the PVDF and fluorinated derivatives used by the battery industry. On the other hand, carboxymethylcellulose/

styrene-butadiene rubber (CMC/SBR) aqueous-based systems have gained significant academic interest and market share as anode binders in the past decades. Over 90% of commercial CE battery anodes uses CMC/SBR binders. Different SBR polymers can have different chemical composition in terms of styrene-to-butadiene monomer ratios, leading to different adhesion and mechanical properties [9]. Other functionalities can also be incorporated into the SBR during the synthesis process to gain additional properties. CMC is a naturally derived product, so its structure varies depending on the source. CMC is also processed into its partial sodium salt form to increase its hydrophilicity [10, 11]. The chemistry underlying the properties of these materials is important to consider for optimum physical and electrochemical performance. JSR Japan is one of the leading suppliers of SBR/CMC material for the battery industry.

Another equally important factor in polymer binders is their processing [2, 12]. Lithium-ion battery electrodes are quite unique in terms of polymer composite systems, typically composed of over 90% micron-sized AM by weight, along with a few percentage of acetylene black (AB) nano-sized conductive additives and only a few percent polymer. As mentioned, the battery industry is pushing further for reduction of the polymer binder to as low as 1%. The discreet size distribution of AM and conductive additive particles will significantly affect final composite electrode properties due to the distribution of polymer binder based on particle surface area. The low polymer content in lithium-ion electrodes makes them uniquely different from traditional polymer composites in both processing and properties. Additionally, during cell operation, ion transport must happen at the particle-electrolyte interface, where the polymer-binder resides. Therefore, binder distribution in the composite can have a large effect on battery performance. Controlled distribution is achieved through process control of slurry mixing, casting and drying of the laminate. Controlled-temperature calendaring, that is, roll-pressing also plays a role in redistribution of binder once the composite is dried [13].

8.2 The functionality of binders in a composite electrode

Li-ion electrodes adopt a porous design, which allows full penetration of the electrolyte throughout the entire thickness of the electrode composite. Therefore, long-range ion transport within the electrode occurs predominantly in the electrolyte phase. Such a design is advantageous for AMs based on Li-ion intercalation, as solid-state ion transport in these materials tends to be rate limiting due to either inherently sluggish diffusion or trapping of ions at structural defects. Although intercalation chemistry generally lends itself well to mechanical stability during energy storage, even the small volume change of such materials during the intercalation and de-intercalation process still poses a major challenge to the long-term integrity of the electrode, producing an impact on battery performance. Volume change can affect both surface chemistry and electrode mechanical integrity. Therefore, bonding the porous composite together with a small percentage of polymer adhesive is necessary to allow the electrode to accommodate volume change during repeated charge and discharge cycles over battery lifetime. When the AM volume change is increased to 300%, the binder effect on the battery performance is significantly amplified. In order to analyze this effect, a detailed understanding of polymer adhesion mechanism becomes necessary.

8.2.1 Fundamentals of adhesion

Three types of binding mechanisms are pertinent to the composite electrode. Mechanical interlocking is one of the most common physical bonding methods between two solid phases, which are intertwined together to provide mechanical stability. Debonding or separation of two interlocked materials involves physically breaking or deforming at least one of the materials to achieve separation. In a composite electrode, the polymer binder infiltrates the micro- and nano-porosity of particles to provide mechanical interlocking adhesion. The second is dispersive adhesion, which refers to the weak interactions, that is "van der Waals forces" that occur between molecules as they approach one another. Dispersive adhesion can be considered to include instantaneous dipole attraction (dispersion) as well as dipole-induced dipole (Debye) attraction, dipole–dipole (Keesom) attraction, and hydrogen bonding between substrate and adhesive. One prominent example of electrode adhesion based on dispersive force is the strong adhesion between non-polar SBR binder and graphite anode surfaces. Breaking of the adhesion normally occurs at the interface of the adhesive and the bonded material's surface, where adhesion force yields to a stronger separation force. Although dispersive adhesion is weak on an individual molecular basis, it can become quite significant when summed over the bulk of a material. Therefore, a well-covered interface between the adhesive and solid surface is critical to improve the number of individual contact points. The third type of binding is covalent binding, which is achieved through the formation of covalent chemical bonds between the substrate and adhesive molecules. Since covalent bonds are an order of magnitude stronger than hydrogen bonds, this is considered to be the strongest form of adhesion. As an example, epoxy glues are based on formation of strong carbon–oxygen bonds at the interface between the epoxy and the surface. Many adhesives use a combination of mechanical interlocking, dispersive forces, and/or covalent binding to achieve very high adhesion. Likewise, in case of PVDF binders for composite electrodes, mechanical interlocking and dispersive adhesion are used to reinforce one another. Since stronger adhesion is needed for Si materials, many developmental binders explore functionalities that include covalent binding in addition to mechanical interlocking and dispersive forces. These combined approaches appear to provide better performance for Si electrodes.

8.2.2 The battery electrode as a unique polymer composite

Normally, a lithium-ion battery electrode is comprised of active materials, electrically conductive additives, and a polymer binder (Figure 8.3) [14]. AB conductive additive, composed of around 40-nm-diameter particles, is commonly used to provide interface electronic conductivity to the usually several-micron-sized AM particles. AB is an essential additive to make the polymer binder phase electrically

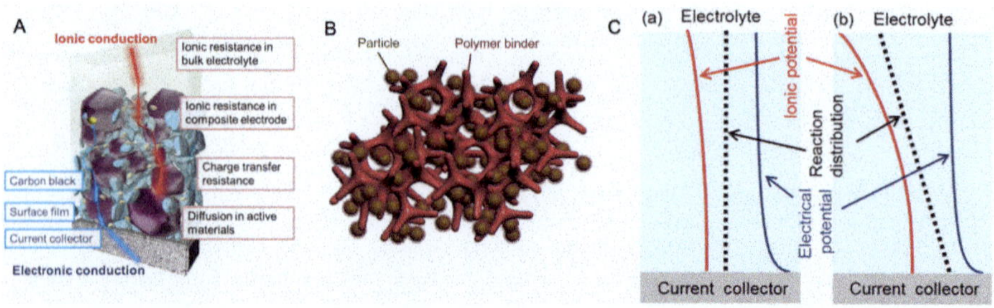

Figure 8.3 Schematic illustration of porous electrode and charge transport properties. (A) Conventional composite electrode with micron size active materials and nano size acetylene black conductive additives bond together by a polymer adhesive binder. (B) 3D porous electrode based on Si nanoparticles and conductive agent have different porosity morphology from conventional electrode. (C) Ionic and electrical potential distribution for (a) high porosity electrode and (b) low porosity electrode.

conductive while maintaining its flexibility to accommodate volume changes in the AM. AB is also low cost and has unique morphology, where primary AB particles are fused into a branched structure that creates a percolation network [15–18]. This electrically conductive medium of polymer and AB functions to provide short-range electron transport to the AM surface in graphite or Si anode systems. On the other hand, most of the long-range current can be carried by the highly conductive graphite or alloy materials themselves. This inherent electrical conductivity is advantageous when compared to cathode materials, which mostly consist of electrically resistive ceramic particles. Thus, multiple conductive additives [19] with nano- to micro-length scale are necessary for cathode design. Some Si nanomaterials, like AB, are also fused into branches, so electrodes made with these Si nanomaterials along with AB and polymer binders tend to form robust electrical percolation pathways [19, 20].

The lithium-ion electrode is a very unique type of polymer composite since the AM content (ceramic for cathode, graphite or Si for anode) is very high compared to regular polymer composite materials, for example, carbon black filler and SBR rubber for tires. In a typical lithium-ion electrode, the AM content is above 90% by weight, along with just a few percentage of conductive additive (AB) and polymer binder [21]. The electrode must also have connected pores comprising about 25%–45% of the total laminate volume for electrolyte penetration and ion transport. The physical properties of binders change when the polymer is close to the particle surface (Figure 8.4) [18, 22–24]. These changes in polymer binder properties are attributable to both chemical reaction of the adhesive with the particle surface functional groups and strong physical interaction of the polymer chains with the surface. The dangling bonds on the surface of a solid particle can chemically bond or physically adsorb polymer adhesives to form a 1- to 5-nm-thick bound polymer

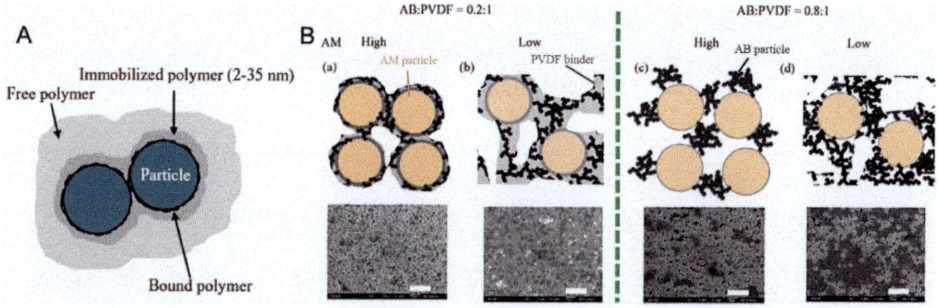

Figure 8.4 *Polymer binder distribution and properties are affected by the surface of the particles. (A) The bulk polymer differentiates into three physical states of bound polymer, immobilized polymer and free polymer, when in contact with a particle surface. The physical properties of bound polymer and immobilized polymer changed significantly. Therefore, the polymer binder in contact with the electrode materials differentiate accordingly, and the polymer binder is distributed based on the surface area and properties of the particles. (B) The schematics and SEM images of binder distribution between AB and AM at different electrode compositions. AB:binder = 0.2:1 at high AM loading and low AM loading; AB:binder = 0.8:1 at high AM and low AM loading. Scale bar: 20 micron.*

layer on the surface of the particle [25]. Due to the chemical bonding, this layer of polymer is insoluble in solvent and can extend its influence to the adjacent polymer chains to immobilize an additional outer layer of the polymer adhesive. The affected polymer layers have higher glass transition temperatures than that of the bulk polymer. In practice, the bound and immobilized layers of polymer can extend well beyond several nanometers. In a well-designed electrode, the adhesive binder is evenly distributed on all particle surfaces; therefore, such a change in physical properties exerts enough influence to affect macroscopic binding and mechanical properties. The overall effect is that the adhesion of the binder and elasticity of the electrode experience a dramatic reduction in the regime where polymer content is limited. Increasing polymer binder content normally enhances the cohesiveness of the electrode and makes the composite electrode more flexible. However, although adding more polymer adhesive improves the mechanical properties of the electrode, it also lowers the overall energy density and slows down ion transport at or near the AM/electrolyte interface. As a result of the latter effect, high binder content also tends to increase charge-transfer impedance, which is a major contributor to overall cell resistance. Therefore, the choice of micron- vs nano-sized Si material not only has an effect on the electrode mechanical properties due to material volume changes, but also has a huge effect on the resulting binder properties and electrode processing due to significantly different particle surface area.

8.3 Surfaces of Si materials

Typically, when discussing polymer binders, the surface that the adhesive must bind to would be introduced and discussed simultaneously. Electrode binders would then be chosen based on existing knowledge from graphite electrode design. However, Si surface chemistry is in fact radically different from that of graphite. Most of the graphite surface consists of basal planes of hexagonal carbon rings or edge planes with dangling carbon bonds; graphite and carbon anode materials are processed to minimize surface carboxylic acid and hydroxide groups, for the purpose of reducing side reactions. However, pure Si materials have a native layer of SiO_2 due to oxide formation during materials synthesis/processing and later exposure to ambient conditions; therefore, Si surfaces are usually either –OH terminated or oxygen terminated (Figure 8.5). The effect of the SiO_2 surface layer on cycling and adhesion was initially studied in 2008 [10, 26, 27]. It was found that these electrodes bind most effectively through the surface –OH functionality in a typical Si particle-based electrode. Therefore, functional binders which can provide strong interaction with

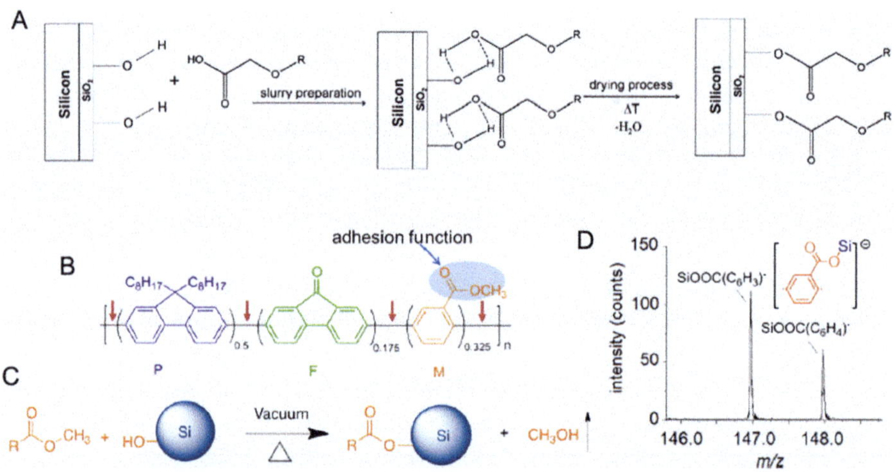

Figure 8.5 (A) Proposed reaction mechanism, condensation of the free carboxylic acid group of CMC and the partially hydrolyzed SiO_2 layer that covers a Si particle. (B) Chemical structure of the PFM functional conductive polymer binder. The red arrows indicate the most likely chemical bonds to be broken during the ionization process in the TOF-SIMS measurement. The blue shadow highlights the ester group that forms adhesion bonds with the SiOH surface. (C) Trans-esterification reaction between the ester functional group of the PFM binder and the SiOH surface group, which provides strong adhesion, during the electrode drying process. (D) TOF-SIMS result of the binder and active materials interface that shows the evidence of chemical bonding between the binder and the active materials.

hydroxyl moieties tend to form stronger electrodes and afford better battery performance [26, 28]. Carbon coating is also widely used in industry to stabilize the Si material's interface and buffer volume expansion. In this case, the surface to be adhered is carbon rather than SiO_2. However, these carbon surfaces can be very different from each other in terms of morphology and surface chemistry, depending on differences in the coating process. This issue further complicates the choice of binders for use in Si-based electrodes.

8.4 Commercial binders: PVDF and CMC/SBR

There have been many attempts to introduced high content (>50% by weight) of Si into the negative electrode in the past two decades. This has been proven challenging, as both storage life and cycle life of the battery become reduced significantly compared to carbonaceous anodes. However, Si has gradually worked its way into commercial high-energy-density Li-ion batteries through inclusion of a small percentage (<10% by weight) of Si along with traditional carbon anode materials. Additionally, both SiO_x and Si/carbon composite materials have made minor inroads into commercial power batteries for EVs and CE batteries within the past few years. Those anodes typically contain less than 10% Si and up to 75% carbonaceous AMs. Commercial adhesive binders for carbon electrodes such as PVDF and CMC/SBR are normally used for these applications and achieve acceptable performance. At low Si content, electrode mechanical integrity is maintained by the graphite particles, and most of the electrode capacity decay comes from the gradual loss of lithium inventory due to Si surface reactions rather than electrode mechanical failure. Traditional binders are preferred due to the well-established electrode slurry and fabrication processes using these polymers. Moreover, most of the Si materials for these applications are carbon coated to improve interface stability, so PVDF and CMC/SBR types of binders are compatible choices for these applications.

PVDF binders are routinely used in literature as a baseline to compare with novel functional binders for Si electrodes. In many cases, the cell performance with PVDF is inferior to that of the specially formulated functional binders. However, PVDF binders are electrochemically stable at low potential, easily processed in N-methyl pyrrolidone (NMP)-based slurries, have a good combination of mechanical properties, and introduce little interfacial impedance. Also, PVDF binders are available in many different molecular weight and copolymer choices, which can be made into different viscosity slurries and electrodes with different mechanical properties. This combination of characteristics makes PVDF a very attractive binder class. On the other hand, PVDF has been gradually losing its dominant position to CMC/SBR binder types, mainly due to the fact that electrode manufacturing is moving toward more environmentally sustainable methods. As a result, aqueous-based slurry formulations have taken over most carbon anode fabrication processes. In response, PVDF suppliers have begun developing latex-based PVDF dispersions in aqueous media; recent reports have found equivalent performance between carbon electrodes made with such PVDF dispersions vs electrodes made with CMC/

SBR-based binders [29]. The other major hurdle for PVDF to successfully compete with CMC/SBR is its cost, given the fluorochemistry required for its synthesis. In order for PVDF to find significant future applications, it must be water-processible and demonstrate superb performance in high-Si-content electrodes.

CMC/SBR is a widely used binder system for carbonaceous anodes. CMC is naturally water-soluble, and SBR is supplied as an aqueous-based latex. Therefore, electrode slurry processes based on CMC/SBR use water as the solvent, making them both low-cost and environmentally friendly. The combination of CMC and SBR also enables excellent performance in graphite electrodes. CMC is also proposed to interact with –OH groups on the native surface of Si, which boosts the effectiveness of this binder system with Si-based AMs. The carboxylic acid (–COOH) groups on CMC may go through a condensation reaction with silanol –OH groups on the Si surface during the high-temperature electrode-drying process, forming a covalent ester bond (Figure 8.5) [26, 28]. As discussed above, covalent bonding is the strongest type of adhesive interaction and is therefore widely considered the most effective strategy to hold Si electrodes together. In current commercial electrodes with low Si content, CMC/SBR has become the preferred binder choice, since there are established manufacturing processes to draw from and most of the Si surface is carbon-coated to minimize its exposure to electrolyte. Much like PVDF binders, SBR binders are also co-evolving with the development of Si-containing electrodes. Many SBR suppliers have begun incorporating proprietary functional groups into their SBR copolymers to adjust both their elasticity and surface adhesion. Of course, PVDF binder suppliers are also doing similar R&D in order to compete for the Si binder market.

8.5 New functional binders

8.5.1 Design principles for Si electrode binders

As the battery community continues its quest for higher energy density, Si content is being pushed increasingly higher. An anode of 50% or greater silicon by weight has become especially desirable, but this goal has proven challenging to achieve in commercial applications. There are many obstacles for this type of electrode: electrode mechanical failure, Si surface reactions, high first-cycle capacity loss, and large volume expansion all play important roles in cell failure. This chapter will focus on binder design and related electrode processing to improve electrode integrity.

Unlike the traditional choices for commercial lithium-ion binders (PVDF and CMC/SBR), which were largely adopted from earlier industry applications, binder development for Si electrodes usually follows a "bottom-up" design approach. New functional binders take advantage of the unique Si surface chemistry, with a focus on addressing the large volume change of Si during charge and discharge. For silicon materials, high adhesion becomes the most important aspect of a binder, as volume-change-related dislocation always leads to debonding of Si and binder at the interface. Thus, covalent binding is routinely explored to enhance adhesion force. Among the various approaches to achieve covalent binding, ester bonds have proven

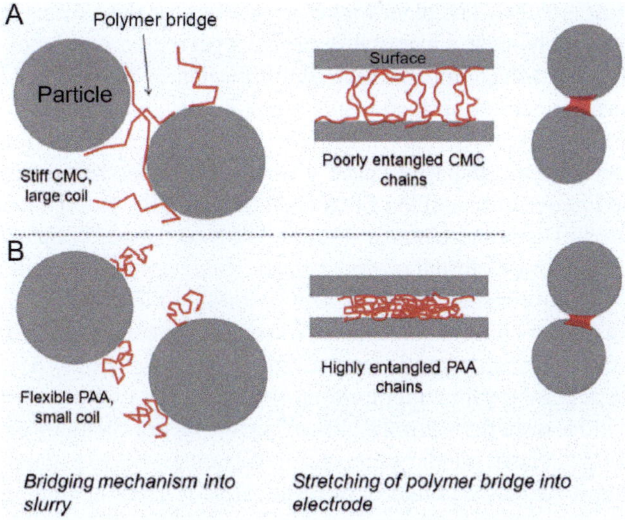

Figure 8.6 *The slurry phase structure of the polymer carries to the composite laminate during fast electrode drying process. (A) The more rigid cellulosic backbone of CMC results in polymer coils of larger size than for the more flexible vinyl type backbone of PAA. Thus, the adsorbed CMC chains extend out into solution more far away from the particle surface than PAA and can more easily form bridges between the particles. (B) Due to its higher flexibility, PAA is more entangled, resulting in a larger toughness of polymer bridges at higher polymer concentration.*

popular, as they are easily formed between organic acid functionality on the binder and hydroxide functionality on the Si surface. Carboxylate groups can also provide strong adhesion with the Cu current collectors typically used on the anode side of a lithium-ion battery, providing an additional advantage for the Si-based electrode [30]. Synthetic poly(acrylic acid) (PAA) has the highest density of carboxylic acid groups out of any common polymer materials; therefore, it has been explored extensively as an electrode-binder for Si materials. Although both CMC and PAA use carboxylate for adhesion, CMC has a much more rigid and chain-extended structure, whereas PAA is more flexible in aqueous solutions and slurries. These different solution properties during the slurry-making process lead to different morphologies in the electrode and hence different electrochemical performance (Figure 8.6) [12, 13].

Multiple strategies have built off the properties of PAA to improve adhesion even further. Crosslinking agents for PAA binders may be used to further improve the mechanical properties of the electrode laminate [31]. There have also been attempts to first modify the Si particle surface with an organic coating, then use the modified surface to further covalently bind with a polymer adhesive. This approach, that is “priming” of the surface has long been used by the coating and adhesive industry before applying a surface layer [32]. The major advantage of covalent adhesion is that the adhesion cannot be easily weakened by swelling of the adhesive with organic electrolyte. Swelling of the adhesive may lead to dislodging of the polymer from the solid surface, weakening dispersive adhesion forces. Most of reported research on novel binders uses uncoated Si materials to demonstrate the strong adhesion between Si and the binders. When the Si is carbon-coated, some of the aforementioned covalent-binding strategies may no longer be relevant, as SiOH groups may no longer be present on the surface. However, there is still no clear best material for high-Si-content electrodes, and research on highly adhesive binders is still warranted. These efforts must evolve, however, to remain aligned with current development efforts in Si materials themselves.

In order to accommodate volume expansion, elasticity of the binder could prove to be another quite important property. If adhesion serves the purpose of grabbing onto the Si surface, elasticity allows the composite electrode to expand and contract without causing mechanical failure of the binder itself. Investigation on binder elasticity has led to some apparently contradictory conclusions, as some research shows that glassy binders give better results, while others report that elastic binders afford better Si electrode performance. However, it is difficult to make a direct comparison of literature results between two different binder materials, as most of the Si materials and electrolytes are different in different publications. Besides these large variables between reports, the fundamental reasons for the contradiction may come down to changes in binder mechanical properties with electrode composition, slurry processing, and electrolyte swelling. More systematic work remains to be done in this field in order to clearly gauge the elasticity of binders under working electrode conditions.

Electrode AMs are both electrical and ionic conductors, and these properties must be maintained throughout the whole of the composite electrode, despite it consisting of many discrete particles. This gap may be bridged *via* electrode engineering. For electron transport, AB nano-conductive additive materials are added to enhance the overall electrical conductivity. These additives form an electrical path which connects all AM particles to the current collector. Ideally, connections occur uniformly across all particle surfaces to ensure uniform electron access to electrode/electrolyte interface. In an Si electrode with low Si content and high graphite content, the need for AB is not so critical, since most of the current is carried by the highly electrically conductive graphite particles, while Si particles are imbedded in the graphite matrix. However, when Si content is increased to 50% and beyond, graphite particles are no longer continuously connected in the composite. Therefore, maintenance of the Si particles’ electric contact to the current collector relies on conductive additives. High concentrations of conductive additive and binder are normally required in a high-Si-content electrode. Recently, intrinsically electron-conductive polymers have been developed to provide both adhesion and molecular-level electrical connection for high-Si-content electrodes (Figure 8.7) [33–35]. These electrically conductive polymers demonstrate superb cycling performance with Si nanoparticles as well as excellent adhesive properties. Up until now, all

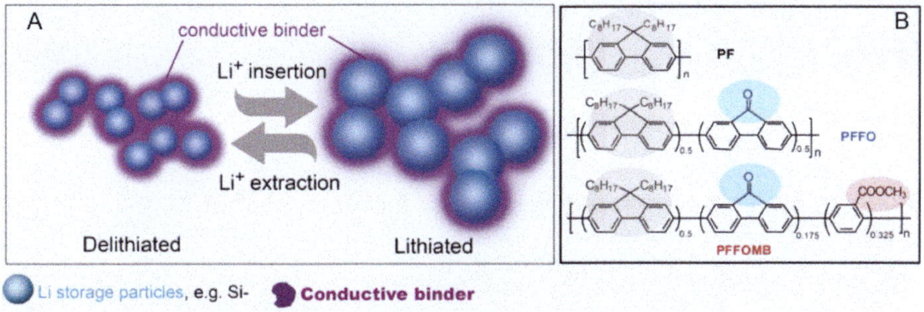

Figure 8.7 *Schematics of the technical approaches to address volume change issue in battery materials. (A) Conductive polymer with dual functionality, as a conductor and binder, could keep both electrical and mechanical integrity of the electrode during the battery cycles. (B) The molecular structure of the PF-type conductive polymers, with two key function groups in PFFOMB, carbonyl and methylbenzoic ester, for tailoring the conduction band and for improving the mechanical binding force, respectively.*

commercial battery electrodes have relied on AB to provide electrical connectivity and an insulating polymer binder to provide mechanical stability. Electrically conductive binders provide an alternative approach to connect the nanomaterials while also accommodating volume change.

Although ion transport in the porous electrode occurs predominantly through the electrolyte, which permeates electrode pores, interfacial ion transport between the AM and electrolyte is normally achieved through an electrolyte-swelled binder layer. If the binder is completely non-swellable in the electrolyte, then this polymer layer will block ion transport to the AM interface (Figure 8.8) [36]. In order to minimize electrode impedance, binders must, therefore, be able to transport lithium ions when swollen by the electrolyte. Alternatively, if the binder is not swellable, the coverage of the AMs needs to be minimized. Due to this trade-off, improving the mechanical properties of the electrode generally tends to reduce the electrode energy and power density. As interfacial impedance is normally the dominant impedance in the cell, enabling effective ion transport to the interface is a major goal in battery-electrode engineering. However, both binder swelling and reduction of binder surface coverage on AMs tend to weaken the polymer's adhesive properties. Bottom-up materials design and new materials development can thus play a large role in developing high-power Si electrodes by allowing one to circumvent this issue. However, manipulating ion transport in the binder is more complicated than it may appear at first glance. Electrolyte swelling normally weakens adhesion and changes the elastic properties of binders; moreover, swelling tends to separate conductive nanocomposites from one other and from the AMs, reducing electrical conductivity at the interface. Interfacial electrical conductivity, binder mechanical

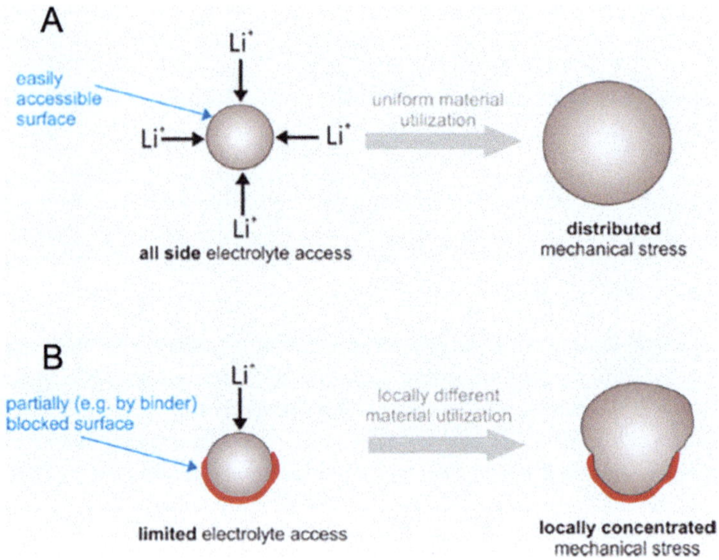

Figure 8.8 *Si material utilization during lithiation and thus material expansion depends on the surface that is accessible to Li^+ and thus the binder distribution covering the surface. (A) An easily accessible surface leads to uniform material utilization and thus homogeneously distributed mechanical stress during the lithiation and thus material expansion process. (B) When the surface is covered by the binders blocking the surface, the surface is partially blocked for the lithiation process and the material utilization is locally concentrated, which leads to strong locally concentrated mechanical stress during the lithiation and thus material expansion process.*

properties, and slurry processing techniques all make a big impact on electrode performance [37].

A bottom-up approach can be used to design and synthesize binders for high-Si-content electrodes as shown in Figure 8.9. In this case, the binder conductivity, adhesion, flexibility, and ion transport are all considered simultaneously for Si electrode [37]. It is very challenging to concurrently optimize all four properties while additionally developing an optimized electrode fabrication process with a novel material. Scientific literature which focuses on merely one or two individual aspects can still have significant value for the Si electrode community when put into proper context. The rest of the chapter will highlight some of the focused research and development efforts that are relevant to enable high-Si-content electrodes.

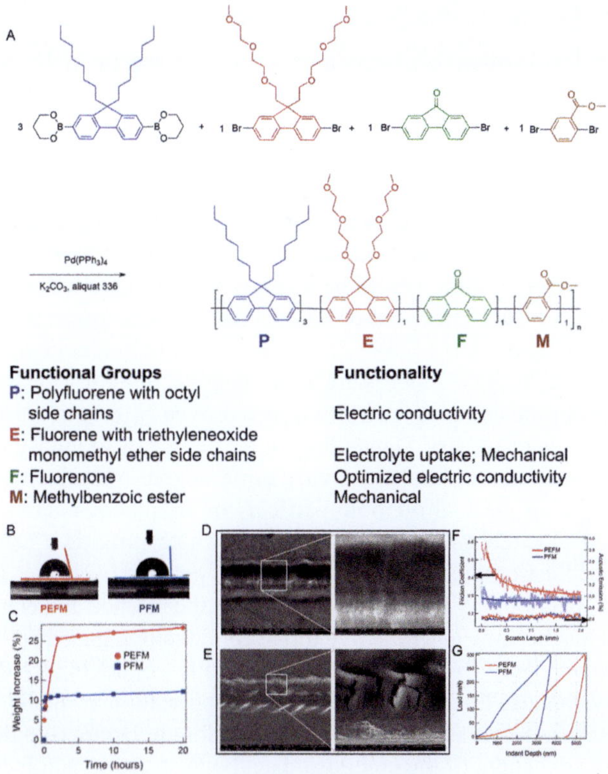

Figure 8.9 *Synthetic scheme and the design purpose of the functional groups incorporated in the polymer binder and their corresponding electrode mechanical properties. (A) Functional groups contribute specific functionalities when the polymer is used as a binder in Li-ion batteries. (B) Water contact angles on the surfaces of PEFM and PFM films. The E side chains in PEFM increase the polarity and thus improve the swelling of the polymer. Therefore the PEFM has a lower contact angle with water. (C) The swelling tests of PEFM and PFM polymer film in the 1 M $LiPF_6$ EC/DEC (1:1) electrolyte. The electrolyte uptake in PEFM is three times higher than that in PFM and is at the same level as that for conventional nonconductive PVDF binder. (D and E) The scratch tests of PEFM/Si (D) and PFM/Si (E) composite electrodes. The SEM image of the scratch morphology of PEFM shows a very smooth scratch path with little crack. (F) The PEFM/Si electrode displays higher friction coefficiency than that of the PFM/Si, indicating the higher binding strength of PEFM. Both have the same level of caustic emission. (G) Nanoindentation force of the PEFM is lower than the PFM at a given indentation depth. PEFM with the ether side chains shows lower modulus, thus higher ductility, than that of the PFM.*

8.5.2 Development of adhesive binders

Silicon binder development has been largely centered around highly adhesive materials in order to compensate for the large volume change of Si. In the early days of this research, the hypothesis that CMC binder forms ester bonds with surface –OH groups was thought to be the major reason for the observed better cycling performance in electrodes with this binder [26]. Much of the subsequent work on carboxylic-acid-functionalized binders was meant to advantage of similar esterification chemistry to improve adhesion between the polymers and Si surfaces. Since CMC is derived from natural products, its early success inspired intense activity to search for other polysaccharide-based binders for Si applications, including alginate and gum arabic (Figure 8.10) [38–41]. This direction in binder development is particularly attractive, as these binders are water-soluble, naturally derived, and biodegradable. Appropriate chemical stability is also proven by the successful application of CMC to graphite chemistry. Depending on the degree of sodium-ion replacement, these natural ionomers could also provide some degree of surfactant properties to help uniformly disperse hydrophobic particles in an aqueous slurry. Additionally, such binders tend to coat the particle surfaces, which provide a certain degree of interfacial modification during the cell formation process [42]. This rich chemistry and multifold functionality is quite interesting for the science and engineering communities, and has become a very active thrust for developing alternative lithium-ion binders. However, only CMC has been successfully commercialized thus far. The rich chemistry can also create confusion for this application as quality control of the binder materials and data verification becomes difficult. Bottom-up development in this area takes advantage of the cellulosic backbone structure and incorporates new functional groups or other moieties to enhance the binder properties. One example is the incorporation of PAA segments onto a chitosan backbone to adjust the morphology and free volume of the carboxylic-acid-containing polymer [43].

Since carboxylic acid groups are an important functionality for adhesion between the Si and binders, poly(acrylic acid) (PAA) has attracted attention as the polymer with the highest density of carboxylic acid groups. For this reason, PAA and its lithium salt derivatives have been extensively researched as binder materials for high-Si-content electrodes [13, 42, 44, 45]. The advantages of this binder are myriad, as it is commercially available, low-cost, and water-soluble. Lithiated poly(acrylic acid) (LiPAA) can also contribute additional lithium-ion inventory to the cell system, which may improve battery performance. Research shows that the degree of lithiation of the PAA impacts the performance of the electrode. However, despite all the obvious advantages, commercial application of LiPAA is still very limited due to high water content, brittleness, and high electrode interfacial impedance.

8.5.3 Development of elastic binders

Although adhesion has been well explored for electrodes based on Si, the factor of elasticity has been relatively less investigated. Since Si materials undergo large volume changes, the more elastic the binder is, the better the composite electrode can withstand repeated cycling. However, both CMC and PAA are brittle binders, yet

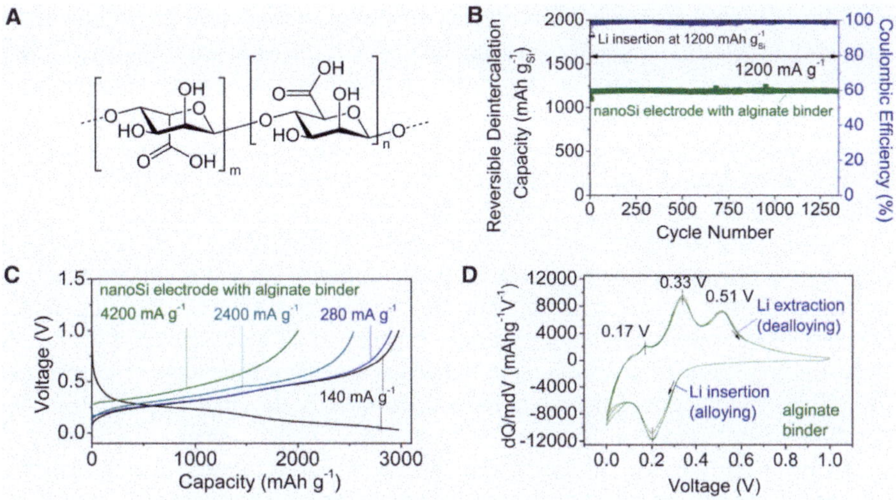

Figure 8.10 *Electrochemical performance of alginate-based nano-Si electrodes (electrode density = 0.50 g/cm^3, weight ratio of Si:C = 3:1). (A) The molecular structure of alginate acid. Partially sodium ion replacement of the proton on the carboxylic acids group forms the Na alginate binder. The chemical composition of the Na alginate binder is very similar to the CMC binders. (B) Reversible Li-extraction capacity and CE of the nano-Si electrodes versus cycle number for Li insertion level fixed to 1 200 mA·hour/g Si (Ah, ampere hour). (C) Galvanostatic discharge profiles of the nano-Si electrode at different current densities between 0 and 1 V. (D) Differential capacity curves of the nano-Si electrode in the potential window of 0 to 1 V versus Li/Li$^+$ collected at the rate of 0.025mV/s after the first galvanostatic charge-discharge cycle. Q/m, specific capacity; V, voltage. All electrochemical measurements were performed at room temperature in two-electrode coin-type half-cells. In (A), the capacity is reported for the Si contribution only. In (D), the capacity is normalized by the total weight of Si and C additives.*

they afford Si electrodes with very good cycling performance. It is difficult to discuss elasticity without considering adhesion as well, as otherwise particles may slide and migrate within the electrode to eventually merge into an aggregate. Therefore, most highly elastic binder designs also incorporate adhesive functionalities. In order to overcome the brittleness of certain polymer binders, functional groups may be included to make the binder more flexible, or hydrogen bonding may be introduced in order to provide strong inter-chain polymer interactions. Binders with mechanical interlocking have also been developed, allowing molecules to slide past each other while maintaining a connection.

Flexible segments can be incorporated into polymer binders to increase compliance and flexibility. Oligo(ethylene oxide) segments provide an example, as their incorporation into rigid polymer chains may reduce overall brittleness. These groups also serve to increase the polarity of the binders, which can improve electrolyte uptake and hence ion transport to the AMs. This design principle is demonstrated in the enhancement of rate and cycling capacity of Si electrodes containing PEFM functional binder as shown in Figure 8.9 [46, 47].

Self-healing polymers have long been used in biological applications. In turn, a self-healing polymer binder with elastic polymer backbone and hydrogen bonding functional groups can accommodate the large volume change of Si materials. When the electrode develops a fracture, hydrogen bonding in the electrode composite may be able to re-establish connections to repair the mechanical damage (Figure 8.11) [48]. Other highly elastic polymers have been recently developed based on incorporation of polyrotaxane into conventional PAA, which imparts extraordinary elasticity to the polymer network from the ring sliding motion of polyrotaxane (Figure 8.12) [49]. The introduction of functional groups to increase binder flexibility and elasticity also leads to changes in other binder properties such as interaction with electrolyte, as well as interaction with the Si surface. In some cases, the overall effect is to improve Si electrode performance, whereas in other cases, the effect is non-obvious or even detrimental to performance. We are unlikely to see many reports of these negative results in literature, although they could prove just as important to understand binder and Si interactions. As the field continues to perform bottom-up binder design, the contribution of elasticity to overall cell performance must be better understood.

8.5.4 Development of conductive binders

In a conventional electrode, electrical connectivity is maintained by percolating conductive additives in the form of acetylene black nanoparticles and micron-sized carbon fiber or graphite particles. The electrical connections of Si AMs to these conductive additives are maintained by a polymer adhesive binder. Such connections are sporadic and point-like on the nanoscale, which is extremely problematic when the AMs are also nanoparticles. Nanoparticles have a small individual surface area per particle with which to make electric connections with (also nano-sized) conductive additives. Moreover, repeated volume change can gradually weaken these connections, leading to loss of overall connectivity among particles. The end result is AM particle isolation and associated capacity loss. A new approach has been developed using an intrinsically electron-conductive polymer as an adhesive binder for Si-based electrodes. There are several apparent advantages of using conductive polymer binders rather than nanoparticle additives. The first advantage is that the Si particles and conductive medium (binder) can be covalently connected to strengthen both the composite electrode and the electronic pathways. The second is the electrical connections between the conductive medium and Si particles are smooth on the molecular level, and connections are evenly distributed across even nano-sized Si as shown in Figure 8.7. In fact, the entire polymer/particle interface is electrically

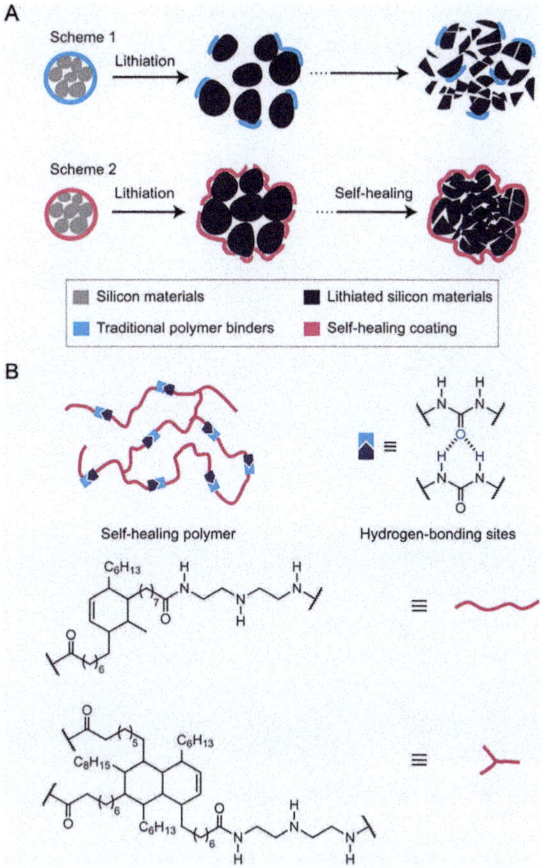

Figure 8.11 *Design and structure of the self-healing electrode. (A) Scheme 1: schematic illustration of the design and behavior of a conventional silicon electrode that shows failure of the electrode because of cracking in particles and polymer binder, which results in loss of electrical contact. Scheme 2: schematic illustration of the design and behavior of our stretchable self-healing electrode that shows the maintaining of electrical contact between the broken particles and no cracks in the polymer binder because of the stretchability and incorporation of self-healing chemistry. (B) Chemical structure of the SHP. Magenta lines, polymer backbones; light-blue and dark-blue boxes, hydrogen-bonding sites.*

conductive, as compared to acetylene black conductive additives which make only a few individual particle-to-particle connections. The third advantage is that the electrical connections and conductive polymer medium are elastic; therefore, the conductive medium can change shape as Si expands and contracts. However, bottom-up

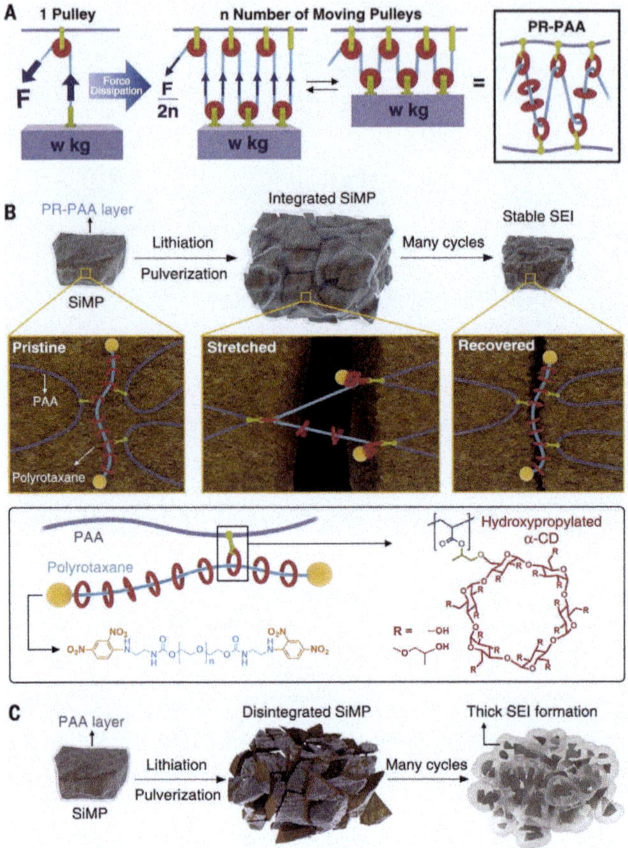

Figure 8.12 *Proposed stress dissipation mechanism of PR-PAA binder for SiMP anodes. (A) The pulley principle to lower the force in lifting an object. (B) Graphical representation of the operation of PR-PAA binder to dissipate the stress during repeated volume changes of SiMPs, together with chemical structures of polyrotaxane and PAA. (C) Schematic illustration of the pulverization of the PAA-SiMP electrode during cycling and its consequent SEI layer growth.*

design of a conductive polymer binder is challenging as it must simultaneously satisfy all the requirements of an electrode binder and a conductive additive, including facile electron transport, stability at very low voltage, ion conductivity, adhesivity, and elasticity (Figure 8.9) [46, 47].

Unlike electron conduction in graphite or metals caused by transport through a conduction band, conducting polymers rely on formation and migration of polarons to achieve conductivity. Formation of a polaron involves a redox reaction; so conductive polymers are by nature also redox-active polymers. Because of the low operational potential of Si anodes, in order for the polymer to be conductive, the

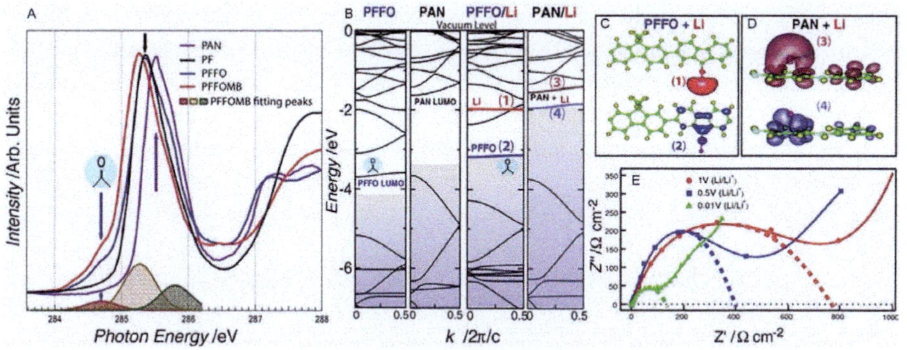

Figure 8.13 *Tailored electronic structure and conductivity of the polymer. (a) Carbon-*1s *XAS spectra collected on a series of polymers. The three peaks at the bottom show the best peak fitting result of the spectrum of PFFOMB (Molecular structures in Figure 8.7B). The carbonyl group leads to a much lower LUMO level at 284.7 eV (blue arrow). (b) Calculated band structure of PFFO (PFFO/Li) in comparison with polyaniline (PAN/Li). The k-point is along the polymer chain direction with* c *as the period length. Shaded area indicates occupied states. Fermi level floats between HOMO and LUMO for undoped materials. (1) and (4) are Li-induced bands, which sits above (2), PFFO LUMO, but below (3), PAN LUMO. (c–d) The wavefunction square of the bands (1), (2), (3) and (4), as marked in (b). The isosurfaces encompass 50% of the wavefunction charge density. The distinct isosurfaces of bands (1) and (2) indicate that the electron charge transfer from Li to PFFO is almost complete, contrasting the weak charge transfer in PAN/Li with close isosurface shape of (3) and (4). (e) The enhanced conductivity of PFFOMB polymer (without Si) at different cathodical doping potentials. The dotted lines are extrapolations of the semi-circle lineshape. The interceptions of the dotted line with the horizontal zero-line indicate the film resistivity, which decreases from 770 to 120 Ω cm^{-2} when doping potential changes from 1 to 0.01 V (Li/Li^{+}).*

polymer is normally cathodically doped from its neutral state to a negatively charged state. A highly conjugated polymer backbone structure such as polyacetylene or polyphenylene allows formation of stable polarons upon redox reaction and allows long-distance polaron migration due to pi-pi electron orbital overlap. An electron-deficient conjugated polymer structure also allows cathodic doping of the polymer to form stable, negatively charged polarons (Figure 8.13) [35].

The PFM polymer design based on this principle allows the formation of a cathodically doped conductive binder during Si electrode operation [35]. A Si composite electrode consisting of 66% Si nanoparticles with 33% PFM polymer can operate without additional conductive additives. The electrode can operate over 700 cycles at greater than 2 000 mAh/g with very little capacity decay, and the electrodes are intact after cycling. These results also underline the important of adhesion, as most conductive polymers are not designed to have good binding properties; similar conducting polymers with the similar band gaps but without adhesive groups can only provide high capacity during the initial few cycles. Inclusion of covalent bonding functionality significantly enhances the cycling stability.

Polyimide (PI) binders have also gained attention for high-Si-content electrodes (Figure 8.14) [50]. In this case, a precursor of poly(amic acid) aqueous solution is used to make a slurry with Si and conductive additive, which is cast into an electrode laminate. The laminate is then cured at 190°C to induce a dehydration and cyclization reaction which forms the final PI structure. Polyimides are high-modulus materials that can withstand the stress generated by the expansion of Si during lithiation. The drying and dehydration process also promotes the formation of covalent bonds between the precursor carboxylic acid groups and Si material, so superb adhesion can also be achieved during the polyimide formation and electrode drying process. Carboxylic acid and imide groups on the poly(amic acid) precursor can additionally form covalent bonds with the Cu current collector surface; therefore, an overall mechanically robust and integrated composite is formed between polyimide binder, Si, and Cu. AB conductive additive is, however, required for the electrode to work properly. Although polyimide is known for its superb thermal and chemical stability, it undergoes extensive electrochemical changes in the Si electrode environment. These changes are associated with the loss of oxygen and protons from the PI and transforms it into an electrically conductive material through partial carbonization, similar to the mechanism of doping in the conductive polymer discussed previously. However, this reaction is more extensive as it consumes a larger amount of charge and lithium ions. Fortunately, after the first lithiation and delithiation process, the electrode capacity becomes fairly stable in subsequent cycles. This process also affords a molecular-level electrical connection between Si materials and the conductive matrix. Some of these modified PI binders are used in pilot-scale production of high-Si-content electrodes.

Conductive polymer binders provide a very different adhesion and connection approach for Si materials vs. traditional binders with conductive additives. Multiple challenges exist for this approach. Most conductive polymers and their precursors are only marginally electrically conductive by themselves, but *in situ* doping by lithium ions enhances their conductivity. This process consumes lithium ions, causing fairly large first-cycle capacity loss. Usually, these conductive polymers are also only soluble in organic solvents, and some require extensive processing after coating. The stability of the conductive polymer interface with electrolyte at very low potential also needs to be evaluated. However, considering the paramount challenge of a high-Si-content electrode and demonstrated successes with PI, further research and development of conductive polymer binders may hold the key to a successful Si electrode design.

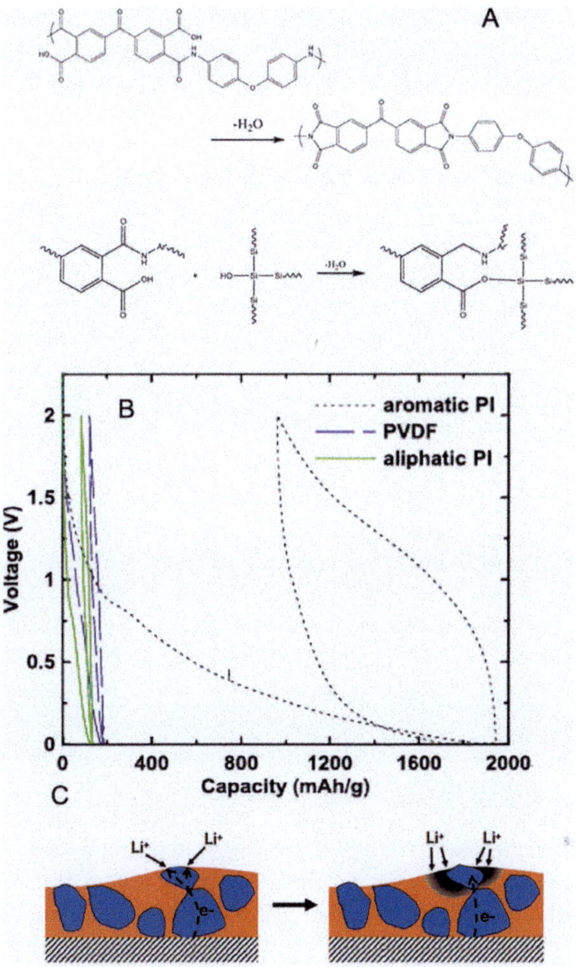

Figure 8.14 Polyimide binder for Si-based electrode. (A) Imidization of poly(amic acid) to polyimide and poly(amic acid) adheres to silicon surfaces. (B) Voltage curves of aro-PI, ali-PI and PVDF binders versus Li-metal. (C) Schematics of how electrochemical aro-PI reduction can occur in a composite coating that provides percolating paths for electrons to reach the coating surface. Aro-PI in contact with the electrolyte and with particles that have access to electrons from the current collector can become reduced and electrically conductive.

8.5 Development of water-soluble or dispersible binders

As the lithium-ion manufacturing industry moves toward environmentally benign processes, the research community continues to work toward phasing out NMP solvent in both anode and cathode slurry processing. There has been a recent explosion in R&D efforts focused on water-soluble binders and natural-material–derived binders for electrode applications. Notable for Si materials are CMC, SBR-based binders, alginate-based binders, LiPAA, and derivatives of these polymers. Most of chemical modifications are aimed at improving the binder properties, while preserving the aqueous solubility of these materials [4, 10, 38, 41].

One significant trend has been the effort to render fluoropolymers water processible. Due to the unique nature of fluorine chemistry, fluoropolymers are not soluble in water, but they can be made into emulsions. Poly(tetrafluoroethylene) (PTFE) was probably one of the earliest-used electrode binders in lithium-ion batteries precisely because PTFE water emulsions are widely available and quite stable, both chemically and electrochemically. PTFE has not seen continued use in lithium-ion anodes, however, mostly due to its high cost. Current research to render PVDF water dispersible is ongoing [29]. Due to the structural variety of PVDF derivatives, including copolymers, PVDF emulsions can be formulated with a wide range of mechanical properties to meet the requirements of Si electrode. Taking into account the superb stability of fluoropolymers as well, aqueous PVDF-based emulsions could very well find application in majority Si electrodes down the road.

Due to the rich chemistry of the conductive polymers, bottom-up design also allows for the development of aqueous-based electrically conductive polymer systems. Modification of conductive polymers into the water-soluble materials may follow one of two different strategies (Figure 8.15). The first strategy is to graft water-soluble functional groups (e.g. carboxylate) onto the conductive polymer backbone to enhance the aqueous solubility of the polymer. These functional groups also promote adhesion of the binder to Si particles and copper current collectors [51]. The drawback of this strategy is the complicated material synthesis required. However, the binder's water solubility and better adhesion with Si and Cu may well compensate for the synthesis process. The second strategy is to use a latex approach to make a binder suspension. Instead of using Suzuki coupling to build carbon–carbon bonds and rely on conduction through a conjugated backbone as in most conducting polymers, simple radical polymerization can be used to synthesize polymers that are side-chain conducting. These kinds of methacrylate-based polymers can be, therefore, be adapted to emulsion polymerization to make an aqueous suspension [52, 53]. Additional copolymerization also allows control over the adhesion and mechanical properties of the latex structure, enabling optimization of such latex binders based on the requirements of the Si surface.

8.6 Additional considerations

While this chapter has covered a broad range of work on functional binders, a few additional points remain to be addressed. The first is the working environment of the binders. Most types of adhesive force are weakened significantly when the joint is wet. Swelling of the adhesive with liquid can lead to the deconstruction of the adhesive interface. However, electrode binders must work while surrounded by an

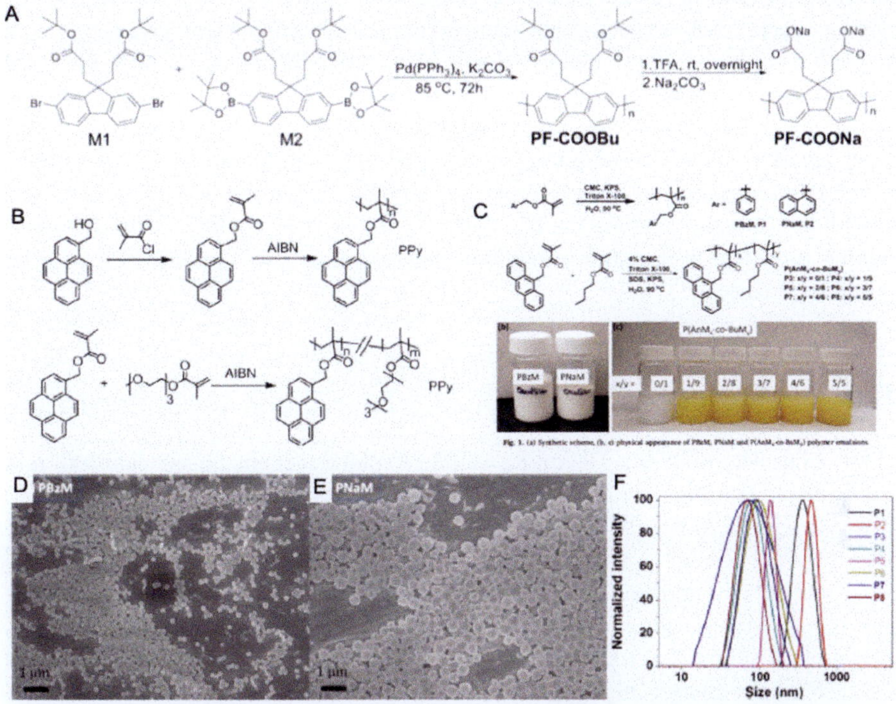

Figure 8.15 *Si functional binders are rendered into aqueous based binders. (A) Synthetic scheme of the water soluble conductive polymer sodium poly(3,3'-(9H-fluorene-9,9-diyl)dipropionic acid) abbreviated as PF-COONa. (B) Generic synthesis of poly(1-pyrenemethyl methacrylate) (PPy, top) and poly(1-pyrenemethyl methacrylate-co-triethylene oxide methyl ether methacrylate) (PPyE, bottom) through radical polymerization process. (C) The electrically conductive polymer binders are synthesized through emulsion polymerization as latex dispersion in water. (D). SEM image of PBzM (P1) particles from emulsion polymerization. (E) SEM image of PNaM (P2) particles from emulsion polymerization. (F) Normalized DLS size distribution of P1–P8 polymer emulsions.*

organic electrolyte, where the binders are fully swelled by the electrolyte to achieve the best ion transport rate and lowest interfacial impedance. Therefore, the adhesion force under real operating conditions is not the same as the force measured in dry electrode laminates. On the contrary, it is likely much reduced when compared to published data. The electrode also operates in a very electrochemically reducing environment: 10–50 mV vs. Li metal. Many functional groups that can provide covalent binding to the Si surface, such as –OH, –COOH, or –COOCH3, may not be stable at that potential and tend to be reduced. Electrochemical transformation of electrode binders has been widely observed using many techniques such as FTIR and first cycle coulomb counting. On the other hand, PVDF binders have been found to exhibit superb stability at low voltage, and CMC/SBR binders also show great stability at low voltage regardless of their –COOH and –COONa moieties. Finally, Si electrodes work best when under bulk external pressure. In fact, even carbonaceous anodes require pressure to operate reliably; generally, 10–20 PSI pressure is needed for optimum graphite performance. Without proper pressure, even a graphite electrode falls apart after only a few cycles. For a high-Si-content electrode, maintaining proper bulk pressure throughout cell operation will doubtless prove even more important.

References

[1] Chen H., Ling M., Hencz L., *et al.* 'Exploring chemical, mechanical, and electrical functionalities of binders for advanced energy-storage devices'. *Chemical Riews*. 2018;**118**(18):8936–82.

[2] Liu G., Zheng H., Song X., Battaglia V.S. 'Particles and polymer binder interaction: a controlling factor in lithium-ion electrode performance'. *Journal of the Electrochemical Society*. 2012;**159**(3):A214–21.

[3] Shi Y., Zhou X., Yu G. 'Material and structural design of novel binder systems for high-energy, high-power lithium-ion batteries'. *Accounts of Chemical Research*. 2017;**50**(11):2642–52.

[4] Li J.T., Wu Z.Y., Lu Y.Q., *et al.* 'Water soluble binder, an electrochemical performance booster for electrode materials with high energy density'. *Advanced Energy Materials*. 2017;**7**(24):30.

[5] Choi N.-S., Ha S.-Y., Lee Y., *et al.* 'Recent progress on polymeric binders for silicon anodes in lithium-ion batteries'. *Journal of Electrochemical Science and Technology*. 2015;**6**(2):35–49.

[6] Zhao H., Yuan W., Liu G. 'Hierarchical electrode design of high-capacity alloy nanomaterials for lithium-ion batteries'. *Nano Today*. 2015;**10**(2):193–212.

[7] Liu G., Zheng H., Kim S., *et al.* 'Effects of various conductive additive and polymeric binder contents on the performance of a lithium-ion composite cathode'. *Journal of the Electrochemical Society*. 2008;**155**(12):A887–92.

[8] Liu G., Zheng H., Simens A.S., Minor A.M., Song X., Battaglia V.S. 'Optimization of acetylene black conductive additive and PVDF composition for high-power rechargeable lithium-ion cells'. *Journal of the Electrochemical Society*. 2007;**154**(12):A1129–34.

[9] Wang L., Fu Y., Battaglia V.S., Liu G. 'SBR–PVDF based binder for the application of SLMP in graphite anodes'. *RSC Advances*. 2013;**3**(35):15022–7.

[10] Jeong S.S., Böckenfeld N., Balducci A., Winter M., Passerini S. 'Natural cellulose as binder for lithium battery electrodes'. *Journal of Power Sources*. 2012;**199**:331–5.

[11] Drofenik J., Gaberscek M., Dominko R., *et al.* 'Cellulose as a binding material in graphitic anodes for Li ion batteries: a performance and degradation study'. *Electrochimica Acta*. 2003;**48**(7):883–9.

[12] Mazouzi D., Karkar Z., Reale Hernandez C., *et al.* 'Critical roles of binders and formulation at multiscales of silicon-based composite electrodes'. *Journal of Power Sources*. 2015;**280**:533–49.

[13] Karkar Z., Guyomard D., Roué L., Lestriez B. 'A comparative study of polyacrylic acid (PAA) and carboxymethyl cellulose (CMC) binders for si-based electrodes'. *Electrochimica Acta*. 2017;**258**:453–66.

[14] Orikasa Y., Gogyo Y., Yamashige H., *et al.* 'Ionic conduction in lithium ion battery composite electrode governs cross-sectional reaction distribution'. *Scientific Reports*. 2016;**6**(1):6.

[15] Heidenreich R.D., Hess W.M., Ban L.L. 'A test object and criteria for high resolution electron microscopy'. *Journal of Applied Crystallography*. 1968;**1**:1–19.

[16] Ayala J.A., Hess W.M., Kistler F.D., Joyce G.A. 'Carbon-black-elastomer interaction'. *Rubber Chemistry and Technology*. 1991;**64**(1):19–39.

[17] Carmona F. 'The electrical-conductivity of polymers filled with carbon particles'. *Annales de chimie (Paris. 1914)*. 1988;**13**(6):395–443.

[18] Dannenberg E.M. 'Bound rubber and carbon black reinforcement'. *Rubber Chemistry and Technology*. 1986;**59**(3):512–24.

[19] Wu H., Yu G., Pan L., *et al.* 'Stable Li-ion battery anodes by in-situ poly merization of conducting hydrogel to conformally coat silicon nanoparticles'. *Nature Communications*. 2013;**4**(1):6.

[20] Xun S., Song X., Grass M.E., *et al.* 'Improved initial performance of Si nanoparticles by surface oxide reduction for lithium-ion battery application'. *Electrochemical and Solid-State Letters*. 2011;**14**(5):A61–3.

[21] Wright R.B., Motloch C.G., Belt J.R., *et al.* 'Calendar- and cycle-life studies of advanced technology development program generation 1 lithium-ion batteries'. *Journal of Power Sources*. 2002;**110**(2):445–70.

[22] Blow C.M. 'Polymer/particulate filler interaction—the bound rubber phenomena'. *Polymer*. 1973;**14**(7):309–23.

[23] Kenny J.C., McBrierty V.J., Rigbi Z., Douglass D.C. 'Carbon black filled natural rubber. 1. Structural investigations'. *Macromolecules*. 1991;**24**(2):436–43.

[24] Medalia A. 'Morphology of aggregates 6 effective volume of aggregates of carbon black from electron microscopy. application to vehicle absorption and to die swell of filled rubber'. *Journal of Colloid and Interface Science*. 1970;**32**(1):115.

[25] Feng J., Li J., Chan C.-M. 'Distribution of carbon black in semicrystalline polypropylene studied by transmission electron microscopy'. *Journal of Applied Polymer Science*. 2002;**85**(2):358–65.

[26] Hochgatterer N.S., Schweiger M.R., Koller S., *et al.* 'Silicon/graphite composite electrodes for high-capacity anodes: influence of binder chemistry on cycling stability'. *Electrochemical and Solid-State Letters*. 2008;**11**(5):A76–80.

[27] Xun S., Song X., Wang L., *et al.* 'The effects of native oxide surface layer on the electrochemical performance of Si nanoparticle-based electrodes'. *Journal of the Electrochemical Society*. 2011;**158**(12):A1260–6.

[28] Zhao H., Wang Z.H., Lu P., *et al.* 'Toward practical application of functional conductive polymer binder for a high-energy lithium-ion battery design'. *Nano Letters*. 2014;**14**(11):6704–10.

[29] Toigo C., Singh M., Gmeiner B., Biso M., Pettinger K. 'A method to measure the swelling of water-soluble PVDF binder system and its electrochemical performance for lithium ion batteries'. *Journal of the Electrochemical Society*. 2020;**167**(2):020514.

[30] Hernandez C.R., Etiemble A., Douillard T., *et al.* 'A facile and very effective method to enhance the mechanical strength and the cyclability of si-based electrodes for Li-ion batteries'. *Advanced Energy Materials*. 2018;**8**(6):1701787.

[31] Mazouzi D., Grissa R., Paris M., *et al.* 'CMC-citric acid Cu(II) cross-linked binder approach to improve the electrochemical performance of Si-based electrodes'. *Electrochimica Acta*. 2019;**304**:495–504.

[32] Jiang S., Hu B., Sahore R., *et al.* 'Tailoring the surface of silicon nanoparticles for enhanced chemical and electrochemical stability for Li-ion batteries'. *ACS Applied Energy Materials*. 2019;**2**(9):6176–83.

[33] Oh J.Y., Rondeau-Gagne S., Chiu Y.C., *et al.* 'Intrinsically stretchable and healable semiconducting polymer for organic transistors'. *Nature*. 2016;**539**(7629):411–5.

[34] Higgins T.M., Park S.-H., King P.J., *et al.* 'A commercial conducting polymer as both binder and conductive additive for silicon nanoparticle-based lithium-ion battery negative electrodes'. *ACS Nano*. 2016;**10**(3):3702–13.

[35] Liu G., Xun S., Vukmirovic N., *et al.* 'Polymers with tailored electronic structure for high capacity lithium battery electrodes'. *Advanced Materials*. 2011;**23**(40):4679–83.

[36] Vogl U.S., Das P.K., Weber A.Z., Winter M., Kostecki R., Lux S.F. 'Mechanism of interactions between CMC binder and Si single crystal facets'. *Langmuir*. 2014;**30**(34):10299–307.

[37] Wu M., Xiao X., Vukmirovic N., *et al.* 'Toward an ideal polymer binder design for high-capacity battery anodes'. *Journal of the American Chemical Society*. 2013;**135**(32):12048–56.

[38] Ling M., Qiu J., Li S., *et al.* 'Multifunctional SA-PProDOT binder for lithium ion batteries'. *Nano Letters*. 2015;**15**(7):4440–7.

[39] Liu J., Zhang Q., Wu Z.-Y., *et al.* 'A high-performance alginate hydrogel binder for the Si/C anode of a Li-ion battery'. *Chemical Communications*. 2014;**50**(48):6386–9.

[40] Kovalenko I., Zdyrko B., Magasinski A., *et al.* 'A major constituent of brown algae for use in high-capacity Li-ion batteries'. *Science*. 2011;**334**(6052):75–9.

[41] Ling M., Xu Y., Zhao H., *et al.* 'Dual-functional gum arabic binder for silicon anodes in lithium ion batteries'. *Nano Energy*. 2015;**12**:178–85.

[42] Parikh P., Sina M., Banerjee A., *et al.* 'Role of polyacrylic acid (PAA) binder on the solid electrolyte interphase in silicon Anodes'. *Chemistry of Materials*. 2019;**31**(7):2535–44.

[43] Cao P.-F., Yang G., Li B., *et al.* 'Rational design of a multifunctional binder for high-capacity silicon-based anodes'. *ACS Energy Letters*. 2019;**4**(5):1171–80.

[44] Kasinathan R., Marinaro M., Axmann P., Wohlfahrt-Mehrens M. 'Influence of the molecular weight of poly-acrylic acid binder on performance of Si-alloy/graphite composite anodes for lithium-ion batteries'. *Energy Technology*. 2018;**6**(11):2256–63.

[45] Hu B., Shkrob I.A., Zhang S., *et al.* 'The existence of optimal molecular weight for poly(acrylic acid) binders in silicon/graphite composite anode for lithium-ion batteries'. *Journal of Power Sources*. 2018;**378**(5):671–6.

[46] Wu M., Xiao X., Vukmirovic N., *et al.* 'Toward an ideal polymer binder design for high-capacity battery anodes'. *Journal of the American Chemical Society*. 2013;**135**(32):12048–56.

[47] Wu M.Y., Song X.Y., Liu X.S., Battaglia V., Yang W.L., Liu G. 'Manipulating the polarity of conductive polymer binders for Si-based anodes in lithium-ion batteries'. *Journal of Materials Chemistry A*. 2015;**3**(7):3651–8.

[48] Wang C., Wu H., Chen Z., McDowell M.T., Cui Y., Bao Z. 'Self-healing chemistry enables the stable operation of silicon microparticle anodes for high-energy lithium-ion batteries'. *Nature Chemistry*. 2013;**5**(12):1042–8.

[49] Choi S., Kwon T.-woo., Coskun A., Choi J.W. 'Highly elastic binders integrating polyrotaxanes for silicon microparticle anodes in lithium ion batteries'. *Science*. 2017;**357**(6348):279–83.

[50] Wilkes B.N., Brown Z.L., Krause L.J., Triemert M., Obrovac M.N. 'The electrochemical behavior of polyimide binders in Li and Na cells'. *Journal of the Electrochemical Society*. 2016;**163**(3):A364–72.

[51] Liu D., Zhao Y., Tan R., *et al.* 'Novel conductive binder for high-performance silicon anodes in lithium ion batteries'. *Nano Energy*. 2017;**36**(5):206–12.

[52] Park S.-J., Zhao H., Ai G., *et al.* 'Side-chain conducting and phase-separated polymeric binders for high-performance silicon anodes in lithium-ion batteries'. *Journal of the American Chemical Society*. 2015;**137**(7):2565–71.

[53] Zheng T., Zhang T., de la Fuente M.S., Liu G. 'Aqueous emulsion of conductive polymer binders for Si anode materials in lithium ion batteries'. *European Polymer Journal*. 2019;**114**(3):265–70.

Chapter 9

Surface modification for silicon anodes

Seoung-Bum Son[1] and Chunmei Ban[2]

9.1 Introduction

The discovery of its superior thermodynamic lithium (Li) storage capacity has successfully introduced silicon (Si) into the electrochemical energy storage field, 40 years after its applications in photovoltaic systems. Compared with Li intercalation into carbonaceous materials, insertion of Li into Si has the advantage of a very high Li packing density; the theoretical packing density in Li–Si alloys is very close to, or slightly higher than, that in pure Li metal, for example, $Li_{22}Si_5$: 0.081 vs Li: 0.0769 mol ml^{-1}. With this supremacy, Si has essentially dominated the efforts to seek higher energy density batteries in the last three decades.

The quest to revealing full potential of Si anodes is lengthy and onerous. Previous research has identified two major challenges that hamper the application of Si materials as an anode in Li-ion batteries, including the unstable electrode–electrolyte interphase and the electrode pulverization caused by the large volume changes upon lithiation. Lithiation of Si materials results in a large volume expansion, which can reach up to ~280%. The consequences of volume expansion are detrimental on all levels: it can cause current collector distortion, crushing of the separator, and cell bulging at the cell level; cause the active materials to lose electrical contact at the electrode level; cause disruption in the solid–electrolyte interphase (SEI) layer at the particle level; cause the increase of internal stress resulting in particle fracture at the subparticle level. The use of X-ray and electron microscopy-based characterization and measurements in late 1990s and 2000s revealed the lithiation mechanism in Si bulk structure, tracked substantial Si volume change, and the phase transformation during cycling. Information from these characterization research studies inspired the research in nanosized Si starting in the late 2000s, where nanosized particles (less than a critical size of ~300 nm for nanowires and ~150 nm for nanoparticles) would limit particle fracture and pulverization and suppress the formation of crystalline $Li_{15}Si_4$ during lithiation. Using nanostructured architecture—including 1D nanowires, 2D thin film, and 3D nanoparticles—greatly

[1]Argonne National Laboratory, 9700 South Cass Avenue, Lemont, IL, USA
[2]University of Colorado Boulder, 1111 Engineering Dr. Boulder, Boulder, CO, USA

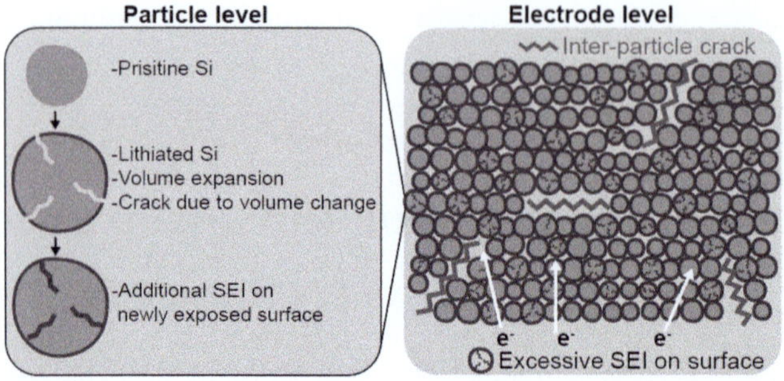

Figure 9.1 *Deterioration of Si anode during continuous lithiation/delithiation in particle and electrode level*

increases the reversible cycling capacity and cycling life. However, decreasing the particle size increases the surface area and catalyzes the interaction between surface species of Si nanomaterials and reactive electrolyte components, resulting in complications in the formation of the SEI layer, which is associated with large irreversible consumption of Li ions.

SEI formed on the graphite anode is indispensable for long-term cyclability of Li-ion battery (LIB) as it stabilizes the graphite surface and prevents further side reaction at the electrolyte/graphite interface. However, structural instabilities such as cracks and pulverizations due to volume changes of Si with continuous lithiation/delithiation expose fresh Si surface to the electrolyte and form additional SEI on the newly formed surface sites. As a result, the excessive amount of SEI formation on the surface of Si anode decreases the electronic conductivity. In addition, continuous volume changes of Si cause inter-particle crack inside Si anodes, and this also considerably limits the electronic conductivity of the Si anodes. Overall, the deterioration of Si anode increases the resistance of electronic/Li-ionic conduction of Si particles and leading to rapid degradation of Si's capability as an energy storage material. Schematics in Figure 9.1 illustrate aforementioned deterioration of the Si during continuous lithiation/delithiation in particle and electrode level.

Original SEI layers immediately form at the electrolyte–anode interface when an anode contacts a carbonate electrolyte, in an uncontrolled and poorly understood process involving complications and uncertainties dependent on native surface species, impurities and composition of electrolyte, cell assembly and operating conditions (cycling protocols, temperature, and pressure). One practical strategy to tailor interphases at the electrolyte–anode interface is to introduce surface coatings acting as an artificial SEI layer prepared before cell assembly. It provides an opportunity to endow new functionalities to artificial interphase, such as facilitating ionic and electron diffusion, stabilizing surface/interface, and offering self-repairing at cracking sites. Instead of relying on electrolyte reduction toward interphases formation, artificial interphase layer can be well tuned in terms of composition, thickness and mechanical and chemical properties, when an appropriate synthesis method is used.

Using surface coatings has shown great promise in enabling graphite or Si anodes, where functional coating layers prevent direct contact from electrolyte and suppress detrimental reduction reactions of electrolytes. Zhang and co-workers applied a thin coating of AlF_3 from chemical vapor deposition (CVD) on graphite powders and observed a 10% capacity increase compared with an uncoated graphite anode. Battaglia's team coated graphite particles with sodium maleate salt via wet process to improve the mechanical properties of graphite anodes. They found that radical polymerization between maleate molecules enhances elasticity of the interphase, leading to improved mechanical integrity. Rudolf and co-workers demonstrated highly reversible and improved rate performance by coating graphite particles with Al_2O_3 through sol-gel method. Later, polymeric materials were used by Long *et al.* and Lee *et al.* to suppress side reactions. Meanwhile, surface coating has also been applied to Si alloy anode materials, to mitigate unfavorable reduction of electrolyte. Among coating materials, carbon has been widely used to promote stable SEI formation and enhance electronic conductivity. On the other hand, coating Si particles with metal oxides, such as Al_2O_3 and TiO_2, has shown improved cycling performance, especially in the suppression of side reactions at Si anode surface. This chapter will discuss the details of the functionalities of the selected coating materials, as well as the coating techniques for scalable surface chemistry modification.

9.2 Materials and techniques for surface modification

9.2.1 Carbonaceous material coatings

Graphitic and carbonaceous materials are good choices to modify Si surface because of their structural flexibility, high electronic conductivity, and good electrochemical stability. Moreover, the use of coatings can effectively suppress the particle–particle interaction, especially for nanoparticles with high surface area. The high surface energy of nanoparticles promotes electrochemical sintering, inducing the particle aggregation and the formation of inert dense blocks [1]. Using coating protects the surface from interacting with other particles, preserving high surface area of nanoparticles.

Using carbonaceous hosts can be dated back to 1995 [2]. It was recognized at that time that the alloying process in Si particles might not be sufficiently reversible, because of the phase transformation and the volumetric changes associated with the alloy process. Direct replacement of the already commercialized graphite-based anodes with Si materials would be quite challenging. Dahn and co-workers introduced Si-carbon composite materials by preparing carbonaceous materials, which contain Si atoms in poorly stacked regions of carbon structures [2]. These composite materials have been formed via CVD with Si precursors such as Si tetrachloride ($SiCl_4$) or dimethyldichlorosilane [$(CH3)_2C_{12}Si$] and carbonaceous precursors such as benzene. The carbonaceous buffer matrix can accommodate volumetric changes while helping the formation of a stable SEI film on the Si surface. Thus, the Si–C composites delivered the reversible cycling behavior with a specific capacity near 500 mAh g^{-1}. Later, they explored the possibility of making similar materials by pyrolysis of siloxane polymers, which contain Si, C, and oxygen elements in one precursor [3, 4]. The compounds exhibited reversible capacity around 600 mAh g^{-1} with an increased amount of Si content (12 atomic % Si), as compared with the capacity of ~340 mAh g^{-1} for the pitch-based carbon. The addition of the Si component increased the reversible capacity, but also the first-cycle irreversible capacity loss. When the carbon content reached more than 70%, the composites showed a much lower reversible capacity. They suggested that sufficient carbon content is required to enable reversible (de)lithiation of the composites. Furthermore, the existence of oxygen impurities is an intrinsic problem of the synthesis by using polysiloxane precursors, which results in large irreversible capacity losses of ~200 mAh g^{-1} [5].

The early studies on Si–C composites confirmed the positive effects of the addition of carbonaceous materials in the Si materials. On the basis of the previous work, systematic investigations including chemical synthesis, characterization, and surface analysis have been conducted on the carbon-coated Si particles, to stabilize the surface of Si materials so that Si materials can be applied to LIB anodes. This section is focused on two main coating approaches including pyrolysis of polymers and CVD, because of their effectiveness, low-cost, and scalable capabilities for future commercialization. Other reported approaches such as hydrothermal carbonization, spray pyrolysis, and sputtering have certain similarities with either pyrolysis of polymers or CVD, but with different reaction temperature and pressure. In addition to discussing the carbon-coating effects on electrochemical properties, one section will be dedicated to characterization of carbon coating and surface analysis, because their results significantly improve understanding the effects of surface coating on electrochemistry of the interphases between the carbon coating and the Si surface.

9.2.1.1 Carbon coatings via pyrolysis of polymers

Polymer pyrolysis has been extensively investigated for preparing carbon coatings on Si particles. Instead of using Si-contained polymers such as polysiloxane for synthesizing Si–C composites, the direct use of Si particles as Si sources largely reduces oxygen impurities while increasing Si content for even higher reversible capacities. Improved cycling performance has been achieved by using the pyrolyzed carbon-coated Si particles via pyrolysis of polyvinylidene fluoride (PVDF) [6], pitch [7], poly vinyl alcohol [8], or poly vinyl chloride [9]. Takeda and co-workers summarized the cycling behavior of the pyrolyzed carbon-coated Si composites under potential cutoff between 20 mV and 1.5 V vs Li/Li+, with the comparable carbon content. The capacity retention after 16 cycles was 45, 44, 47, 62, 47, 63, 46, and 67% for the pyrolyzed Si–sucrose, Si–resin, Si–poly ethylene oxide (PEO), Si–PVC, Si–PE, Si–CPE, Si–pitch, and Si–PVDF composites, respectively. It was found that the pyrolyzed Si–PVDF electrode showed most stable cycling behavior. Note that Si particles used in this work have an average size of 1 μm, which was not small enough to mitigate fracture and cracking in Si particles. Later, J. Huang's team and Y. Cui's team identified the particles' critical sizes for crystalline Si particles (150 nm) and amorphous Si particles (870 nm), below which cracking was not observed [10, 11]. After replacing the micro-sized particles with nanoscale particles, the cycling performance was greatly improved with much higher capacities of 1328.8 mAh

g^{-1} with a capacity retention of 76.8% over 150 cycles [12]. The data also suggested that the acetylene black and pyrolyzed carbon materials had small reversible capacities: 90 and 65 mAh g^{-1}, respectively, indicating a tiny contribution to the total capacity of the pyrolyzed carbon-coated Si anodes—which also used the acetylene black as the conductive additive. Without the coating, the Si anodes exhibited a relatively low reversible capacity of 364.7 mAh g^{-1} and decay sharply within 30 cycles.

Derived from the polymer pyrolysis, hydrothermal carbonization followed by annealing is an alternative method to create hydrophilic carbon materials. The process includes a hydrothermal process of a water-soluble carbohydrate heated at mild temperatures (180–200 °C) in an autoclave. During the hydrothermal process, carbohydrates undergo dehydration, subsequent polymerization, and finally carbonization. For example, Maier and co-workers coat Si nanoparticles (20–50 nm) by using hydrothermal carbonization of glucose to form a Si@SiO_x/C nanocomposite [13]. Although the annealing at 750 °C was performed to improve the structural order of the carbon coating, the large I_D/I_G ratio indicated from Raman spectroscopy indicated the low-graphitic degree in the nanocomposites. In addition to the growth of carbon coating, Si oxide (SiO_x) coatings were also formed on the surface of Si nanoparticles during the hydrothermal process and subsequent annealing processes, as shown in Figure 9.2 [13]. Slightly increased cycling performance was observed for the Si@SiO_x/C nanocomposite anode than the pure Si nanoparticles. Rapid capacity decay was shown for the carbon-coated nanocomposites when using a typical Li-ion electrolyte (1M $LiPF_6$ in EC/DMC solution) suggesting that the double-layered SiO_x/C coating seems insufficient for obtaining good cycling performance. However, when the Si@SiO_x/C nanocomposite electrode was cycled in the vinylene carbonate (VC)-containing electrolyte, an excellent cycling performance was achieved as indicated in Figure 9.3. Note that VC is widely regarded as the best agent for the formation of passivating films on the surface of electrode materials, especially for carbon materials. The poly alkyl Li carbonate species found on the SEI layer of graphite anodes are expected to be more cohesive and flexible and thus provide better passivation than surface films comprised of only Li salt [14].

Alternatively, aerogel carbonization has been used to host Si particles in a carbonaceous porous matrix. Wang *et al.* used resorcinol and formaldehyde precursors for preparing carbon aerogel for hosting Si nanoparticles, followed by carbonization at 650 °C [15]. The carbon-coated Si anode delivered a highly reversible capacity of about 1 450 mAh g^{-1} but with an irreversible capacity of about 550 mAh g^{-1} for the first cycle. When introducing covalent bonding between Si particles and carbon coating by hydrosilylation, slightly improved capacities were observed but again with large irreversible capacity loss [16]. The preparation schematic in Figure 9.4 shows that the H-terminated Si (HSi) nanoparticles were hydrosilylated with 2-allylphenol (2AP) prior to resorcinol-formaldehyde polymerization. The resulting gel was then carbonized at 700 °C in argon atmosphere. Without carbon encapsulation, the Si nanoparticles anode experienced drastic capacity degradation with over 90% of the initial capacity lost over 10 cycles. With the carbon matrices, the Si anodes in this work exhibited an initial reversible capacity of 590 mAh g^{-1} at 600 mA g^{-1} (~1C, based on

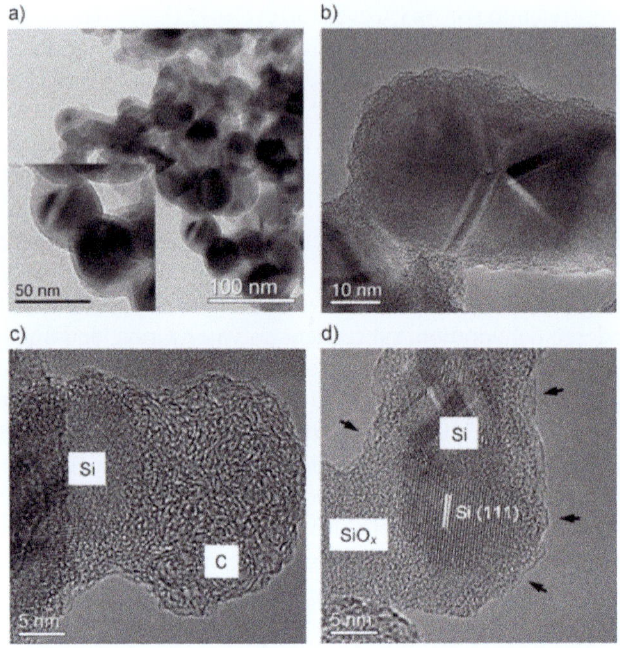

Figure 9.2 *Morphology and composition of the carbon coating produced by hydrothermal carbonization and further carbonization at 750 °C under N_2. (a) Overview of the Si@SiOx/C nanocomposites and a TEM image at higher magnification (in the inset) showing uniform spherelike particles; (b) HRTEM image clearly showing the core/shell structure; (c, d) HRTEM image displaying details of the silicon nanoparticles coated with SiOx and carbon. Reproduced with permission from [13]. Copyright 2008 WILEY-VCH Verlag GmbH & Co. KGaA, Weinheim.*

the weight of Si-C composites) and a capacity retention of ~76% over 50 cycles. The authors attributed these improvements in cycling stability and rate capability to the flexible carbon matrix obtained at relatively lower carbonization temperature, such as 700°C used in their work. When using higher carbonization temperatures (>700 °C), the carbon matrix tends to become more rigid after further dehydrogenation and less capable to accommodate volume changes caused by the embedded Si particles [13, 15].

In addition to the formation of amorphous carbon coating via polymer pyrolysis, Piper *et al.* cyclized polyacrylonitrile (PAN) polymer to form hybrid graphitic-polymer coating for Si nanoparticles [17]. When limiting the cyclization temperature below 350 °C in an inert environment, PAN proceeds initial cyclization to form a pyridine-based conjugated polymer without carbonization, thus the obtained coating still maintains polymeric properties while introducing

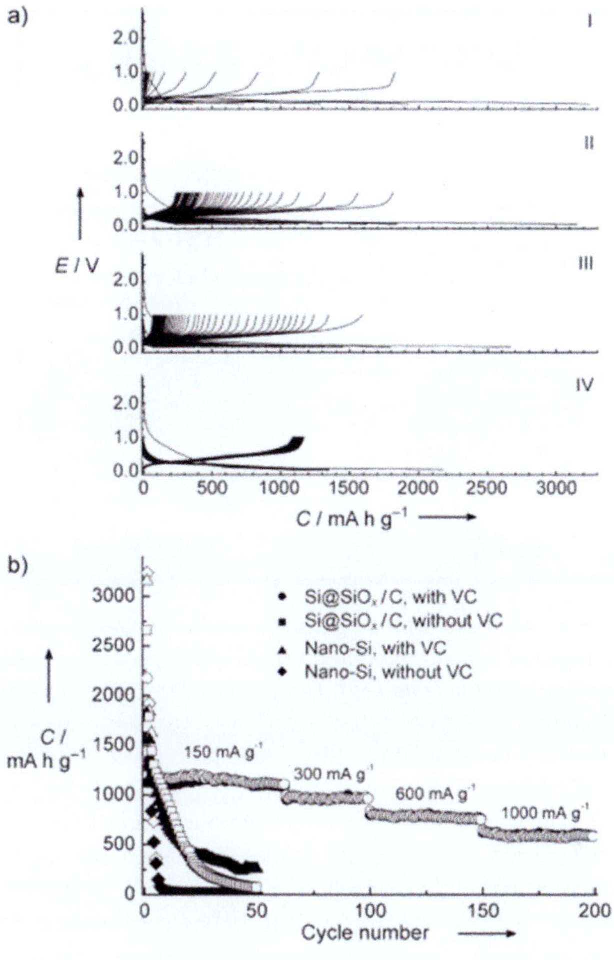

Figure 9.3 *(a) Galvanostatic discharge/charge curves (I, II) of pure Si nanoparticle and carbon-coated Si nanocomposite electrodes (III,IV) cycled at a current density of 150 mA g^{-1} between the voltage limits of 0.05^{-1} V in vinylene carbonate (VC)-free (I, III) and VC-containing 1 M $LiPF_6$ in ethylene carbonate/dimethyl carbonate (EC/DMC) solution (II, IV). (b) Cycling and rate performance of pure Si nanoparticles and Si@SiOx/C nanocomposite electrodes cycled in VC-free and VC-containing 1 M LiPF6 in EC/DMC solutions (solid symbols: charge; empty symbols: discharge). Reproduced with permission from [13]. Copyright 2008 WILEY-VCH Verlag GmbH & Co. KGaA, Weinheim.*

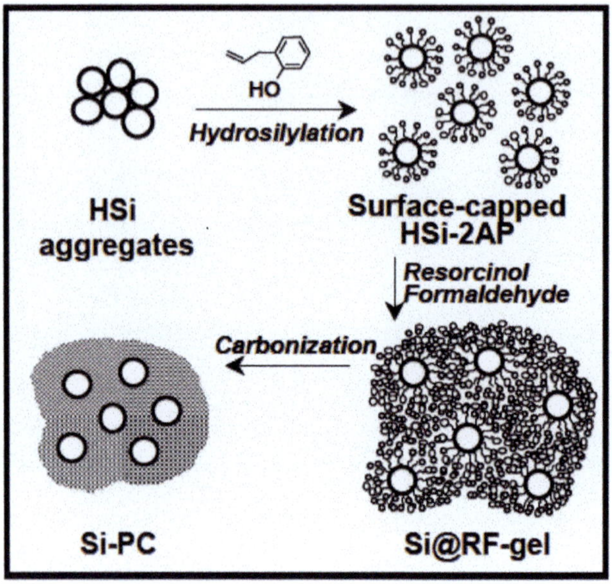

Figure 9.4 *Synthetic scheme of the carbon-coated Si anodes by using the carbon aerogel from resorcinol and formaldehyde precursors. Reproduced with permission from [15]. Copyright 2008 American Chemical Society.*

delocalized sp2 π bonding for intrinsic electronic conductivity. Results from advanced characterization and analysis studies of interphases suggest that both chemical composition and surface morphology determine the properties of the interphase, ultimately dictating the electrochemical cycling performance. The in situ Si-PAN cyclization enables uniform distribution of Si nanoparticles into the flexible cyclized-PAN matrix while protecting the nanoparticles with the hybrid coating, as displayed in Figure 9.5. As a result, there was no cracking observed for the cyclized-PAN coating and the Si nanoparticles, nor signs of delamination detected at the interface between the cyclized-PAN and Si. The authors suggested that the superior resiliency of the cyclized-PAN coating successfully accommodated large volumetric changes where previously reported dense carbon matrices would have experienced brittle failure. Furthermore, the existence of ordered sp2 graphitic structure in the cyclized-PAN coating offered good electronic conductivity, resulting in improved rate capability.

But due to the instantaneous nature of the process, pyrolysis of organic carbon precursors leads to the formation of disordered carbon, porosity, and the presence of impurities such as hydrogen, oxygen, or sulfur in the final product, resulting in reduced conductivity and stability, which causes large initial capacity loss and lower reversible capacity than the theoretical capacity of Si particles.

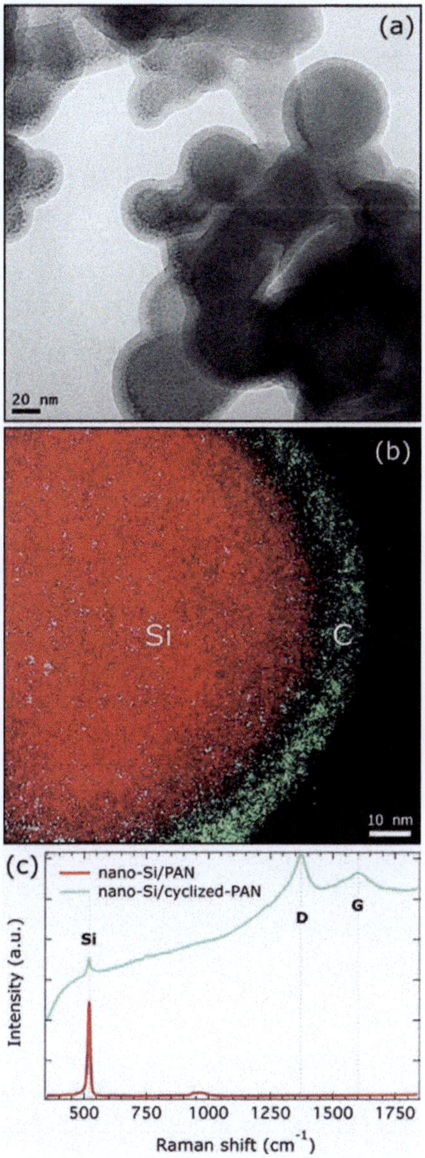

Figure 9.5 *(a) TEM micrograph of the uncycled nano-Si/cyclized-PAN electrode with (b) EELS elemental mapping of the conformal 5-nm-thick cyclized-PAN carbon coating. (c) Raman spectra for an untreated electrode (red) and for an electrode treated at 300 °C (cyan). The delocalized sp^2 π D and G bands present in the spectra for the heat-treated sample confirm cyclization of PAN coatings. Reproduced with permission from [17]. Copyright 2013 WILEY-VCH Verlag GmbH & Co. KGaA, Weinheim.*

9.2.1.2 Carbon coating via chemical vapor deposition

In contrast to solution-based approaches including polymer pyrolysis and sol-gel method, vapor deposition techniques such as sputtering, pulsed laser deposition (PLD), and CVD offer better control over the thickness at the nanometer level. As compared with sputtering and PLD, the CVD technique is a non-line-of-sight deposition technique with the capability of providing conformal coating on 2D substrates. CVD with vapor precursors (benzene, methane, acetylene, propylene, or toluene) has been wildly used to coat Si-based particles and films with carbonaceous materials.

Ogumi *et al.* are among the first groups to apply CVD carbon coating to Si particles in the early 2000s [18]. The CVD carbon coating enabled reversible cycling of Si microparticles (with an average particle size larger than 5 μm), although the reversible capacities were less than 1 000 mAh g^{-1}, much lower than the theoretical capacity of Si particles [18–20]. They attributed the improved cyclability to the fact that the carbon coating can suppress the decomposition of electrolyte and provide integral and continuous electric contact networks around Si particles [20, 21]. However, the cycling performance often faded after extended cycling (>50 cycles), in large part because of Li-ion trapping and the failure of the entire electrode structure exacerbated by cracking in micro-Si particles [1, 19, 21]. The results, while confirming the positive effects of carbon coating for Si anodes, suggest that the strategies to mitigate mechanical failures of electrode structures should be also considered to ultimately stabilize cycling performance.

Research into the root cause of mechanical fracturing and cracking of Si particles shows using Si nanoparticles less than the critical size would largely increase fracture toughness of the lithiated alloys [22–24]. Significant improvement in cycling properties has been well demonstrated by using nano-engineered Si materials, such as Si nanowires, Si nanotubes, interconnected hollow Si nanospheres, and Si core/hollow carbon nanostructures [25, 26]. The development of nanoscale Si materials as the anode materials is detailed in Chapter 7. The use of nano-structured Si particles has been proven to increase the overall capacity due to the improved mechanical stability against fractures and cracking, as well as short diffusion path for Li^+ ions.

When combining the morphological design and the carbon-coating strategies, further improved performance has been achieved with negligible capacity fade over hundreds of cycles. Yang *et al.* reported that the CVD carbon-coated Si nanowires with an average diameter of 90 nm offered better cycling performance including higher capacity retention and better rate capability than those from the uncoated Si nanowires [27]. Ma *et al.* reported that the CVD carbon-coated yolk-shell-structured Si anodes showed a specific capacity of 796 mAh g^{-1} with a high retention of 88.3% after 100 cycles at a high current density of 1.0 A g^{-1}, while the capacity retention of the uncoated electrode was only ~ 48.5% after 100 cycles [28].

Combining high electrical conductivity with strong mechanical integrity, graphene coating has been rapidly adopted to enhance the structure integrity and provide electrical conductivity. The graphene layers anchored onto the Si surface

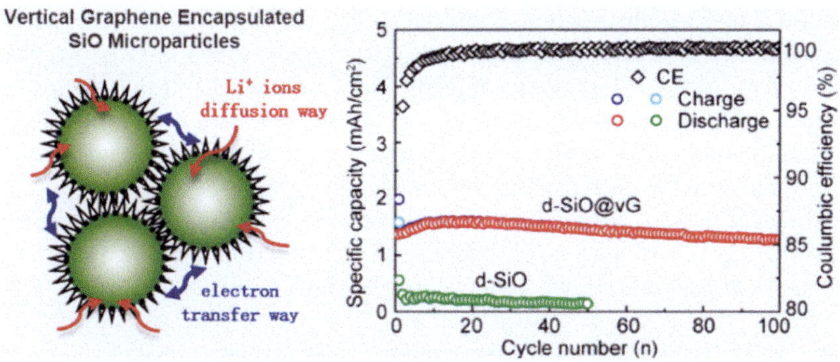

Figure 9.6 Schematics of direct CVD growth of vertical graphene nanosheets on SiO (left); and the comparison of electrochemical cycling performance for both pristine SiO and the graphene-encapsulated SiO anodes (right). Reproduced with permission from [30]. Copyright 2017 American Chemical Society.

accommodate the volume expansion of Si via a sliding process between adjacent graphene layers [29]. Son *et al.* used a mild oxidant (using a mixture of CO_2 and CH_4, instead of solely using CH_4) to avoid the formation of SiC during CVD process, as well facilitate the growth of graphene coating at a relatively low temperature of 1 000 °C [29]. When paired a commercial Li cobalt oxide cathode with the graphene-coated Si anode, the full cell delivers a volumetric energy density of 972 and 700 Wh l^{-1} at the 1st and 200th cycle, respectively, which are 1.8 and 1.5 times higher than those of current commercial Li-ion batteries. Later, Liu and coworkers reported the growth of vertical graphene on surface of Si particles via CVD process. The process is unique because of the simultaneous disproportionation reaction and CVD graphene growth on Si monoxide particles, as shown in Figure 9.6 [30]. During the CVD process, the SiO microparticles were transformed into a SiO_x matrix where Si nanoparticles were immersed into, while forming the amorphous SiO_2 surface that provides oxygen sites for graphene nucleation. They found that the silica skin layer would act as a barrier layer to prevent the formation of inert SiC, which was proven critically challenging for the direct growth of carbon coating for Si particles under CVD process [18, 29]. Consequently, the resultant Si-based materials showed impressive stability (100 cycles, 93% retention) even at a high areal mass loading of 1.5 mg cm^{-2}. When adding only 5 wt.% of the coated composites in the graphite anode, the mixture anode boosted the capacity of graphite/NCA 18 650 cell to 3 200 mAh (vs. 2 800 mAh in the commercial graphite/NCA 18 650 cell), manifesting great potential for industrial application.

Furthermore, to accommodate the inevitable volumetric changes during (de) lithiation, porous structures may be a viable solution to conserve the volume needed for Si expansion, which is critical for industrial applications because commercial battery cells have very little volume available for anode expansion. However, nanoscale porous structures dramatically increase the surface reaction sites for the unfavored parasitic reactions between Si and electrolyte solutions, as well as reduce the overall electrical conductivity. The former results in large irreversible capacity loss and the formation of thick SEI layers which further increase cell impedance. Therefore, the impact of carbon coating becomes even more important for the porous structures, to suppress the surface reactions and ensure the conductive matrix for electrochemical reactions.

Excellent cycling performance was demonstrated on the porous Si composites coated with CVD carbon [31–33]. Yushin and co-workers designed a two-step CVD synthesis of Si-carbon porous composites, in which Si nanoparticles were firstly deposited on the surface of annealed carbon-black particles followed by the carbon deposition on the self-assembled large porous spherical granules [34]. As a result, the carbon-coated Si-carbon porous structure offered a specific capacity of 1 590 and 870 mAh g^{-1} at the fast discharge rates of 1 and 8C, respectively. Magnesiothermic reduction process has been also used to produce porous Si structures, followed by CVD carbon coating [32, 33]. The coated Si anodes demonstrated highly reversible capacity of 1 835 mAh g^{-1} and outstanding cycle stability. When mixed with graphite materials, the Si-graphite composite anode offered a specific capacity of 850 mAh g^{-1} and 3.4 mAh cm^{-2} areal capacity with 92% capacity retention after 200 cycles [33]. Full cells assembled with the commercial Li iron phosphate cathode and the composite anode also demonstrate excellent performance.

Most of the above efforts in CVD depositing carbon coating are based on static deposition design, in which only partial surface areas of particles are exposed to vapor precursors, and powder particles are prone to aggregate during the CVD process. Thus, the deposition thus obtained might be unevenly distributed, and the mass production is challenging. To address these issues, rotatable furnace and fluidized thermal CVD techniques have been recently developed to achieve uniform coating while avoiding particle aggregation, and most importantly, it paved the way for large-scale carbon coating in commercial applications [32, 35–38]. As compared to conventional CVD under static deposition mode, the new technologies enable to-be-coated particles to be fluidized and suspended in the precursor gas phase for uniform carbon coating [38]. Figure 9.7 that exemplifies the design of a fluidized thermal CVD illustrates the synthesis procedure for the carbon-coated Si–SiO_x composites [36]. The thermal disproportionation occurred during CVD induced the formation of Si nanoparticles embedded in SiO_x matrix, which reacted in a manner similar to the design used in [30, 31].

9.2.1.3 Effects of carbon coating on conductivity and SEI composition

As compared to amorphous carbon coating, the graphitic carbon coating further improves the cycling performance. Nava *et al.* found that the graphitic carbon-coated Si anodes exhibited average R_{SEI} and R_{CT} values that are 40 and 67% lower than the impendences of the amorphous carbon-coated Si nanoparticles anodes [39]. The authors attributed the reduction of the SEI layer impedance R_{SEI} to the higher electrical conductivity and the slightly lower coating thickness and impurities of the graphitic carbon-coating than that of the amorphous carbon

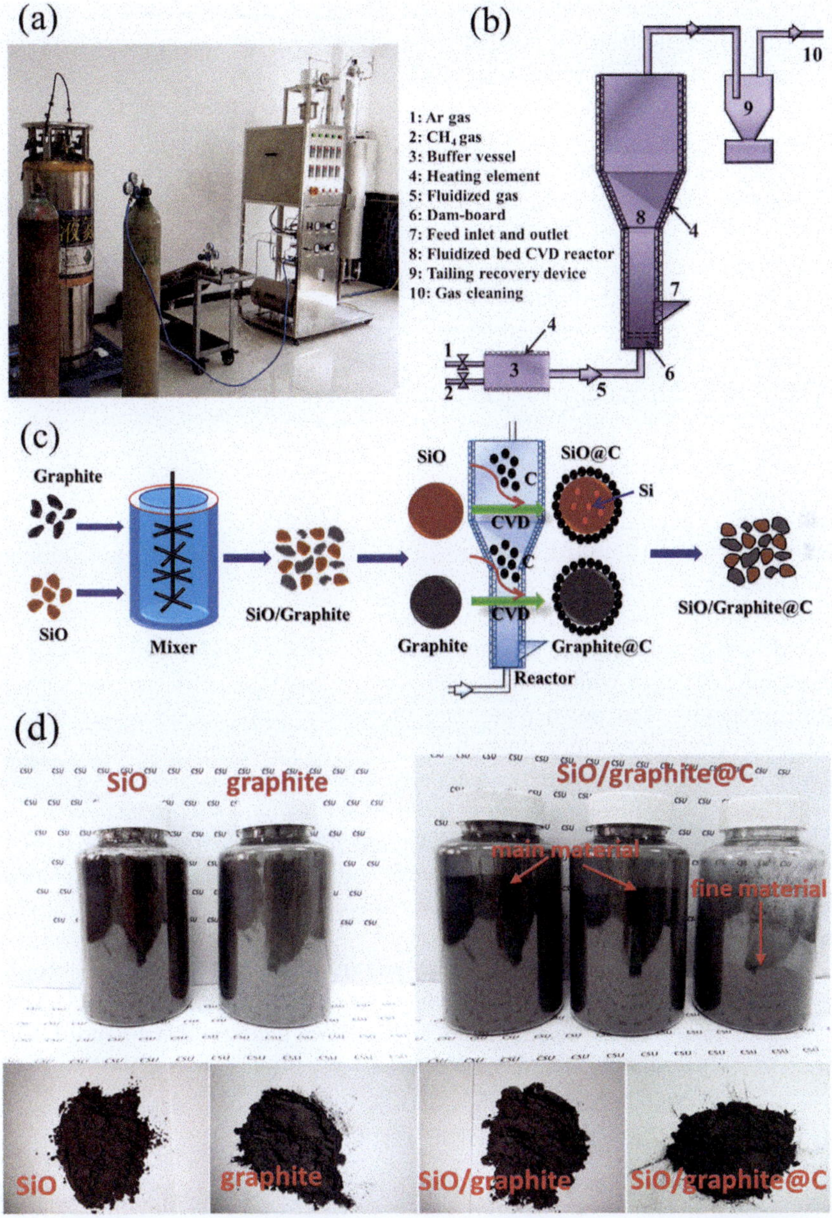

Figure 9.7 *Illustration of (a) the fluidized bed reactor and (b) its schematic diagram for large-scale carbon coating for silicon particles. (c) Schematic of formation mechanism for carbon layer and Si crystal. (d) Photograph of SiO, graphite, SiO/graphite, and SiO/graphite@C product.). Reproduced with permission from [36]. Copyright 2018 Elsevier Ltd and Techna Group S.r.l.*

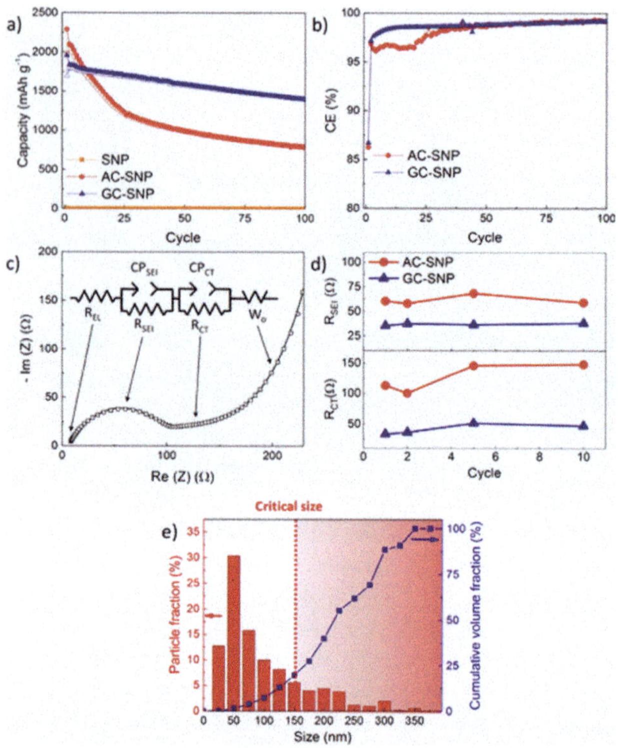

Figure 9.8 *Effects of conductive and nonconductive carbon coatings on electrochemical cycling performance. (a) and (b) show charge/discharge curves and Coulombic efficiency of pristine SNP, AC-SNP, and GC-SNP anodes. (c) displays the electrochemical impedance plot and fitting procedure used for the analysis of the impedance curves. (d) compares* R_{SEI} *and* R_{CT} *obtained from the fitting of the impedance curves. (e) the observed SNP size distribution and cumulative volume distribution. The critical size for crack formation during lithiation of crystalline silicon is indicated with a dotted red line to guide the eye of the reader. Reproduced with permission from [39]. Copyright 2019, American Chemical Society.*

coating layer. They stated that the reduction in charge transfer impedance R_{CT} suggests that the highly graphitic carbon coating can also better facilitate the transport of Li ions to the Si nanoparticles core area compared to the amorphous carbon coating. Consequently, the graphitic carbon-coated Si anodes (GC-SNP) retain 71% of their initial reversible capacity after 100 cycles, as compared to the retention of only 34% for the amorphous carbon-coated Si anodes (AC-SNP), as shown in Figure 9.8a and b. Note that both the amorphous and graphitic carbon coatings enabled reversible cycling of Si nanoparticles. Gradual increases in R_{SEI} and R_{CT} values were observed for both carbon coatings (Figure 9.8c and d), which might be related to the locally uneven stress/strain distribution and corresponding loss of electric contact within the electrode layer. Figure 9.8e depicted the particle size distribution and their contribution to volume fraction of the sample. Surprisingly, the small number of large particles (≥150 nm, critical size for fracture during lithiation) actually accounts for a significant volume fraction of the sample. Thus, the inevitable mechanical fracture caused by the large particles lead to localized mechanical and electrical disconnections, hence explaining the gradually increased R_{CT} values and capacity fade.

Raman spectroscopy was often used to identify the amorphous carbon structure and indicate the graphitization level in the coating layers. Two obvious characteristic peaks of D and G bands were usually observed in the typical amorphous carbon [40]. The G-bands are resonantly enhanced tangential bands between 1 500 and 1 600 cm^{-1}, which are strong first-order bands, consisting of six components with 2A + 2E1 + 2E2 symmetries arising from curvature-induced splitting of the tangential E2g mode of graphite. The D-bands are shown in a broad band at ~1 350 cm^{-1} attributed to a convolution of the disorder-induced band of carbon impurities and defects in the hexagonal sp2 carbon and edges. The Raman spectrum is a characterizing signature of carbon-based materials, and the intensity ratio of D and G bands (I_D/I_G) can be also used to evaluate the degree of graphitization in carbon-based materials. The I_D/I_G ratio is inversely proportional to graphitization degree. A lower I_D/I_G ratio means a higher degree of graphitization of the carbon component in the composites.

Wilcox *et al.* pointed out that using a fit with two carbon D and G lines did not give accurate results [41]. Four Gaussian bands are necessary to account for the observed Raman feature with minimum error. As shown in Figure 9.9, the bands at ~1 518 and 1 190 cm^{-1} have uncertain origins, but they have already been observed in disordered carbon black and diamond-like carbons. That observation could imply that short-range vibrations of sp3-coordinated carbons contribute to the disordered spectra. The sp2/sp3 ratios have also been used to compare the content of sp2 graphene fraction. When the graphene domain sizes are maintained similar, the D/G ratios are not significantly changed. In this case, the sp2/sp3 ratios can be used to indicate the proportion of carbon with a graphitic nature and the electronic conductivity. The improved rate capability was observed in the carbon-coated Si materials annealed at high temperature (≥800 °C) [31]. The authors attributed the improved rate performance to the improved carbon quality confirmed with low D/G ratios and high sp2/sp3 ratios, indicating the enhanced electronic conductivity.

In addition to the impact on the electrical conductivity, the carbon coating on the Si particles has a profound effect on the surface morphology and SEI composition. Yen *et al.* found that after prolonged cycling, the Si electrode made with Si microparticles (~3 μm) exhibits a very rough surface, and its SEI contains an abundance of fluorinated C and Si species [21]. The C-coated Si electrode (coating via CVD of benzene) exhibits a smooth SEI layer that does not contain the fluorides but some siloxane species. Other than these differences, however, these two electrodes

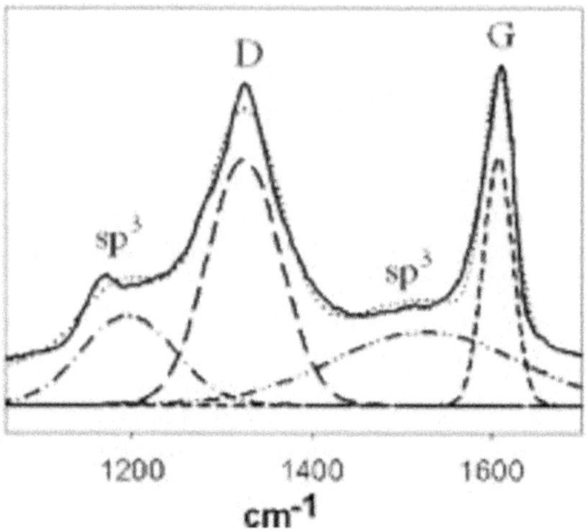

Figure 9.9 *Raman spectrum of the residual carbon is resolved into four bands including D and G bands, and the short-range vibrations of sp3-coordinated carbons. Reproduced with permission from [41]. Copyright the electrochemical society (published by IOP publishing).*

do show some common SEI species, including, Li_2CO_3, $ROCO_2Li$, LiF, ROLi, and polyethylene oxides, which have long been known to the SEI of graphite electrode. The SEI layer on the Si electrode vividly shows the presence of LiF after one cycle, while LiF was not detected on the C–Si counterpart. The data suggest that the C coating helps to reduce decomposition.

9.2.2 Si-metal alloy modification

Si-metal alloy systems can be categorized into two sections based on whether the metal element does or does not participate in electrochemical reactions with Li. It should be noted that these two have dissimilar electrochemical mechanism with Li. This chapter will first describe the Si-metal compounds which are readily active to Li, and then describe Si-metal compounds where only Si phase actively reacts with Li. Readers will note that most of Si-metal alloys were synthesized by using high energy ball milling (HEBM) method as it provides a low-cost and relatively easy fabrication process to produce massive amount of alloy powders. Many examples can be found dealing with the preparation of ball milling configurations combining with the various material composition [42–46].

9.2.2.1 Si-metal as a Li-active material

Initial studies on Si-metal alloy were reported in the early 1990s by Anani and Huggins at Stanford University [47, 48]. Ternary alloy systems of Li-Si-Mg were investigated to use in thermal battery normally operating at around 400 °C. The binary system of Li_7Si_3-$Li_{13}Si_4$ was the active anode materials for the thermal batteries, and a phase diagram approach revealed that addition of Mg to the binary system improved the electrode potentials as well as the theoretical capacity compared to the pure Li_7Si_3-$Li_{13}Si_4$ binary system. As the ternary system was investigated to use in thermal battery, there is no evidence that the Li-Si-Mg system could be helpful to mitigate the issues of current Si anode mentioned earlier in this book.

Later, Mg_2Si was investigated to use in current LIB systems for the purpose of reducing the amount of irreversible capacity during electrochemical cycles [49]. Authors proposed an electrochemical reaction scheme as follow based on XRD and differential capacity analysis.

$$\text{First step}: \; Mg_2Si + xLi^+ + e^- \rightarrow Li_xMg_2Si \tag{9.1}$$

$$\text{Second step}: \; Li_xMg_2Si + Li^+ + e^- \rightarrow Li_{critical}Mg_2Si \rightarrow Li_{saturated}Mg_2Si + Mg + Li - Si\text{ alloy} \tag{9.2}$$

$$\text{Final step}: \; Mg + Li^+ + e^- = Li - Mg\text{ alloy} \tag{9.3}$$

Mg_2Si electrochemically reacts with Li and converts to a mixture of Li_xMg_2Si, Li-Mg and Li-Si alloy. Those Li-Mg and Li-Si alloys are, however, also inhibited by volume changes with lithiation/delithiation and shown as poor cycling stability. The study revealed that 1 mol of Mg_2Si reacts with 3.9 mol of Li ion, which results in the specific capacity of 1 370 mAh g^{-1}, but the discharge capacity decreases dramatically within 10 cycles. Additional XRD analysis revealed that rapid degradation attributed to the change of starting materials from Mg_2Si to residual Mg_2Si, Li-Mg and Mg upon cycling, which led to the large volume expansion and electrodes disintegration. $CaSi_2$ is also investigated as an active anode material by Wolfenstine [50]. The proposed reaction mechanism based on the Ca-Si-Li ternary phase diagram is

$$10.8\,Li + CaSi_2 \rightarrow Li_2Ca + 2Li_{4.4}Si \tag{9.4}$$

The theoretical reversible capacity associated with the above reaction would be 3015 mAh g^{-1}. The observed specific capacity of $CaSi_2$ depends on the particle size. It is shown that the specific capacity of the 20-h ball-milled $CaSi_2$ is higher than that of the as-received $CaSi_2$. The ball-milled $CaSi_2$ partilces show an initial specific capacity up to 1 900 mAh g^{-1} but degrade rapidly. This instability is due to the formation of $Li_{4.4}Si$ phase upon lithiation. The phase suffered from volume changes and mechanical failures which inhibit reversible phase recombine and electrochemical reactions. $FeSi_2$ has been also studied, but its electrochemical performances have not drawn any attention due to aforementioned reasons [51]. A common feature of those Li-active Si-metal alloy is that they all showed unsatisfactory electrochemical stability mostly due to the Li-Si phases formation.

9.2.2.2 Si as a Li-active material

To cope with those disadvantages with Li-active Si-metal alloys, researchers have worked to develop Si-metal alloys where only the Si phase electrochemically reacts with Li and the metal alloy is inactive to Li have been performed.

These approaches take advantage of inactive metal alloys as a mechanically stable matrix that alleviates the volume changes of Si upon lithiation/delithiation.

One of the earlier studies follows this coping strategy was a Si-Ag composite material reported by Hwang *et al.* [52]. Authors expected that nano-Si composite material, in which nanometer-scale Si particles are finely dispersed within a solid, mixed-conducting, ductile metallic matrix, would demonstrate a good electrochemical performance. This makes Ag a good choice of materials as the Si–Ag system is immiscible in addition to serving as a mixed conducting matrix. The $Si_{50}Ag_{50}$ powders were obtained with HEBM using various time periods. No evidence of SiAg alloy was observed with XRD and SEM/EDS study. The study revealed that longer ball milling time provides improved cycle stability in LIB owing to well dispersed Si phase in ductile Ag matrix proven by microstructural observations. Initial discharge capacity of around 800 mAh g^{-1} was reported and only 400 mAh g^{-1} was retained after 10 cycles. It should be noted that the Ag itself becomes electrochemically active to Li below 0.07 V, thus the utilization of lower voltage range needs to be avoided for the $Si_{50}Ag_{50}$ alloys. Later, Kim *et al.* deposited Cu on Si particle surface by an electroless deposition method to enhance the electrical conductivity [53]. The Si–Cu anode exhibits improved cycle performance as a result of conductivity enhancement, but the long-term stability is still unsatisfactory due to detachment of Cu from the Si surface that is caused by a severe volume change of the Si during cycling. The detachment of Cu was greatly suppressed by annealing at 400 °C in an inert atmosphere. Furthermore, longer annealing time at 400 °C results in better cycle stability with lower initial discharge capacity (from 2 250 mAh g^{1} to 1 500 mAh g^{1}). However, the capacity level of Si-Cu was still far from commercial level anode for LIB.

Later, a NiSi phase was synthesized using a HEBM method and confirmed with XRD analysis [54]. The proposed electrochemical reaction of NiSi with Li is

$$xLi^+ + xe^- + NiSi \rightarrow Li_xSi + Ni \leftrightarrow Ni + Si + xLi^+ + xe^- \quad (9.5)$$

In this reaction, Ni does not react with Li but acts as a buffering matrix for the Li_xSi phase in the subsequent cycles. This results in initial discharge capacity of 1 180 mAh g^{-1} with 80% CE, but the electrochemical reaction is only partially reversible and its capacity degraded with each cycle. Improved electrochemical performances were expected with further optimization of nanocrystalline NiSi synthesis. The Ni–Si alloy was revisited later by Wang *et al.* and synthesized using hydrogen plasma-metal reaction (HPMR) method [55]. In general, the HPMR method is applied to synthesize commercial-level nanosized particles. Initial discharge capacity of 1 304 mAh g^{-1} was reported but rapidly lost its reversibility with cycles. The authors indicated that Si electrochemically reacts with Li, while the Si_2Ni phase plays the role of matrix which buffers Si volume expansion and facilitate charge transfer among Si particles. However, the poor cycling stability implicates that the volume-expansion-mitigating property of the matrix is not retained for long-term cycles. Then, Liu *et al.* developed porous NiSi–Si alloy particles with a pore size distribution peaked at 200 nm and an intraparticle porosity around 40% by using HEBM and subsequent dissolution of unreacted Ni [56]. Pre-existing voids were expected to help accommodate volume expansion arising from alloying of the Si component. As a result, the composite electrode exhibits reduced expansion, as compared with pure Si electrode, and hence significantly reduced capacity fading upon cycling. Synchrotron XRD further indicates that the NiSi component of the composite undergoes reversible transformation to Ni_2Si and Li_xSi during cycling, which was not a previously not considered mechanism of this material.

Then, the nanocomposites alloys of Si/TiB_2 and Si/TiN have been developed by Kim *et al.* using HEBM [57, 58]. This study investigated the impact of milling time on its electrochemical properties and observed that higher milling times deliver lower initial specific capacity but better stability. The achieved specific capacity is around 300–700 mAh g^{-1}. Neither Si/TiB_2 nor Si/TiN achieved theoretical capacity assuming complete reaction of Si with 4.4 Li. (It is now generally accepted that fully lithiated Si phase is $Li_{15}Si_4(Li_{3.75}Si)$ corresponding to 3 579 mAh g^{-1}) Obrovac and Christensen [59] mentioned that small particle size of the inactive component provides better stability because the stress induced by volume expansion of the active phase during cycling is distributed more homogeneously in the composite containing pre-milled inactive component. This appears to be a good way to improve the electrochemical stability of the active–inactive nanocomposite anodes, but the reduction in initial discharge capacity and further improvement in cycling stability were still issues to be overcome. Later, Si/TiC was synthesized using HEBM by Patel *et al.* [60]. XRD analysis after HEBM only showed TiC peaks indicating Si particles were amorphized and distributed inside the TiC matrix. Initial capacity of 999 mAh g^{-1} was expected considering 40 mol% Si in the composite but only 340 mAh g^{-1} was achieved. The capacity loss might be due to the inability of either Li ions or electrons to diffuse to the Si phase enclosed by TiC matrix. The authors discussed that other matrix materials should be tried to exploit the full capacity of Si inside the matrix. The study on Si-Mn composite is another good example, which reports less initial discharge capacity than expected [45]. Excess ball milling of Si and Mn produces a phase of Si_3Mn_5 and results in specific capacity of around 450 mAh g^{-1}. The metallic phase mitigates volume expansion of Si upon lithiation by encapsulating and preventing Si from directly contacting the liquid electrolyte. However, excess amounts of inactive phases with Li limits the specific capacity level of composites. In general, mass of metal alloys is much heavier than that of other surface modification materials such as carboneous materials; therefore, mass ratio of Si and metal alloys should be carefully designed to achieve enough gravimetric capacity. Considering gravimetric capacity of the graphite, which is commercially being used in LIBs, is 372 mAh g^{-1} [61]. Si-based composite should aim for decidedly higher specific capacities than that of graphite. Multi-component alloys such as Fe-Cu-Si, Ti-Si-Al, Si-Co-Mn-Al, and Si-Co-Cu-Cr-Al were developed by HEBM, but electrochemical performances were not as stable as needed [62–65].

An example of powder morphology achieved with HEBM method can be found in Figure 9.10 [66]. Xilin *et al.* fabricated a core-shell structured Si/B_4C composite with graphite coating [66]. The B_4C provides high electronic conductivity and was used as millers to break down micron Si. The graphite coating on the Si/B_4C composite further improves the conductivity and favors the SEI formation on its surface. The Si/B_4C/Graphite composite exhibits a specific capacity

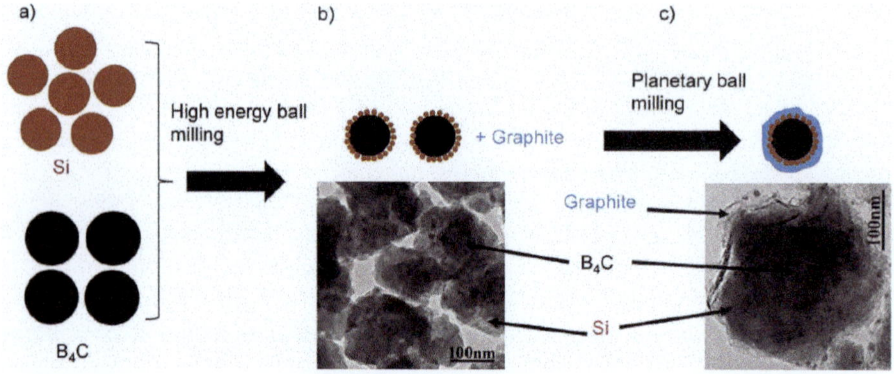

Figure 9.10 *Schematic diagram of the synthesis process. (a) Starting materials of micron-sized B_4C and Si. (b) Schematic diagram of the Si/B_4C core–shell structure and TEM image. (c) Schematic diagram of the Si-B_4C-Graphite structure and TEM image.*

of 822.5 mAh g^{-1} at 0.63 A g^{-1} and only results in 6% capacity loss over 100 cycles. The B_4C has also been used in other types of battery electrodes such as Li-sulfur and Li-oxygen [67, 68].

After many attempts to develop the Si-metal alloy with a HEBM method, researchers have used rapid quenching methods such as arc melting and melt-spinning to accomplish a better homogeneous and well-distributed Si-metal alloy system. The Al-Mn-Si alloys with various compositions were prepared by melt spinning by Sun *et al.* [69]. TEM images revealed that fine Si grains (less than 1 μm) are randomly distributed in an alloy matrix. A specific capacity higher than 500 mAh g^{-1} was achieved in $Al_{55}Si_{40}Mn_5$ and $Al_{53}Si_{40}Mn_7$ alloys after 10 cycles. In addition, an ordered phase of Al_3Li was also formed in the alloys after Li insertion, whereas no other Li-related phases have been detected. The authors claimed that refining alloy structures and the formation of metastable Al-Si-Mn compounds play important roles in limiting volume expansion upon Li insertion and improves the Li storage capability.

Then, Ni-Ti alloy (Nitinol), one of the most well-known shape-memory alloys, was utilized as an alloy material [70]. Jung *et al.* have developed a nanosize Si embedded in Nitinol alloy matrix composite using arc melting followed by a rapid quenching method [71]. The authors confirmed that approximately 50-nm Si crystallites were surrounded by the Ni–Ti matrix using XRD and high-resolution transmission electron microscope (HRTEM) with energy dispersive spectroscopy (EDS) analyses. Initial capacity of 1 047 mAh g^{-1} was reported and degradation of electrochemical stability was alleviated. Son *et al.* also synthesized an Si-Ti-Ni alloy using a melt spinning method and found that nano-Si phases were embedded in the $Ti_4Ni_4Si_7$ phases [72]. Figure 9.11 shows the materials preparation procedures and the microstructures of achieved materials. Figure 9.11d displays that Si grains with sizes of 20–50 nm are well distributed in $Ti_4Ni_4Si_7$ matrix phase. Son

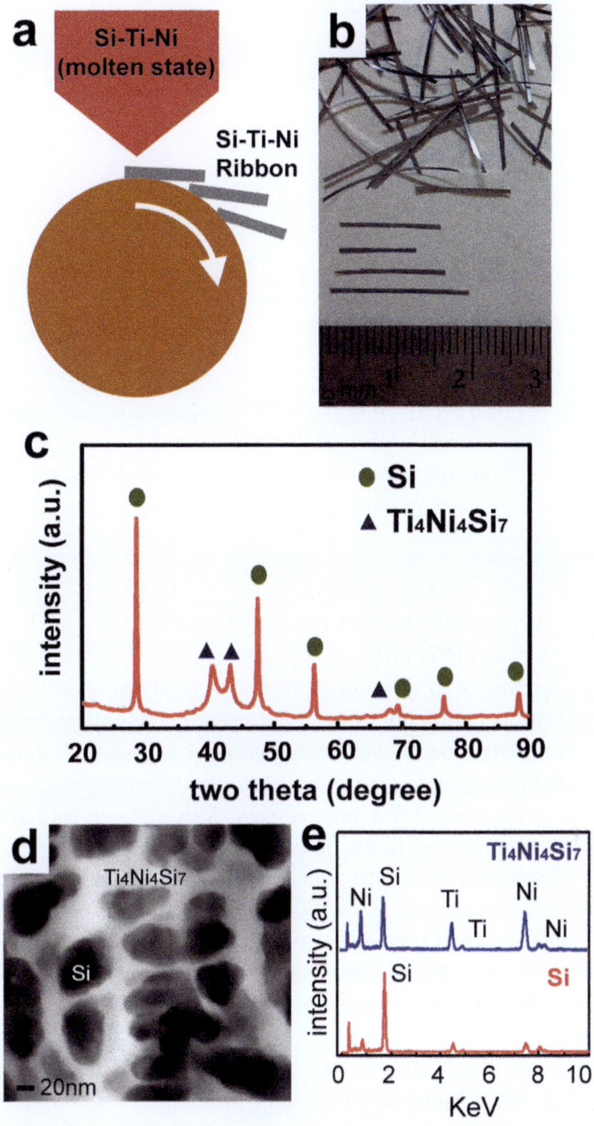

Figure 9.11 *(a) Simplified schematics of melt spinner. (b) Achieved melt spun Si-Ti-Ni ribbon. (c) XRD pattern of ball-milled melt spun Si-Ti-Ni powder that matched with Si phase (JCPDS, 27−1 402) and $Ti_4Ni_4Si_7$ phase (JCPDS, 80−1 015). (d) TEM image with HAADF mode of nano-Si and $Ti_4Ni_4Si_7$. (e) EDS results of given area in d.*

et al. also studied the lithiation mechanism of the Si-Ti-Ni alloy by synthesizing lithiated $Ti_4Ni_4Si_7$ phases and measured the ionic conductivity. The study found that the lithiated $Ti_4Ni_4Si_7$ phases have an ionic conductivity around 2.5×10^{-5} S cm^{-1},

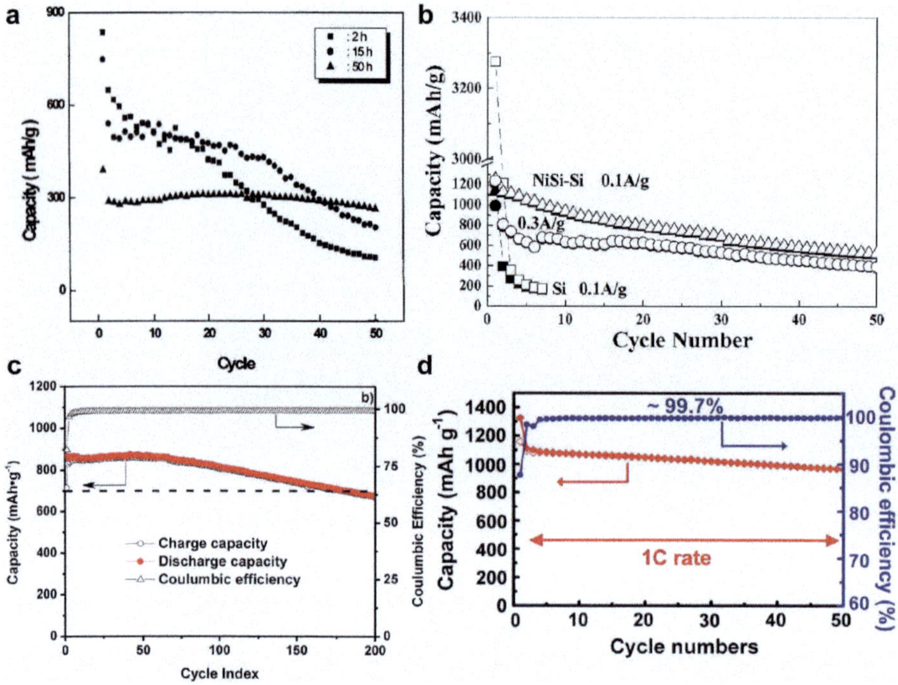

Figure 9.12 *(a) Specific capacity as a function of cycle counts for Si–TiC composite obtained after ball milling. (b) Specific capacity as a function of cycle counts for $CaSi_2$. (c) discharge−charge profiles of Si/B_4C/Graphite composite. (d) Specific capacity and CE of STN anode versus cycle numbers.*

which is lower compared to the general Li-ion electrolytes but enough to conduct Li ions to the Si phases. Then the Si-Ti-Ni system demonstrated a specific capacity in excess of 900 mAh g^{-1} after 50th cycles with a CE of 99.7%. Additional studies on this system in terms of compositions and compatibility with solid state electrolyte have been done as well [73–76]. Figure 9.12 summarized the cycling stability of some Si-metal alloys introduced in this chapter for clear comparison [44, 50, 66, 72]. Figure 9.12 displays the cycling performances of Si-TiC, $CaSi_2$, Si/B_4C/Graphite composite, and Si-Ti-Ni alloy, respectively. While earlier study with Si-TiC shows lower specific capacity around 400 mAh g^{-1} after 35 cycle, the specific capacity has been increased to about 2 000 mAh g^{-1} with the $CaSi_2$ anode but rapidly degrades. Both Si/B_4C/graphite composite and Si-Ti-Ni alloy showed improved initial capacity and cycling stability but could not prevent continuous degradation of capacity as cycle number increases. Degradation could be caused by structural instability, a consequence of the high lithiation degree of Si and excessive SEI formation on the surfaces. Besides, all cycling performances displayed in Figure 9.12 are achieved from half cell using Li metal, which provide infinite Li sources to Si-based anodes.

It is suggested to have CE of higher than 99.99% to be cycled long enough in full cell with commercially used cathode such as $Li(Ni_xMn_yCo_{1-x-y})O_2$.

9.2.3 Surface modification via ALD and MLD

9.2.3.1 Introduction

Atomic layer deposition (ALD) is a sequential, self-limiting surface reaction and has proven to be the best method to deposit continuous, conformal, and pinhole-free films with superb control over film thickness, composition, and conformality at the atomic level [77, 78]. ALD has been used to fabricate various inorganic materials, such as binary oxides, sulfides, and nitrides [77]. Previous works have demonstrated that ALD is an important tool to modify the surface of positive electrodes [79–81]. ALD coatings can mitigate the parasitic side reactions between the liquid electrolyte and the surface of electrodes, and lead to the improved performance of Li-ion electrodes. Comparatively, limited number of papers have been reported for electrochemical performance improvement of Si with ALD coating [82]. Unlike the typical Li metal oxide cathodes, Si-based electrodes require that the coating materials are chemically stable and also sufficiently robust to allow volumetric expansion. Rigid inorganic coatings such as Al_2O_3 do not satisfy the new needs for high-capacity, high-volume-expansion Si electrodes. Coating materials with improved mechanical properties are required to realize these promising electrode materials for next-generation LIB.

Successive development of ALD techniques has broadened the range of possible materials from inorganic materials to organic materials. In the 1990s, organic polymers were deposited by a variant of ALD which is now generally known as molecular layer deposition (MLD) [83]. Continuous achievements were made to combine typical ALD precursors and all-organic precursors to grow organic–inorganic hybrid materials [84, 85]. The hybrid organic–inorganic materials achieved by MLD bring new class of materials with versatile characteristics which cannot be achieved with typical ALD [82]. Figure 9.13 displays the generic growth sequences of ALD and MLD [85]. Figure 9.13a shows Al_2O_3 ALD layer with precursors of trimethylaluminum (TMA) and H_2O. The two self-limiting surface reactions that define Al_2O_3 ALD growth are [86, 87]

$$\text{A. AlOH}^* + \text{Al(CH}_3)_3 \rightarrow \text{AlO} - \text{Al(CH}_3)_2^* + \text{CH}_4 \tag{9.6}$$

$$\text{A. AlO} - \text{Al(CH}_3)_2^* + \text{H}_2\text{O} \rightarrow \text{AlO} - \text{Al} - \text{OH}_* + \text{CH}_4 \tag{9.7}$$

where the * denotes the surface species. When reaction is performed in an ABAB sequence, these sequential reactions produce linear, atomic-layer-controlled Al_2O_3 growth. A single reaction-cycle of ALD yields a thickness of around 1 to 1.5 Å and two cycles of ALD layer were generally used to improve the electrochemical performances of cathodes materials. While recent work has proven ALD to be an important tool in improving the performance of cathodes via surface coating [79, 80], attempts to attenuate volume expansion in high-capacity electrodes such as Si-based anodes have not been successful as much as cathodes have been. ALD coatings may help reduce the effect of parasitic side reactions between the liquid electrolyte

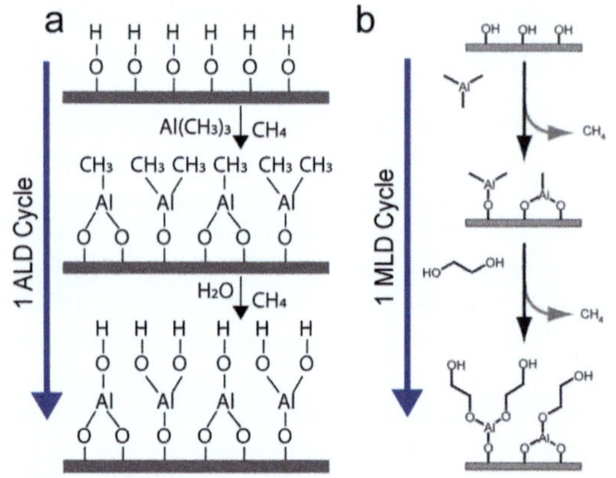

Figure 9.13 *(a) Schematic representation of Al_2O_3 ALD on $LiCoO_2$. (b) Schematic illustration of AlGL MLD.*

and electrode surface, but the coatings are not robust enough to provide structural support when applied to materials such as Si, which undergoes volumetric expansion of up to 300%. Figure 9.13b shows the schematic growth sequence of MLD using the reaction of an aluminum containing metal alkyl, $Al(CH_3)_3$ (trimethylaluminum (TMA)), with HO-$(CH_2)_2$-OH (ethylene glycol (EG)). The sequential reactions of TMA and EG during MLD yield a polymeric film that can be approximately described as poly(aluminum ethylene glycol). The MLD also takes the advantages of ALD with respect to uniformity, conformality, and controllability analogous to ALD [77]. Moreover, MLD significantly expands the choices of precursors and final products. Various organic polymeric materials, including polyimide, polyamide, polyurea, and cross-linked carbosiloxane, have been synthesized using all-organic precursors based on condensation polymerization reactions [88–92]. By combining all-organic precursors and the inorganic precursors used in the ALD process, MLD also enables the fabrication of a huge variety of hybrid organic–inorganic materials [83, 84]. Developing durable electrode materials with high energy density requires functional and stable interphases that mitigate deleterious side reactions, reduce the loss of Li inventory, and also ensure mechanical integrity through electrochemical reactions. As surface modification via MLD has been recognized as one of the most effective methods to evolve the surface chemistry on porous battery electrodes with high surface area and aspect ratio, applying the emerging MLD technique for surface modification may offer significant improvements for electrodes with high capacity for future Li-based batteries. And, in the following chapter, new strategies of using the various MLD methods to develop the advanced high energy dense Si-based anodes will be introduced.

9.2.3.2 Surface modifications via MLD on Si-based anodes

Figure 9.14 displays the surface microstructures of uncoated and aluminum glycerol (AlGL)-MLD-coated Si particles using HRTEM [93]. The surface morphology has

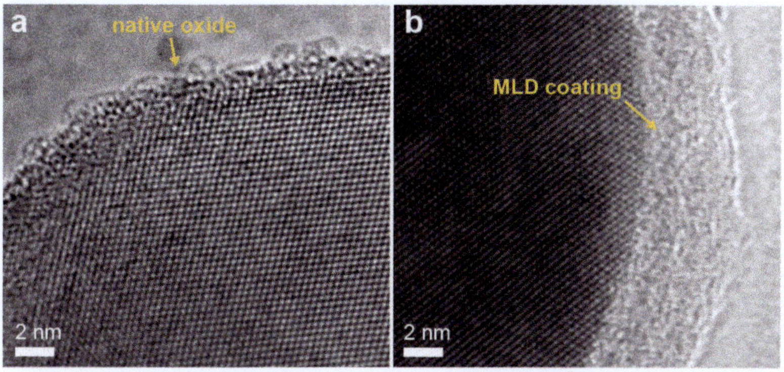

Figure 9.14 TEM images of uncoated Si and AlGL-MLD-coated Si

been largely changed after the MLD coating, and the particles were conformally covered by the AlGL coating. Intimate contact at the interface of the AlGL-MLD coating and Si phase was clearly observed with HRTEM, confirming that the coating is well adhered on the Si particle with the MLD technique. One of the featured characteristics that MLD dominates over ALD is mechanical resiliency of the coating material. Unlike the ALD coating generally composed of oxide or sulfide materials, MLD coating hybridized with organic–inorganic provides unique physical properties. It is also known that native oxide on the Si surface is removed during the MLD process [78]. The XPS results in Figure 9.15 elucidate how the surface chemistry evolves with MLD coating as well as electrochemical reaction [94]. Before cycle, both uncoated and MLD-coated samples show Si (100.1 eV) and SiO_2 peaks

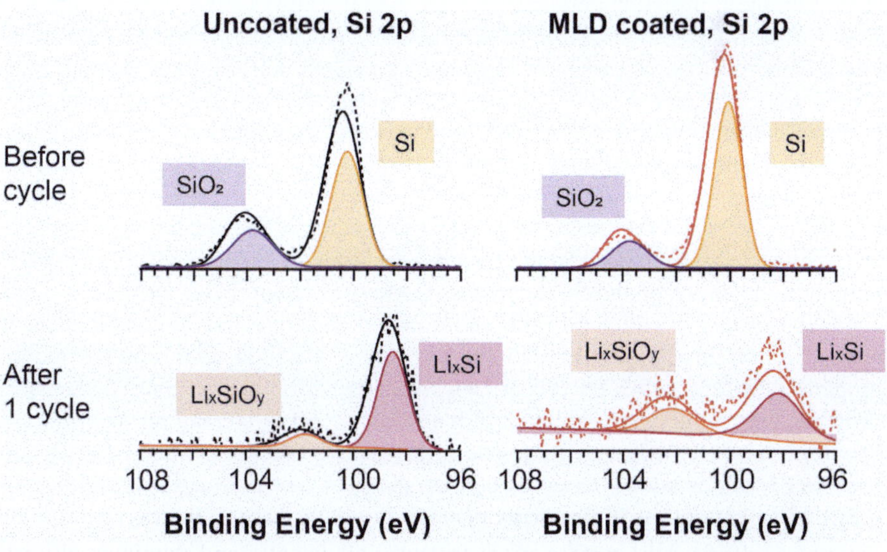

Figure 9.15 XPS, Si 2p spectra of uncoated and alucone-MLD-coated silicon

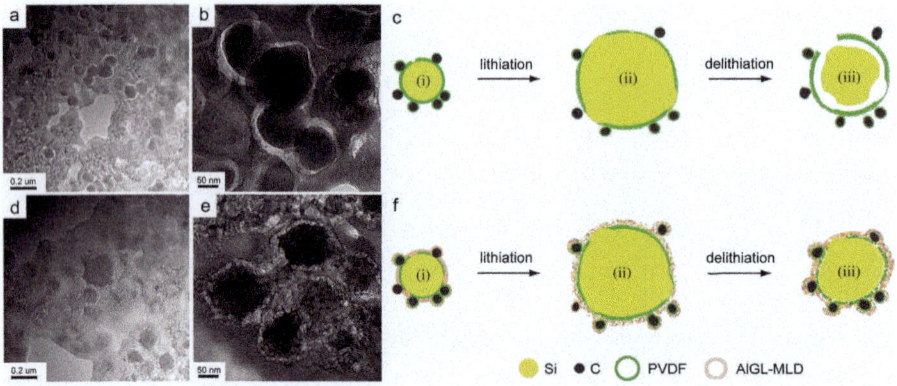

Figure 9.16 *(a, b) TEM images of the uncoated Si electrode. (c) Detailed schematic of the uncoated Si electrode's lithiation–delithiation mechanism. (d, e) TEM images of the AlGL-MLD-coated Si electrode. (f) Detailed schematic of the AlGL-MLD-coated Si electrode's lithiation–delithiation mechanism.*

(103.7 eV) in Si 2p region. The peak fitting results revealed that SiO_2 intensity for the MLD-coated Si (17.03%) is notably lower than uncoated Si (29.76%) referring removal of SiO_2 on the Si surface after MLD coating. After the first cycle, both Si and SiO_2 peaks disappear and are replaced by peaks of Li_xSi and Li_xSiO_y, which are caused by irreversible Li loss in Si. The intensity of Li_xSi peak is much lower for the MLD coated Si indicating lower irreversible Li loss in Si.

Piper *et al.* reported how MLD coating recovers and preserves the expansion of electrodes upon lithiation/delithiation with in-depth TEM images [95]. Figure 9.16 shows TEM images of uncoated Si electrode and AlGL-MLD-coated Si electrodes after 20 charge–discharge cycles. The uncoated electrode shows a severance of the connections between the Si particles and the electrode network once delithiation occurs. This behavior isolates active Si and prevents its utilization, thus contributing to rapid capacity degradation. On the other hand, the AlGL-MLD-coated electrode shows an intimate adherence between the Si particles and the rest of the electrode network (Figure 9.16e) during cycling. This intimate contact between particles and coatings provides robust and resilient properties to accommodate the extreme volumetric changes of the Si electrodes, helping to maintain an intimately linked conductive network which prevents the isolation of active material and enables its full utilization throughout electrochemical cycling. Detailed studies on mechanical properties evolution after MLD coating were also performed. Son *et al.* used nanoindentation to quantitatively investigate the effect of MLD coating on the mechanical property of Si electrodes [93]. The study revealed that MLD-coated electrodes have H/E* values almost twice as high as those of the uncoated electrodes, indicating more resistance to the permanent plastic deformation. In other words, the microstructure of MLD-coated electrodes would be more recoverable and will be well-preserved during electrochemical cycling. Remarkable improvement in electrochemical performance using AlGL-MLD coating on Si-based electrodes was reported [93–95]. While uncoated Si electrodes undergo rapid degradation of a typical Si-based electrodes, AlGL-MLD-coated Si electrodes exhibit around 1 000 mAh g^{-1} with a CE in excess of 99%. These results confirm that enhancements in mechanical properties, mechanical robustness, and elastic reversibility contribute to the improved cycling performance of Si-based electrode. Computational study helps to understand the compositional evolution of AlGL-MLD layer after lithiation [96]. Ma *et al.* insists that AlGL is composed of Al–O complexes with 3-O or 4-O coordination. During lithiation, Li atoms bind very strongly to the O atoms in the most energetically favorable sites. After the film is irreversibly saturated with Li atoms, it becomes electronically conductive. The ethylene carbonate molecules in liquid phase are found to be reduced at the surface of the Li-saturated alucone film following similar electron transfer mechanisms as found previously for lithiated Si anodes. Detailed in situ TEM observation on lithiation process of AlGL-MLD-coated Si particles has also been performed [97]. It shows that lithiation starts from the AlGL layer and gradually diffuses into Si particles, leading to two-phase formation of Si (lithiated amorphous Si and unlithiated crystalline Si). Upon lithiation, the thickness of the AlGL-MLD coating changes from ~1.7 to ~3 nm. This intimate contact between particle and coating provides mechanical strength that maintains the original electrode network, prevents the isolation of active material, and enables its full utilization throughout cycling. Most importantly, the swelling of the AlGL coating along with lithiation helps the AlGL coating to expand with the Si particle. Upon delithiation, the coated Si particles together with the AlGL coating contract without pore formation or coating delamination.

Then, Piper *et al.* have reported new reaction precursor—an aromatic organic diol, hydroquinone—combined with trimethylaluminum [98]. Alumination of hydroquinone in liquid has been studied to produce polydentate Lewis acids for strong binding with nucleophilic substrates by Kaul *et al.* [99]. The knowledge has been successfully transferred to produce a cross-linked aluminum dioxybenzene compound via MLD. The hybrid coating (denoted by AlHQ) was deposited conformally onto the Si-based anodes using the sequential, self-limiting reaction of TMA ($Al(CH_3)_3$) and HQ ($C_6H_4(OH)_2$) according to:

$$\text{A. } Si - OH^* + Al(CH_3)_3 - Si - O - Al(CH_3)_2^* + CH_4 \tag{9.8}$$

$$\text{B. } Si - O - Al(CH_3)_2^* + HOC_6H_4OH - Si - P - Al - OC_6H_4OH^* + CH_4 \tag{9.9}$$

where * indicates surface species and Si represents the underlying Si electrode. Figure 9.17 shows a schematic of the controlled structure of the AlHQ. The growth rate of this MLD reaction is measured to be 7.5 Å per AB cycle at a substrate temperature of 150 °C. The laminated electrode was used as a substrate for the hybrid coating as illustrated in Figure 9.17a. During the MLD reaction, the aluminum oxide was connected with dioxybenzene via a conjugated chain in the form of $(Al–O–benzene–O–Al–)_n$ [100]. Further dehydrogenation occurs when annealing the as-deposited film under both inert and air atmosphere, as shown in Figure 9.17b. Dehydrogenation induces a longitudinal crosslinking reaction of the

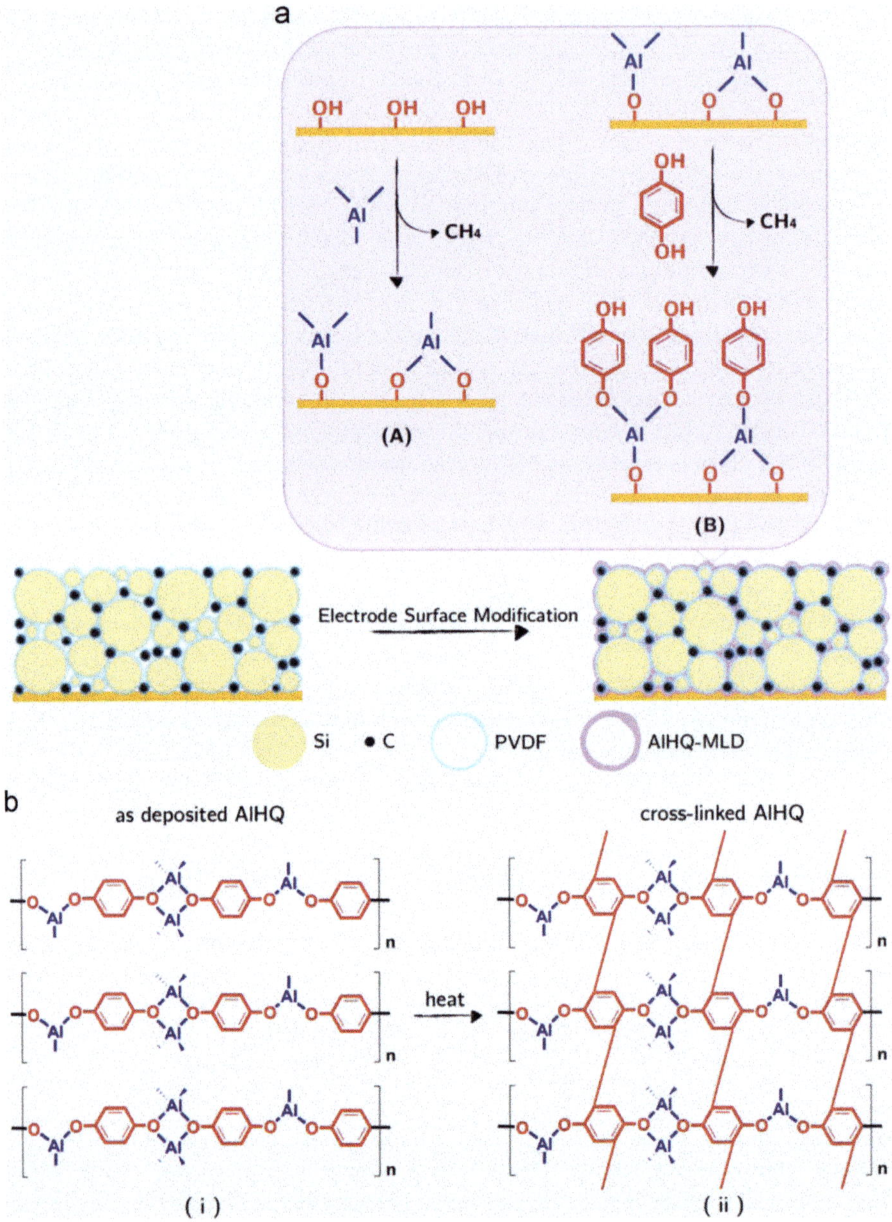

Figure 9.17 *(a) The controlled layered chemistry in MLD AlHQ coating and the schematic of MLD coating for the laminated Si electrode. (b) Schematic of the crosslinking process of the AlHQ chains after post-deposition heat treatments. Dehydrogenation occurs during the post-annealing process, which induces a longitudinal crosslinking reaction of the aluminum-dioxybenzene chains.*

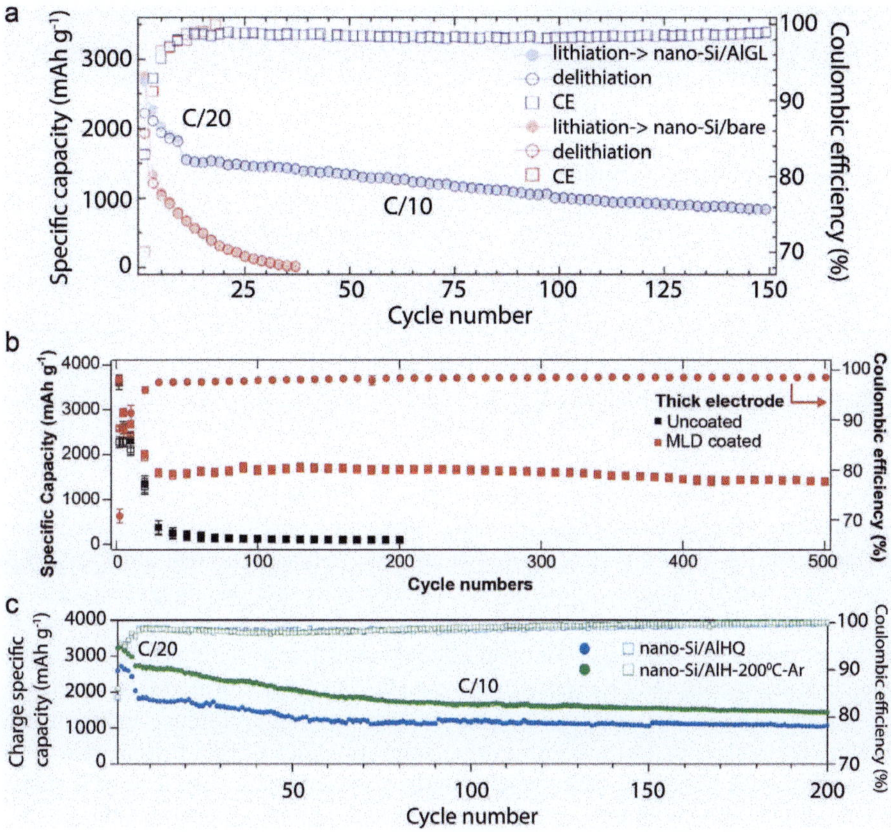

Figure 9.18 *(a) Cyclic capacity and CE (blue symbols) of a nano-Si/AlGL anode in comparison with the capacity and CE (red symbols) of a nano-Si/bare composite electrode. (b) Cycling performances of (a) thin and (b) thick electrodes with and without Alucone-MLD coating. (c) Cyclic capacity and CE of a Si anode coated with as-deposit AlHQ (blue symbols) compared to the cyclic capacity and CE of a Si anode coated with Ar-treated AlHQ (green symbols).*

aluminum-dioxybenzene chains [101]. The three-dimensional cross-linked structure leads to improved conductive and mechanical properties, as discussed in a later section. Furthermore, the crosslinking structure enabled by the polydentate Lewis acid not only creates a strong covalent bonding to the surface of Si particles, but also provides chemical stability of the coating film.

Figure 9.18 compares the cycling performances of Si-based anodes with MLD coating [93, 95, 98]. Figure 9.18a and b shows the Si-based anodes with AlGL-MLD, and Figure 9.18c shows the Si-based anode with AlHQ MLD. Although Figure 9.18a and b utilizes the identical AlGL-MLD on the Si electrodes surface, but Figure 9.18b exhibits the improved initial capacity and cycling stability, which might be caused by optimization of MLD processes. As a result, AlGL-coated Si electrodes in Figure 9.18b delivers the specific capacity of 1 410 mAh g^{-1} even after 500 cycles. The reason for lower initial capacity with AlGL-MLD-coated Si anode could be kinetic limits for Li ions with minor contribution from additional mass of MLD layers. Figure 9.18c displays the cycling performances of Si anode with AlHQ-MLD coating. Specific capacity of around 1,500 mAh g^{-1} was achieved, which is similar level to the results from AlGL-MLD in Figure 9.18b, but initial CE was improved from 70% (with AlGL-MLD) to ~85% (with AlHQ-MLD).

9.2.4 In situ Li–M–Si phase formation using additional salt

Recently, a new type of approach using mixed salts in electrolyte to stabilize the Si anode surface chemistry has been reported by Han *et al.* [102]. In this work, additional salt of 0.1M $M(TFSI)_x$ (M = Mg, Zn, Al and Ca) was added into the electrolyte to stabilize the Si anode chemistry through the in situ formation of Li–M–Si ternary phase during charging process. Here, only M of inert to Li were chosen to avoid any anodic or cathodic activity other than insertion into Si to form ternary Zintl phases. These ternary phases which presumably formed on the surface could protect the underlying lithiated Si, reduce side reactions with the electrolyte and Si surface. The authors revealed that reactions between common electrolyte solvent and model compounds (Li-Mg-Si) are dramatically suppressed with controlled Mg doping using solid-state magic angle spinning (MAS) nuclear magnetic resonance (NMR) spectroscopy. Authors also performed electrochemical test in full-cell combining with commercially available cathodes and reported that higher capacities and superior cyclabilities with improved CE were observed with additional $M(TFSI)_x$ salts when compared to standard electrolyte system. For example, Mg containing electrolyte in NMC full cells improves the CE by 0.15% and in turn the capacity retention rates by ~40% after 90 cycles. The Zn is the only exception due to the activity of Zn in the cathode, which motivates authors to explore more Zintl phases with other cations, particularly the multivalents that can form Li–M–Si Zintl phases with minimum activity in the cell expecting to provide in situ stabilization of the Si surface with the Zintl phases.

References

[1] Dimov N., Kugino S., Yoshio M. 'Carbon-coated silicon as anode material for lithium ion batteries: advantages and limitations'. *Electrochimica Acta.* 2003;**48**(11):1579–87.

[2] Wilson A.M., Dahn J.R. 'Lithium insertion in carbons containing nanodispersed silicon'. *Journal of the Electrochemical Society.* 1995;**142**(2):326–32.

[3] Wilson A.M., Zank G., Eguchi K., Xing W., Dahn J.R. 'Pyrolysed silicon-containing polymers as high capacity anodes for lithium-ion batteries'. *Journal of Power Sources.* 1997;**68**(2):195–200.

[4] Xing W., Wilson A.M., Eguchi K., Zank G., Dahn J.R. 'Pyrolyzed polysiloxanes for use as anode materials in lithium-ion batteries'. *Journal of the Electrochemical Society.* 1997;**144**(7):2410–6.

[5] Larcher D., Mudalige C., George A.E., Porter V., Gharghouri M., Dahn J.R. 'Si-containing disordered carbons prepared by pyrolysis of pitch/polysilane blends: effect of oxygen and sulfur'. *Solid State Ionics*. 1999;**122**(1–4):71–83.

[6] Liu Y., Wen Z.Y., Wang X.Y., Hirano A., Imanishi N., Takeda Y. 'Electrochemical behaviors of Si/C composite synthesized from F-containing precursors'. *Journal of Power Sources*. 2009;**189**(1):733–7.

[7] Wen Z.S., Yang J., Wang B.F., Wang K., Liu Y. 'High capacity silicon/carbon composite anode materials for lithium ion batteries'. *Electrochemistry Communications*. 2003;**5**(2):165–8.

[8] Guo Z.P., Jia D.Z., Yuan L., Liu H.K. 'Optimizing synthesis of silicon/disordered carbon composites for use as anode materials in lithium-ion batteries'. *Journal of Power Sources*. 2006;**159**(1):332–5.

[9] Liu Y., Hanai K., Yang J., Imanishi N., Hirano A., Takeda Y. 'Morphology-stable silicon-based composite for Li-intercalation'. *Solid State Ionics*. 2004;**168**(1–2):61–8.

[10] Liu X.H., Zhong L., Huang S., Mao S.X., Zhu T., Huang J.Y. 'Size-dependent fracture of silicon nanoparticles during lithiation'. *ACS Nano*. 2012;**6**(2):1522–31.

[11] McDowell M.T., Lee S.W., Harris J.T., *et al.* 'In situ TEM of two-phase lithiation of amorphous silicon nanospheres'. *Nano Letters*. 2013;**13**(2):758–64.

[12] Xu Y., Yin G., Ma Y., Zuo P., Cheng X. 'Nanosized core/shell silicon@carbon anode material for lithium ion batteries with polyvinylidene fluoride as carbon source'. *Journal of Materials Chemistry*. 2010;**20**(16):3216–20.

[13] Hu Y.-S., Demir-Cakan R., Titirici M.-M., *et al.* 'Superior storage performance of a Si@SiOx/C nanocomposite as anode material for lithium-ion batteries'. *Angewandte Chemie*. 2008;**47**(9):1645–9.

[14] Aurbach D., Gamolsky K., Markovsky B., Gofer Y., Schmidt M., Heider U. 'On the use of vinylene carbonate (VC) as an additive to electrolyte solutions for Li-ion batteries'. *Electrochimica Acta*. 2002;**47**(9):1423–39.

[15] Wang G.X., Ahn J.H., Yao J., Bewlay S., Liu H.K. 'Nanostructured Si–C composite anodes for lithium-ion batteries'. *Electrochemistry Communications*. 2004;**6**(7):689–92.

[16] Lee J.-K., Kung M.C., Trahey L., Missaghi M.N., Kung H.H. 'Nanocomposites derived from phenol-functionalized Si nanoparticles for high performance lithium ion battery anodes'. *Chemistry of Materials*. 2009;**21**(1):6–8.

[17] Piper D.M., Yersak T.A., Son S.-B., *et al.* 'Conformal coatings of cyclized-PAN for mechanically resilient Si nano-composite anodes'. *Advanced Energy Materials*. 2013;**3**(6):697–702.

[18] Yoshio M., Wang H., Fukuda K., Umeno T., Dimov N., Ogumi Z. 'Carbon-coated Si as a lithium-ion battery anode material'. *Journal of the Electrochemical Society*. 2002;**149**(12):A1598.

[19] Dimov N., Fukuda K., Umeno T., Kugino S., Yoshio M. 'Characterization of carbon-coated silicon: structural evolution and possible limitations'. *Journal of Power Sources*. 2003;**114**(1):88–95.

[20] Liu W.-R., Wang J.-H., Wu H.-C., Shieh D.-T., Yang M.-H., Wu N.-L. 'Electrochemical characterizations on Si and C-coated Si particle electrodes for lithium-ion batteries'. *Journal of the Electrochemical Society*. 2005;**152**(9):A1719.

[21] Yen Y.-C., Chao S.-C., Wu H.-C., Wu N.-L. 'Study on solid-electrolyte-interphase of Si and C-coated Si electrodes in lithium cells'. *Journal of the Electrochemical Society*. 2009;**156**(2):A95.

[22] Son S.-B., Kappes B., Ban C. 'Surface modification of silicon anodes for durable and high energy lithium-ion batteries'. *Israel Journal of Chemistry*. 2015;**55**(5):558–69.

[23] Wang J.W., He Y., Fan F., *et al*. 'Two-phase electrochemical lithiation in amorphous silicon'. *Nano Letters*. 2013;**13**(2):709–15.

[24] Gordon M.J., Baron T., Dhalluin F., Gentile P., Ferret P. 'Size effects in mechanical deformation and fracture of cantilevered silicon nanowires'. *Nano Letters*. 2009;**9**(2):525–9.

[25] Chan C.K., Peng H., Liu G., *et al*. 'High-performance lithium battery anodes using silicon nanowires'. *Nature Nanotechnology*. 2008;**3**(1):31–5.

[26] Wu H., Chan G., Choi J.W., *et al*. 'Stable cycling of double-walled silicon nanotube battery anodes through solid-electrolyte interphase control'. *Nature Nanotechnology*. 2012;**7**(5):310–15.

[27] Chen H., Dong Z., Fu Y., Yang Y. 'Silicon nanowires with and without carbon coating as anode materials for lithium-ion batteries'. *Journal of Solid State Electrochemistry*. 2010;**14**(10):1829–34.

[28] Ma Y., Tang H., Zhang Y., Li Z., Zhang X., Tang Z. 'Facile synthesis of Si-C nanocomposites with yolk-shell structure as an anode for lithium-ion batteries'. *Journal of Alloys and Compounds*. 2017;**704**:599–606.

[29] Son I.H., Hwan Park J., Kwon S., *et al*. 'Silicon carbide-free graphene growth on silicon for lithium-ion battery with high volumetric energy density'. *Nature Communications*. 2015;**6**(1):1–8.

[30] Shi L., Pang C., Chen S., *et al*. 'Vertical graphene growth on SiO microparticles for stable lithium ion battery anodes'. *Nano Letters*. 2017;**17**(6):3681–7.

[31] Yi R., Dai F., Gordin M.L., Sohn H., Wang D. 'Influence of silicon nanoscale building blocks size and carbon coating on the performance of micro-sized Si-C composite Li-ion anodes'. *Advanced Energy Materials*. 2013;**3**(11):1507–15.

[32] Xiao K., Tang Q., Liu Z., *et al*. '3D interconnected mesoporous Si/SiO_2 coated with CVD derived carbon as an advanced anode material of Li-ion batteries'. *Ceramics International*. 2018;**44**(4):3548–55.

[33] Yan Z., Guo J. 'High-performance silicon-carbon anode material via aerosol spray drying and magnesiothermic reduction'. *Nano Energy*. 2019;**63**:103845.

[34] Magasinski A., Dixon P., Hertzberg B., Kvit A., Ayala J., Yushin G. 'High-performance lithium-ion anodes using a hierarchical bottom-up approach'. *Nature Materials*. 2010;**9**(4):353–8.

[35] Iwamura S., Nishihara H., Kyotani T. 'Fast and reversible lithium storage in a wrinkled structure formed from Si nanoparticles during lithiation/delithiation cycling'. *Journal of Power Sources*. 2013;**222**:400–9.

[36] Xia M., Li Y., Zhou Z., *et al*. 'Improving the electrochemical properties of SiO@C anode for high-energy lithium ion battery by adding graphite through fluidization thermal chemical vapor deposition method'. *Ceramics International*. 2019;**45**(2):1950–9.

[37] Ko M., Chae S., Ma J., *et al*. 'Scalable synthesis of silicon-nanolayer-embedded graphite for high-energy lithium-ion batteries'. *Nature Energy*. 2016;**1**(9):1–8.

[38] Shi H., Zhang H., Li X., *et al*. 'In situ fabrication of dual coating structured SiO/1D-C/a-C composite as high-performance lithium ion battery anode by fluidized bed chemical vapor deposition'. *Carbon*. 2020;**168**(23):113–24.

[39] Nava G., Schwan J., Boebinger M.G., McDowell M.T., Mangolini L. 'Silicon-core-carbon-shell nanoparticles for lithium-ion batteries: rational comparison between amorphous and graphitic carbon coatings'. *Nano Letters*. 2019;**19**(10):7236–45.

[40] Ban C., Wu Z., Gillaspie D.T., *et al*. 'Nanostructured Fe_3O_4/SWNT electrode: binder-free and high-rate Li-ion anode'. *Advanced Materials*. 2010;**22**(20):E145–9.

[41] Wilcox J.D., Doeff M.M., Marcinek M., Kostecki R. 'Factors influencing the quality of carbon coatings on $LiFePO_4$'. *Journal of the Electrochemical Society*. 2007;**154**(5):A389.

[42] Ladam A., Bibent N., Cénac-Morthé C., *et al*. 'One-pot ball-milling synthesis of a Ni-Ti-Si based composite as anode material for Li-ion batteries'. *Electrochimica Acta*. 2017;**245**:497–504.

[43] Wang G.X., Sun L., Bradhurst D.H., Zhong S., Dou S.X., Liu H.K. 'Innovative nanosize lithium storage alloys with silica as active centre'. *Journal of Power Sources*. 2000;**88**(2):278–81.

[44] Guo Z.P., Zhao Z.W., Liu H.K., Dou S.X. 'Lithium insertion in Si–TiC nanocomposite materials produced by high-energy mechanical milling'. *Journal of Power Sources*. 2005;**146**(1):190–4.

[45] Zuo P., Yin G. 'Si–Mn composite anodes for lithium ion batteries'. *Journal of Alloys and Compounds*. 2006;**414**(1):265–8.

[46] Yoon S., Park C.-M., Kim H., Sohn H.-J. 'Electrochemical properties of Si–Zn–C composite as an anode material for lithium-ion batteries'. *Journal of Power Sources*. 2007;**167**(2):520–3.

[47] Anani A., Huggins R.A. 'Multinary alloy electrodes for solid state batteries II. A new LiSiMg alloy negative electrode material for use in high energy density rechargeable lithium cells'. *Journal of Power Sources*. 1992;**38**(3):363–72.

[48] Anani A., Huggins R.A. 'Multinary alloy electrodes for solid state batteries I. A phase diagram approach for the selection and storage properties determination of candidate electrode materials'. *Journal of Power Sources*. 1992;**38**(3):351–62.

[49] Kim H., Choi J., Sohn Hun-Joon., Kang T. 'The insertion mechanism of lithium into Mg_2Si anode material for Li-Ion batteries'. *Journal of the Electrochemical Society*. 1999;**146**(12):4401–5.

[50] Wolfenstine J. 'CaSi2 as an anode for lithium-ion batteries'. *Journal of Power Sources*. 2003;**124**(1):241–5.

[51] Santos-Peña J., Brousse T., Schleich D.M. 'Mg_2Si and MSi_2 (M=Ca, Fe) silicon alloys as possible anodes for lithium batteries'. *Ionics*. 2000;**6**(1):133–8.

[52] Hwang S.-M., Lee H.-Y., Jang S.-W., *et al*. 'Lithium insertion in SiAg powders produced by mechanical alloying'. *Electrochemical and Solid-State Letters*. 2001;**4**(7):A97–100.

[53] Kim J.W., Ryu J.H., Lee K.T., Oh S.M. 'Improvement of silicon powder negative electrodes by copper electroless deposition for lithium secondary batteries'. *Journal of Power Sources*. 2005;**147**(1):227–33.

[54] Wang G.X., Sun L., Bradhurst D.H., Zhong S., Dou S.X., Liu H.K. 'Nanocrystalline NiSi alloy as an anode material for lithium-ion batteries'. *Journal of Alloys and Compounds*. 2000;**306**(1):249–52.

[55] Wang Z., Tian W.H., Liu X.H., Li Y., Li X.G. 'Nanosized Si–Ni alloys anode prepared by hydrogen plasma–metal reaction for secondary lithium batteries'. *Materials Chemistry and Physics*. 2006;**100**(1):92–7.

[56] Liu W.-R., Wu N.-L., Shieh D.-T., *et al*. 'Synthesis and characterization of nanoporous NiSi-Si composite anode for lithium-ion batteries'. *Journal of the Electrochemical Society*. 2007;**154**(2):A97–102.

[57] Kim I.-seok., Blomgren G.E., Kumta P.N. 'Nanostructured Si/TiB_2 composite anodes for Li-ion batteries'. *Electrochemical and Solid-State Letters*. 2003;**6**(8):A157–61.

[58] Kim I.-seok., Kim I., Kumta P.N., Blomgren G.E. 'Si/TiN nanocomposites novel anode materials for Li-ion batteries'. *Electrochemical and Solid-State Letters*. 1999;**3**(11):493–6.

[59] Obrovac M.N., Christensen L. 'Structural changes in silicon anodes during lithium insertion/extraction'. *Electrochemical and Solid-State Letters*. 2004;**7**(5):A93–6.

[60] Patel P., Kim I.-S., Kumta P.N. 'Nanocomposites of silicon/titanium carbide synthesized using high-energy mechanical milling for use as anodes in lithium-ion batteries'. *Materials Science and Engineering: B*. 2005;**116**(3):347–52.

[61] Yazami R., Touzain P. 'A reversible graphite-lithium negative electrode for electrochemical generators'. *Journal of Power Sources*. 1983;**9**(3):365–71.

[62] Li M., Yu Z., Qu M. 'Preparation of SixCo0.3Cu0.3Cr0.6Al0.2/modified graphite sphere composites and their electrochemical performance as anode materials for Li-ion batteries'. *Journal of Alloys and Compounds*. 2010;**491**(1):643–8.

[63] Doh C.-H., Shin H.-M., Kim D.-H., *et al*. 'A new composite anode, Fe–Cu–Si/C for lithium ion battery'. *Journal of Alloys and Compounds*. 2008;**461**(1):321–5.

[64] Lee K.-M., Lee Y.-S., Kim Y.-W., Sun Y.-K., Lee S.-M. 'Electrochemical characterization of Ti–Si and Ti–Si–Al alloy anodes for Li-ion batteries produced by mechanical ball milling'. *Journal of Alloys and Compounds*. 2009;**472**(1–2):461–5.

[65] Mingqi L., Ying Z. 'Electrochemical properties of Si–Co–Mn–Al-modified graphite sphere composites as negative electrode materials for Li-ion batteries'. *Journal of the Electrochemical Society*. 2010;**157**(6):A729–34.

[66] Chen X., Li X., Ding F., *et al*. 'Conductive rigid skeleton supported silicon as high-performance Li-ion battery anodes'. *Nano Letters*. 2012;**12**(8):4124–30.

[67] Song S., Xu W., Cao R., *et al*. 'B4C as a stable non-carbon-based oxygen electrode material for lithium-oxygen batteries'. *Nano Energy*. 2017;**33**(4):195–204.

[68] Song N., Gao Z., Zhang Y., Li X. 'B4C nanoskeleton enabled, flexible lithium-sulfur batteries'. *Nano Energy*. 2019;**58**:30–9.

[69] Wang X.D., Li X.P., Zhao M.S., Li Y. 'Electrochemical properties of melt-spun Al-Si-Mn alloy anodes for lithium-ion batteries'. *Journal of Power Sources*. 2008;**182**:353–8.

[70] Eckelmeyer K.H. 'The effect of alloying on the shape memory phenomenon in nitinol'. *Scripta Metallurgica*. 1976;**10**(8):667–72.

[71] Jung H., Kim Y.-U., Sung M.-S., *et al*. 'Nanosize Si anode embedded in super-elastic nitinol (Ni–Ti) shape memory alloy matrix for Li rechargeable batteries'. *Journal of Materials Chemistry*. 2011;**21**(30):11213–16.

[72] Son S.-B., Kim S.C., Kang C.S., Yersak T.A. 'A highly reversible nano-Si anode enabled by mechanical confinement in an electrochemically activated LixTi4Ni4Si7 matrix'. *Advanced Energy Materials*. 2012;**2**(10):1226–31.

[73] Yersak T.A., Son S.-B., Cho J.S., *et al*. 'An all-solid-state Li-ion battery with a pre-lithiated Si-Ti-Ni alloy anode'. *Journal of the Electrochemical Society*. 2013;**160**(9):A1497–501.

[74] Kwon H.J., Sohn K.Y., Park W.-W. 'Electrochemical properties of rapidly solidified Si-Ti-Ni(-Cu) base anode for Li-ion rechargeable batteries'. *Electronic Materials Letters*. 2013;**9**(6):859–63.

[75] Loka C., Yu H., Lee K.-S., Cho J. 'Nanocomposite Si/(NiTi) anode materials synthesized by high-energy mechanical milling for lithium-ion rechargeable batteries'. *Journal of Power Sources*. 2013;**244**:259–65.

[76] Suh S.-S., Yoon W.Y., Lee C.-G., *et al*. 'Implementation and characterization of silicon anode with metal alloy inactive matrix for lithium-ion secondary batteries'. *Journal of the Electrochemical Society*. 2013;**160**(4):A751–5.

[77] George S.M. 'Atomic layer deposition: an overview'. *Chemical Reviews*. 2010;**110**(1):111–31.

[78] Ban C., George S.M. 'Molecular layer deposition for surface modification of lithium-ion battery electrodes'. *Advanced Materials Interfaces*. 2016;**3**(21):1600762.

[79] Jung Y.S., Cavanagh A.S., Dillon A.C., Groner M.D., George S.M., Lee S.-H. 'Enhanced stability of $LiCoO_2$ cathodes in lithium-ion batteries using surface

modification by atomic layer deposition'. *Journal of the Electrochemical Society*. 2010;**157**(1):A75–81.

[80] Jung Y.S., Lu P., Cavanagh A.S., *et al.* 'Unexpected improved performance of ALD coated $LiCoO_2$ graphite Li-ion batteries'. *Advanced Energy Materials*. 2013;**3**(2):213–19.

[81] Sharma K., Routkevitch D., Varaksa N., George S.M. 'Spatial atomic layer deposition on flexible porous substrates: ZnO on anodic aluminum oxide films and Al_2O_3 on Li ion battery electrodes'. *Journal of Vacuum Science &Amp; Technology. A, Vacuum, Surfaces, and Films an Official Journal of the American Vacuum Society*. 2015;**34**(1):146.

[82] Bai Y., Yan D., Yu C., *et al.* 'Core-shell Si@TiO_2 nanosphere anode by atomic layer deposition for Li-ion batteries'. *Journal of Power Sources*. 2016;**308**:75–82.

[83] Yoshimura T., Tatsuura S., Sotoyama W. 'Polymer films formed with monolayer growth steps by molecular layer deposition'. *Applied Physics Letters*. 1991;**59**(4):482–4.

[84] George S.M., Yoon B., Dameron A.A. 'Surface chemistry for molecular layer deposition of organic and hybrid organic-inorganic polymers'. *Accounts of Chemical Research*. 2009;**42**(4):498–508.

[85] Dameron A.A., Seghete D., Burton B.B., *et al.* 'Molecular layer deposition of Alucone polymer films using trimethylaluminum and ethylene glycol'. *Chemistry of Materials*. 2008;**20**(10):3315–26.

[86] Dillon A.C., Ott A.W., Way J.D., George S.M. 'Surface chemistry of Al_2O_3 deposition using $Al(CH_3)_3$ and H_2O in a binary reaction sequence'. *Surface Science*. 1995;**322**(1):230–42.

[87] Groner M.D., Fabreguette F.H., Elam J.W., George S.M. 'Low-temperature Al2O3 atomic layer deposition'. *Chemistry of Materials: a Publication of the American Chemical Society*. 2004;**16**(4):639–45.

[88] Zhou H., Bent S.F. 'Highly stable ultrathin carbosiloxane films by molecular layer deposition'. *The Journal of Physical Chemistry C*. 2013;**117**(39):19967–73.

[89] Ivanova T.V., Maydannik P.S., Cameron D.C. 'Molecular layer deposition of polyethylene terephthalate thin films'. *Journal of Vacuum Science &Amp; Technology. A, Vacuum, Surfaces, and Films : an Official Journal of the American Vacuum Society*. 2011;**30**(1):121.

[90] Loscutoff P.W., Lee H.-B.-R., Bent S.F. 'Deposition of ultrathin polythiourea films by molecular layer deposition'. *Chemistry of Materials*. 2010;**22**(19):5563–9.

[91] Adamczyk N.M., Dameron A.A., George S.M. 'Molecular layer deposition of poly(p-phenylene terephthalamide) films using terephthaloyl chloride and p-phenylenediamine'. *Langmuir*. 2008;**24**(5):2081–9.

[92] Du Y., George S.M. 'Molecular layer deposition of nylon 66 films examined using in situ FTIR spectroscopy'. *The Journal of Physical Chemistry C*. 2007;**111**(24):8509–17.

[93] Son S.-B., Wang Y., Xu J., *et al.* 'Systematic investigation of the alucone-coating enhancement on silicon anodes'. *ACS Applied Materials & Interfaces*. 2017;**9**(46):40143–50.

[94] Son S.-B., Cao L., Yoon T., *et al.* 'Interfacially induced cascading failure in graphite-silicon composite anodes'. *Advanced science*. 2019;**6**(3):1801007.

[95] Piper D.M., Travis J.J., Young M., *et al.* 'Reversible high-capacity Si nano-composite anodes for lithium-ion batteries enabled by molecular layer deposition'. *Advanced Materials*. 2014;**26**(10):1596–601.

[96] Ma Y., Martinez de la Hoz J.M., Angarita I., *et al.* 'Structure and reactivity of alucone-coated films on Si and LixSiy surfaces'. *ACS Applied Materials & Interfaces*. 2015;**7**(22):11948–55.

[97] He Y., Piper D.M., Gu M., *et al.* 'In situ transmission electron microscopy probing of native oxide and artificial layers on silicon nanoparticles for lithium ion batteries'. *ACS Nano*. 2014;**8**(11):11816–23.

[98] Molina Piper D., Lee Y., Son S.-B., *et al.* 'Cross-linked aluminum dioxybenzene coating for stabilization of silicon electrodes'. *Nano Energy*. 2016;**22**:202–10.

[99] Kaul F.A.R., Tschinkl M., Gabbai F.P. 'Alumination of resorcinol and hydroquinone: an easy access to bifunctional lewis acids'. *Journal of Organometallic Chemistry*. 1997;**539**(1):187–91.

[100] Yoon B., Lee Y., Derk A., Musgrave C., George S. 'Molecular layer deposition of conductive hybrid organic-inorganic thin films using diethylzinc and hydroquinone'. *ECS Transactions*. 2011;**33**(27):191–5.

[101] DuMont J.W., George S.M. 'Pyrolysis of alucone molecular layer deposition films studied using in situ transmission fourier transform infrared spectroscopy'. *The Journal of Physical Chemistry C*. 2015;**119**(26):14603–12.

[102] Han B., Liao C., Dogan F., *et al.* 'Using mixed salt electrolytes to stabilize silicon anodes for lithium-ion batteries via in situ formation of Li-M-Si ternaries (M = Mg, Zn, al, Ca)'. *ACS Applied Materials & Interfaces*. 2019;**11**(33):29780–90.

Chapter 10

Mechanical characterization of silicon-based electrodes

Yikai Wang[1] and Yang-Tse Cheng[1]

The substantial volume change (≈300%) of silicon (Si) during repeated lithiation/delithiation cycling can cause large mechanical stresses that fracture the Si-based electrodes, resulting in electrical isolation of Si particles and consuming Li and electrolytes by continuously forming solid electrolyte interphase (SEI). These factors lead to the fast capacity fading of Si-based electrodes. To mitigate the electromechanical degradation of Si electrodes, it is necessary to understand the mechanical property evolution of Si electrodes during cycling. At room temperature, Si transforms to amorphous Li_xSi during electrochemical lithiation. Amorphous Li_xSi becomes crystalline $Li_{3.75}Si$ as the electrochemical potential (vs. Li/Li^+) drops to 50 mV [1, 2]. The composition and structure of Li_xSi determines its mechanical properties. Density functional theory (DFT) calculations show that the strong covalent Si–Si bonds are gradually replaced by weaker ionic Li–Si bonds during lithiation [3]. As lithium concentration increases, the Young's modulus of Li_xSi decreases from 95 GPa ($x = 0$ for amorphous Si) to 38 GPa ($x = 3.75$) [3]. For Si composite electrodes consisting of Si particles, polymeric binders, and carbon black, the adhesion between the binders and Si particles can greatly influence the cycling stability of Si composite electrodes. Recently, several mechanical characterization techniques, such as nano- and micro-indentation, scratch tests, and peel tests, have been used to measure the mechanical properties of Si-based electrodes. In addition, stress sensors have been employed for in situ measurements of the lithiation-/delithiation-induced stresses. Together with the fracture behavior investigated by microscopic methods, these measurement techniques have helped provide a deeper understanding of the capacity fading mechanisms of Si-based electrodes. This chapter reviews several mechanical measurement methods and their applications in developing strategies for making high capacity and durable Si composite electrodes.

[1]Department of Chemical and Materials Engineering, University of Kentucky, Lexington, KY 40506-0046, USA

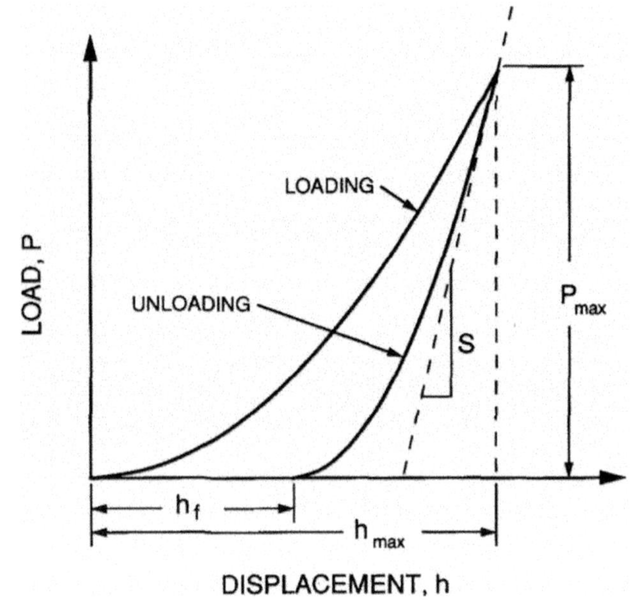

Figure 10.1 *A schematic diagram of a typical load-displacement curve obtained by nanoindentation [4]*

10.1 Nanoindentation

10.1.1 Instrumented nanoindentation technique

Common LIB electrodes consist of an electrode layer on a current collector foil. Since the thickness of electrodes is usually at the micrometer scale, conventional tensile and compression tests are unsuitable for their mechanical property measurements. In contrast, depth-sensing nano- and micro-indentation, which can quantitatively measure the mechanical properties of a small volume of materials, have been used for mechanical characterization of battery electrodes and electrode materials. During nanoindentation measurements, the tip, for example a Berkovich indenter, is pressed into the sample, held at the maximum load briefly (usually 10 s), and then removed from the sample, as shown in Figure 10.1. The nanoindentation instrument records the load, penetration depth, and time during the test. Based on the Oliver–Pharr method [4], the reduced modulus is given by

$$E_r = S \cdot \frac{\sqrt{\pi}}{2\beta\sqrt{A}} \tag{10.1}$$

where S is the initial slope (dP/dh) of the unloading curve, β is a geometry constant (β = 1.034 for a Berkovich indenter), and A is the projected area of the indent. The Young's modulus of the sample, E_s, is calculated by [4]

$$E_s = \frac{1 - v_s^2}{\frac{1}{E_r} - \frac{1 - v_i^2}{E_i}} \tag{10.2}$$

where υ_s is Poisson's ratio of the sample, υ_i and E_i are the Poisson's ratio and Young's modulus of the indenter, respectively. For indenters made of diamond, E_i = 1141 GPa and υ_i = 0.07.

The indentation hardness of the sample is calculated by [4]

$$H = \frac{P_{max}}{A} \tag{10.3}$$

where P_{max} is the maximum load. Although the Oliver–Pharr method is widely used to calculate Young's modulus and hardness from nanoindentation tests, it can be inaccurate, especially when pile-up occurs around indents [5].

10.1.2 Mechanical property evolution of Li_xSi alloys during electrochemical cycling

Since the dimension of indents can be as small as several micrometers and even tens of nanometers, nanoindentation has been used to measure the mechanical properties of Si thin film and wafer electrodes after electrochemical lithiation/delithiation. To avoid the substrate effect, the maximum indentation depth should be less than 1/10 of the thickness of the Si or Li_xSi layer. Because Li_xSi compounds obtained by electrochemical reactions are unstable in the air [6], sample preparation, transfer, and nanoindentation measurements of Si electrodes need to be conducted in an inert environment. Researchers either put nanoindenters inside argon-filled gloveboxes or fill the indentation chamber with argon [7–10]. For ex situ indentation measurements, Si thin films and wafers can be electrochemically lithiated to different states of charge (SOC). A constant voltage holding or holding at the open circuit voltage (OCV) after the galvanostatic cycling helps eliminate the lithium concentration gradient in electrodes [10]. After cycling, Si thin films and wafers are rinsed by electrolyte solvents, such as diethyl carbonate (DEC) and dimethyl carbonate (DME), to remove Li salts before indentation measurements. Hertzberg *et al.* found that the indentation depth corresponding to the same indentation load increases as Li content increases (Figure 10.2a), indicating a softening tendency [10]. In addition, the unloading slope decreases with increasing Li concentration. The Young's modulus of the pristine nanocrystalline Si film was determined as 92 GPa. After lithiation, the Young's modulus decreases to 58 GPa for $Li_{0.6}Si$, slightly increases at higher lithiation concentrations ($Li_{1.08}Si$ and $Li_{2.06}Si$), and remarkably decreases to about 12 GPa after the full lithiation ($Li_{3.75}Si$), as shown in Figure 10.2b. Similar softening tendencies of Li_xSi during lithiation were also found in nanoindentation measurements by Berla *et al.* [9], Wang *et al.* [11], and Vasconcelos *et al.* [8] despite of some discrepancies among the reported values.

The hardness of Li_xSi also decreases with increasing Li concentration, as shown in Figure 10.2c. Due to distinct sample preparation procedures (e.g., current density, which

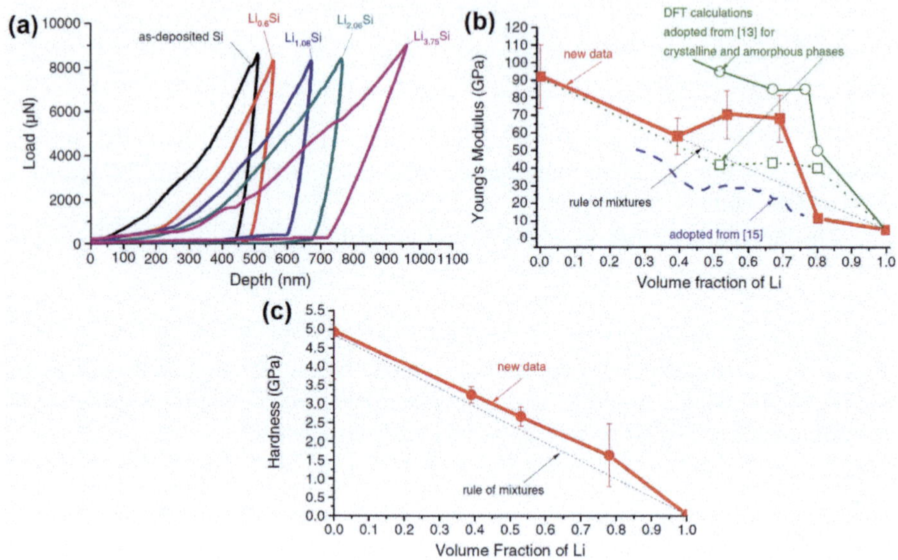

Figure 10.2 *(a) The load-displacement curves of Si thin films at different SOCs, (b) the Young's modulus of Li_xSi alloys as a function of the volume fraction of Li, and (c) nanoindentation hardness of Li_xSi alloys as a function of the volume fraction of Li [10].*

affects the Li concentration profiles in Si thin films) and the time-dependent behavior of Li_xSi, the reported hardness of Li_xSi differs in literature [8–11]. Nevertheless, the hardness of $Li_{3.75}$ Si measured by different groups distributes in a narrow range between 1.2 and 1.5 GPa. According to the Tabor relationship [12], that is, nanoindentation hardness (*H*) is about 3 times of the yield strength (σ_y) of materials, σ_y of $Li_{3.75}Si$ is estimated to be about 450 MPa.

Nanoindentation technique has also been used to measure the fracture toughness and creep behavior of Li_xSi compounds. Based on the indentation-induced radial cracks [13, 14], Wang *et al.* found that the fracture toughness of *a*-Si film ($0.51 \pm 0.014MPa\sqrt{m}$ for the as-prepared Si film) decreases slightly upon lithiation ($Li_{0.31}Si$) as Li insertion weakens the bonds among Si and reduces the surface energy [15]. However, the fracture toughness increases substantially after further lithiation ($0.77 \pm 0.03MPa\sqrt{m}$ for $Li_{1.09}Si$) as Li_xSi significantly softens and plastic deformation dominates the deformation. For indentation creep measurements, the stress exponent can be derived from the indentation strain rate and hardness during the indentation holding period with a constant load. Vasconcelos *et al.* found the stress exponent of Si is about 50, implying a rate-insensitive deformation behavior. The stress exponent decreases abruptly upon lithiation and reaches a stable value of 22 as the Li concentration increases above 0.5 (i.e., $Li_{0.5}Si$) [8], which is consistent with the brittle-to-ductile transition from Si to Li_xSi during lithiation [15, 16].

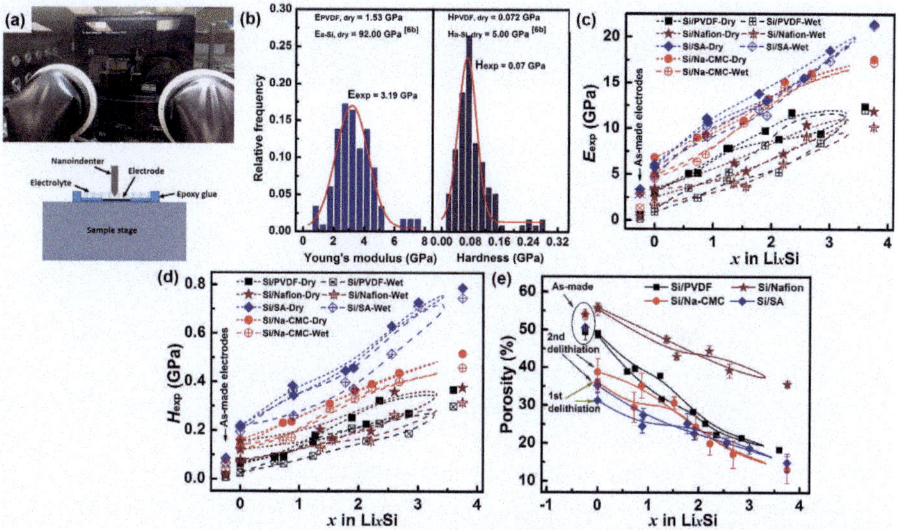

Figure 10.3 *(a) An environmental nanoindentation system inside an argon-filled glovebox. The lower figure shows the schematic diagram of the liquid cell for nanoindentation measurements under wet conditions. (b) The distribution histograms of Young's modulus and hardness of the Si/PVDF composite electrode after the first delithiation under dry conditions. The evolution of the expectation values of (c) the Young's modulus and (d) hardness of Si composite electrodes measured under both dry and wet conditions. (e) The evolution of the porosity of Si composite electrodes [7, 19].*

10.1.3 Mechanical property evolution of Si composite electrodes

The Young's modulus and hardness of Si composite electrodes have a distinct dependence on Li concentration since their porosity changes as Si particles expand and contract during electrochemical cycling. In addition, the mechanical properties of composite electrodes are influenced by the testing environment, that is, measurements under dry conditions are different from that under wet conditions (e.g., in the organic liquid electrolyte) since polymeric binders swell and soften in the organic electrolyte [17, 18]. Since SEI, Li_xSi compounds, and electrolytes are sensitive to air and moisture [6], an environmental nanoindentation system was installed inside an argon-filled glovebox [7], as shown in Figure 10.3a. The liquid cell made of electrolyte-resistant glue was used for nanoindentation measurements in the liquid electrolyte (denoted as under wet conditions). Because the disassembling process of coin cells causes deformation, introduces external stress, and thus changes the mechanical properties of the cycled electrodes, Swagelok cells were used to cycle Si composite electrodes to different SOCs for environmental nanoindentation measurements. The disassembly process of Swagelok cells affects little the microstructure and mechanical properties of electrodes. The Si composite electrodes were made of Si nanoparticles (30–50 nm), carbon black, and different binders, including polyvinylidene fluoride (PVDF), Nafion, sodium carboxymethyl cellulose (Na-CMC), and sodium alginate (SA) [7]. The indentation depth was 1500 nm to reduce the influence of surface roughness and tiny cracks. Since the indent covers thousands of Si nanoparticles, nanoindentation measures the average mechanical properties of Si composite electrodes instead of individual particles.

The measured Young's modulus and hardness of Si composite electrodes at each SOC distribute over a range. For example, the Young's modulus of the Si/PVDF electrode at the first delithiation state ranges from 0.59 to 7.76 GPa. The hardness is between 0.01 and 0.29 GPa, as shown in Figure 10.3b. Gaussian function was used to analyze the histograms of Young's modulus and hardness. The expectation values (mean values) of Gaussian function, that is, Young's modulus (E_{exp}) and hardness (H_{exp}), were used to plot against the Li concentration (i.e., x in Li_xSi), as shown in Figure 10.3. In contrast to the lithiation-induced softening of Si thin film electrodes [8–11], both E_{exp} and H_{exp} increase with increasing Li concentration during lithiation, reach the peak values at the lithiation state ($Li_{3.75}Si$), and then decrease with decreasing Li concentration during delithiation. For example, E_{exp} of the Si/SA electrode is 5.74 GPa at the first delithiation state. It increases almost linearly to the peak value of 21.40 GPa at the second lithiation state and then decreases to 6.04 GPa at the second delithiation state [19].

Although the lithiation-induced hardening of Si nanoparticle electrodes has been found in previous studies [7, 19, 20], lithiation-induced softening of Si composite electrodes could also happen if the indentation depth is smaller than individual Si (or Li_xSi) particles. For example, if the ratio of the particle diameter to the maximum indentation depth (d/h_{max}) is larger than 50, the indent only covers a single Si (or Li_xSi) particle. The indentation response will be dominated by the mechanical behavior of individual Si (or Li_xSi) particles, showing a lithiation-induced softening tendency similar to Si thin films [3, 8, 10, 11]. To measure the average mechanical properties (incorporating the effects of porosity and SEI) of Si composite electrodes, nanoindentation measurements should be conducted with a maximum indentation depth much larger than Si particle size, for example, $h_{max}/d > 30$ in previous studies [7, 19, 20]. Hereinafter, we focus on the average mechanical properties of Si composite electrodes.

The lithiation-induced "hardening" of Si composite electrodes can be attributed to the porosity change during cycling. As shown in Figure 10.3e, the porosity of Si composite electrodes decreases continuously during lithiation due to the volumetric expansion of Li_xSi particles [19]. It reaches the lowest value after the full lithiation and then gradually increases to a level close to that at the first delithiation state as Li_xSi particles contract during the second delithiation process. The porosity change is consistent with the cross-sectional microstructure of electrodes at different SOCs. As shown in Figure 10.4, many voids are visible in the as-made and delithiated electrodes, while only smaller and fewer voids can be found in the lithiated electrode (Figure 10.4c) [7].

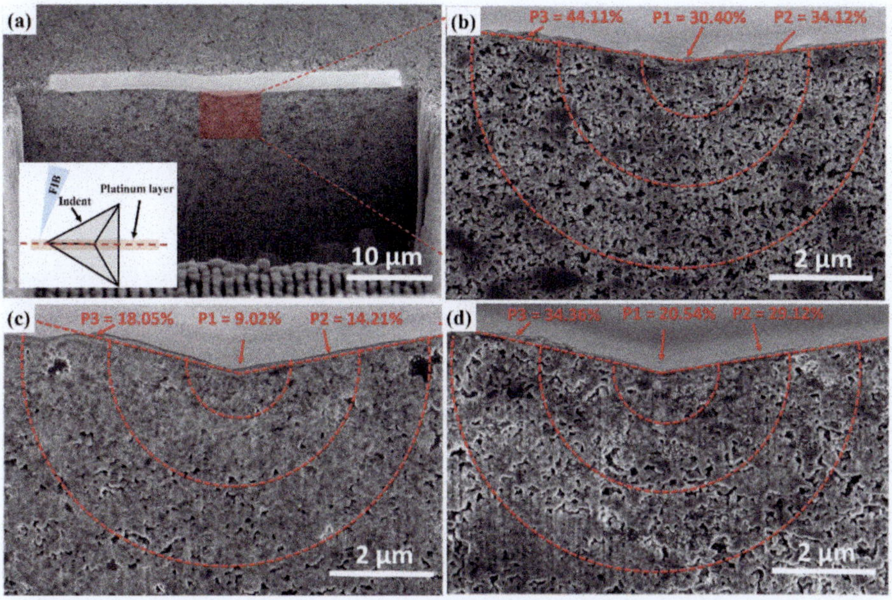

Figure 10.4 *(a) A typical cross-sectional SEM image of an indent in the as-made Si/PVDF composite electrode. The cross section was prepared by a focus ion beam (FIB) along the median of each triangular indent, as shown in the inserted schematic diagram. High magnification cross-sectional SEM images of (b) the as-made electrode; (c) the electrode after the second lithiation and (d) the electrode after the second delithiation. The regional porosity (P1, P2, and P3 regions) underneath the indent increases with increasing distance from the apex of the indent [7].*

During nanoindentation measurements, the indenter densifies the porous composite electrode. As indicated by the dashed lines in Figure 10.4b–d, there is a porosity gradient underneath the indents. The porosity increases with increasing distance from the apex of the indenter. Densification is the dominated deformation mechanism in the highly porous as-made electrodes and electrodes at the first delithiation and second delithiation states since most Si particles in the indents were not deformed, as shown in Figure 10.5. In contrast, Li_xSi particles in the second lithiated electrode are clearly deformed by the indenter (Figure 10.5g). Based on these observations, a porosity-dependent indentation deformation mechanism was proposed to understand the lithiation-induced hardening of Si composite electrodes. The as-made and delithiated Si composite electrodes have high porosity; nanoindentation induces the densification of the highly porous composites instead of deforming the small and stiff Si particles (see Figure 10.5). Compared with crystalline and amorphous Si, the highly porous binder/CB matrix is much more compliant. Therefore, the measured Young's modulus and hardness of the as-made and fully delithiated electrodes are small. As the lithium concentration increases, active particles expand, and the

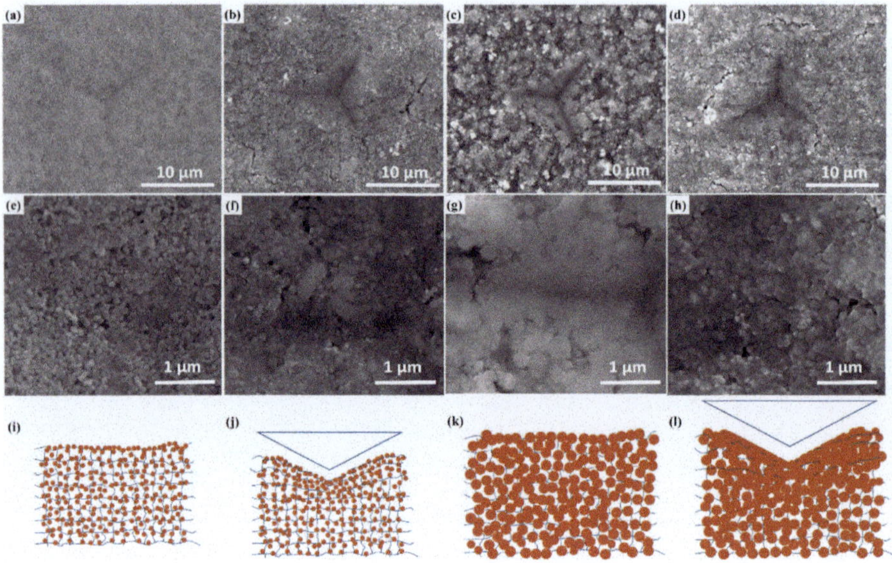

Figure 10.5 *Typical indent morphologies in Si/PVDF composite electrodes at different SOCs: (a) as-made, (b) the first full delithiation, (c) the second full lithiation, and (d) the second full delithiation. (e)–(h) are enlarged microstructure in indents corresponding to (a)–(d), respectively. (i) and (j) are schematic diagrams of the indentation deformation of highly porous electrodes at the delithiation state. (k) and (l) are schematic diagrams of the indentation deformation of electrodes with low porosity at the lithiation state [7].*

porosity of electrodes decreases. Densification completes at a shallow indentation depth (Figure 10.5). The deformation of Li_xSi particles increasingly contributes to nanoindentation responses, resulting in higher values of the Young's modulus and hardness although Li_xSi particles soften as the lithium concentration increases. Overall, the lithiation-induced hardening of Si composite electrodes is a coupling result of the porosity-dependent deformation mechanism (densification vs. deformation of Li_xSi particles) and the lithiation-induced softening of Li_xSi particles.

The Young's modulus and hardness of Si composite electrodes at the same SOC are affected by polymeric binders, as shown in Figure 10.3. Their values follow generally the sequence of Si/Nafion < Si/PVDF < Si/Na-CMC < Si/SA. The effects of binders on the mechanical properties of Si composite electrodes may be attributed to two factors: (1) mechanical properties of binders and (2) the adhesion between binders and Si. Nanoindentation measurements (Figure 10.6a–c) show that the magnitudes of Young's modulus and hardness of binders, under both dry and wet conditions, follow the sequence: SA > Na-CMC > PVDF > Nafion, which is consistent with that of Si composite electrodes made of different binders. Lap joint shear tests (Figure 10.6d) show that the adhesive strength between binders and Si follows the sequence: Nafion/Si > SA/Si > Na-CMC/Si

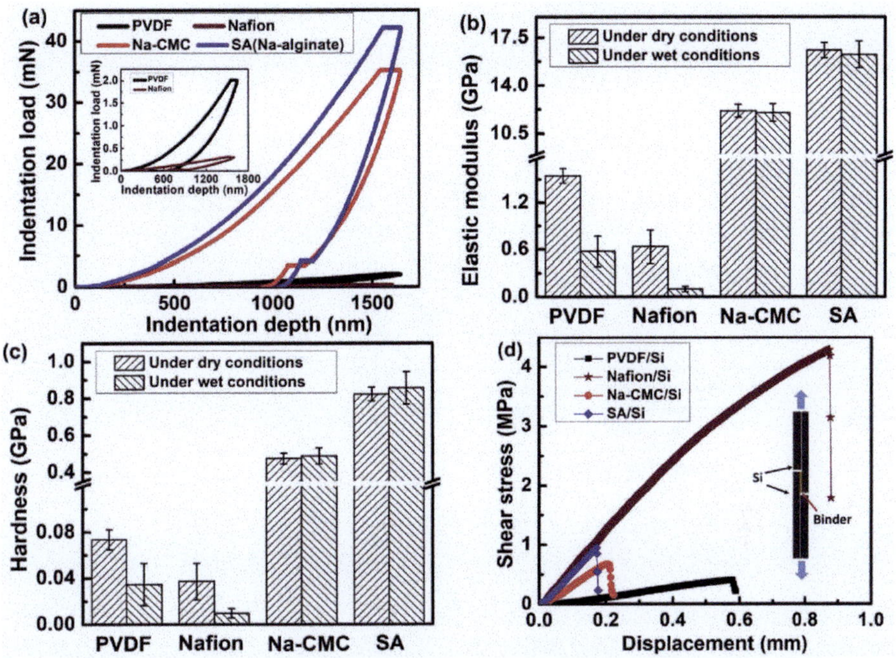

Figure 10.6 *(a) The indentation load-displacement curves of PVDF, Nafion, Na-CMC, and SA thin films. (b) The Young's modulus. (c) Hardness of binder films under dry and wet conditions. (d) The shear stress-displacement profiles of Si/binder/Si lap joints [19].*

> PVDF/Si. Because (1) the nanoindentation response is dominated by the porosity of Si composite electrodes and (2) the porosity of Si electrodes at different SOCs follows: Si/Nafion > Si/PVDF > Si/Na-CMC > Si/SA (as shown in Figure 10.3e, although Si/SA and Si/Na-CMC electrodes have very similar porosity when $x \geq 1.8$), it is concluded that the mechanical properties of binders, instead of the adhesion between binders and Si, strongly affect the porosity change as well as the Young's modulus and hardness of Si composite electrodes. Compared with soft binders (Nafion and PVDF), stiff binders (Na-CMC and SA) can effectively suppress the irreversible volume change and strengthen Si composite electrodes.

Electrolytes also affect the mechanical properties of Si composite electrodes. As shown in Figure 10.3, the Young's modulus and hardness of Si composite electrodes measured under wet conditions are smaller than that measured under dry conditions, which can be attributed to the softening of binders in the organic electrolyte. As shown in Figure 10.6, PVDF and Nafion films significantly soften in the electrolyte. For example, E and H of the Nafion film under wet conditions are 16% and 26% of that under dry conditions, respectively. In contrast, the Young's modulus and hardness of Na-CMC and SA under wet conditions are very close to that under dry conditions. The softening behavior of binders is consistent with their swelling in the electrolyte. Swelling ellipsometry

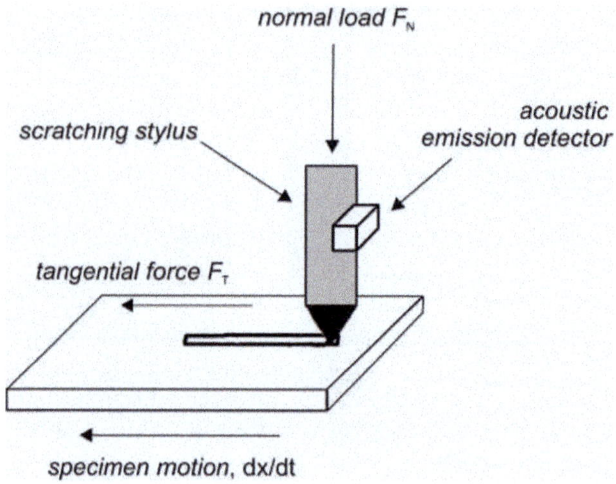

Figure 10.7 A schematic diagram of scratch tests [21]

measurements indicated that the thickness of a PVDF thin film increased by 18% after immersed in the DEC vapor for 1 000 s, while Na-CMC and SA films swelled only 0.5% and negligibly under the same condition, respectively [17, 18]. The softening behavior of binders in the electrolyte is "inherited" by Si composite electrodes as Si/PVDF and Si/Nafion electrodes soften more than Si/Na-CMC and Si/SA electrodes in the electrolyte (Figure 10.3). Cautions should be taken when selecting mechanical properties of Si composite electrodes for electromechanical modeling.

Although nanoindentation has been used to measure the mechanical properties of electrodes and electrode components, it cannot measure the adhesive strength between binders and Si particles, the adhesion between Si electrode layers and copper (Cu) current collectors, and the cohesive strength of Si composite electrodes. Recently, scratch and peel tests have emerged as popular approaches to evaluate the adhesion and cohesion of Si composite electrodes.

10.2 Scratch tests

During scratch tests, a hard stylus with known geometry and size (usually a conical indenter) moves on the surface of the coating at a constant rate and with a progressively linearly increasing normal load (Figure 10.7). As the normal load increases, coating failures, such as spalling, buckling, and tensile cracks, can occur. A "scratch critical load" (L_c) is the minimum load that would significantly damage the coating [22]. The magnitude of L_c can be used to compare qualitatively the adhesive strength between coatings and substrates. Microscope observations are often used to identify scratch failures, which is usually accompanied with an abrupt change of the frictional coefficient between the tip and the coating [23].

To evaluate the adhesion between polymeric binders and Si, the polymeric binder@Si sample can be prepared by casting binder solutions on Si wafers. The binder@Si wafer samples are completely dried before scratch tests. For fair comparisons, the thickness of binder coatings should be the same. Figure 10.8 shows the scratch depth-distance curves of poly(acrylic acid) (PAA) films and lithium-substituted PAA films (with 75% H in the carboxyl groups substituted by Li, denoted as PAA-0.75Li) on Si wafers. Severe fluctuations are found in the depth profile of PAA@Si after 700 μm scratch distance. Microscope observations in Figure 10.8 show that PAA@Si has discrete delamination along the scratch. In contrast, the depth profile of PAA-0.75Li@Si is quite smooth and without abrupt changes in the scratch depth within the 1500 μm scratch. In addition, the PAA-0.75Li has shallower and narrower scratches than PAA. Therefore, PAA-0.75Li has stronger adhesion with Si as well as higher strength than PAA, indicating that PAA-0.75Li is likely a better binder for Si electrodes than PAA.

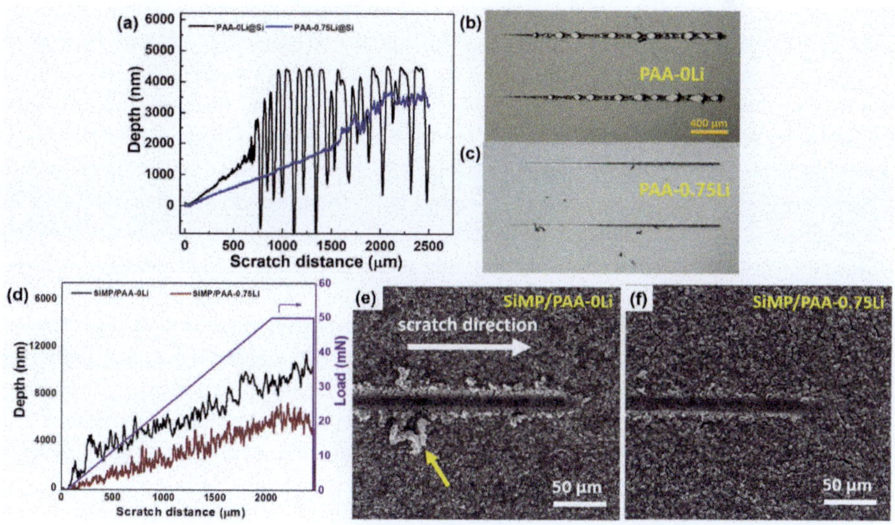

Figure 10.8 *(a) The scratch depth vs. distance profiles of PAA@Si and PAA-0.75Li@Si films. The PAA and PAA-0.75Li films have the same thickness of 5 μm. (b) and (c) are the scratch morphology of PAA@Si and PAA-0.75@Si films, respectively. (d) The scratch depth vs. distance profiles of Si/PAA and Si/PAA-0.75Li electrodes. (e) and (f) are the scratch morphology of PAA@Si and PAA-0.75@Si films, respectively (from unpublished results)[1].*

[1]These results have been published in our recent paper, Dang, Dingying, Yikai Wang, Ming Wang, Jiazhi Hu, Chunmei Ban, and Yang-Tse Cheng. "Lithium Substituted Poly (acrylic acid) as a Mechanically Robust Binder for Low-Cost Silicon Microparticle Electrodes." *ACS Applied Energy Materials* 3, no. 11 (2020): 10940–49.

Scratch tests were carried out in Si composite electrodes made of Si microparticles with respective PAA and PAA-0.75Li as binders (denoted as Si/PAA and Si/PAA-0.75Li electrodes, respectively) to further evaluate the effect of binders on the mechanical integrity of Si composite electrodes. As shown in Figure 10.8, the scratch depth profiles of both electrodes slightly fluctuate due to the surface roughness and microstructural discontinuity of the porous composite electrodes. No delamination between the electrode and the Cu current collector was found at the maximum normal load, as shown in Figure 10.8. Under the same normal load, the Si/PAA electrode have wider and deeper scratches than the Si/PAA-0.75Li electrode. Moreover, many electrode fragments were peeled off from the Si/PAA electrode after scratch (Figure 10.8), while only several small fragments were found in the Si/PAA-0.75Li electrode. Based on the scratch morphology and depth profiles, the Si/PAA-0.75Li electrode has better cohesive strength than the Si/PAA electrode, which can be attributed to the robust mechanical properties of the PAA-0.75Li binder and the strong adhesive strength between PAA-0.75Li and Si.

Scratch tests have also been used to investigate the effects of surface modifications on the mechanical integrity of Si composite electrodes. To stabilize the interface between Si nanoparticles and the binder/carbon black matrix, Son *et al.* deposited a thin alucone coatings on Si nanoparticle composite electrodes by using molecular layer deposition (MLD) [24]. High-resolution transmission electron microscopy (HRTEM) confirmed that the alucone coating, despite with a thickness of about 2 nm, has intimate contact with Si particles. Energy dispersive spectroscopy (EDS) analysis detected a uniform Al signal from the electrode surface to the interface between electrode and the current collector, indicating that the alucone coating disperses through the porous Si composite electrodes. The MLD coating imposes significant effects on the scratch behavior of the composite electrodes, as shown in Figure 10.9. For example, at the scratch distance of 900 μm, the uncoated Si electrodes delaminate from the Cu current collector, as indicated by the exposed Cu in

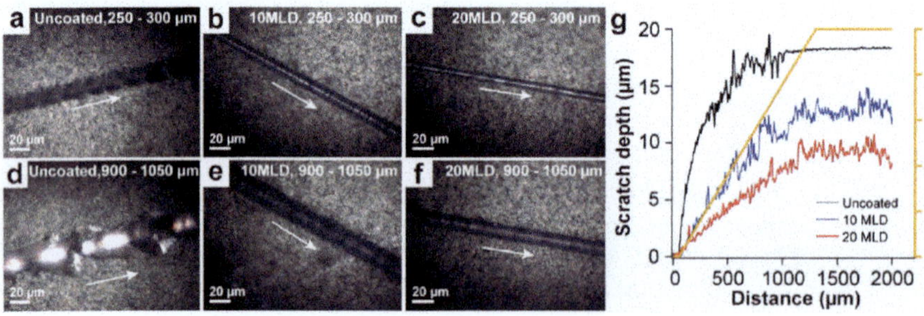

Figure 10.9 Microstructure of the uncoated and coated electrodes corresponding to different scratch distances: (a) and (d) uncoated electrodes, (b) and (e) electrodes coated with 2.5 nm alucone (10 MLD), (c) and (f) electrodes coated with 5 nm alucone (20 MLD). (g) Depth profiles of electrodes as a function of scratch distance [24].

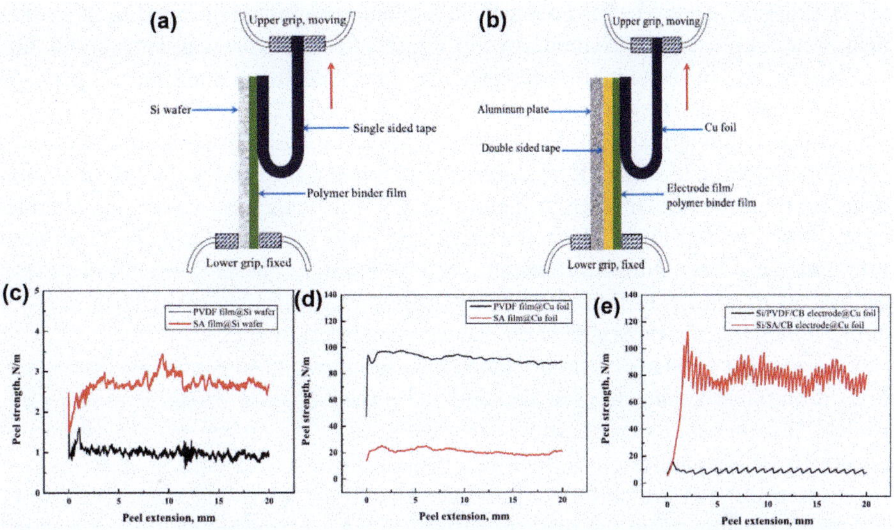

Figure 10.10 *Schematics of 180° peel test set-ups for (a) binder@Si wafer samples and (b) electrode@Cu or binder@Cu samples. (c)–(e) are the peel strength vs. peel extension profiles of binder@Si wafer, binder@Cu, and electrode@Cu samples, respectively [26].*

Figure 10.9. In contrast, the MLD coated electrodes keep intact and show smaller scratches than the uncoated ones. The scratch profile (Figure 10.9) shows that the scratch depth of the coated electrodes (denoted as 10- and 20-MLD) is smaller than that of the uncoated ones corresponding to the same normal load. These results show clearly that the alucone coating effectively improves the cohesion of Si composite electrodes, which is one of the critical reasons to understand the beneficial effect of the MLD coating on the cycling stability of Si composite electrodes [24].

10.3 Peel tests

Peel tests are one of the standard approaches to measure the adhesion in films/substrate systems. There are various types of peel tests with different angles, including 90° peel test, 180° peel test, and T-type peel test [25]. The angle influences the stress distribution at the bonding interface as the peeling front moves through the specimens. The 180° peel test (ASTM D903 standard) has been used to study the cohesive and adhesive strength of battery electrodes. During the tests, the bonded parts are separated at a constant angle of 180°, as shown in a typical configuration in Figure 10.10. The peel strength is measured as the load per unit width of the bond when the peeling speed becomes constant. After observing the failure mode, the peel strength is considered as a measure of the work of delamination or the work of cohesion.

The applications of 180° peel tests in Si electrodes include evaluating the adhesion between polymeric binders and Si (binder@Si), the adhesion between binders and the Cu current collector (binder@Cu), and the cohesion or adhesion between the electrode layer and the Cu current collector (electrode@Cu). To measure the adhesion between binders and Si, Hu *et al.* cast polymeric binder solutions, including PVDF and SA on Si wafers [26]. After drying, binder strips with a width of 13 mm were cut from the binder films. A piece of single-sided tape was attached to a binder strip. As shown in Figure 10.10, the upper grip pulls the tape at a speed of 50 mm/min and peels the binder film apart from the Si wafer. The peel strength of SA@Si (~2.7 N/m, shown in Figure 10.10) is about three times that of PVDF@Si (~1.0 N/m), which is consistent with shear lag measurements [19].

To conduct 180° peel tests of binder@Cu samples and Si composite electrodes, a rigid aluminum plate is used to bond the binder coating or the laminate electrodes. The uncoated side of the Cu strip is pulled by the upper grips, as shown in Figure 10.10. The adhesion of the PVDF/Cu interface is measured to be about 91.3 N/m (Figure 10.10), which is over four times that of the SA/Cu interface (20.5 N/m). Figure 10.10 shows that the peel strength of Si/SA electrodes (78.3 N/m) is about ten times that of Si/PVDF electrodes (8.7 N/m). The peel profiles of both electrodes consist of many serrations due to the curvature change of the peeling front during peel tests. After peel tests, the Cu current collector was divided into dark and bright regions, as shown in Figure 10.11. The bright regions have a few electrode component clusters (Figure 10.11), while the dark regions are still covered by electrodes (Figure 10.11). Since the peel failure is dominated by cohesive fractures within the electrode layer, these peel tests measure the cohesion of the Si composite electrodes. The adhesion of Si/SA and Si/PVDF electrode layers to the Cu current collector is stronger than the cohesive strength between electrode layers and Cu foils. As shown in Figure 10.10, the peel strength of Si composite electrodes is much higher than the adhesion between the corresponding binder films and Si wafers. This observation can be explained by the different microstructure configurations of samples. For binder@Si samples, peeling-induced separations take place between polymer chains and a smooth Si wafer surface without the influence of solid particles. In electrode@Cu samples, binders connect Si and carbon black particles. The total area of cohesive fractures at the binder/Si interface in composite electrodes is much larger than that of binder@Si samples. In addition, the energy barrier to separate binders from carbon black particles, though unknown, can also increase the peel strength of composite electrodes.

10.4 In situ stress measurements

Stress in Si electrodes is primarily caused by lithium diffusion and the phase transformation among Li_xSi compounds with different molar volumes during cycling. It affects not only the mechanical integrity of Si electrodes but also the electrochemical reaction kinetics during cycling [27, 28]. Real-time measurements of the lithiation-/delithiation-induced stress, though quite challenging, is imperative for

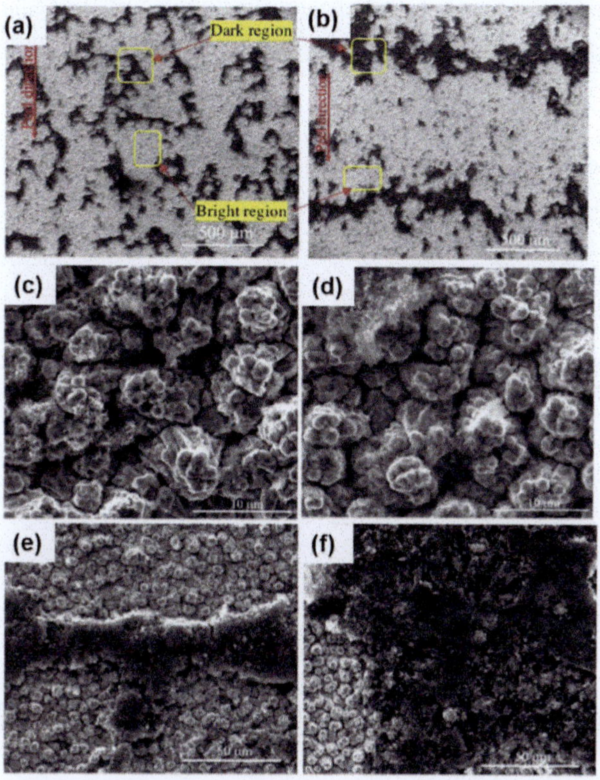

Figure 10.11 *Microstructure of Si composite electrodes after 180° peel tests: (a) Si/SA and (b) Si/PVDF electrodes. (c) and (d) are microstructure of the bright regions in the Si/SA and Si/PVDF electrodes after peel tests, respectively. (e) and (f) are microstructure of the dark regions in the Si/SA and Si/PVDF electrodes after peel tests, respectively [26].*

understanding the fracture behavior and the electrochemical-mechanical coupling effects of Si-based electrodes. To date, two major techniques have been used for in situ stress measurements of Si-based electrodes, including optical stress sensors and curvature measurements of cantilever electrodes.

10.4.1 Multi-beam optical stress sensor (MOSS)

Figure 10.12 shows a typical MOSS system configurated with a custom-made electrochemical cell [29]. The working electrode can be double-side polished Si wafers or Si thin films deposited on an electrochemically inert wafer substrate with smooth surface on both sides [29–31]. Si wafers with thick thermal oxide layers can also be used as the substrate. To prepare Si thin film samples, Ti and Cu layers are usually sputtered on the substrate to improve the adhesion between Cu and the substrate and

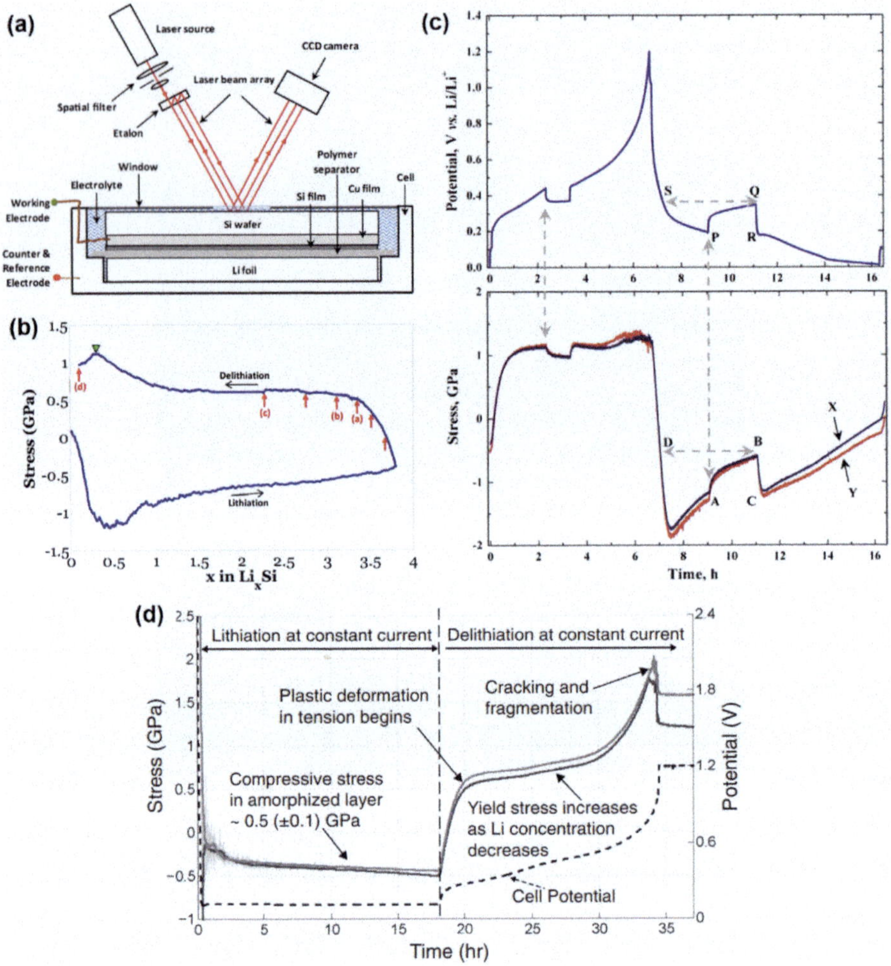

Figure 10.12 *(a) The schematic of an electrochemical cell for MOSS [29]. (b) The stress vs. lithium concentration profile of a 300 nm amorphous Si thin films [30]. (c) The potential and stress profiles of a Si thin film electrode during lithiation/delithiation and open-circuit-potential relaxation. The relaxation starts from P and A and ends at Q and B [29]. (d) The stress and potential profiles of a Si wafer electrode in the first cycle [31].*

to serve as the current collector, respectively [30]. Upon electrochemical cycling, the lithiation/delithiation induced volume changes of Si (thin films or wafers) cause the curvature change ($\Delta\kappa$) of the substrate, which is monitored by a parallel array of laser beams through the transparent quartz window of the cell. The biaxial film stress, σ, can be calculated by Stoney equation [32],

$$\sigma = \sigma^r + \frac{E_s h_s^2 \Delta\kappa}{6h_f(1-\nu_s)} \tag{10.4}$$

where E_s and ν_s are the Young's modulus and Poisson's ratio of the substrate, respectively. σ^r is the residual stress after sputtering Si films. σ^r can be measured by the curvatures of the substrate before and after film preparation. h_f and h_s are the thickness of the film and substrate, respectively. Based on the assumption that 370% volumetric expansion of Si corresponds to the theoretical capacity of 3579 mAh g^{-1}, σ^r is assumed as a linear function of SOC, that is, h_s [29], where z is the Li concentration and h_f is the initial thickness of Si thin films.

The excessive formation of SEI may influence the stress measurement in the initial cycle. Pharr *et al.* found that the stress profiles of Si thin films with different thickness (100 and 300 nm) are almost the same, implying that the SEI formation has negligible influence on stress [30]. A typical stress vs. lithiation concentration profile is shown in Figure 10.12. The tensile stress (70 MPa) of the film before electrochemical cycling is caused by the sputtering process of Si thin films. During lithiation, the expansion of the Si thin film is restricted by the rigid substrate, resulting in compressive stress. The rapidly linear increase of compressive stress indicates the elastic response during the initial lithiation stage. When the lithiation concentration increases to 0.4, the compressive stress reaches to the maximum value of 1.2 GPa and the stress decreases during further lithiation, indicating continuous plastic flow. At the full lithiation state (corresponding to $Li_{3.75}Si$), the compressive stress reaches the minimum value of about 450 MPa, which is consistent with the yield strength of $Li_{3.75}Si$ [10]. During delithiation, the initial unloading is with a large slope and elastic. The compressive stress transforms to tensile stress linearly until it reaches a tensile plateau (500 MPa corresponding to $Li_{3.4}Si$), when plastic deformation takes place in the Si film to accommodate the volume contraction. During further delithiation, the tensile stress increases to the maximum value of 1.2 GPa. Similar stress evolution tendency has also been found by other studies [29, 33–35]. In addition, stress relaxation happens simultaneously with the potential relaxation when galvanostatic lithiation is interrupted by a short period of delithiation, as shown in Figure 10.12 [29]. The stress of the Si thin film decreases rapidly upon the interruption (open circuit) at A and P. Both the potential and stress increase gradually toward an equilibrium value during the 2-h interruption period. Upon re-lithiation at B and Q, both the potential and the stress quickly resume to that prior to the interruption, which suggests that Li concentration changes are not responsible for the stress relaxation. Sethuraman *et al.* proposed that the stress relaxation is driven by viscoplastic relaxation of the amorphous Li_xSi [29].

Si wafers show a different stress profile from Si thin films. As shown in Figure 10.12 [31], the compressive stress reaches a plateau of about 0.5 GPa. Due to the small lithiation current density (12.5 μA/cm^2), it is speculated that the phase boundary between amorphous Li_xSi and crystalline Si moves slowly. The Li concentration in the amorphous layer can be assumed the same. Based on the thickness of the lithiated layer and the lithiation capacity, the lithium concentration is 3.5. The stress plateau at 0.5 GPa suggests that the flow stress of $Li_{3.5}Si$ is at about the same level.

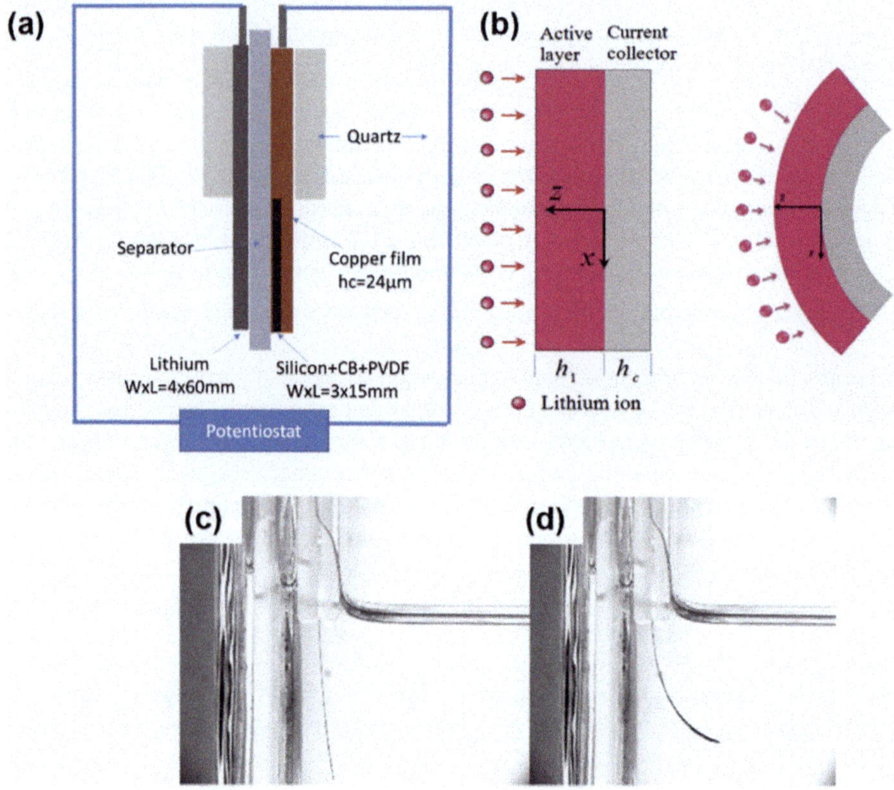

Figure 10.13 *(a) A schematic of the electrochemical cell for curvature measurements of cantilever electrodes. (b) A schematic of the deformation of an electrode active layer bonded to a current collector. Photos of a composite electrode cantilever before (c) and after lithiation (d) [36].*

10.4.2 In situ curvature measurements of cantilever electrodes

The volume expansion of Si also drives the bending of composite electrodes during electrochemical cycling. The stress can be derived from the curvature change of cantilever electrodes based on elastic bending of bi-layer plates. Figure 10.13 shows the transparent quartz cell configuration of this method [36]. A Si composite electrode stripe (cantilever) is used as the working electrode. The uncoated side of the Cu current collector is fixed. The Si electrode layer faces the counter electrodes, for example a piece of Li foil, separated by a separator film. The cell is filled with electrolytes and sealed to prevent the evaporation of the electrolyte solvent. Upon lithiation, the electrode layer expands, while the current collector restricts its free expansion. As a result, the electrode bends to a circle, as shown in Figure 10.13. During delithiation, the curvature decreases as the electrode layer contracts. The curvature change of the cantilever electrode is recorded by a video camera. Based on the elastic deformation theory, the in-plane normal strain (along the z direction in Figure 10.13) varies in the thickness direction by

$$\varepsilon = \varepsilon_0 + \kappa z \tag{10.5}$$

where ε_0 is the in-plane strain at the plane of $z = 0$, representing the elongation strain of the electrode. κ is the curvature of the electrode cantilever. Equation (10.5) is valid with the assumption that the electrode layer bonds to the current collector during cycling. At a low C-rate, the Li distribution is assumed as homogeneous in the thickness direction. The constitutive equation of the active layer is [37]

$$\sigma_e = E_e(\bar{c})(\epsilon_0 + KZ) - \frac{1}{3}E_e(\bar{c})\epsilon^*_{max}\bar{c} \tag{10.6}$$

where $E_e\left(\bar{c}\right)$ is the Young's modulus of the electrode as a function of the normalized Li concentration, $\bar{c} = c/c_{max}$ (c_{max} is the maximum concentration of Li in the electrode). ε^*_{max} is the maximum volumetric strain of the electrode after fully lithiated. $\varepsilon^*_{max} = f\Omega c_{max}$, where f and Ω are the volume fraction and the partial molar volume of the electrode, respectively. The Li diffusion induced stress is represented by $\frac{1}{3}E_e\left(\bar{c}\right)\varepsilon^*_{max}\bar{c}$ since only the z direction can expand freely. The stress in the Cu current collector is [37]

$$\sigma_{Cu} = E_{Cu}(\epsilon_0 + KZ) \tag{10.7}$$

where E_{Cu} is the Young's modulus of Cu. As the cantilever electrode is free of mechanical constrain, the force and torque balances follow

$$\int_{-hc}^{h1} \sigma_x dz = 0 \tag{10.8}$$

$$\int_{-hc}^{h1} \sigma_x z dz = 0 \tag{10.9}$$

With the normalized Li concentration and the curvature of the cantilever electrode, the stress and Young's modulus as a function of Li concentration can be derived by the above equations. Li *et al.* have used this method to measure the stress evolution of Si composite electrodes cycled with a partial lithiation/delithiation protocol, as schematically shown in Figure 10.14 [36, 37]. The curvature of Si electrodes gradually increases during lithiation and decreases during delithiation. The curvature, at the same SOC, increases with the cycle number in the first three cycles and the curvature does not completely reverse back to zero after the second and third delithiation (Figure 10.14). The derived stress vs. SOC profiles (Figure 10.14) show that compressive stress is generated in Si composite electrodes upon lithiation and increases linearly with increasing the Li concentration during lithiation. The maximum compressive stress of Si composite electrodes is tens of MPa, much smaller than that of Si wafer and thin film electrodes. The stress rapidly decreases linearly upon delithiation, implying elastic recovery. During further delithiation, the stress reaches a plateau of several MPa, which can be attributed to the plastic deformation of Si composite electrodes, including the rearrangements of particles, detachments among particles, and cohesive fractures.

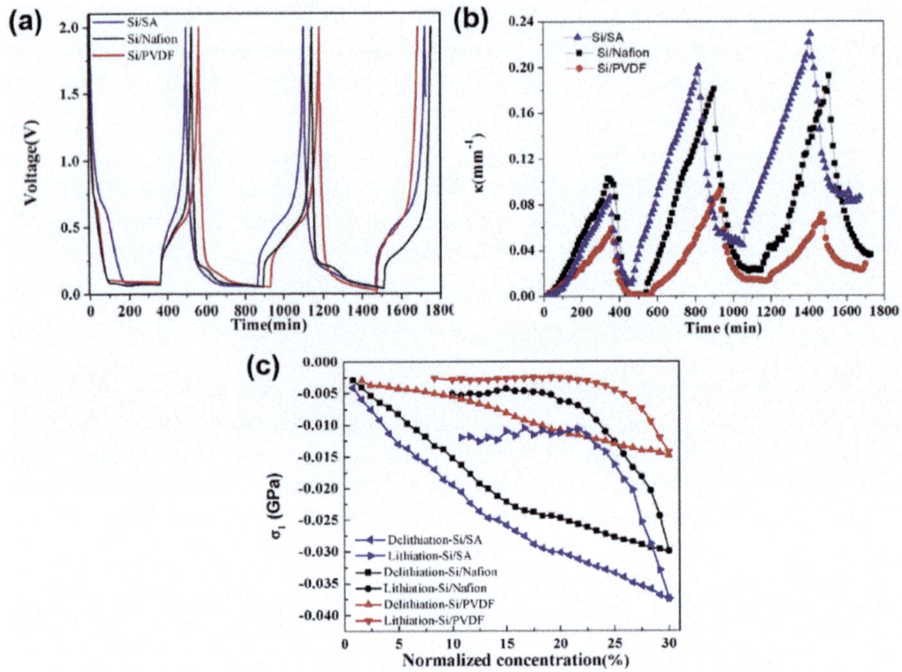

Figure 10.14 *(a) Voltage and (b) curvature evolution of composite Si/binder electrodes in the first three cycles. (c) The stress vs. Li concentration profiles of the electrodes in the second cycle [37]. Si composite electrodes were cycled at a low C-rate of C/20 to homogenize the Li distribution.*

Polymeric binders affect the stress evolution of Si composite electrodes. The magnitude of curvature and stress follows the sequence of Si/SA > Si/Nafion > Si/PVDF. The binder-dependent stress of Si composite electrodes can be attributed to the adhesion between binders and Si and the cohesion of Si composite electrodes. Both SA and Nafion have strong adhesion with Si [19] and hence can secure the connections between Si particles and the electrode matrix. In addition, the robust mechanical properties of SA also strengthen the composite electrode. In contrast, PVDF is a soft binder and has quite weak Van der Waals interaction with Si. The mechanical integrity of Si/PVDF electrode is much poorer than that of Si/SA and Si/Nafion electrodes. For instance, the crack density of the cycled Si/PVDF electrode (after three cycles) is about six times that of the pristine one, while the crack increments of Si/SA and Si/Nafion electrodes are about three times [37]. The extensive fracture may relax the electrochemical stress in the Si/PVDF electrode. The mechanically robust Si/SA and Si/Nafion electrodes can endure high stress without significant fractures, which is one of the reasons for their better electrochemical stability than the Si/PVDF electrode.

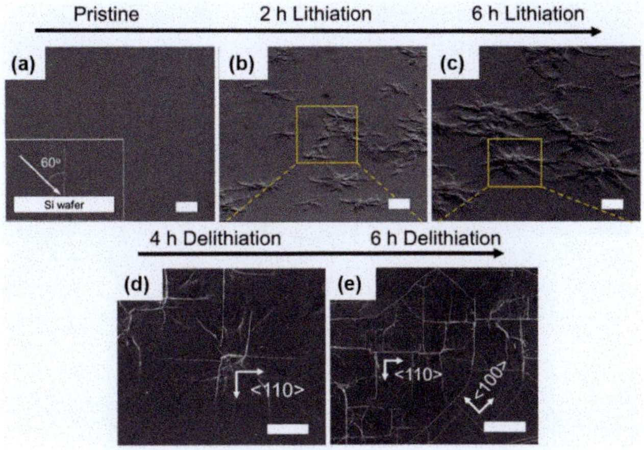

Figure 10.15 *Surface morphology changes of Si (100) wafers during electrochemical cycling: (a) the pristine wafer, (b) and (c) are wafers after lithiation at 100 μA/cm² for 2 and 6 h, (d) and (e) are wafers after delithiation at 100 μA/cm² for 4 and 6 h. Scale bars are 50 μm [38].*

10.5 Fracture behavior of Si-based electrodes

10.5.1 Si wafer electrodes

Upon the initial lithiation, amorphous Li_xSi (*a*-Li_xSi) layers form in wafer electrodes. As the volume expansion of Li_xSi is constrained by the wafer substrate, compressive stress is generated in *a*-Li_xSi layers and drives the formation of buckles, as shown in Figure 10.15 [38]. Buckles propagate preferably along the <110> direction due to the larger lithium ion diffusion coefficient in this direction than that of the <100> and <111> directions [39]. The microstructure of buckles consists of multiple *a*-Li_xSi and crystalline Si layers. Some areas in the wafer keep the same intact and smooth morphology as the pristine wafer though they also participate into lithiation [38]. During delithiation, the *a*-Li_xSi layer contracts and its stress state transforms gradually from compressive to tensile. After delithiation for 4 h, the tensile stress induces the fracture of buckles along the <110> direction, while the flat areas keep intact, as shown in Figure 10.15. Further delithiation increases the dimension of the <110>-oriented cracks. Cross-sectional images show that <110>-oriented cracks can penetrate over 5 μm into the wafer (Figure 10.16). Some cracks parallel to the wafer surface were also found underneath the buckles, which can be attributed to (1) the delamination between different layers (*a*-Li_xSi and *c*-Si) during lithiation and (2) the deflection of crack tips at the *a*-Li_xSi/*c*-Si interface during delithiation.

After full delithiation, the fracture of the flat area is dominated by the <100>-oriented cracks. There is no <110>-oriented crack in the flat area, as shown in Figure 10.15. The preferred orientation of cracks in the flat area can be explained

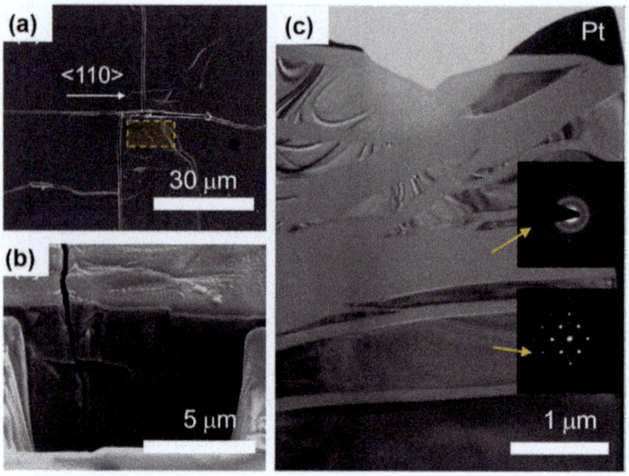

Figure 10.16 *Microstructure of <110>-oriented cracks. (a) is the top-view of a <100>-oriented crack. (b) is a cross-sectional image of the <100>-oriented crack. (c) is a HRTEM image and two SAED patterns of the indicated layers [38].*

by a linear elastic fracture model [38]. For a coating/substrate system, the critical fracture thickness, larger than which coatings start to fracture, is [40]

$$h_c = \frac{2}{\pi}\frac{E_f\Gamma_c}{(1-\nu_f^2)\sigma_f^2 g(\alpha,\beta)} \tag{10.10}$$

where ν_f and E_f are the Poisson's ratio, elastic modulus, and thickness of the a-Li_xSi layer, respectively. σ_f is the delithiation-induced tensile stress in the a-Li_xSi layer. $g(\alpha, \beta)$ is a function of the Dundurs parameters [41]

$$\alpha = \frac{\bar{E}_f-\bar{E}_s}{\bar{E}_f+\bar{E}_s} \text{ and } \beta = \frac{1}{2}\frac{\mu_f(1-2\nu_s)-\mu_s(1-2\nu_f)}{\mu_f(1-\nu_s)+\mu_s(1-\nu_f)} \tag{10.11}$$

where ν_s and E_s are the Poisson's ratio and elastic modulus the wafer substrate, respectively. $\mu = E/2(1+\nu)$ is shear modulus. $\bar{E} = E/(1-\nu^2)$ is the plane strain modulus. $g(\alpha, \beta)$ increases with increasing α but has a weak dependency on β. For coatings under the same stress (e.g., in the same area of a thin film), h_c is determined by the orientation-dependent plane strain modulus of the wafer substrate. The plane strain moduli of Si are 141.2 and 169.8 GPa along the <100> and <110> orientations, respectively [42]. Based on (10.10) and (10.11), the <100> orientation has a larger $\mu = E/2(1+\nu)$ and thus a smaller h_c than the <110> orientation. Therefore, the <100>-oriented crack propagate in the flat area prior to the appearance of the <110>-oriented crack.

The fracture behavior of Si wafers is affected by the cycling protocol, for example, current density and areal capacity. Some randomly oriented cracks were found between large <100> cracks in the Si (100) wafer cycled at 100 $\mu A/cm^2$ with an areal capacity of 600 $\mu Ah/cm^2$ [43], as shown in Figure 10.17. In addition, the

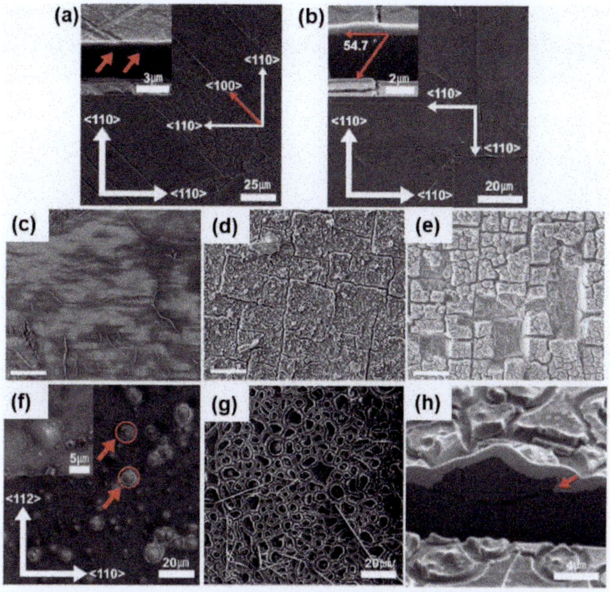

Figure 10.17 *Cracks in a Si (100) wafer after cycling with an areal capacity of 600 μAh/cm^2: (a) and (b) show random cracks and <110>-oriented cracks, respectively [43]. (c)–(e) are the surface microstructure of Si (100) wafers cycled by using cyclic voltammetry after different cycle numbers: (c) three cycles, (d) eight cycles, and (e) 50 cycles. The scale bars are 20 μm [44]. (f)–(h) are surface morphology of Si (111) wafers: (f) at the full lithiation state (600 μAh/cm^2), (g) the full delithiation state (600 μAh/cm^2). (h) is a cross-sectional image between* c-*Si and* a-*Si after delithiation of a Si (111) wafer [43].*

<110>-oriented cracks form an angle of 54.7° to the surface (Figure 10.17), implying that cracks propagate inside the wafer along the {111}/<110> slip plane. Shi *et al.* cycled Si (100) wafer using cyclic voltammetry (CV) technique and they did not observe extensive <100>- and <110>-oriented cracks in the wafer after three cycles, as shown in Figure 10.17 [44]. Cracks with two orthogonal directions initiate and propagate after additional CV cycling. After 50 cycles, the cracks divided the wafer into square islands. Small cracks also formed in individual islands. Some islands detached from the wafer substrate, as shown in Figure 10.17.

The morphology changes of Si wafers also rest with their orientation. As shown in Figure 10.17, irregular triangular humps form on the surface of Si (111) wafers after lithiation due to the spatial directionality of the <110> orientation. During delithiation, randomly oriented cracks formed along the edge of humps (Figure 10.17). Cracks also generated along the interface between *a*-Li_xSi and *c*-Si underneath the humps, as shown by the cross-sectional images in Figure 10.17.

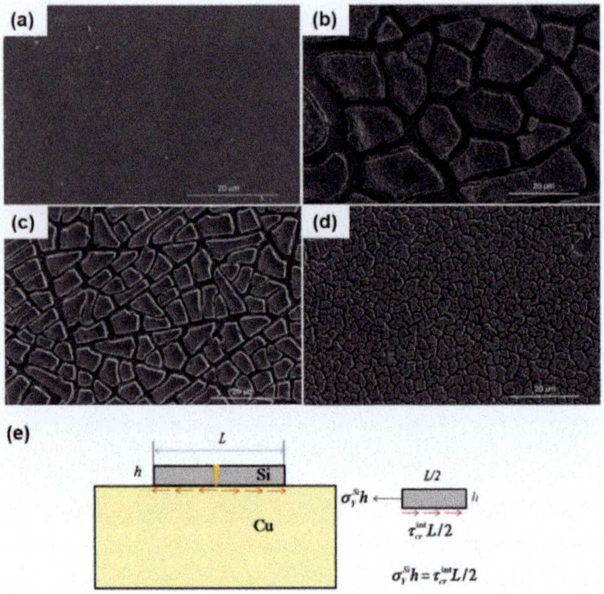

Figure 10.18 *Cracking patterns of Si thin films after electrochemical cycling with different thickness. (a) The as-prepared Si thin film with a thickness of 500 nm, (b) 1000 nm thick, after five cycles, (c) 500 nm thick after five cycles, and (d) 200 nm thick, after 10 cycles [45]. (e) is a schematic of fracture in lithiated Si film on Cu substrate by localized plastic strain [46].*

10.5.2 Si thin film electrodes

Different from crystalline Si wafers, the mechanical properties and Li diffusion coefficient are isotropic in amorphous Si thin films. There is no preferred orientation of cracks in amorphous Si thin films after electrochemical cycling. Random primary wide cracks interact with each other and divide the film into islands. The fracture morphology varies with the thickness of Si thin films [45], as shown in Figure 10.18. Compared with thick films (e.g., 1 000-nm thick), cracks in thinner films (e.g., 200-nm thick) have more branches and wiggles. The cracked area and cracking spacing decrease with decreasing the thickness of thin films. It is therefore inferred that Si thin films would not crack if the film thickness reduces to a critical value, h_c. Based on Griffith's criterion, cracking happens if the stored strain energy is larger than the surface energy required for the newly formed surface after fracture. h_c can be calculated by [47]

ν (10.12)

where E, ν, and σ are the Young's modulus, Poission's ratio, and stress of the film, respectively. $G_c = \frac{K_{Ic}}{E'}$, where K_{Ic} is the fracture toughness and $E' = E$ for the plane stress condition. By taking $K_{Ic} = 1MPam^{1/2}$, $\nu = 0.2$, and $\sigma = 2GPa$, $\nu = 0.2$ was estimated in the range between 100 and 200 nm [45]. Li *et al.* confirmed that an amorphous Si thin film (on a stainless steel substrate) with a thickness of 100 nm did not crack after 10 cycles [45].

The cracks in Si thin films also show that the average size of the islands decreases as the thickness of Si thin film decreases. It is speculated that fracture would not happen in a film patterned with islands smaller than that divided by cracks. By assuming that additional cracks form by the plastic strain between two adjacent cracks, the force equilibrium when a Si island starts to yield is [46]

$$\frac{\tau_{Cr}^{int} L_{Cr}}{2} = \sigma_Y^{Si} h \tag{10.13}$$

where σ_Y^{Si} is the yield strength of Si; τ_{cr}^{int} is the interfacial shear strength between Si and the substrate. τ_{cr}^{int} is bounded by the flow stress of the substrate and the interfacial frictional strength, that is, $\tau_{cr}^{int} = \min(\tau_Y^{Sub}, \tau_f^{int})$. The critical length of the island is

$$L_{cr} = \frac{2\sigma_Y^{Si} h}{\tau_{cr}^{int}} \tag{10.14}$$

If Cu is used as the substrate, τ_Y^{Cu} and τ_f^{int} are closed to 40 MPa [48]. σ_Y^{Si} is 1~1.75 GPa. For a Si film with a thickness of 100 nm (on a Cu substrate), the value of L_{cr} is calculated to be between 5.1 and 8.9 µm. As confirmed by Xiao *et al.*, a Si film made of Si patches of $7 \times 7\mu m^2$ and with a thickness of 100 nm, remained intact on the Cu substrate and the islands did not fracture after cycling [46]. In Si films made of larger patches (e.g., $17 \times 17\mu m^2$), many patches delaminated from the substrate after electrochemical cycling. With the improved mechanical integrity, the thin film patterned by $7 \times 7\mu m^2$ islands showed better cycling stability and capacity retention than the films patterned with larger islands.

10.5.3 Si composite electrodes

Despite that their porous structure could accommodate, to some extent, the repeated volume expansion/contraction of Si particles, some Si composite electrodes fracture after electrochemical cycling [20, 49, 50]. Only narrow cracks were found in the as-prepared Si/SA composite electrodes made of 50 wt% Si nanoparticles (30–50 nm), 25 wt% carbon black, and 25 wt% SA [20], while extensive random cracks form in the electrodes after the first cycle, as shown in Figure 10.19. The cracks separate the electrode into many irregular shape islands. The dimension of primary cracks is much larger than the Si nanoparticles in the electrode. Therefore, the fracture of the binder/carbon black matrix, instead of individual Si particles, causes the fracture of Si composite electrodes. Cracks gradually close during lithiation and reopen during the subsequent delithiation [20]. The island size and the crack spacing at the second delithiation state are similar to that at the first delithiation state. Ex situ observations showed that cracks in the as-made Si/SA electrode reappear with larger crack spacing after the first delithiation, as shown in Figure 10.19. Some additional secondary cracks form during electrochemical cycling. All primary cracks at the first, third, and tenth full delithiation states occur at the same locations. In situ digital microscope observations show that cracks open and close periodically at the same locations

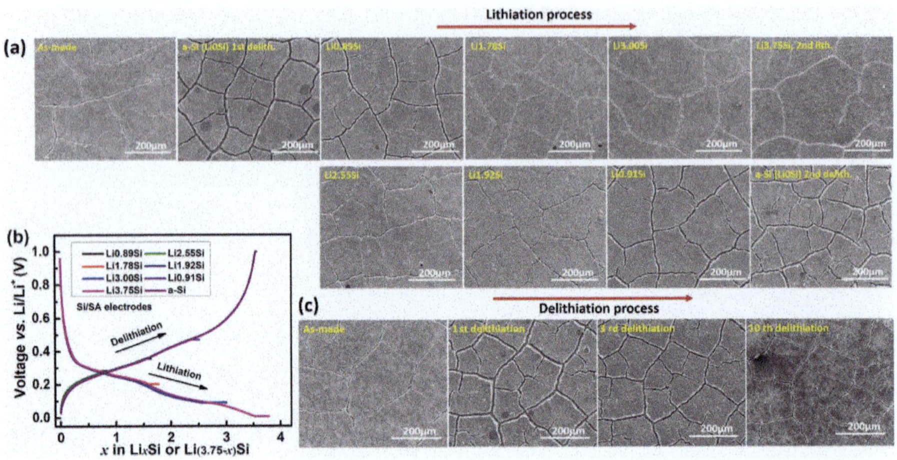

Figure 10.19 *(a) The evolution of cracks in Si/SA electrodes in the second cycle. (b) Voltage profiles of Si/SA electrodes at different SOCs in the second cycle. The first and second delithiated Si electrodes were denoted as* a-*Si. (c) Cracks in the Si/SA electrode at the same locations during multiple cycles [20].*

during multiple cycles [20]. This also suggests that individual islands are electronically connected to the Cu foil and remain electrochemically active during cycling.

Similar periodic cracking behavior was also found in Si composite electrodes made with other binders, such as Na-CMC and Nafion. As shown in Figure 10.20, the primary cracks in the Si/Na-CMC electrode are smaller than those in the Si/SA and Si/Nafion electrodes. The Si/Nafion electrode has larger crack gaps than the Si/SA and Si/Nafion electrodes. Cross-sectional SEM images show that primary cracks in Si/SA, Si/Na-CMC, and Si/Nafion electrodes reach the electrode/Cu current collector interface and propagate along the interface, as shown in Figure 10.20. In contrast, the Si/PVDF electrode remains intact after several initial cycles, as shown in Figure 10.20c and g. These results indicate that the cracking behavior of Si composite electrode depends on polymeric binders.

Since binders bind Si particles with the electrode matrix, the adhesion between binders and Si may be a dominated factor for the cracking behavior of Si composite electrodes. Lap joint shear tests showed that the adhesion strength of the Nafion@Si interface is about four times of the SA@Si and Na-CMC@Si interfaces and about 40 times of the PVDF@Si interface [19]. The strong adhesion of the SA@Si, Na-CMC@Si, and Nafion@Si interfaces results from the hydrogen bond between Si and these binders, while the weak adhesion of the PVDF@Si interface can be attributed to the weak Van der Waals interaction between PVDF and Si [17, 18, 26]. Moreover, PVDF has large electrolyte up-take capability. Excessive SEI forms between PVDF and Si particles and weakens their bonding strength. As a result, the delithiation-induced contraction and stress of Li_xSi particles cannot be transferred to

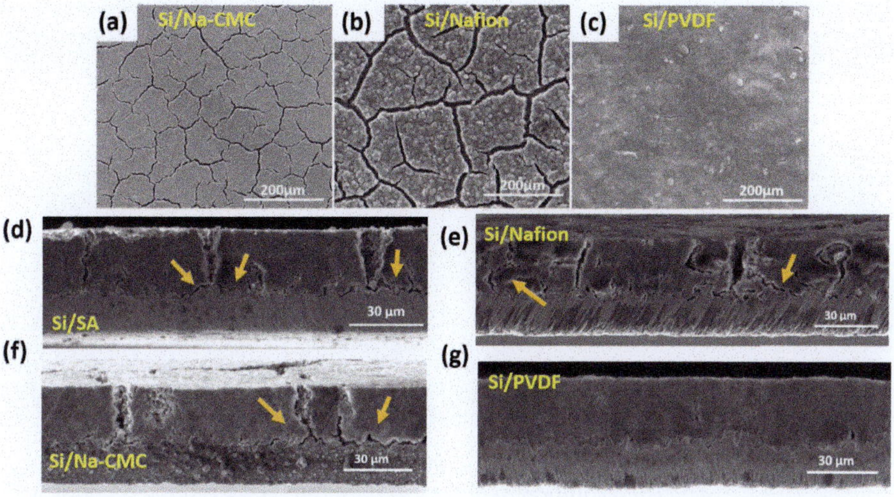

Figure 10.20 *The microstructure of (a) Si/Na-CMC, (b) Si/Nafion, and (c) Si/PVDF electrodes after the first delithiation. Cross-sectional images of (d) Si/SA, (e) Si/Nafion, (f) Si/Na-CMC, and (g) Si/PVDF electrodes at the second delithiation state [20].*

the electrode matrix, which makes the cracking of the flexible Si/PVDF electrode unlikely. For the other binders (i.e., SA, Na-CMC, and Nafion), their strong adhesion with Si is efficient to transfer the delithiation-induced tensile stress from Si particles to the electrode matrix, which drives the formation of cracks in the electrode.

The mechanism for the periodic cracking behavior of Si/SA, Si/Na-CMC, and Si/Nafion electrodes is proposed as follows. During the first lithiation, the Si particles expand, fills the pores and crack spacings of the as-made electrodes. Cracks close by physical contacts, instead of chemical bonding, among particles. The expansion of composite electrodes is restricted by the Cu current collector. Compressive stress is generated in the Si composite electrodes and will not induce fractures of composite electrodes during the first lithiation. In the first delithiation process, the progressive contraction of Li_xSi particles "pulls" the electrode matrix through binders and causes the shrinkage of the composite electrodes. The localized stress state gradually transforms from compression to tension, leading to the separation of electrode islands along the cracks in the as-made electrodes due to the weak physical contact among particles at these locations. As delithiation proceeds, cracks gradually widen, grow deeper, connect with each other, and separate the electrodes into islands, as shown in Figure 10.21. Secondary cracks propagate within individual islands, when the energy release rate in individual islands exceeds the fracture toughness of composite electrodes. Once extensive cracks form after the first delithiation, they periodically close and open at the same locations synchronously with the expansion and contraction of Li_xSi particles during subsequent cycling.

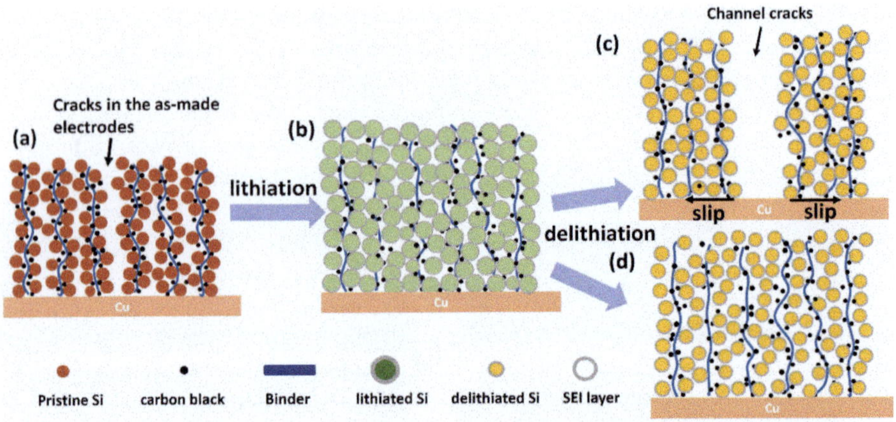

Figure 10.21 *Schematic diagrams of the cracking behavior of Si composite electrodes during cycling. (a) As-made Si composite electrodes, (b) electrodes after the first lithiation, (c) Si electrodes made with SA, Na-CMC, and Nafion after the first delithiation, and (d) Si/PVDF electrodes after the first delithiation [20].*

Although Si/Nafion, Si/Na-CMC, and Si/SA electrodes experience severe fractures during cycling, they all have much better capacity retention than Si/PVDF electrodes [20]. The fast capacity degradation of Si/PVDF electrode is due to the weak adhesion between PVDF and Si particles. As shown in Figure 10.21, many Si particles may detach from the PVDF/carbon black matrix and lose electronic conductivity. In contrast, Si particles adhere well with the SA, Na-CMC, and Nafion binders and maintain reliable electronic connectivity with the fractured electrode matrix. Since electrode islands are still well connected with the current collector under external compressive pressures in coin cells, electrons can transport in and out of Si particles. Therefore, most Si particles in Si/Nafion, Si/Na-CMC, and Si/SA electrodes remain electrochemical activity and help retain the cycling capacity.

Similar to Si thin film electrodes, the cracks in Si composite electrodes can be inhibited by reducing the electrode thickness and patterning with patches smaller than the separated islands. However, these two approaches are not practical from the consideration of energy density and manufacturing cost. Alternatively, cracks can be suppressed or "healed" if polymeric binders can be reversibly deformed to relax the localized tensile stress [51, 52]. Developing low cost, highly recoverable binders with robust adhesion with Si will provide a facile solution to improve mechanical integrity of Si composite electrodes.

10.6 Summary

Mechanical and structural degradations threaten the cycling stability of high capacity Si electrodes. Measuring the mechanical properties, quantifying the stress evolution, and analyzing the fracture behavior of Si electrodes form a basis for understanding the rapid capacity fading of Si. To this end, nanoindentation measurements have been used to quantify the Young's modulus and hardness of Si electrodes; scratch and peel tests have been used to characterize the adhesion between Si electrode components and the cohesion of Si composite electrodes; MOSS and curvature measurements have been used to measure the stress evolution of Si electrodes during electrochemical cycling. These measurements provide indispensable mechanical properties for developing strategies to overcome the coupled mechanical and electrochemical degradation of Si-based electrodes. Specifically, mechanical properties have been used in the design of Si electrodes, such as using nanoindentation and peel test results to select polymeric binders for Si composite electrodes [26, 53, 54]. Despite the popularization of these mechanical characterization methods in Si electrode research, more state-of-the-art techniques are in demand to quantify (1) the interactions among Si particles, binders, and carbon black at the particle level, (2) the heterogeneous microstructure and mechanical properties of SEI, and (3) the interactions between SEI and Si electrode components in realistic liquid organic electrolyte environment. These measurements will deepen the existing understanding of degradation mechanisms and accelerate the commercialization of Si-based electrodes for LIB applications.

Acknowledgments

We would like to thank colleagues, collaborators, and former group members at University of Kentucky for their contributions to the research reviewed in this chapter, in particular, Jiazhi Hu, Dawei Li, Dingying Dang, Xingcheng Xiao, Juchuan Li, Jiagang Xu, Jie Pan, Qinglin Zhang, Bo Lu, Junqian Zhang, Taeho Yoon, Seoung-Bum Son, and Jasmine M. Wallas. Special acknowledgments are due to Chunmei Ban and Dingying Dang for their valuable comments and suggestions on the manuscript. We would like to acknowledge the support from the National Science Foundation Award No. 1355438 and the Vehicle Technologies Office of the U.S. Department of Energy Battery Materials Research (BMR) Program under Contract Number DE-EE0007787.

References

[1] Obrovac M.N., Christensen L. 'Structural changes in silicon anodes during lithium insertion/extraction'. *Electrochemical and Solid-State Letters*. 2004;7(5):A93–6.

[2] Li J., Dahn J.R. 'An in situ X-ray diffraction study of the reaction of Li with crystalline Si'. *Journal of the Electrochemical Society*. 2007;**154**(3):A156–61.

[3] Shenoy V.B., Johari P., Qi Y. 'Elastic softening of amorphous and crystalline Li–Si phases with increasing Li concentration: a first-principles study'. *Journal of Power Sources*. 2010;**195**(19):6825–30.

[4] Oliver W.C., Pharr G.M. 'An improved technique for determining hardness and elastic modulus using load and displacement sensing indentation experiments'. *Journal of Materials Research*. 1992;**7**(6):1564–83.

[5] Cheng Y.-T., Cheng C.-M. 'Scaling, dimensional analysis, and indentation measurements'. *Materials Science and Engineering: R: Reports*. 2004;**44**(4-5):91–149.

[6] Schroder K.W., Celio H., Webb L.J., Stevenson K.J. 'Examining solid electrolyte interphase formation on crystalline silicon electrodes: influence of electrochemical preparation and ambient exposure conditions'. *The Journal of Physical Chemistry C*. 2012;**116**(37):19737–47.

[7] Wang Y., Zhang Q., Li D., *et al.* 'Ameliorating the interfacial problems of cathode and solid-state electrolytes by interface modification of functional polymers'. *Advanced Energy Materials*. 2018;**8**:1702578.

[8] de Vasconcelos L.S., Xu R., Zhao K. 'Operando nanoindentation: a new platform to measure the mechanical properties of electrodes during electrochemical reactions'. *Journal of the Electrochemical Society*. 2017;**164**(14):A3840–7.

[9] Berla L.A., Lee S.W., Cui Y., Nix W.D. 'Mechanical behavior of electrochemically lithiated silicon'. *Journal of Power Sources*. 2015;**273**:41–51.

[10] Hertzberg B., Benson J., Yushin G. 'Ex-situ depth-sensing indentation measurements of electrochemically produced Si–Li alloy films'. *Electrochemistry Communications*. 2011;**13**(8):818–21.

[11] Wang X., Singh S.S., Ma T., Lv C., Chawla N., Jiang H. 'Quantifying electrochemical reactions and properties of amorphous silicon in a conventional lithium-ion battery configuration'. *Chemistry of Materials*. 2017;**29**(14):5831–40.

[12] Tabor D. 'The hardness of solids'. *Reviews of Physics in Technology*. 1970;**1**(3):145–79.

[13] Anstis G.R., Chantikul P., Lawn B.R., Marshall D.B. 'A critical evaluation of indentation techniques for measuring fracture toughness: I, direct crack measurements'. *Journal of the American Ceramic Society*. 1981;**64**(9):533–8.

[14] Cuadrado N., Seuba J., Casellas D., Anglada M., Jiménez-Piqué E. 'Geometry of nanoindentation cube-corner cracks observed by FIB tomography: implication for fracture resistance estimation'. *Journal of the European Ceramic Society*. 2015;**35**(10):2949–55.

[15] Wang X., Fan F., Wang J., *et al.* 'Confining energy migration in upconversion nanoparticles towards deep ultraviolet lasing'. *Nature communications*. 2015;**6**:1–7.

[16] Zhao K., Wang W.L., Gregoire J., *et al.* 'Lithium-assisted plastic deformation of silicon electrodes in lithium-ion batteries: a first-principles theoretical study'. *Nano Letters*. 2011;**11**(7):2962–7.

[17] Magasinski A., Zdyrko B., Kovalenko I., *et al.* 'Toward efficient binders for li-ion battery si-based anodes: polyacrylic acid'. *ACS Applied Materials & Interfaces*. 2010;**2**(11):3004–10.

[18] Kovalenko I., Zdyrko B., Magasinski A., *et al.* 'A major constituent of brown algae for use in high-capacity Li-ion batteries'. *Science*. 2011;**334**(6052):75–9.

[19] Wang Y., Dang D., Li D., Hu J., Cheng Y.-T. 'Influence of polymeric binders on mechanical properties and microstructure evolution of silicon composite electrodes during electrochemical cycling'. *Journal of Power Sources.* 2019;**425**:170–8.

[20] Wang Y., Dang D., Li D., Hu J., Zhan X., Cheng Y.-T. 'Effects of polymeric binders on the cracking behavior of silicon composite electrodes during electrochemical cycling'. *Journal of Power Sources.* 2019;**438**:226938.

[21] Brostow W., Kovačevic V., Vrsaljko D., Whitworth J. 'Tribology of polymers and polymer-based composites'. *The Journal of Materials Education.* 2010;**32**:273–90.

[22] Sharifi H., Aliofkhazraei M., Darband G.B., Shrestha S. 'A review on adhesion strength of PEO coatings by scratch test method'. *Surface Review and Letters.* 2018;**25**(03):1830004.

[23] Li X., Bhushan B. 'Micro/nanomechanical and tribological characterization of ultrathin amorphous carbon coatings'. *Journal of Materials Research.* 1999;**14**(6):2328–37.

[24] Son S.-B., Wang Y., Xu J., *et al.* 'Systematic investigation of the alucone-coating enhancement on silicon anodes'. *ACS Applied Materials & Interfaces.* 2017;**9**(46):40143–50.

[25] Kawashita L.F., Moore D.R., Williams J.G. 'Comparison of peel tests for metal–polymer laminates for aerospace applications'. *The Journal of Adhesion.* 2005;**81**(6):561–86.

[26] Hu J., Wang Y., Li D., Cheng Y.-T. 'Effects of adhesion and cohesion on the electrochemical performance and durability of silicon composite electrodes'. *Journal of Power Sources.* 2018;**397**:223–30.

[27] Pan J., Zhang Q., Li J., Beck M.J., Xiao X., Cheng Y.-T. 'Effects of stress on lithium transport in amorphous silicon electrodes for lithium-ion batteries'. *Nano Energy.* 2015;**13**(4):192–9.

[28] Lu B., Song Y., Zhang Q., Pan J., Cheng Y.-T., Zhang J. 'Voltage hysteresis of lithium ion batteries caused by mechanical stress'. *Physical Chemistry Chemical Physics.* 2016;**18**(6):4721–7.

[29] Sethuraman V.A., Chon M.J., Shimshak M., Srinivasan V., Guduru P.R. 'In situ measurements of stress evolution in silicon thin films during electrochemical lithiation and delithiation'. *Journal of Power Sources.* 2010;**195**(15):5062–6.

[30] Pharr M., Suo Z., Vlassak J.J. 'Measurements of the fracture energy of lithiated silicon electrodes of Li-ion batteries'. *Nano Letters.* 2013;**13**(11):5570–7.

[31] Chon M.J., Sethuraman V.A., McCormick A., Srinivasan V., Guduru P.R. 'Real-time measurement of stress and damage evolution during initial lithiation of crystalline silicon'. *Physical Review Letters.* 2011;**107**(4):045503.

[32] Stoney G.G. 'The tension of metallic films deposited by electrolysis'. *Proceedings of the Royal Society of London. Series A, Containing Papers of a Mathematical and Physical Character.* **82**; 1909. pp. 172–5.

[33] Soni S.K., Sheldon B.W., Xiao X., Bower A.F., Verbrugge M.W. 'Diffusion mediated lithiation stresses in Si thin film electrodes'. *Journal of the Electrochemical Society.* 2012;**159**(9):A1520–7.

[34] Bucci G., Nadimpalli S.P.V., Sethuraman V.A., Bower A.F., Guduru P.R. 'Measurement and modeling of the mechanical and electrochemical response of amorphous Si thin film electrodes during cyclic lithiation'. *Journal of the Mechanics and Physics of Solids*. 2014;**62**(4):276–94.

[35] Pharr M., Suo Z., Vlassak J.J. 'Variation of stress with charging rate due to strain-rate sensitivity of silicon electrodes of Li-ion batteries'. *Journal of Power Sources*. 2014;**270**:569–75.

[36] Li D., Wang Y., Hu J., Lu B., Cheng Y.-T., Zhang J. 'In situ measurement of mechanical property and stress evolution in a composite silicon electrode'. *Journal of Power Sources*. 2017;**366**(8):80–5.

[37] Li D., Wang Y., Hu J., *et al*. 'Role of polymeric binders on mechanical behavior and cracking resistance of silicon composite electrodes during electrochemical cycling'. *Journal of Power Sources*. 2018;**387**(2):9–15.

[38] Yoon T., Xiao C., Liu J., *et al*. 'Electrochemically induced fractures in crystalline silicon anodes'. *Journal of Power Sources*. 2019;**425**:44–9.

[39] Lee S.W., McDowell M.T., Choi J.W., Cui Y. 'Anomalous shape changes of silicon nanopillars by electrochemical lithiation'. *Nano Letters*. 2011;**11**(7):3034–9.

[40] Ye T., Suo Z., Evans A.G. 'Thin film cracking and the roles of substrate and interface'. *International Journal of Solids and Structures*. 1992;**29**(21):2639–48.

[41] Beuth J.L. 'Cracking of thin bonded films in residual tension'. *International Journal of Solids and Structures*. 1992;**29**(13):1657–75.

[42] Chow L.A., Xu Y.H., Dunn B., Tu K.N., Chiang C. 'Cracking behavior of xerogel silica films on silicon substrates'. *Applied Physics Letters*. 1998;**73**(20):2944–6.

[43] Kang C.S., Son S.-B., Kim J.W., *et al*. 'Electrochemically induced and orientation dependent crack propagation in single crystal silicon'. *Journal of Power Sources*. 2014;**267**:739–43.

[44] Shi F., Song Z., Ross P.N., Somorjai G.A., Ritchie R.O., Komvopoulos K. 'Failure mechanisms of single-crystal silicon electrodes in lithium-ion batteries'. *Nature Communications*. 2016;**7**:1–8.

[45] Li J., Dozier A.K., Li Y., Yang F., Cheng Y.-T. 'Crack pattern formation in thin film lithium-ion battery electrodes'. *Journal of the Electrochemical Society*. 2011;**158**(6):A689–94.

[46] Xiao X., Liu P., Verbrugge M.W., Haftbaradaran H., Gao H. 'Improved cycling stability of silicon thin film electrodes through patterning for high energy density lithium batteries'. *Journal of Power Sources*. 2011;**196**(3):1409–16.

[47] Gross D., Seelig T. *Fracture Mechanics*. New York: Springer; 2006.

[48] Li Q., Kim K.-S. 'Static friction at fractal interfaces'. *Proceedings of the Royal Society A: Mathematical, Physical and Engineering Sciences*. 2008;**464**:1319–43.

[49] Nguyen C.C., Yoon T., Seo D.M., Guduru P., Lucht B.L. 'Systematic investigation of binders for silicon anodes: interactions of binder with silicon particles and electrolytes and effects of binders on solid electrolyte interphase formation.'. *ACS Applied Materials & Interfaces*. 2016;**8**(19):12211–20.

[50] Reale Hernandez C., Karkar Z., Guyomard D., Lestriez B., Roué L. 'A film maturation process for improving the cycle life of Si-based anodes for Li-ion batteries'. *Electrochemistry Communications*. 2015;**61**:102–5.
[51] Wang C., Wu H., Chen Z., McDowell M.T., Cui Y., Bao Z. 'Self-healing chemistry enables the stable operation of silicon microparticle anodes for high-energy lithium-ion batteries'. *Nature Chemistry*. 2013;**5**(12):1042–8.
[52] Xu Z., Yang J., Zhang T., Nuli Y., Wang J., Hirano S.-ichi. 'Silicon microparticle anodes with self-healing multiple network binder'. *Joule*. 2018;**2**(5):950–61.
[53] Wei L., Chen C., Hou Z., Wei H. 'Poly (acrylic acid sodium) grafted carboxymethyl cellulose as a high performance polymer binder for silicon anode in lithium ion batteries'. *Scientific Reports*. 2016;**6**:19583.
[54] Ryou M.-H., Kim J., Lee I., *et al*. 'Mussel-inspired adhesive binders for high-performance silicon nanoparticle anodes in lithium-ion batteries'. *Advanced Materials*. 2013;**25**(11):1571–6.

Chapter 11

Practical implementation of silicon-based negative electrodes in lithium-ion full-cells—challenges and solutions

Tobias Placke[1], Gebrekidan Gebresilassie Eshetu[2,3], Martin Winter[1,4], and Egbert Figgemeier[2,4]

11.1 Introduction

Despite the striking improvements in terms of specific energy and energy density of commercial lithium-ion batteries (LIBs), that is, of up to ≈260 Wh kg^{-1} and ≈700 Wh L^{-1} (cell level) and ≈140 Wh kg^{-1} and ≈210 Wh L^{-1} (battery pack level), based on lithium nickel manganese cobalt oxide (NMC) || graphite (Gr) cell chemistries, the LIB of today (i.e., year 2020) is not yet adequate to meet all the challenging requirements for different applications, such as electromobility (electric vehicles, xEVs) [1–4]. To enable massive xEV market penetration, technological breakthroughs in terms of increased specific energy/energy density (>235 Wh kg^{-1} and >500 Wh L^{-1}) to enable driving ranges of at least 500 km and in terms of reduced cost per energy (<125 US$ kWh^{-1}), both at battery pack level, are promptly needed [2, 5]. The most suitable way to fulfill these requirements is the development of new high-capacity negative (anode) and positive (cathode) LIB-active materials [6, 7].

Presently, silicon (Si) has stood out as the most appealing replacement for state-of-the-art Gr-based anodes [2], owing to its peculiar advantageous features such as the high theoretical capacity of ≈3 590 mAh g^{-1} and ≈8 360 mAh cm^{-3} ($Li_{3.75}Si$) at room temperature (in the de-lithiated state) [8], the relatively low operating potential (average de-lithiation potential of ≈0.4 V vs. Li|Li^+) enabling a high cell voltage, environmental benignity, and the high abundance and low cost of precursor materials

[1]University of Münster, MEET Battery Research Center, Institute of Physical Chemistry, Correnssstraße 46, 48149 Münster, Germany

[2]ISEA, RWTH Aachen, Jägerstr. 17-19, 52066 Aachen, Germany

[3]Department of Chemistry, College of Natural and Computational Sciences, Mekelle University, Mekelle, Ethiopia

[4]Helmholtz-Institute Münster, IEK-12, Forschungszentrum Jülich GmbH, Correnssstraße 46, 48149 Münster, Germany

such as SiO_2 [9–14]. However, despite all these expedient aspects, the broad practical deployment of Si-based anode materials in commercial LIB cells is hindered due to the presence of multiple and interwoven hurdles, such as colossal volume changes at Si particle level (≥280%), an ineffective solid electrolyte interphase (SEI) formation, low electronic conductivity, severe electrode swelling, electrolyte consumption and "drying," and so forth [14–18]. Thus, due to these challenges, the large-scale implementation of Si in particular as well as of other metallic storage hosts forming reversibly intermetallic phases with Li, so-called alloy-type anode materials (e.g., tin or aluminum) in general remains indescribable so far. Moreover, despite superior performance reported for various Si-based materials in the scientific battery community for laboratory cells, these often "outstanding" results may not be directly translated into "practical" electrodes and cells as well as commercial cell formats, because the conditions under which they are typically evaluated are usually not aligned with those for realistic "practical" applications. In this chapter, we will provide the fundamental insights for the practical implementation of Si-based negative electrode materials in LIB full-cells, address the major challenges and give guidance for future approaches to achieve the targets in terms of the battery's key performance metrics in commercial cell formats.

11.1.1 Lessons learned from commercialized battery cells based on lithium-alloying anodes

The first systematic research studies on lithium-alloys (*e.g.*, Li-Al, Li-Si) were associated with their use in high-temperature battery cells, operating in molten salt electrolytes at ≈400–450 °C [19, 20]. The major approach was the replacement of Li metal anodes, due to their low melting point of 180 °C. Lithium alloys first appeared in room temperature lithium cells when they were introduced as substrates for electrodeposition of Li metal. In this respect, much effort was put into the replacement of Li metal anodes in rechargeable lithium cells in the late 1970s and early 1990s to overcome performance and safety issues of Li metal electrodes [19, 20]. Lithium alloys have been used for several years in high-temperature "thermal batteries" (*e.g.*, FeS_2 || Li-alloy (Li-Al or Li-Si), molten chloride electrolyte), which were produced commercially for military purposes. These cells could be stored at ambient temperature (low self-discharge rate) and had to be heated (above the melting point of the electrolyte) for operation [20]. The first use of lithium alloy anodes in commercial cells was the employment of the so-called Wood's metal alloy, that is, an alloy of Bi, Pb, Sn, and Cd, in button-type cells developed by *Matsushita-Panasonic* in the early 1980s [19, 20].

In the early 1990s, Sn and related compounds were the preferred metallic Li storage hosts both in academic and industrial research and quite a few findings on tin at that time were simply transferred to Si later on [21–26]. However, there are also distinct differences of the two metals. In 1997, *Fuji Photo Film Celltec Co.* (Japan) announced the commercialization of their so-called *Stalion* LIB cell based on an alloy material as negative electrode [27], in which a tin (Sn)-based amorphous composite oxide was applied. In the composite material, the active SnO was embedded in an amorphous oxide matrix based on P_2O_5, Al_2O_3, and B_2O_3 phases [19]. However, the attempt was aborted at the end of the 1990s, attributed to challenges associated with a large irreversible capacity, voltage-dependent accretion of Sn, and somewhat low volumetric energy improvements, cumulatively leading to a poor electrochemical performance [9]. In particular, the energy density of these cells was much smaller than previously expected, as not the lithiated, but the de-lithiated state was considered in the energy calculation and, further, the irreversible capacity over cycling was assumed to be zero. Overall, these cells only showed an energy gain of <10% compared to Gr-based LIB cells, which was most likely not worth to pursue for commercialization, considering the significant challenges and costs to achieve a high cycle life [9]. In 2005, *Sony Corporation* announced the commercialization of a LIB cell based on a carbon-tin-transition metal composite anode, that is, a Sn–Co–C hybrid anode, with the trade name *Nexelion* ("Next Era Lithium Ion") for use in hand-held camcorders [9, 28]. The cylindrical cells (14430 (AA) format, 900 mAh, 2.5–4.2 V) offered energy densities of 478 Wh L^{-1} and 158 Wh kg^{-1} and exhibited a capacity retention of 80% after 200–250 cycles. However, the development was once again abandoned due to its short lifetime in terms of cycle life and high cost of Co, prohibiting its adoption in large cells for xEVs and stationary storage applications [29, 30].

Nevertheless, despite their shortcomings, the commercialization of the abovementioned lithium-alloys has revitalized another wave of appealing anode materials for LIBs and brought further fundamental understanding with respect to the challenges for the practical application of alloying materials in LIB cells.

Which lessons can be drawn from these first trials for commercialization of alloying anodes? In a first attempt, it has to be kept in mind that not only one single or a few, but various performance metrics and parameters of novel Si-based materials and cells have to be addressed and fulfilled along the battery value chain, that is, from material to the cell level, to become suitable for industrial purposes. Addressing only a few metrics or neglecting certain parameters might lead to an overestimation of the true "practical" cell performance, for example, unrealistic high values for the specific energy or energy density, which cannot be reached/realized in practical LIB cells [8, 31]. Therefore, it is also of utmost importance to improve the reporting of experimental parameters [8]. Only in such cases, a suitable and fair comparison of new Si-based materials to state-of-the-art Gr-based anode materials in terms of cell performance, safety, and so forth can be guaranteed and allow an evaluation of their practicality. The challenges related to the multi-requirement optimization will be comprehensively introduced and discussed in the following sub-chapters.

11.1.2 Can we directly transform the knowledge gained from graphite anodes to Si anodes?

Synthetic or natural Gr is the state-of-the-art negative electrode material in commercial LIB cells and has been firstly introduced in the third generation of *Sony*'s LIB cell more than two decades ago [32, 33]. Over the past decades, comprehensive fundamental knowledge could be gained for further performance and stability improvements of Gr-based electrodes in LIB cells, in particular with respect to material improvements (*e.g.*, particle design), optimized electrolyte (salt anions, additives, etc.) formulations for effective interphase design, as well as the impact of electrochemical evaluation in LIB full-cells. Can this knowledge (at least partially) be transferred to the development of Si-based negative electrodes?

For Gr-based materials, it is well known that the particle size distribution and specific surface area play a major role in terms of SEI formation and irreversible losses in the first charge/discharge cycle(s), that is, for electrolyte and active lithium consumption, which is expressed by the Coulombic efficiency (C_{Eff}). In general, Gr materials are optimized with respect to a low specific surface area, typically ≤ 4 m^2 g^{-1}, resulting in a high initial C_{Eff} of ≈90–94%. In contrast, a major approach to improve the performance of Si-based materials focuses on sophisticated nano-designs, which offer a high specific surface area and, thus, reduce the initial C_{Eff} (see Section 11.2.1.2). Furthermore, for Gr-based anodes, SEI formation is mainly limited to the first few cycles, which is a result of the volume stability during lithium intercalation/de-intercalation. In contrast, Si, as Li storage metal, suffers from severe volume changes within each cycle, which results in continuous SEI re-formation [34].

Considering the design of Si-based active materials, it can clearly be stated that the nanodesign approach alone will not result in a successful implementation of Si-based materials. Si-based materials need to be optimized toward low specific surface areas and high tap densities [35, 36], as it has been shown for the optimization of Gr-based anode materials. In terms of electrochemical testing procedures, various concepts, which have been systematically applied to evaluate Gr-based materials, can relatively easily be converted to Si-based anode materials, as both materials operate in a similar potential range, even though the average de-lithiation potential of Si is shifted toward slightly higher values (≈0.2 V vs. ≈0.4 V vs. Li|Li^+) [36]. However, when it comes to electrode composition, for example, in terms of finding suitable binders, as well as to the electrolyte formulation and additives, the lessons learned from Gr anodes cannot easily be converted to Si-based materials, which is primarily related to the different requirements in terms of mechanical stability (volume stability vs. severe volume changes) and dynamic nature of the SEI. These challenges and possible solutions will be addressed in the following chapters.

11.2 Impact of Si-addition to lithium-ion cells: performance and application

11.2.1 Key metrics for practical Si anode-based lithium-ion cells

Several key metrics in terms of performance and cost have to be addressed and fulfilled to achieve a high practicality of Si-based materials to be employed in commercial LIB cells. In this respect, a major challenge is the transfer of Si material and Si-based electrode properties and their electrochemical characteristics measured in laboratory cells (at non-optimized conditions) to practical LIB cell formats (with

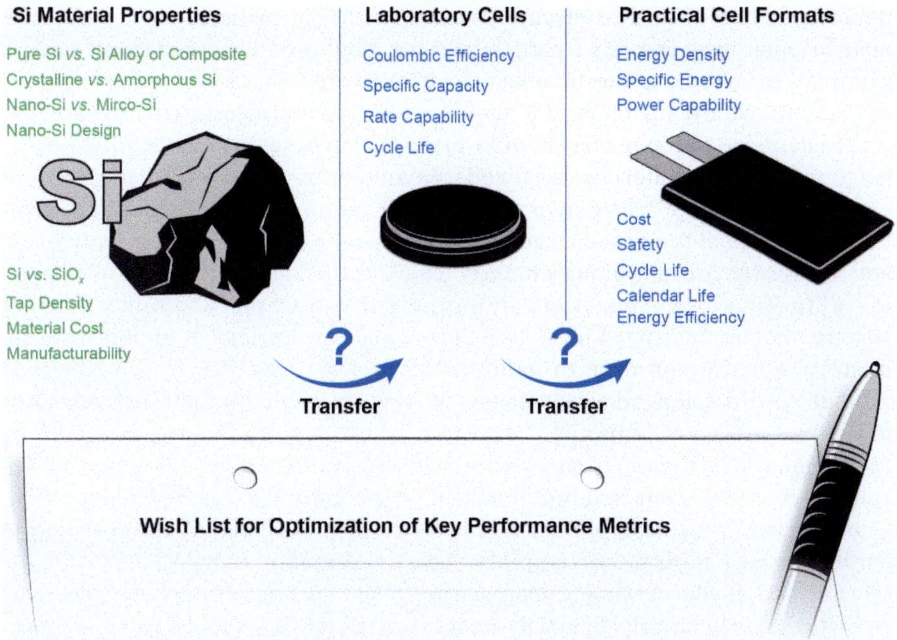

Figure 11.1 *Major challenge for the practical application of Si-based materials is to achieve the most suitable balance of key metrics in terms of performance, safety, and cost in practical LIB cell formats. For application in practical cells, optimized conditions and various requirements have to be fulfilled (wish list for optimization of key performance metrics). Thereby, the transfer of Si material properties and their electrochemical results measured in laboratory cells (at non-optimized conditions) to practical LIB cells is the most critical step.*

optimized conditions and many further requirements) to achieve the optimum balance of these metrics suitable for certain application purposes (Figure 11.1).

11.2.1.1 Cycle and calendar life

The lifetime in terms of cycle life is the first major disquiet for the practical application of Si-based LIB cells. Most importantly, it becomes more critical for Si anode-based LIB cells having high Si contents (=higher anode capacity) owing to their serious challenges in terms of volume changes and corresponding capacity decay. The increasing demands for batteries that should last ≥10 service life years (calendar life) for electromobility [4], conforming to ≈1 300–2 000 charge/discharge cycles (cycle life), or even 20–40 years lifetime for grid applications [30], calls the critical improvement of the lifetime along with superior capacity retention of Si anode-based LIB cells. Thereby, a high cycle life needs not only to be guaranteed at the material and electrode levels (*e.g*., prevention of Si particle cracking and Si particle detachment), which is often evaluated in non-optimized laboratory cells, but also in practical LIB cells with optimized conditions, such as limited contents in lithium and electrolyte (see Section 11.3.3). Therefore, the right and optimum design of Si materials and electrodes and the "translation" of their properties to electrochemical characteristics is a major challenge (Figure 11.1). For example, various nano-design approaches (Si nanoparticles, nanowires, etc.) [12, 13] have led to remarkably improved performance with high cycle life in non-optimized laboratory cells ("half cells," see Section 11.3.1) and have thus been praised as most suitable strategies toward commercialization of Si [17, 37]. However, it has to be stated that the nano-approach alone does not solve the cycle life challenge [29], which is a result of a huge electrochemically active surface area, leading to parasitic side reactions (low Coulombic efficiency, high active lithium losses) [38] and, therefore, to a very limited cycle life in practical LIB cells (see Section 11.3.2). Further, nano-Si materials are often evaluated by use of composite electrodes having relatively low mass loadings, which can also result in an overestimation of the electrochemical performance, as much higher mass loadings are needed in practical LIB cells. In particular, novel nano-Si design approaches, such as yolk-shell or pomegranate-like structures [39, 40], are highly elegant concepts from fundamental perspective, however, seem to be poorly practical so far [29].

11.2.1.2 Efficiencies of redox reactions: impact on cycle life and power (rate capability)

For practical application, in particular for large format battery cells, the Coulombic (C_{Eff}), voltage (V_{Eff}), and energy (E_{Eff}) efficiencies are of paramount importance and need to be seriously considered [38, 41]. Defined as the quotient of discharge capacity by charge capacity, C_{Eff} is a crucial parameter to evaluate the reversibility of the accompanying redox reactions and should ideally be close to 100%, inferring all the charge from the cathode in a fully charged state can return to the cathode in a fully discharged state [12, 38]. However, this is not possible under practical conditions, especially for Si anode-based LIB cells due to various parasitic side reactions such a continuous breakage and (re-)formation of the SEI at Si. Customarily, a retention of 80% of its original capacity is marked as a criterion for the life evaluation of an energy storage device for the xEV industry. For example, a C_{Eff} of 99.96% is required to achieve a cycling stability of up to 500 cycles for commercialization (see also Section 11.3.2) [42]. Therefore, very small changes in the C_{Eff} can lead to severe capacity fading in practical LIB full-cells [11, 37, 43]. One major challenge is particularly the first cycle C_{Eff} of Si anode-based LIB cells, which is typically in the range of ≈65–85% (depending on the Si anode material/electrode design), far below that of state-of-the-art Gr anodes (≈90–94%) [12].

Analogously to C_{Eff}, the energy efficiency (E_{Eff}) refers to the ratio of the discharge to charge energy and is coined as key performance indicator of a battery cell [38, 41]. In general, E_{Eff} is defined as the product of C_{Eff} and voltage efficiency (V_{Eff}), while the V_{Eff} of a cell is determined by the voltage hysteresis between the charge and discharge reaction, so that a low-voltage hysteresis results in a high V_{Eff}. Main parameters influencing V_{Eff}, C_{Eff}, and E_{Eff} of the individual electrode active materials are the charge/discharge current (or rate), state-of-charge (SOC), internal resistance, operating temperature, and state-of-health (SOH). In general, EV batteries should be able to allow high power charging (fast charging), which is strongly affected by the battery cell's E_{Eff}. Therefore, it is of utmost importance not only to study the current rate capability of new Si-based materials (capacity retention in dependence of charge/discharge current), but more importantly the power rate capability (energy retention in dependence of the charge/discharge current) [41, 44]. Moreover, low energy efficiency also leads to higher internal temperature inside the cell, presenting a challenge for the thermal management, thus, the lifetime of the cell. In this respect, new Si material concepts, for example, "pure Si anode designs," might face additional challenges in terms of their power capability.

11.2.1.3 Specific energy and energy density

Energy content can be defined as the amount of energy stored per unit mass (=specific energy or gravimetric energy density) or volume (=(volumetric) energy density) and can be calculated for different levels along the battery value chain, that is, from the material level to the battery cell or battery pack level [31]. For a typical LIB cell, the energy density (E_{Cell}) is collectively determined by both the cell voltage (U_{Cell}) between the employed redox couples of the positive and negative electrodes and storage capacity of the cell (Q_{cell}), as given by $E = Q_{Cell} \cdot U_{Cell}$ (or $E = \int_a^b U_{Cell}\, dQ_{Cell}$) [1, 42]. In principle, this implies that increasing the overall energy density requires maximizing the cell voltage (=potential difference between the cathode and anode) and/or the charge storage capacity of both electrode materials [42]. Considering the negative electrode, the potential of Gr- and Si-based anodes is relatively flat over the course of the charge/discharge processes (average de-lithiation potentials of ≈0.2 V (Gr) and ≈0.4 V (Si) vs. Li|Li$^+$), therefore, the obtainable energy density is rather uttered by the reversible capacity than by the voltage/potential.

In general, target energy density values can be calculated by a top-down approach from the battery pack level to the electrode/material level, as reported by Andre *et al.* [14], or by a bottom-up approach starting from theoretical values as reported by Betz *et al.* [31]. To report realistic values for specific energy or energy density, various parameters need to be included in these calculations [8, 14, 31], while—in case of negative electrodes, such as Si-based electrodes—it is of utmost importance to consider the lithiated state and, thus, the displaced electrolyte volume, to achieve the true energy density [8]. In general, there are various discounting factors when translating the Si electrode specific capacity to a practical energy density of a Si-based LIB cell, while calculations indicate that an improvement of ≈20% in terms of energy density can be achieved for Si-based LIB cells compared to state-of-the-art Gr-based LIB cells [9, 45].

Overall, a combination of suitable measures is needed to improve the energy density of Si-based LIB cells from the material design to the practical cell level design (Figure 11.1), including various measures at the material level (*e.g.*, materials with high tap density [35, 46]), at the electrode level (*e.g.*, high active Si content and low amount of inactive materials, high mass loading and areal capacity, lower porosity) and at the cell level (*e.g.*, optimized electrolyte amount, reduction of electrode swelling). However, refining these different parameters for energy density improvements will result in limitations of other key metrics such as power capability and cycle life (*e.g.*, by fracture and delamination of Si materials from the current collector). In particular, realizing high Si contents within the electrode as well as high areal capacities is rather difficult and focused research efforts aiming at optimizing materials/electrodes need to be strengthened for the practical application of Si anodes.

11.2.1.4 Cost: from material cost to cell cost

Though Si is an abundant element (most likely in the form of its precursor materials such as SiO_2), cost related to the large-scale production of practical Si-based anode cells could pose a major concern. In general, the costs of various Si materials and resulting various electrodes and cells will be rather scattered. From an industrial perspective, Si material design must be scalable and highly reproducible at production costs comparable to currently used Gr materials and allow a direct incorporation in the large-scale manufacturing process of negative electrodes [2, 47]. In principle, the overall price of a given battery cell involves the cost of the starting reactants (*i.e.*, sources of Si raw material), synthesis, processing, electrolytes, binders, current collector, conductive additives, formation, conditioning steps, and so forth [48]. For instance, nano-Si materials, in particular sophisticated designs such as yolk-shell or pomegranate-like structures [39, 40], are produced *via* complex processes and sophisticated techniques, making them more costly and more challenging toward upscaling than upstream materials such as micro-Si, metallurgical Si, Si alloys, and SiO_x [13]. Hence, a careful assessment of the overall cost of the whole Si anode-based LIB production is needed, while a fair comparison should be based on the cell's cost per energy content.

11.2.1.5 Safety

The large-scale adoption of LIB cells necessitates the appraisal of safety-induced risks enlisting both thermal and chemical threats [49, 50]. Legitimate questions regarding the practical application of Si anodes could be: How safe are Si anode-based LIB cells? What is the impact of Si in Gr/Si blend electrodes in terms of safety? These and other related questions need a proper answer to have a complete appraisal of the practicality of Si anode-based LIB cells. Except few studies based on differential scanning calorimetry and related material level devices [51, 52], there is a lack of comprehensive investigations on the effect of incorporating Si toward the safety aspects of Si anode-based cells. Allcorn *et al.* reported on a material safety study of practical nano-Si/Gr anodes and found several parameters influencing exothermic reactions [53]. With increasing Si content, they found an increased peak heating rate and a decreased onset temperature for exothermic reactions. In addition, they found that the exothermic reactivity can be mitigated by reducing the electrolyte ratio. Further, the Si surface area had a major impact on the reaction rate, that is, smaller Si particles showed an enhanced thermal reactivity [53]. Of course, unfavorable issues related to an ineffective SEI layer and electrode swelling might pose more severe safety concerns on Si-based LIB cells, compared to its Gr-based counterparts. Similar to Gr anodes, all the cascading thermal reactions and, thus, the worst case scenario of a thermal runaway, are initially triggered by SEI decomposition at temperatures even lower than 90 °C [54, 55]. This infers how much the quality of the SEI layer is important for the thermal threats of LIB cells, and an ineffective SEI layer accompanied with large volume changes could be a major challenge for large format Si-based LIB cells.

11.2.2 Impact of Si-addition on the performance characteristics of lithium-ion cells

The major approach for adding Si to the negative electrode is the beneficial impact on the LIB cell energy density. As explained above, when seriously considering the various discounting factors for translating the Si electrode-specific capacity to a practical energy density of a Si-based cell, calculations indicate that an improvement of ≈20% in terms of energy density can be achieved for Si-based LIB cells compared to state-of-the-art Gr-based LIB cells, which, however, significantly depends on the Si amount and the cathode capacity [9, 45].

Furthermore, Si addition to a Gr-based anode also impacts capacity-balancing of negative-to-positive electrode, as schematically illustrated in Figure 11.2. For

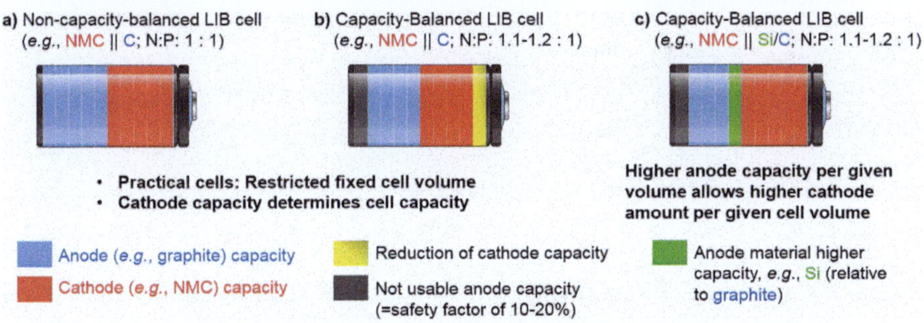

Figure 11.2 *Impact of Si addition to a Gr anode with regard to capacity balancing in the full-cell (considerations are based on a restricted fixed cell volume as in practical LIB cells). (a) Non-capacity balanced LIB cell with Gr anode; (b) Capacity-balanced LIB cell with a Gr anode capacity excess of 10–20% to reduce the risk of Li metal plating; (c) Capacity-balanced LIB cell in which a high-capacity anode (Si) has been added (e.g., in the form of a Si/Gr blend): A higher anode capacity per given volume can be realized, which allows the addition of a higher cathode amount per volume, resulting in a higher cell capacity.*

practical cells having a fixed cell volume, the cathode capacity needs to be reduced to avoid the risk of Li metal plating at the anode, resulting in an anode capacity excess of ≈10–20% (=not usable anode capacity; Figure 11.2b). When adding a certain amount of Si to the anode, a higher anode capacity per given volume can be realized, which results from the high capacity of Si. In turn, also a higher cathode amount per given volume can be realized, while the anode capacity remains in excess due to the addition of Si (Figure 11.2c). Overall, a higher cell capacity can be achieved.

In addition, as the incorporation of Si will have a significant impact on the negative electrode thickness (=thinner electrodes), a positive influence on the rate capability, for example, during fast charging, can be expected. Therefore, suitable approaches, such as specifically designed Si materials or structured electrodes [56], need to be followed and verified in practical cells. The assumption of better and safer fast charging capabilities of negative electrodes with high Si loadings, without the risk of Li metal plating, is also further supported by the enhanced lithiation potential of Si compared to Gr [57]. In summary, an optimum performance of Si-based cells can only be achieved by carefully designing Si negative electrode in combination with suitable electrolytes, while considering the various trade-offs, for example, between high capacity and limited cycle life or safety, to find the best compromise.

11.2.3 Silicon in commercial LIB cells: state of the art and markets for Si-based cells

Since the late 1990s, Si-based materials are considered as the most promising choice for high-capacity negative electrodes in LIB cells, which is displayed by the enormous number of patents (≥2 720) and scientific publications (≥4 530) in the field (Figure 11.3), covering a broad range of various types of Si materials and related processing and application strategies. Among others, first patents on Si-based materials have been published at the end of the 1990s and in the early 2000s, for example, by *3M* (US 6255017 B1, US 7498100 BB) for Si alloys/composite, by *Sanyo Electric* (WO 2001029913 A1) for pure thin film Si electrodes or by *Samsung SDI* (US2002/0164479 A) for Si/C composites. The transition from science to industry within the technology life cycle, that is, from basic research to applied research and development and to commercialization/industrial application, can also be captured by the establishment of start-up companies indicating an early phase of application [58]. In the recent decade and particularly, recent years, various start-ups and established companies announced the production of Si-based materials, Si-based electrodes, or completely designed Si-based cells for different LIB application purposes with enhanced energy contents (*i.e.,* from Si/C and SiO_x composites or Si-alloy anodes to pure Si anodes), including *Sila Nanotechnologies* [59], *Enovix* [60], *Enevate* [61], *Amprius* [62], *Group14* [63], *Leyden Jar* [64], *Nexeon* [65], *EoCell* [66], *Ecellix Inc* [67], *Nanotek Instruments* [68], *StoreDot* [69], *Enwires* [70], *CalBattery* [71], *Zeptor* [72], *OneD Material* [73], *E-magy* [74], *Elkem* [75] and *Cenate* [76], among others.

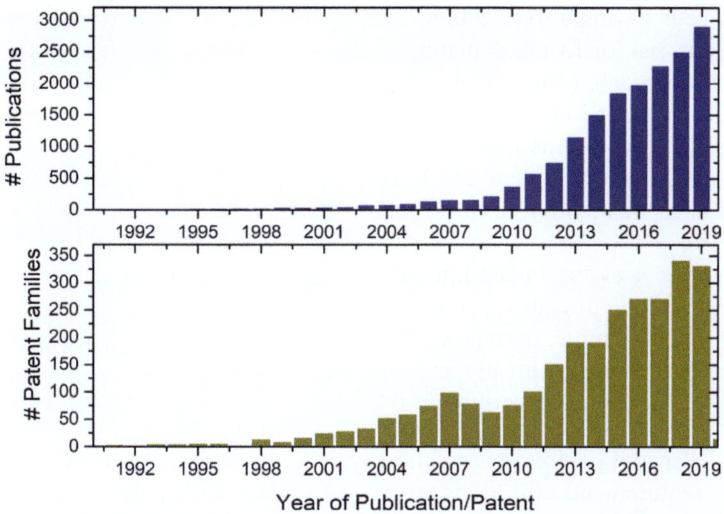

Figure 11.3 *Statistical analysis of on the trend of publications and patents on Si-based negative electrodes. Last updated on 26th of November 2019. Keywords used in the Scopus search were "Lithium-battery + Silicon anode".*

For the practical integration of Si into the current commercial LIB ecosystem, the co-utilization of Gr and Si in form of Gr-blended Si or Gr/Si composites has resurfaced as the most viable approach for realizing high-energy LIB cells [13]. This strategy combines the best and synergistic features of Gr (*i.e.,* long cycle life, high Coulombic efficiency, low volume changes, facile calendaring process, high electronic conductivity, and industrial applicability to conventional electrode manufacturing technologies) and Si (i.e., high specific capacity) [13]. Nowadays, Si is already added in small amounts (up to ≈3–8 wt.%; often in form of SiO_x) to Gr-based electrodes within some commercial LIB cells; however, the incorporation of higher amounts of Si (≥10–20 wt.%) is still hampered by the insufficient cycle life of such cells so far. For example, *Tesla* (*Panasonic*) claimed in 2015 that they had increased the Si content to 6% in the *Model S* car batteries and very recently *BMW* announced to adopt 10–15% Si in EV cars by 2023 [77]. Besides application of Gr-blended Si (nano)particles, the use of SiO_x, for example, in form of sub-μm SiO_x particles blended with Gr [78], seems to be a viable approach for practical LIB cells and has been patented by *Shin Etsu Chemical Co Ltd* (US 7776473 B2, US 6759160 B2), among others. An example of a cross-sectional SEM image of a negative electrode, obtained from a commercial 18650 cylindrical LIB cell, is shown in Figure 11.4. It clearly shows that the anode is composed of Gr and SiO_x particles. Despite the above attempts, much effort is needed to bring the academic and industrial sectors closer to each other. The research activity on Si-based anodes has received huge celebrity, particularly in the last decade (Figure 11.3).

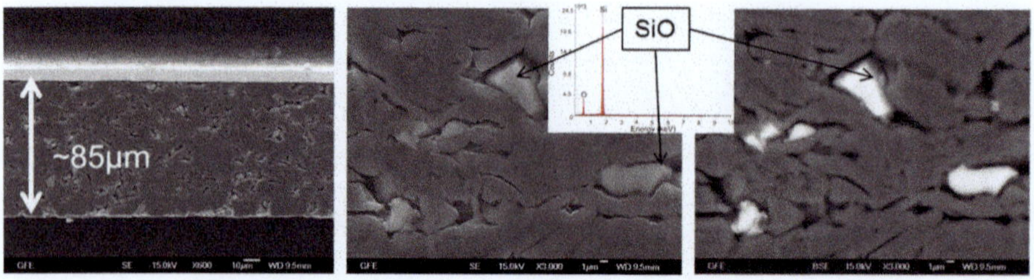

Figure 11.4 *SEM and EDX analysis of negative electrode extracted from high-energy 18650 LIB cell. Source: F. Rohmann, L. Willenberg, J. Meyer, E. Figgemeier (Gescheinschaftslabor Elektronenmikroskopie und ISEA, RWTH Aachen University).*

However, the progress on practicality of Si anode-based LIB cells in view of the basic requirements still needs more work. When further illuminating the commercial aspect of high-energy Si-based LIBs cells, that is, most likely for use in EV applications, the tradeoff and, therefore, right "balancing" between high energy density and long cycle life can be considered as one of the most important factors for successful implementation. In this respect, the practical usage of EV batteries might significantly differ from the laboratory testing conditions of high-energy LIB cells. For example, it is well known that the cycle life of LIB cells can be significantly increased by "shallow" cycling, that is, by charge/discharge cycling with a limited depth-of-discharge (DoD) [20]. Considering real-life drive cycles of EVs (Figure 11.5), it is known that a complete discharge of the battery will only very randomly be needed. Considering the electrochemical properties of Si-/Gr-based negative electrodes, it also means that Si will predominantly remain in the charged

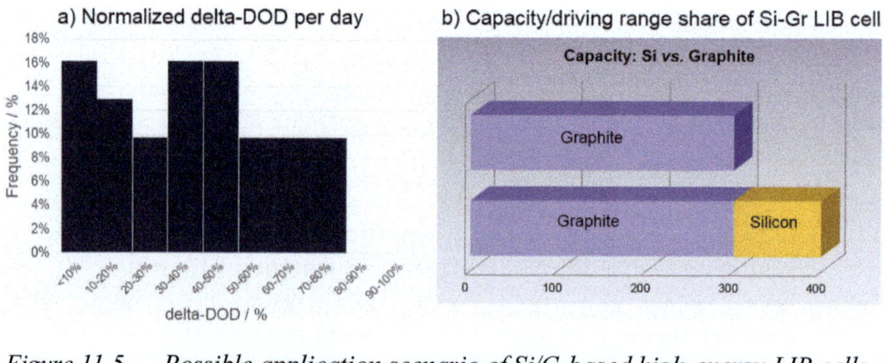

Figure 11.5 *Possible application scenario of Si/C-based high-energy LIB cells in EVs considering real-life drive cycles. (a) Normalized delta-depth-of-discharge (DOD) per day (data from: ISEA Institut für Stromrichtertechnik und Elektrische Antriebe, RWTH Aachen) and (b) capacity and driving range (e.g., in miles) share of Si-Gr-based LIB cells.*

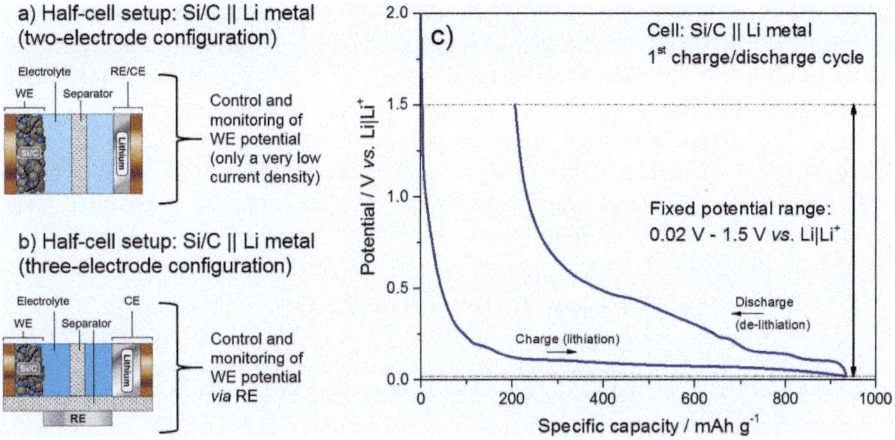

Figure 11.6 *Studies of silicon-based electrodes in a half-cell setup (Si/C || Li metal) with an "excess" Li metal counter electrode (CE) in (a) two-electrode configuration (Li metal cell) or (b) in three-electrode configuration (with Li metal reference electrode (RE)). (c) Representative potential profiles for the first cycle lithiation/de-lithiation of a silicon/carbon (Si/C) composite electrode in a Si/C || Li metal cell (half-cell setup) within a fixed potential range of 0.02–1.50 V vs. Li|Li$^+$.*

state for partial discharge of the battery/LIB cells, which is a result of the different operation potentials of Gr and Si (Gr will be discharged first; see Section 11.3.2). The extreme volume changes of the Si-based anode could be strongly reduced and be limited to those occasions when a complete discharge (=driving distance) is needed. In turn, this capacity and driving range share of Si/Gr-based LIB cells would allow for both, an enhanced cycle life and a high-energy/driving range when needed (Figure 11.5). On the other hand, this usage approach would also result in a higher share of the highly reactive cathode in the charged state (at high SOC) and, thus, might result in stability and lifetime challenges.

11.3 Challenges for practical implementation of Si-based negative electrodes in lithium-ion battery full-cells

A very common approach to study novel Si-based negative electrode materials in terms of their electrochemical performance, in particular with respect to their charge/discharge cycling or rate performance, is the use of a so-called half-cell setup (either in two- or three-electrode configuration), as illustrated in Figure 11.6a and b [37, 79]. Such a half-cell setup typically implies two major requirements: First, the counter electrode (CE) typically consists of Li metal, which means that a nearly unlimited amount of active lithium is available in the cell system. Second, the potential range of the working electrode (WE), for example, Si or a Si/carbon composite (further abbreviated as "Si/C"), is fixed during the whole electrochemical study of a Si/C || Li metal cell. Typically, Si-based materials/electrodes are studied in potential ranges of, for example, 1.5 V to 0.05–0.01 V vs. Li|Li$^+$, as exemplarily shown by the charge/discharge WE potential profile of a Si/C composite electrode in Figure 11.6c. However, it has to be considered that there are several issues and limitations when Li metal is used as CE and even more when it is used in a two-electrode configuration without reference electrode (RE). It has to be noted that the latter cell setup should correctly be termed "Li metal cell," as the cell voltage is controlled (Figure 11.6a). The term "half-cell" can only be used for two-electrode configurations in which polarization effects at the Li metal RE/CE can be neglected (cell voltage ≈ CE potential), that is, at a very low current density at the Li metal RE/CE. The challenges related to the electrochemical studies in Li metal cells will be comprehensively discussed in Section 11.3.1, as there is a high risk of misinterpreting the electrochemical performance and related structure-property relations for Si-based negative electrode materials.

In addition, the use of Li metal will inevitably result in accelerated and continuous electrolyte decomposition and, thus, excess electrolyte is necessary. However, with the use of excess electrolyte, continuous electrolyte decomposition will lead to continuous accumulation of electrolyte decomposition products that will affect cell performance.

Unfortunately, in comparison to the large number of publications for electrochemical studies of Si (Figure 11.3), there is only a minor number of publications available in which Si-based negative electrode materials are studied in a "full-cell setup" (either in two- or three-electrode configuration), as illustrated in Figure 11.7a and b [37]. In such a full-cell setup, the cell is operated by control of the cell voltage, which is defined as the difference between the positive and the negative electrodes' potential, as exemplarily shown for an NMC-111 || Si/C full-cell in Figure 11.7c [79–82]. In contrast to a half-cell setup (Figure 11.6), the individual electrode potentials will change over operation, which depends on the used materials and electrochemical operation conditions. The changes of the negative and/or positive electrode might be monitored *via* a RE (*e.g.*, Li|Li$^+$). Furthermore, the active lithium content, also often referred to as "lithium inventory" or "lithium reservoir," is limited in a full-cell setup [37, 38, 43, 83], which in turn has a significant impact on the cycling performance, especially with respect to the capacity retention. The requirements and challenges related to the electrochemical studies in LIB full-cells will be discussed in Section 11.3.2. Furthermore, there are specific challenges for the implementation of Si-based materials in practical/commercial lithium-ion full-cells, which are related to practical mass loadings/areal capacities (mAh cm^{-2}), limited amount of electrolyte and issues linked to "drying," as well as negative electrode swelling. These aspects are typically not primarily addressed in literature, as most studies are performed in "laboratory cells," such as coin cells or single-layer pouch-bag cells. Such laboratory cells typically use high amounts of electrolyte, which makes the knowledge transfer

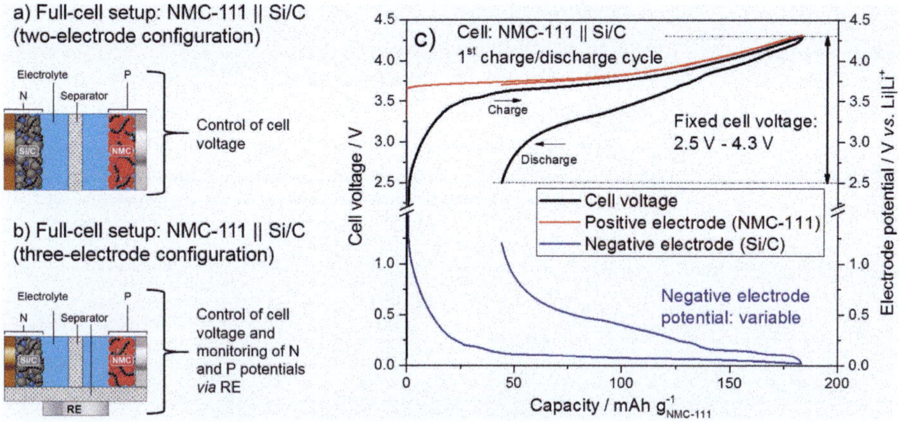

Figure 11.7 *Studies of silicon-based electrodes in a full-cell setup (e.g., NMC-111 || Si/C) in (a) two-electrode configuration or (b) in three-electrode configuration (with Li metal reference electrode (RE)). (c) Representative cell voltage and electrode potential profiles for the first charge/discharge cycle of an NMC-111 || Si/C cell within a fixed cell voltage range of 2.50–4.30 V.*

to practical cells and cell formats (*e.g.*, multilayer pouch cells or cylindrical cells) rather difficult. In Section 11.3.3, these practical aspects and challenges will be comprehensively introduced and discussed.

11.3.1 Studies of Si-based materials in Li metal cells: opportunities, pitfalls, and lessons learned

In general, there are two major approaches to study Si-based materials within Si || Li metal cells, either in a two-electrode configuration (Li metal cell) in which the cell voltage (in V) is controlled (Figure 11.6a) or in a three-electrode configuration ("half-cell" setup) in which the WE potential (in V vs. Li|Li^+) is controlled (Figure 11.6b) [37]. These cells are typically investigated with a constant current charging/discharging procedure within a defined voltage/potential range, that is, a lower cut-off voltage/potential slightly > 0 V (vs. Li|Li^+) for lithiation of the Si-based electrode and an upper cut-off voltage/potential of typically 1–3 V (vs. Li|Li^+). This quite simple and effective approach to study novel Si-based materials can provide important information, but might also lead to misinterpretations, as there are several issues, which need to be considered when using Li metal as CE [37].

In a first step, electrochemical studies of Si-based materials in Li metal cells (two-electrode configuration), which are often performed in a coin cell setup, will be discussed. In these cells, Li metal is termed as negative electrode and the Si-based electrode is designated as positive electrode, according to their electrode potentials. For such experiments, it has to be considered that Li metal is not an ideal negative electrode, which is related to its polarizability, as high overpotentials of up to

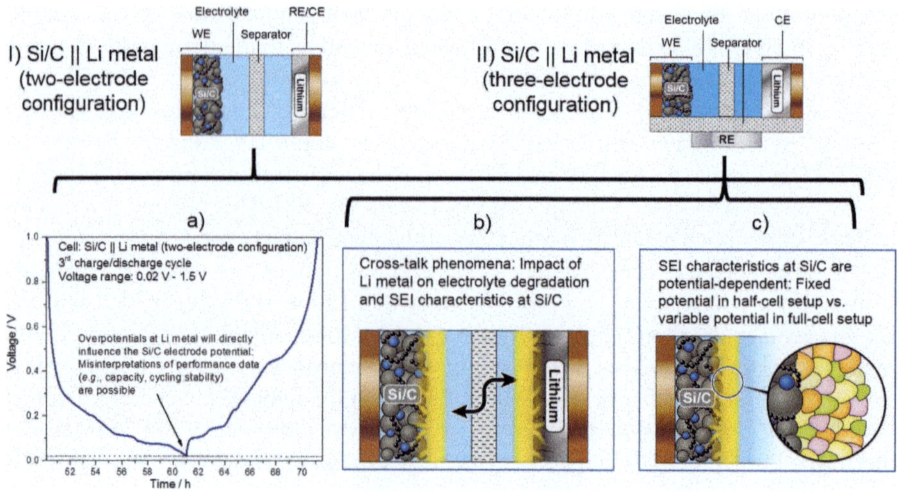

Figure 11.8 *Challenges and possible misinterpretations of Si-based materials and electrodes in Si/C || Li metal cell setup in (I) two- or (II) three-electrode configuration using an "excess" Li metal counter electrode: (a) Impact of overpotential at the Li metal electrode on the Si electrode potential (setup I); (b) Impact of Li metal on electrolyte/additive consumption and degradation and SEI formation at Si (setups I and II); (c) Impact of fixed WE potential on SEI formation and characteristics at Si (setup I and II).*

hundreds of mV can occur during dissolution/deposition of Li metal [37, 84]. The overpotentials are significantly influenced by the applied current density and will have a direct impact on the measured cell voltage. In turn, it is not possible to determine the exact potential of of the Si-based electrode at the applied cut-off voltage within this setup, especially at high current densities or C-rates (Figure 11.8Ia). This is a major drawback of this cell setup, as the lithiation degree of Si is determined by its electrode potential and vice versa [37]. An exact determination of the potential and, therefore, of the specific capacity of Si-based materials will only be possible at very low current densities (at Li metal) or when applying an appropriate constant-current-constant voltage procedure. Misinterpretations in terms of specific capacity and cycling stability might easily take place for this cell setup, as minor differences of the applied voltage can have a significant impact on the lithiation degree, on the reversible capacity and, thus, on the cycling stability [37]. Second, the electrochemical evaluation of Si-based materials in a half-cell setup using Li metal as CE and RE (three-electrode configuration) will be discussed. In general, the application of REs ($Li|Li^+$) is more viable to study novel Si-based materials (WE), especially in terms of their lithiation/de-lithiation capacity. In this setup, the WE can be cycled in a defined cut-off potential range and overpotentials at the Li metal CE will not have any significant impact on the WE potential, in contrast to the two-electrode configuration. However, both of the mentioned cell setups display several limitations, which are related to the use of Li metal as well as to the fixed potential range of the investigated Si-based materials.

It is well known that Li metal electrodes suffer from the formation of "high surface area lithium" (HSAL), often denoted as "dendrites," during the ongoing dissolution/deposition process, which is one of the major reasons hindering the commercial breakthrough of Li metal batteries to date [85–88]. This process is accompanied by continuous electrolyte reduction at the continuously formed fresh Li metal surfaces and severe consumption and decomposition of the electrolyte species, which in turn deteriorate the performance of Si || Li metal cells (Figure 11.8I/IIb). In this respect, the modification of the electrolyte, for example, by using electrolyte additives for improving the SEI formation at Si electrodes, is a well-known strategy to improve the performance of Si-based LIB cells [3, 54, 89–93]. Accordingly, there are various literature studies on the use of electrolyte additives for Si that have been performed in Li metal cells, in particular focusing on performance and SEI formation at the Si electrode. The results obtained in these cells may not be transferred to practical Si-based full-cells, due to the significant impact of the Li metal electrode, that is, its highly reducing ability throughout the cycling process. Thus, it is not possible to differentiate whether the positive effect of a certain electrolyte formulation in terms of an improved cell performance is related to beneficial SEI formation and/or positive influences on the charge/discharge reactions at the Si or at the Li metal electrode (Figure 11.8I/IIb). Thus, due to the strong impact of Li metal on electrolyte consumption and degradation, the effect of certain electrolyte additives on cell performance and SEI formation at Si (*e.g.*, SEI composition) can be completely different in a practical Si-based full-cell. A similar dependency has been reported for the study of positive electrode materials and corresponding "cathode electrolyte interphase" (CEI) formation in cells with and without Li metal CE or negative electrode [94, 95].

Another major issue of electrochemical studies in a half-cell setup is that the potential of the Si-based WE is fixed during prolonged cycling, while typically low electrode potentials (*e.g.*, 10–50 mV vs. $Li|Li^+$) are applied as lower cut-off potential. In this respect, it is often disregarded that the precise control of a certain cut-off potential during charge/discharge cycling may provoke the formation of a unique SEI layer, which may differ in composition and thickness compared to the SEI layer formed in a full-cell setup in which the potential is not constant over cycling (Figure 11.8I/IIc) [96–98]. For example, Vogl *et al.* showed that SEI formation and composition on single crystal Si electrodes strongly depends on the electrode potential [96, 97]. In contrast, in a full-cell setup, the low potentials of Si-based electrodes are typically not reached, due to the negative/positive electrode capacity balancing [82], which will be discussed in Section 11.3.2. Therefore, a potential-dependent impact on SEI composition and thickness cannot completely be excluded. As a result, it is hard to distinguish the predominant aging origins of potential-controlled cells using Li metal, which might be either related to the Li metal electrode or the investigated Si-based electrode, especially for long-term cycling studies. Therefore, it is recommended not to over-interpret the long-term cycling stability of Si-based electrodes obtained in Li metal cells, for example, when comparing different binder materials or distributions [99, 100]. Such results might give a first estimation in terms of performance, but should be verified in a LIB full-cell setup without "auxiliary" Li metal electrode [37]. One of the most important aspects for electrochemical studies in cells using an "auxiliary" Li metal electrode is the fact that an almost unlimited amount of active lithium (lithium reservoir) is available in these cells, as typically quite thick (up to ≈500 mm) Li metal foil is used. This excess of active lithium can certainly help to understand fundamental material and/or electrode-intrinsic degradation mechanisms. Nevertheless, parasitic reactions, which consume active lithium, for example, by continuous SEI (re-)formation due to the significant volume changes, are neglected in such investigations [37, 38]. The impact of active lithium losses, which is often expressed by the Coulombic efficiency (C_{Eff}), on the cell performance in Si-based full-cells without an "auxiliary" Li metal electrode is discussed in Section 11.3.2.

In summary, the electrochemical evaluation of novel silicon-based materials and electrodes within Si || Li metal cells using an "auxiliary" Li metal electrode can deliver important information about material- or electrode-intrinsic properties. These studies can either be conducted in a half-cell setup with three-electrode configuration (potential control) or in a full-cell setup ("Li metal cell," cell voltage control). However, these cell setups also display certain limitations and drawbacks for the determination of battery performance metrics. It can be recommended to perform initial studies of Si-based materials in Si || Li metal cells to determine:

- Material-intrinsic characteristics (reversible and irreversible capacity; Coulombic efficiency; structural and mechanical material changes, *e.g.*, with respect to particle cracking and amorphization behavior)
- Electrode-intrinsic characteristics (suitability of electrode composition and processing, *e.g.*, disconnection of Si particles; impact of conductive additives, binder materials)

The obtained results from Si || Li metal can be used as a first estimation for performance evaluation of the Si-based material and/or electrode. However, follow-up studies should be performed in a full-cell setup without excess Li metal electrode (Figure 11.7). Furthermore, to avoid misinterpretation of the electrochemical or analytical data (*e.g.*, in terms of SEI composition and structure) obtained from cells using an "auxiliary" Li metal electrode, as illustrated in Figure 11.8, the following issues need to be considered:

- Due to the high reactivity of Li metal, it is not recommended to perform electrolyte studies (*e.g.*, on novel electrolyte additives) only in a Si || Li metal cell setup. These studies should be always performed also in a full-cell setup without excess Li metal CE.
- Due to the Li metal electrode (high reactivity) and the WE potential control (fixed potential) in Si || Li metal cells, analytical data obtained for SEI composition and structure cannot directly be correlated to Si-based full-cells (no Li

metal electrode, varying Si electrode potential). Thus, it is recommended to perform SEI studies in Si-based full-cells without excess Li metal electrode. Thus, electrolyte degradation at Li metal can be excluded and the potential of the Si negative electrode will be different compared to the potential-controlled cell.

- When studying Si-based materials in Si || Li metal cells, it is not possible to give any statement about the impact of active lithium losses on cell performance. Nevertheless, the C_{Eff} of Si || Li metals cell can give a rough estimation for the performance in a practical full-cell without excess Li.

11.3.2 Studies of Si-based materials in lithium-ion full-cells: considerations for anode/cathode capacity balancing and challenges with respect to active lithium losses

To evaluate the practicality of Si-based materials for LIB cells, electrochemical studies have to be conducted in a cell setup in which the (negative) impact of an "excess" Li metal electrode can be excluded. Accordingly, different approaches have been proposed and discussed in literature to overcome the issues of the Li metal electrode.

One strategy is the use of a capacity-oversized positive electrode, such as $LiFePO_4$ (LFP), within an LFP || Si/C cell setup (two- or three-electrode configuration) in which the potential of the negative Si/C electrode can be controlled *via* a Li metal RE (Figure 11.6b) [38, 83, 101–103]. LFP is well suited as positive electrode for such studies, as it exhibits a very flat lithiation/de-lithiation potential. Thus, LFP can be also used to monitor the potential of the Si/C negative electrode during cell voltage controlled operation within a two-electrode configuration. These types of cells are typically named "pseudo full-cells," as a "practical" full-cell setup exhibits a 10–15% capacity-oversized negative electrode (N:P ratio >1) to prevent safety issues due to Li metal plating on top of the negative electrode, which is critical for high charging rates as well as for operation at low temperatures [104, 105]. Nevertheless, pseudo full-cells can serve well as a diagnostic tool to study the performance of Si/C negative electrodes, as it is possible to implement a defined lithium reservoir within the cell to exclude capacity fading related to the depletion of active lithium [37]. In addition, by exclusion of a Li metal electrode, the challenges related to continuous electrolyte consumption and degradation can be prevented and fundamental insights into the degradation mechanism of novel Si/C materials can be obtained [38, 83, 101, 103].

In a classical LIB cell setup, the amount of active lithium (lithium reservoir) within the cell is limited by the positive electrode, for example, by layered transition metal oxides based on the formula $LiMO_2$ (M = Co, Ni, Mn, etc.) such as $LiNi_xMn_yCo_zO_2$ (NMC, with $x + y + z = 1$), resulting in an N:P ratio >1 [37, 82]. Therefore, lithium-consuming parasitic reactions [38, 43], which cannot be identified in a half-cell setup with an "excess" Li metal CE, as discussed above, will lead to capacity fading of the full-cell. It is well known from literature that there are several reactions which can be attributed to active lithium losses (= consumption of electrons and lithium), including in particular SEI formation [9, 18, 38, 106],

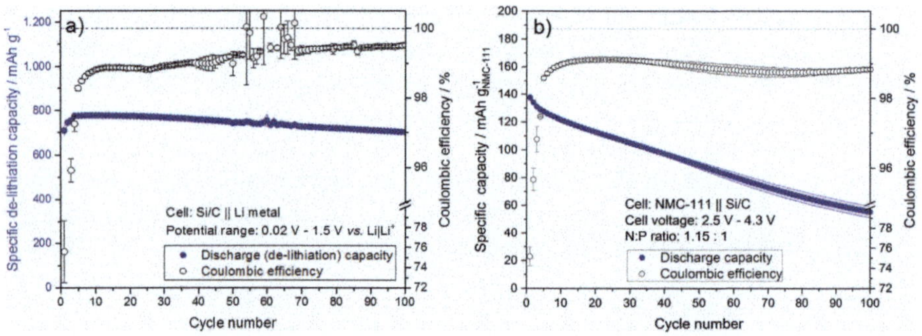

Figure 11.9 *Comparison of the discharge capacity and Coulombic efficiency during charge/discharge cycling of a Si/C-based electrode (a) within a half-cell setup (Si/C || Li metal cell; potential range: 0.02–1.50 V vs. Li|Li*$^+$*) and (b) within a full-cell setup (NMC-111 || Si/C; cell voltage range: 2.5–4.3 V; N:P capacity ratio: 1.15).*

active material loss such as mechanical detachment of lithiated Si particles [38], lithium trapping within Si particles [107] as well as irreversible Li metal plating [104]. Even though various Si-based materials have been proposed to show an outstanding capacity retention within a Li metal cell, the good performance might not be transferred to a LIB full-cell without "excess" Li metal electrode. Exemplarily, Figure 11.9a displays the cycling performance of a Si/C negative electrode within a Si/C || Li metal cell (half-cell setup). While a relatively good cycling stability is achieved within this cell setup, the good performance cannot be transferred to the full-cell setup, in which the same material is evaluated in an NMC-111 || Si/C cell (Figure 11.9b), showing severe capacity fading.

Several publications already pointed out that the major failure mechanism of full-cells comprising a Si-based negative electrode is due to the ongoing loss of active lithium, which is primarily caused by the initial formation and continuous breakage/(re-)formation of the SEI layer at Si during each lithiation/de-lithiation cycle [47, 90, 108–116]. Irreversible parasitic reactions within a LIB cell are typically expressed by the Coulombic efficiency (C_{Eff}), which is defined as the quotient of the discharge capacity and the charge capacity [38, 42]. An ideal C_{Eff} should be close to 100%, which would indicate that parasitic side reactions, such as electrolyte decomposition or active lithium loss, are reduced to a minimum [38, 42].

"Irreversible capacity" (Q_{irr}) is another term, which is often used in literature to describe parasitic reactions. It is defined as the difference between charge capacity and discharge capacity of one cycle. In turn, the accumulated Q_{irr} is defined as the sum of the Q_{irr} over a certain number of charge/discharge cycles, and, thus, should also be minimized to achieve a high capacity retention in a LIB full-cell [38]. In this respect, it is important to know that the accumulated Q_{irr} is often equated to active lithium losses. However, this assumption is only a rough estimation, as there can also be non-lithium consuming parasitic reactions within the cell, which do not reduce the active lithium content [38]. A simple approach to differentiate between

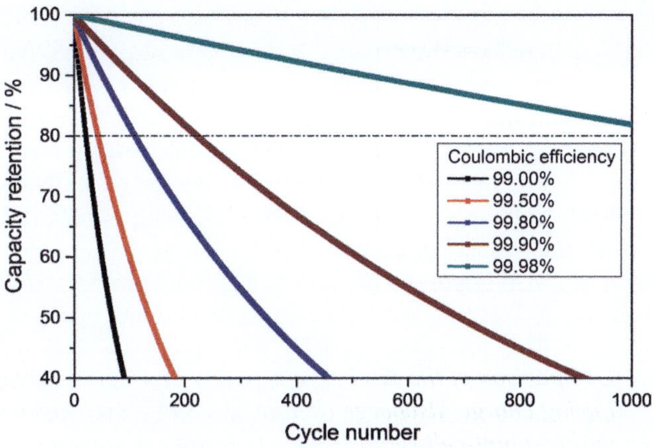

Figure 11.10 *Calculations on the impact of Coulombic efficiency (C_{Eff}) on capacity retention in a LIB full-cell without "excess" Li metal electrode, for example, for an NMC || Si/C cell. Different C_{Eff} values are considered, i.e., from 99.00% to 99.98%, while in each case the respective values are considered from the 1st to the 1000th cycle. For the theoretical calculation, it is assumed that the active lithium loss is equal to the irreversible capacity. Example: A C_{Eff} of 99.90% indicates that the remaining active lithium content in the cell is only 0.999^{200} 117 = 81.86% after 200 cycles. Calculations have been conducted as reported in [42].*

Q_{irr} and detrimental active lithium losses has been recently proposed by using LFP || Si/C pseudo full-cells in a three-electrode configuration with Li metal RE. This cell setup allows to state whether capacity fading of such Si-based cells is related to active lithium losses or material/electrode failure [38]. Nevertheless, in a first approximation, C_{Eff} gives a decent indication for the suitability of Si-based materials for practical application in a full-cell.

As shown, for the example of a Si/C || Li metal cell in Figure 11.9a, C_{Eff} of the Si/C material is only ≈75% in the first cycle and only slowly increases up to ≈99.5% after 100 cycles. When the same Si/C material is evaluated in an NMC-111 || Si/C cell (Figure 11.9b), a C_{Eff} of only ≈99.0% is achieved over cycling, which results in relatively strong capacity fading. Unfortunately, many research studies still primarily focus on improvements of the cycling stability of Si/C materials in Li metal cells, while neglecting the importance of a high C_{Eff}, for example, by insufficient reporting or by an inappropriate presentation of the C_{Eff} data. In particular, we recommend to report both, the first cycle C_{Eff} as well as average C_{Eff} over cycling, while also including error bars to demonstrate the reproducibility of the results. As an example, Figure 11.10 shows the impact of C_{Eff} on capacity retention of a theoretical LIB full-cell, while it is assumed that a C_{Eff} below 100% is completely related to active lithium loss. It becomes clearly evident that a C_{Eff} of 99.00% (over the whole cycling range) is still far too low to achieve a reasonable capacity retention (state-of-health (SOH) below 80%: <100 cycles), thus, at least a C_{Eff} of ≥99.90% should be targeted to demonstrate a commercial relevant cycle number (>200 cycles for SOH >80%). In this respect, Lin *et al.* reported that a C_{Eff} of 99.96% is required to achieve a cycling stability up to 500 cycles for commercialization [42]. Nevertheless, the calculations shown in Figure 11.10 do not even consider the typically low C_{Eff} values (<90% or even <80%) obtained in the first charge/discharge cycle(s), as illustrated in Figure 11.9a. In this respect, high precision Coulometry, as reported in several works of Dahn *et al.*, is an important method to gain fundamental insights into degradation mechanisms of LIB full-cells related to active lithium losses [43, 118, 119].

The negative/positive electrode capacity balancing of a LIB full-cell, often described by the N:P capacity ratio (>1), is essential for safe cell operation and for a high capacity retention. However, it also has to be accurately adjusted, as the capacity-oversizing of the negative electrode results in a decrease of the cell's energy density [80, 82, 112, 114, 116, 120]. Ideally, the negative electrode potential can be indirectly adjusted by precise selection of the cell cut-off voltages and of the N:P capacity ratio, as exemplarily shown in Figure 11.7c. However, this adjustment is only precisely possible for the first charge/discharge cycle, as parasitic reactions, in particular active lithium loss, will lead to an undesired shift of both the negative and positive electrode potentials in the following charge/discharge cycles, as reported by Krueger *et al.* [80] and in the review of Nölle *et al.* [37]. Due to SEI formation and active lithium losses (C_{Eff} < 100%), less lithium is released during discharge of the negative electrode than accepted in the previous charge, as exemplarily shown in Figure 11.7c for the first charge/discharge cycle of an NMC-111 || Si/C cell. As a result, the discharge reaction of the full-cell will be terminated by the complete de-lithiation of the negative Si/C electrode when the charge cell cut-off voltage is reached. In turn, the positive electrode will not be completely lithiated during discharge (Figure 11.7c), even though it could theoretically accept more lithium [37]. As a consequence of continuous active lithium losses during cycling, the individual end-of-charge (EOC) potentials of the negative and positive electrodes will gradually shift toward higher values [17, 37, 108]. Figure 11.11 depicts the development of the negative and positive electrode potentials in dependence of the time of an NMC-111 || Si/C full-cell, which is operated in a cell voltage range of 2.5–4.3 V. The potentials have been monitored *via* an Li metal RE. It is obvious that the negative EOC potential of Si/C is gradually shifted to higher values during the ongoing cycling. As a result, the positive electrode is also forced to cycle at higher EOC potentials (as seen in Figure 11.11), thus, resulting in a higher degree of de-lithiation (=higher Li extraction ratio), which in turn might cause (additional) capacity fading of the full-cell due to instability of the positive electrode, for example, by structural instability [121].

The gradual shift of the negative electrode potential toward higher values during cycling also has a further impact on the overall cycling stability. In particular, this potential shift needs to be considered for the evaluation of Si/carbon (Si/C) or Si/Gr composite materials, as the individual materials display different operation potential regions. In this context, Yao *et al.* reported on an operando method for the

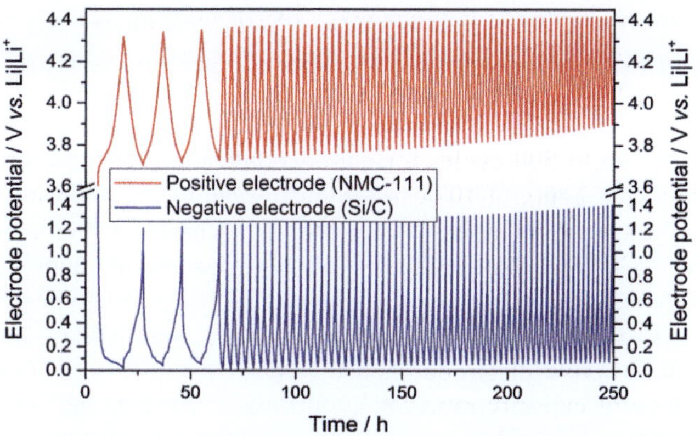

Figure 11.11 *Negative and positive electrode potentials (vs. Li|Li⁺) of an NMC-111 || Si/C cell (full-cell setup, three-electrode configuration; with Li metal RE). Cell voltage range: 2.50–4.30 V; N:P capacity ratio: 1.15.*

quantification of the lithiation and de-lithiation amounts of Si/Gr blends, which is of high importance [57]. In the charge process (=lithiation of the negative electrode), Si-based materials will be first lithiated compared to Gr, followed by Li^+ intercalation into Gr at potentials below 0.2 V vs. Li|Li^+. During discharge, the de-lithiation reaction preferentially starts from Gr in a potential range from ≈0.01 to 0.23 V vs. Li|Li^+, before de-lithiation of Si takes place in a potential range from ≈0.23 to 1.00 V vs. Li|Li^+. As a consequence, the above-mentioned potential shift of the negative electrode can lead to EOC potential values, at which the Gr material is less active toward lithiation or even not at all involved in the lithiation process (>200 mV vs. Li|Li^+). Hence, in such a case, only Si would be predominantly cycled, which might induce further parasitic side reactions and capacity fading. It also needs to be mentioned that other carbon materials, such as amorphous carbon, exhibit a different operation potential region (*i.e.*, 0–1.2 V vs. Li|Li^+) [33], thus, will still be active toward lithiation at higher potentials.

The above-introduced challenges and examples clearly demonstrate that direct transfer of results for Si/C materials obtained from a half-cell setup (two- or three-electrode configuration) with an "excess" Li metal CE to a LIB full-cell setup with limited lithium inventory is not possible. It is highly recommended that researchers do not only focus on improvements in terms of cycling stability within Li metal cells but also consider design strategies of novel Si-based materials (*e.g.*, with low specific surface area [10, 36, 47] for minimizing active lithium losses by evaluating in LIB full-cells [79, 114, 116], that is, to develop Si-based materials with the goal to achieve a high C_{Eff} during the initial and ongoing charge/discharge cycling.

11.3.3 Challenges for implementation Si-based materials in practical lithium-ion cell formats: electrolyte drying-out and electrode swelling

To achieve a successful transfer of novel battery materials, such as Si/carbon composites, and close the gap between academia (*e.g.*, laboratory studies in coin cells) and industry (*e.g.*, practical studies in cylindrical or wound pouch cells), a broad series of different performance metrics needs to evaluated and quantified [14, 17, 42, 122].

In general, there are rising concerns in the battery community about how new battery materials and emerging battery technologies are reported in literature, while various of such publications report on a "promising," "outstanding," or even "ultra-high" performances [123]. The "practicality" of new materials and technologies is often neglected or not at all addressed in such publications, resulting in concerns about the "performance accessibility" for practical applications [8, 31, 42]. As an example, various emerging cell chemistries (*e.g.*, S || Li metal or O_2 || Li metal) have been reported to exhibit a very high theoretical gravimetric or volumetric energy. However, the calculation of the energy values often excludes numerous relevant parameters for practical battery cells, such as the practical mass utilization of the active material or practically achievable discharge voltages [31]. Therefore, theoretical energy values, which are commonly reported for evolving battery technologies, might drastically overestimate their realistic and practical potentiality compared to state-of-the-art LIB cells [8, 31]. Several researchers already pleaded on more transparency for reporting performance data as well as on a more balanced view, reporting not only advantages but also disadvantages/challenges, for those novel materials and cell chemistries [8, 31, 42, 123–127].

Lin *et al.* reported on a list of performance metrics that need to be quantified to allow for a realistic evaluation of the practicality of novel materials and emphasized that reporting only a limited number of these metrics will result in an incomplete picture [42]. Typically, material developments in laboratory scale only focus on a few performance metrics such as specific reversible capacity, Coulombic efficiency, energy efficiency [41], or cycling stability. In general, the optimization of just a few of these parameters seems to be relatively easy, especially when tested in Li metal cells as discussed above (Figure 11.12). However, when it comes to practical application, there are many more parameters, which need to be optimized. These particularly include an increase of electrode mass loading (mg cm^{-2}) and areal capacity (mAh cm^{-2}), an optimized electrode composition (high active material content; low contents of binder/conductive carbon), a reduced electrolyte amount per cell capacity, as well as practically relevant testing conditions. As a result, this multi-requirement optimization will directly have a (negative) impact on different performance parameters such as cycling stability, rate capability, or safety (Figure 11.12).

The practical application of Si-based negative electrodes in commercial cell formats is not only hindered by ongoing active lithium losses, as discussed in Section 11.3.2. Other challenges particularly include the continuous consumption of electrolyte during operation, resulting in accumulation of

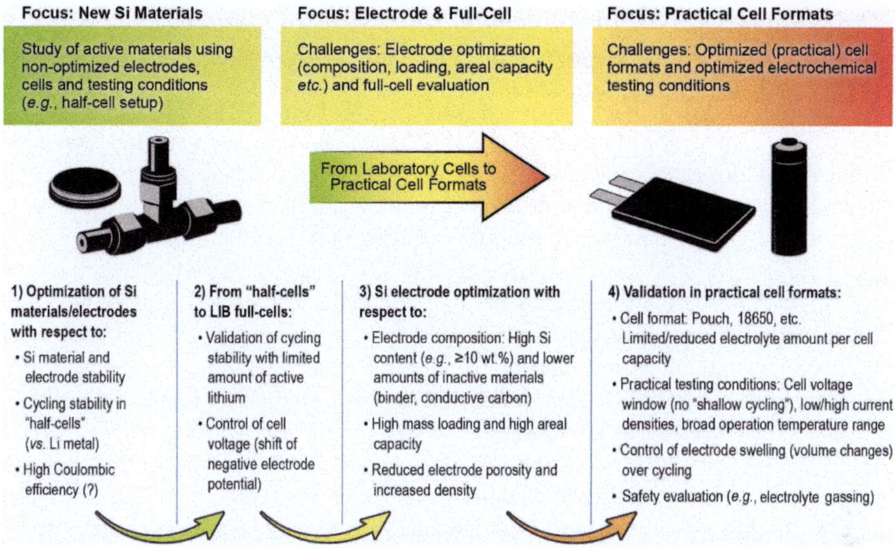

Figure 11.12 *Challenges and requirements for optimization of performance metrics from the material level (green) to the cell level (red) and from laboratory cells (left) to "practical," commercially relevant cell formats (right). The improvement of single- or few-metrics in laboratory cells can be achieved relatively easy, while the improvement of multi-requirements in "practical" cells is rather difficult. Various challenges in terms of optimized parameters at the electrode level, cell level and in terms of electrochemical testing conditions have to be fulfilled for "practical" cells.*

decomposition products within the porous electrode, that is, SEI thickening, as well as a significant "swelling" (= thickness increase) of the composite electrode structure [17, 103, 113, 128–130]. These effects might not be observed in lab scale experiments/cells, in which typically much high(er) amounts of electrolyte are used than in commercial cells [14].

In general, there are several parameters affecting the swelling of Si/C electrodes, such as the Si content in the electrode, the electrode composition and porosity, the N:P ratio as well as the state-of-charge (SOC) [13, 17]. Electrode swelling will have a negative impact on the volumetric energy density and can further result in severe performance and safety issues, for example, by deforming the separator, by blocking pores of the separator, or by squeezing out the electrolyte from the separator [17]. Various techniques have been reported for the analysis of the volume expansion of cell stacks, such as in situ pressure measurements for cylindrical cells and specialized in situ force-displacement cells [131, 132] as well as *operando* pressure measurements in pouch cells [113, 133]. The swelling behavior of Si-based electrodes has been particularly studied by ex situ cross-sectional SEM image studies [102, 103] and in situ electrochemical dilatometry [17, 134, 135].

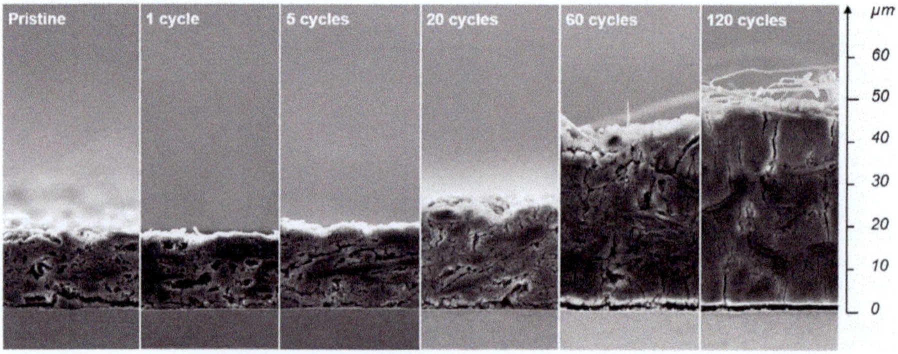

Figure 11.13 *Cross-sectional SEM images of Si/Gr electrodes in the pristine state and after different numbers of cycles, as reported by Wetjen* et al. *Reproduced from [103] with permission from The Electrochemical Society, Copyright 2018.*

Wetjen *et al.* reported on the quantification of lithium-containing electrolyte decomposition products in Si/Gr-based negative electrodes, combined with studies of the mass changes and coating thickness changes during charge/discharge cycling [103]. They studied an Si/Gr composite electrode (35 wt.% nano-Si, 45 wt.% Gr, 10 wt.% binder, 10 wt.% carbon fibers) and found a major electrode thickness and mass increase within the first 60 cycles, that is, from ≈19 μm to ≈49 μm and from ≈1.4 mg cm^{-2} to ≈3.9 mg cm^{-2}, respectively (Figure 11.13). Furthermore, they found that the porosity remained nearly unchanged during cycling. In this respect, the mass increase is due to the ongoing SEI growth and accumulation of decomposition products within the composite electrode (Figure 11.14).

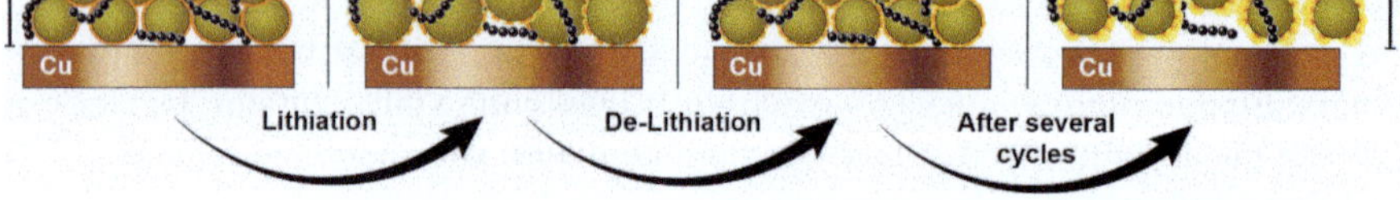

Figure 11.14 *Repeated volume expansion and contraction of Si particles (a–c) during charge/discharge cycling causes SEI damage and (re-) formation, resulting in a continuous thickening of the SEI layer and, thus, an irreversible volume expansion of the entire electrode (d).*

Louli *et al.* reported *operando* pressure measurements to study the reversible and irreversible volume expansion of pouch cells based on Gr and Si-/C-based negative electrodes [113]. They observed irreversible pressure growth with ongoing cycling and found an inverse correlation to the capacity retention and polarization growth, attributed to SEI growth at the negative electrode (Figure 11.14). In another work, Louli *et al.* reported an asymmetric nature of the volume expansion profiles of LIB cells with Si-based negative electrodes, which is independent of the type of Si [133]. Furthermore, it is known that aggregated Si contributes even stronger to the electrode swelling/expansion than homogeneously dispersed Si particles [13]. Müller *et al.* reported on the application of mechanical pressure on NMC811 || Si/C pouch cells and the impact of fixed and flexible compression (0.08–0.84 MPa) on the cycle life, which was evaluated by in situ analysis of the pressure distribution [136]. They found an improved cycling stability for a fixed compression at low pressure (0.08 MPa) and flexible compression at medium pressure (0.42 MPa). De Sutter *et al.* published a comprehensive study on the aging of volumetric constrained LIB pouch cells using Si alloy anodes at various cycling conditions and give fundamental insights on the impact of different stress factors on the cycle life [137]. Recently, Willenberg *et al.* presented a novel method for monitoring the volume expansion of LIB cells using strain gauges [138]. In their study, a commercial cylindrical 18 650 cell, having a small fraction of Si in the negative electrode, was evaluated. They observed irreversible diameter changes as well as reversible diameter changes over cycling, which could be correlated to an increase of the negative electrode thickness [138]. Overall, swelling of Si-based negative electrodes plays a critical role for the application in commercial cells, thus, further material optimization and electrode/cell engineering as well as suitable analysis techniques are needed to elaborate the practicality of Si-based LIB cells. Furthermore, Bommier *et al.* presented an *operando* acoustic technique, allowing the monitoring of SEI formation for NMC || Si/C LIB full-cells, providing fundamental insights into the capacity fading mechanism and protection of Si particles over cycling [79].

Another major challenge for the commercial application of Si-based electrodes is related to the consumption of electrolyte during charge/discharge cycling. Most of the studies reported in literature focus on lab-scale cells in which a large excess of electrolyte is used, while the electrolyte volume is strongly reduced for commercial cells, for example, fixed by a electrolyte volume to electrode pore volume ratio of ≈2.5–3.0, as reported by Chae *et al.* [17] or by the electrolyte mass per cell capacity, as reported by Obrovac and Chevrier [9] and Liu *et al.* [86]. In this respect, an electrolyte to capacity ratio of <3 g Ah^{-1}, most likely in the range of 1.3–1.5 g Ah^{-1}, has been reported for high-energy LIB cells, while laboratory cells typically have a large excess of electrolyte, that is, >13 g Ah^{-1}, which in turn can significantly increase the cell's cycle life [9, 86].

Electrolytes containing certain additives (typically <10 wt.%) have proven beneficial for the capacity retention of LIB cells with Si-containing negative electrodes (see Section 11.4.4) [3, 89, 139–141]. In this respect, fluoroethylene carbonate (FEC) has become the most common electrolyte additive utilized with Si-based electrodes and in combination with state-of-the-art carbonate solvents [141]. FEC has been shown to significantly improve the capacity retention of Si-based LIB cells, as a result of an effective SEI formation and reduction of active lithium losses [140, 141]. However, it was also shown that the ongoing SEI (re-)formation and growth also results in continuous consumption of the electrolyte (additives), which can lead to a rapid cell failure when the electrolyte or additive is depleted [10, 113, 141, 142]. For example, Jung *et al.* reported the sudden cell failure of Si-based cells after complete FEC consumption [141]. They found that in the presence of FEC, FEC gets almost exclusively reduced, while the reduction of other electrolyte components such as ethylene carbonate is prevented to a large extent. Therefore, future works should focus on the optimization and development of novel electrolyte chemistries including new additives or combinations to form an effective SEI layer for Si-based negative electrodes.

11.4 Strategies for the development of advanced Si-based lithium-ion full-cells

To actualize the practical implementation of Si-based anodes in commercial LIB cells, numerous strategies at the material level, electrode level, and cell level have been extensively investigated [3, 17, 36]. Among various approaches, an optimized Si material design offering a high tap density and low specific surface area, an advanced electrode formulation (*e.g.*, by novel binder concepts) [3], an improved electrolyte formulation for effective SEI formation [3], suitable pre-treatment and/or pre-lithiation concepts [143], etc. are among the most promising strategies in terms of practicality. Thus, a summarized account of the major improvement strategies is provided as below.

11.4.1 Material concepts toward practical usage of Si-based anodes

11.4.1.1 Silicon/graphite blends as anode material

As detailed in the introduction section of this chapter, the co-utilization of Si and Gr can be regarded as the most viable and near-term strategy of integrating Si into practical LIB cells [13, 122, 144]. Si/Gr electrodes can be designed either by blending of the two anodes or creating Si/Gr composite, in which both requires the optimization of the Si/Gr ratio and fine distribution of Si in the whole electrode. In this regard, the ratio of Si to Gr dictates the obtainable specific capacity, and thus incorporation of Si can improve the volumetric energy density by reducing the use of active materials and thus electrode thickness while still keeping equivalent electrode densities. Owing to the large volume variation, an improvement in volumetric energy density of pure Si is considered as theoretically almost impossible. However, increase in capacity with the addition of Si could result in severe electrode swelling. Moreover, due to the different (de-)lithiation potentials of Si and Gr [36, 57], the capacity ratio of Si to Gr determines the actual potential of the blended electrode. This shows that both proper material and cell design of Si/Gr blends and/or composites is necessary to achieve high volumetric capacities in commercial LIBs. For the same ratio of Si to Gr, the electrode swelling heavily depends on the distribution of Si. An uneven distribution will create an intensified volume expansion leading to electrode swelling and decay in life cycle.

11.4.1.2 Major recent results on Si-based LIB full-cells

As already briefly discussed above, in comparison to the large number publications for electrochemical studies of Si, there is only a minor number of publications available in which Si-based negative electrode materials are studied in a full-cell setup and/or in practical cell formats, such as pouch or cylindrical cells. A brief summary of full-cell data (among others) is given in Table 11.1.

A fair comparison of Si-based full-cells seems to be very difficult, which is related to the fact that nearly all publications display differences in the used cell setup, cell chemistry (type of Si-based anode, cathode material, etc.), electrode composition (Si amount vs. inactive material content), mass loadings and areal capacities, N:P capacity balancing, operating conditions, electrolyte composition and amount, etc. (Table 11.1). In addition, some Si-based anodes were pre-lithiated before usage in full-cells to compensate the high active lithium losses (see Section 11.4.3). Furthermore, on a critical note, it has to be mentioned that various performance metrics (*e.g.,* areal capacities, capacity balancing, Coulombic efficiency data) are not sufficiently discussed or even not mentioned in several publications. Unfortunately, it is often not clearly stated how specific energies and energy densities are calculated, for example, if the calculation only considers the cathode capacity or both anode and cathode capacities, and which inactive materials are included in the calculation [31]. Therefore, a satisfactory and fair evaluation cannot be guaranteed, and we strongly recommend researchers to improve the reporting of experimental parameters to allow their own comparison of performance data.

11.4.2 Electrode formulation and binder development

In general, the electrode composition (active material(s):binder:conductive additive(s)) needs to be adopted to the total amount of Si and type of Si species (Si, SiO_x, Si alloy, Si/C composite, etc.) used in the negative electrode, as well as with respect to the targeted mass loading and areal capacity to fulfill the requirements in terms of mechanical and electrochemical stabilities.

Polymeric binders play key role in maintaining the physical integrity among Si particles and between Si particles and the metallic current collectors, both during electrode fabrication and throughout the cycling process [161, 162]. Owing to the above-mentioned drastic conditions associated with Si anodes, that is, the dynamic SEI and huge volume changes, the required quality of polymeric binders for Si anode is different from that of Gr [3]. Moreover, contrary to the initial consensus that polymeric binders forming strong bonding lead to better mechanical stability of electrodes, supramolecular interactions of intermediate bond strength are turned out to be more convenient for the long-term cycle life of Si anode-based LIB cells (Figure 11.15b), compared to covalently cross-linked (Figure 11.15c) or simple van der Waals forces (Figure 11.15a). This results from the fact that supramolecular

Table 11.1 *Overview of a variety of reported silicon full-cell data, including cell format, electrode materials, electrolyte formulation, operating conditions, capacity and energy density, capacity retention, and Coulombic efficiency*

	Electrode materials							
Cell format	**Anode (areal capacity)**	**Cathode**	**Electrolyte formulation**	**Operating conditions (cell voltage; N:P ratio; rate or current; temperature)**	**Capacity/Specific energy**	**Capacity retention (%)**	**Coloumbic efficiency**	**Ref.**
Swagelok-type T cell	Fe_xSi_y/C materials (≈2.7 mAh cm^{-2})	NMC-111	1M $LiPF_6$ in EC:EMC 3:7 (w:w) + 5 wt.% FEC	2.5–4.3 V; N:P = 1.15:1; 0.1C (3 cycles), 0.5C (charge and discharge) + CV step at 4.3 V	≈150 mAh g^{-1} (cathode); ≈400 mWh g^{-1} (anode + cathode)	77%@150 cycles	≈81% (1st cycle) ≈99.8% (100th cycle)	[35]
Coin cell (CR2032)	Si microparticles (D50 ≈ 4.5 µm)	NMC-622	1M $LiPF_6$ in FEC:DEC 1:4 (v:v)	3.0–4.2 V; 0.1C (1st cycle), 0.5C (cycling) + CV step during charge	≈2 mAh cm^{-2}	80%@200 cycles (25-35% lithiation degree)	≈90% (1st cycle; ≈25-60% lithiation degree)	[47]
Wound pouch cell (800 mAh)	Si alloy (17.5% Si) (≈4-4.5 mAh cm^{-2})	NMC-622	1 M $LiPF_6$ in EC:DEC 1:1 (w:w) + 5 wt.% (VC/FEC/TEOSCN (1.67:1.67:1.67 w:w:w)	2.5–4.2V; 1C charge + CV step, 0.5C discharge; 45 °C	≈800 mAh	58% @417 cycles	99.99% after 417 cycles	[54]
Single-layer pouch cell	Nano-Si/Gr (≈2.2 mAh cm^{-2})	NMC-532	1.2M LiFSI in EC:EMC 3:7 (w:w) + 10 wt.% VC	3.0–4.1 V; 0.5C (formation after tap charge) C/3 (cycling)	≈140 mAh g^{-1} (cathode)	≈60% @100 cycles	99.0% (at C/20, after 100 cycles)	[109]
Pouch cell	SiO_x/Gr	NCA	1.2M $LiPF_6$ in EC:EMC:FEC 3:6:1 (w:w:w) + 1 wt.% LiDFOB	2.75–4.25 V; N:P = 0.8:1 to 1.8:1; cycling: 0.5C + CV step (charge), 0.5C (discharge)	≈180 mAh g^{-1} (cathode)	≈80% @500 cycles (N:P = 1.03:1)	>80% (1st cycle; N:P = 1:1 to 1.1:1	[112]
Multilayered pouch cell	Columnar Si thin film (≈3.3 mAh cm^{-2})	NMC-622 or NMC-811	1M $LiPF_6$ in EC:DMC 1:1 (v:v) + 10 wt.% FEC	3.0–4.3 V; N:P ≈1.4:1 to 1.65:1; 0.1C (2x), 0.5C (2x) + CV step, 0.5C (charge and discharge)	≈180 mAh g^{-1} (NMC-622); ≈210 mAh g^{-1} (NMC-811); ≈135 Wh kg^{-1} (NMC-622); ≈183 Wh kg^{-1} (NMC-811);	70%@50 cycles (NMC-622); 67%@50 cycles (NMC-622)	≈92-94% (1st cycle); ≈99.2-99.4% (cycling)	[114, 116]
21 700 cell, pouch cell	Li-doped SiO-C (≈6 mAh cm^{-2})	NMC-532, NMC-622, NMC-811	1M $LiPF_6$ in EC:DMC:EMC 1:1:1 (v:v:v) + 5 wt.% FEC	3.0–4.2 V; 0.2C + CV step	≈173 mAh g^{-1} (NMC-622 cathode, pouch cell); up to 265 Wh kg^{-1} (21700 cell)	≈90%@300 cycles	≈89% (1st cycle)	[122]
Stacked pouch cell (1.22 Ah)	Si alloy (≈3.7 mAh cm^{-2})	NMC-532	1M $LiPF_6$ in EC:EMC 3:7 + 10 wt.% FEC	3.0–4.35 V; N:P ≈ 1.28:1; charge: CC-CV 0.25-0.05, discharge: 0.25	≈1.22 Ah (stacked pouch cell); ≈4.2 Wh (cell energy)	≈80%@290 cycles	≈75% (1st cycle); ≈99.88% (from 4th cycle)	[129]

(Continues)

Table 11.1 Continued

	Electrode materials							
Cell format	Anode (areal capacity)	Cathode	Electrolyte formulation	Operating conditions (cell voltage; N:P ratio; rate or current; temperature)	Capacity/Specific energy	Capacity retention (%)	Coloumbic efficiency	Ref.
Pouch cell	a)SiO/Gr b) Si alloy/ Gr c) Nano Si/C	a)NCA b) LCO c) NCA	a)1.2M $LiPF_6$ in EC:EMC: DMC 25:5:70 (v:v:v) + 2 wt.% VC; b, c) 1.2M $LiPF_6$ in EC: EMC:FEC 27:63:10 (v:v:v)	3.0–4.2 V; C/6 (cycling), 40 °C	a)260 mAh (4.2 V) b) 230 mAh (4.35 V) c) 165 mAh (4.4 V)	a)N/A b) ≈86% @90 cycles c) ≈93% @90 cycles	N/A	[133]
Stacked pouch cell	Si/C composite	NMC-811	1M $LiPF_6$ in EC:DEC 3:7 + 10 wt.% FEC	2.7–4.2 V; 0.1C (formation), cycling: C/3 + CV step (charge), 1C (discharge)	≈160 mAh g^{-1} (active material)	≈45% @100 cycles	≈84% (1st cycle)	[136]
Pouch cell (1360 mAh)	Si alloy/Gr	NMC-622	1M $LiPF_6$ + 0.1M LEA in EC:EMC 3:7 (v:v) + 10 wt.% FEC	2.7–4.4 V; C/3 + CV step (charge), C/3 (discharge); various DoD and temperatures	≈1360 mAh (cell)	Various results dependent on applied pressure, rate, DoD, temperature, etc.	N/A	[137]
Pouch cell (200 mAh)	Si alloy/Gr (≈3.0 mAh cm^{-2})	LCO	1M $LiPF_6$ in EC:EMC:FEC 27:63:10 (w:w:w) + 2 wt.% additive	3.0–4.35 V; N:P ≈ 1.24:1; 0.5C + CV step at end of charge (cycling)	≈200 mAh (cell)	≈80%@300 cycles	N/A	[142]
Cylindrical cells	Si alloy (≈4.3 mAh cm^{-2}; pre-lithiated)	NMC-622	0.95M $LiPF_6$ + 0.05M LiBOB in FEC:EMC:DMC:PC 3:3:3:1 (w:w:w:w) + 2 wt.% VC	2.75–4.2 V; N:P (safety factor) ≈1.3:1; 0.05C (1st cycle), 0.5C + CV step (cycling)	≈1850 mAh (cell)	≈80%@500 cycles	≈99.9% to 99.95%	[145]
Coin cell (CR2025)	CNTs-Si-CNTs (pre-lithiated)	$LiFePO_4$	1M $LiPF_6$ in EC:EMC:DMC 1:1:1 (v:v:v)	2.5–3.7 V; N:P = 6:1; 0.1C	139 mAh g^{-1}; 374 Wh kg^{-1}	92%@200 cycles	N/A	[146]
Coin cell (CR2032)	Si@BGra/Ni (≈4.5 mAh cm^{-2})	$LiFePO_4$	1M $LiPF_6$ in EC: DMC 1:1 (v:v)	2.0–3.6 V; N:P = 1.15:1; 0.3 and 1.5 mA cm^{-2}	3.5 mAh cm^{-2}	90%@260 cycles	N/A	[147]
Coin cell (CR2016)	Microscale porous Si (pre-lithiated) (≈5.1 mAh cm^{-2})	NMC-111	1M $LiPF_6$ in EC:DEC 1:1 + 6 wt.% VC or FEC	2.8–4.25V; N:P = 1.1:1; 0.5C	≈150 mAh g^{-1} (cathode); 502 Wh kg^{-1}	84%@400 cycles	N/A	[148]
Wound pouch cell	Si alloy	$LiCoO_2$, NMC-622, and blends	1M $LiPF_6$ in EC:EMC 3:7 (w:w) + additives (FEC, DEPC, CO_2)	3.0–4.35 V; C/11	≈220-250 mAh (cell)	e.g., 90%@250 cycles	N/A	[149]
Pouch cell	MGS (≈3.8 mAh cm^{-2})	$LiCoO_2$	1.3M $LiPF_6$ in EC:EMC:DEC 3:5:2 (v:v:v) + 10 wt.% FEC + various other additives	2.7–4.35 V; N:P = 1.1:1; 1C discharge/0.5C charge	494 mAh cm^{-3}; 1826 Wh L^{-1} after 100 cycles	N/A	N/A	[150]

(Continues)

Table 11.1 Continued

	Electrode materials							
Cell format	Anode (areal capacity)	Cathode	Electrolyte formulation	Operating conditions (cell voltage; N:P ratio; rate or current; temperature)	Capacity/Specific energy	Capacity retention (%)	Coloumbic efficiency	Ref.
Coin cell (CR2016)	Nano-Si/Gr blend (≈1.47 mAh cm^{-2})	NMC-532	1M $LiPF_6$ in FEC:DEC 1:1 (v:v) + 5 wt.% FEMC + 3wt.% VC	3.0–4.55 V; N:P ≈ 1.15:1; C/6 (formation cycles), C/3 (cycling)	173 mAh g^{-1} (cathode); 622 Wh kg^{-1} (based on cathode)	71% @100 cycles	75% (1st cycle)	[151]
Coin cell (CR2032)	C/SiNW and C/SiNW/Li (pre-lithated)	$LiNi_{0.5}Mn_{1.5}O_4$	1M $LiPF_6$ in EC:DEC 1:1 (v:v)	3.5–4.9V; 1C and 5C (each charge and discharge)	≈80 mAh cm^{-2} (cathode)	≈75% @500 cycles (1C); ≈65% @2000 cycles (5C)	98% after 1st activation	[152]
Pouch cell	Si-graphene composite (≈2 mAh cm^{-2})	NCA	1M $LiPF_6$ in EC:DEC 1:1 (w:w) + 10 wt.% FEC	2.75–4.2 V; N:P ≈1:1; C/15 (5 cycles), 0.2C (5 cycles) 0.5C (40 cycles)	≈133 mAh g^{-1} (cathode)	≈80% @50 cycles	98.9-99.5% after ten cycles	[153]
Coin cell (CR2016)	Si/Gr blend (pre-lithiated)	NCA	1M $LiPF_6$ in EC:PC:DEC 1:1:3 (w:w:w) + 1 wt.% VC + 5 wt.% FEC	2.2–4.4 V; 0.5C (cycling)	≈150 mAh g^{-1} (cathode)	≈80% @245 cycles	≈99% (cycling)	[154]
Reference electrode cell setup	Si/Gr blend (pre-lithiated: ≈45%)	NMC-532	1.2M $LiPF_6$ in EC:EMC 3:7 (w:w) + 10 wt.% FEC	3.1–4.1 V; N:P ≈ 1.75:1 (non-pre-lithiated); 5 x [0.05C (3x), 0.33C (94x), 0.05C (3x)]	≈160 mAh g^{-1} (cathode)	≈70% @500 cycles	≈99.5-99.8%	[155]
Coin cell (CR2016)	Si nanoparticles (pre-lithiated)	NMC-532	1M $LiPF_6$ in EC:EMC 3:7 (v:v) + 10% FEC	3.0–4.2 V; N:P = 1.6; C/3 (cycling)	N/A	90% @20 cycles; 72% @200 cycles	99.7% (average) after 240 cycles	[156]
Coin cell (CR2032)	Si/C composite	NMC-532	1M $LiPF_6$ in EC:DMC:FEC 45:45:10 (v:v:v)	2.8–4.2 V; 14 mA g^{-1} (1st cycle), 47 mA g^{-1} (charge), 70 mA g^{-1} (discharge)	≈50 mAh (cell)	72% @100 cycles	N/A	[157]
Coin cell + reference electrode	Si/Gr	NMC-532	1.11M $LiPF_6$ in EC:EMC 3:7 (w:w) + 10 wt.% FEC (or + 10 wt.% VC)	3.0–4.2 V; N:P = 1.39:1; ≈C/3 (cycling); 30 °C	≈1.5-1.6 mAh cm^{-2}	≈80% @20 cycles	N/A	[158]
Coin cell (CR2032)	Si/C composite	$Li_{1+x}Ni_{0.5}Mn_{1.5}O_4$	1M $LiPF_6$ in EC:DMC 1:1 (w:w)	3.0–4.78 V	≈130 mAh g^{-1} (cathode)	≈75% @100 cycles	>99.7% (cycling)	[159]
Three-electrode cell	Si/C composite	$Li_{1+x}Ni_{0.5}Mn_{1.5}O_4$	1M $LiPF_6$ in EC:EMC 3:7 (w:w) + 5 wt.% FEC	2.4–4.95 V; N:P ≈ 1.3:1 to 1.8:1; ≈0.1C (3 cycles), ≈0.5C (cycling)	≈134 mAh g^{-1} (cathode)	≈60% @103 cycles	≈98.0-98.5 (at 0.1C after 106 cycles)	[160]

BGra: bilayer graphene;CNTs: carbon nanotubes; C/SiNW: Si nanowires arrays-covered carbon cloth; CV: constant voltage; DCS-Si: double carbon shells-coated Si nano-particles; DoD: depth-of-discharge; FEC: fluoroethylene carbonate; FEMC: methyl (2,2,2-trifluoroethyl) carbonate; LEA: lifetime electrolyte Arkema [137]; MGS: macropore-coordinated graphite-Si; TEOSCN: (2-cyanoethyl)triethoxysilane; VC: vinylene carbonate.

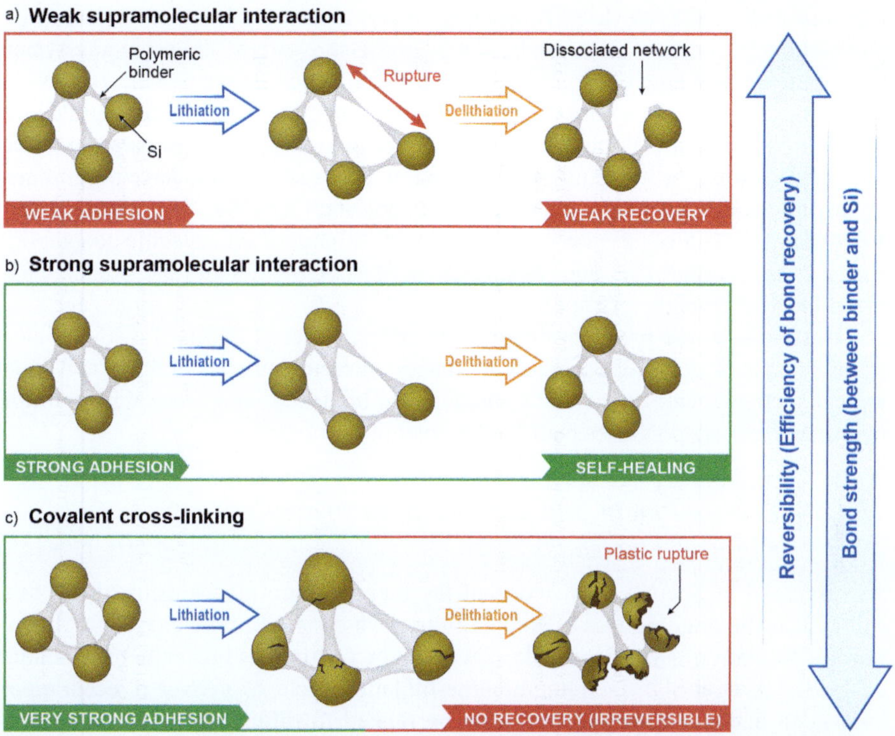

Figure 11.15 *Classification of (a) and (b) strong supramolecular interactions and (c) chemical bonds, as well as their adhesion behaviors for Si-based anodes.*

interactions can be reversibly broken and re-formed (a phenomenon called self-healing) under an applied mechanical stress. Hence, binders endowed with abundant —O— or —N— and highly polarized moieties such as —COOH, —OH, and —NH_2 — are known to provide superior binding properties, characterized by an improved performance of Si-embedded negative electrodes. Such functional groups which bestow abundantly moderately strong, but elastic (reversible/self-healing) supramolecular interactions such as hydrogen bonding, ion–ion, ion–dipole, dipole–dipole, π–π, cation–π, anion–π and dative bonds are of utmost importance for the practical realization of highly reactive/high capacity anodes based on Si [99, 163, 164].

A comprehensive overview on the different classes of polymeric binders used for Si-based anodes has been reported by Eshetu and Figgemeier [3]. In particular, it has been shown that the use of binders having a high density of carboxylic acid moieties can improve the long-term cycling performance of Si-based anodes. In this respect, polyacrylic acid (PAA) binders possess more carboxylic groups than classical carboxymethyl cellulose (CMC) binders, resulting in a higher amount of grafting sites on the Si surface through strong and reversible hydrogen bonds, which in turn can improve the cycling performance [3]. Among other binder systems, PAA or lithium substituted PAA (LiPAA) alone or in combination with other binders (polymer blending), such as CMC, can be considered as the state-of-the-art binder system up to date, showing the most promising results for Si-based negative electrodes [36, 129, 145, 165–168]. In addition, it must be mentioned that not only the right choice of the binder system is important but also an optimized electrode preparation procedure, for example, by adjusting the pH value, which was shown to have a significant impact on the cycling performance and electrode stability [168, 169]. In the case of Si/Gr blends, it is also noteworthy to use blends of binders with specific adaptability both to Si and Gr particles, such as LiPAA for Si and CMC for Gr (and LiPAA/CMC blend for Si/Gr blend electrode).

In summary, the practical deployment of high-energy Si-based negative electrodes demands polymeric binders of tuned supramolecular chemistry and topologically cross-linked, conductive, moderately branched, and those with multiple hydrogen bonding polymers could be of potential candidates.

11.4.3 Pre-treatment and pre-lithiation strategies for Si-based negative electrodes

As discussed in Section 11.3.2, one of the major failure mechanisms of Si-based full-cells is the ongoing loss of active lithium, which is primarily caused by the initial SEI formation in the first cycle as well as the continuous breakage/re-formation of the SEI layer at Si in each lithiation/de-lithiation cycle. Moreover, a recent report evidenced that lithium losses can also be related to diffusion-controlled Li trapping in Si anodes, which in turn results in an accelerated capacity decay during subsequent cycling [170], as has been previously reported for Sn-based negative electrodes [22, 34]. In general, the loss of active lithium results in an irreversible decrease of the available LIB cell energy content by consuming (= immobilizing) the lithium supplied by the cathode-active material [38]. Therefore, the compensation of the initial irreversible capacity within the first cycle seems to be a viable approach to improve the electrochemical performance of Si-based full-cells in terms of energy density and cycling stability [143].

In this regard, the pre-treatment or "pre-lithiation" of the Si-based negative electrode is considered as an appealing strategy of increasing significance for building practical LIB cells of high energy density and long cycle life [122, 143, 145, 155, 171]. In general, several different synonyms are used in literature to describe the pre-lithiation approach, such as "pre-doping (of lithium ions)," "pre-formation," "pre-conditioning," "pre-alkaliation," "mitigating lithium losses," etc. Up to date, more than 200 scientific publications and more than 1000 patents have been published on pre-lithiation for various types of materials, components, or electrochemical storage devices. Furthermore, pre-lithiation is already commercially used for lithium-ion capacitors (LICs), to significantly increase the LIC's cell voltage by a potential drop of the negative electrode [143].

The consequences and benefits of pre-lithiation on the electrochemical performance of Si-based full-cells in terms of an energy gain have been quantitatively explained by the work of Chevrier *et al.* [145] using the cell model of Obrovac and

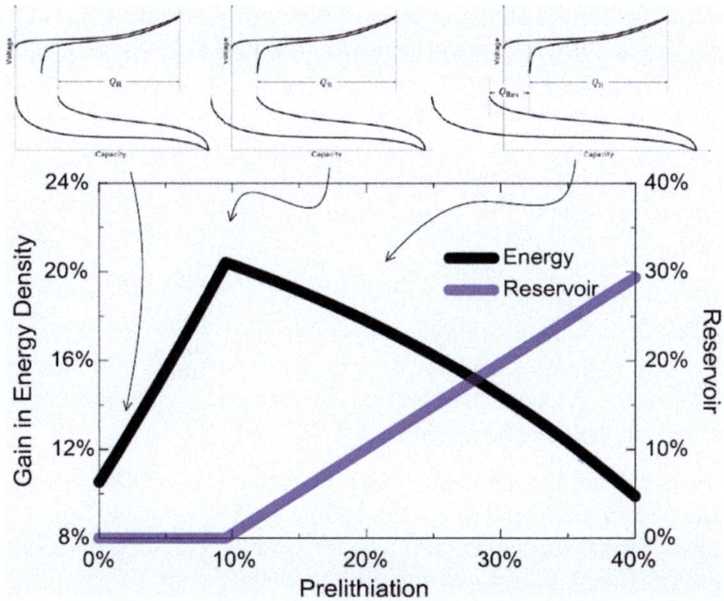

Figure 11.16 *Impact of the pre-lithiation degree for a LIB full-cell stack containing a 5:3 Si alloy:Gr negative electrode on the gain in energy density (relative to a reference Gr-based full-cell stack) and on the creation of a lithium reservoir (relative to the first lithiation capacity of the negative electrode). Insets show schematic potential curves. The calculations are based on the first charge/discharge cycle. Reproduced from [145] with permission from The Electrochemical Society, Copyright 2018.*

Chevrier [9], which was extended to include pre-lithiation. In general, pre-lithiation cannot only be used to compensate the active lithium losses in the first cycle, which directly results in an increase of the LIB cell's energy density (Figure 11.16). In addition, pre-lithiation can also be used to create a lithium reservoir in the negative electrode, which can be used to gradually release lithium for further compensation of active lithium losses due to the ongoing SEI re-formation. However, the creation of such a lithium reservoir will result in a decrease of the cell energy density (Figure 11.16), as the thickness/capacity of the negative electrode must be increased to achieve a safe N:P ratio to avoid Li metal plating at the negative electrode, while this safe behavior is considered as happening for an N:P ratio for a non-pre-lithiated cell [145]. For example, a reversible lithium reservoir of ≈20% can be achieved by pre-lithiation of ≈30%, when assuming that ≈10% of active lithium losses take place during the first charge [145].

In general, there are various strategies and methods for pre-lithiation of the Si negative electrode, which include several physical (contact), chemical, or electrochemical routes, that can either be pursued before cell assembly (*e.g.*, on Si material or electrode level) or by in situ approaches. Following approaches have been reported in literature and can be briefly categorized as follows [143]:

- Pre-lithiation with the help of the positive electrode
  - Positive electrode additives (in situ pre-lithiation): Electrode additives, such as Li_3N or Li_2O_2, have to be introduced in the positive electrode processing and should be able to release lithium in a suitable potential range in the first charge cycle [172, 173].
  - Overlithiated positive electrode materials (in situ pre-lithiation): One example is the high-voltage spinel $LiNi_{0.5}Mn_{1.5}O_4$, which can be overlithiated by (electro-)chemical routes and can be used for pre-lithiation of the negative electrode [160, 174].
- Direct pre-lithiation of the negative electrode
  - Chemical pre-lithiation by active reactants (before cell assembly or in situ): Examples are the use of organometallic compounds (*e.g.*, n-BuLi) or lithium-organic complex solutions for pre-lithiation of Si materials or electrodes [175, 176]. An example for the in situ pre-lithiation of the negative electrode within the LIB cell includes the use of sacrificial electrolyte salts (pre-lithiation solutions) [177].
  - Use of lithiated active materials as negative electrode additives (before cell assembly): Examples are the use of (partially) lithiated powder materials (*e.g.*, Li_xSi, Li_xSn, Li_xGe) or lithiated polymers, which need to be incorporated in the negative electrode processing [178, 179].
  - Pre-lithiation by direct contact to Li metal (before cell assembly or in situ): Various approaches have been presented for pre-lithiation by use of Li metal, including the use of Li metal foils, Li metal powder (such as "stabilized lithium metal powder" (SLMP)), as well as sputtered or evaporated Li thin films [180–182]. Pre-lithiation of Si negative electrodes can either be performed before cell assembly or by an in situ pre-lithiation during operation by including an auxiliary Li metal electrode within the LIB cell (Li metal reservoir) [183].
  - Electrochemical pre-lithiation (before cell assembly): A suitable approach, besides lab-scale methods, is the pre-lithiation *via* a continuous roll-to-roll process (electrochemical bath) [184–186]. Pre-lithiation of the negative electrode can be either achieved by a Li metal electrode (external short circuit) or by an electrolysis process, in which low-cost Li salts (*e.g.*, LiCl) can be used as precursor, resulting in pre-lithiation of the negative electrode and gas evolution at the counter electrode.

From all of the above introduced pre-lithiation approaches, only some are potentially suitable for industrial manufacturing processes, while many approaches suffer from a high reactivity, poor stability, and/or high safety issues and would require changes of the production process (*e.g.*, use of different solvents or binders), which particularly includes the use of Li metal. The pre-lithiation by a continuous roll-to-roll process (electrochemical bath) seems to be one of the most promising approaches, that is, due to the avoidance of reactive salts, the use of low-cost Li salts, and most likely a homogeneous pre-lithiation. However, up to date, the implementation into the manufacturing process and the cost-competitiveness needs to be verified.

The positive impact of pre-lithiation on the electrochemical performance of Si-based LIB cells, in terms of an improved cycling stability and energy density, has already been evaluated in several publications [145, 155, 160, 187, 188]. Chevrier *et al.* concluded that the improved performance of highly pre-lithiated cells (with Li reservoir) is not only related to the compensation of active lithium losses, but also due to a lower "capacity swing" of the negative electrode (*i.e.*, lithiation/delithiation at a lower and more constant electrode potential) [145]. Furthermore, they assumed that parasitic reactions will be reduced in the presence of a Li reservoir, which could be validated by a significantly enhanced C_{Eff} for the pre-lithiated Si electrodes in cylindrical cells. Recently, Rodrigues *et al.* reported that the relative contribution of Gr (of a Si/Gr negative electrode) to the reversible capacity increases for high levels of pre-lithiation, which in turn results in an enhanced cycling stability [155]. They also observed a higher C_{Eff} for the pre-lithiated cells compared to non-pre-lithiated cells, even after consumption of the Li reservoir. Besides these aspects, pre-lithiation also provides a pre-volume expansion, reducing the excessive volume change of Si during upcoming cycling [189]. This reduces the extent of cracking and pulverization of Si, yielding to improved cycling performance. Moreover, pre-lithiation is reported to decrease the elastic and shear moduli of the electrodes, and increase of the diffusion coefficient of lithium. This, on the other hand, reduces the strain compared to the non-pre-lithiated Si and thus boosting the cycling performance of Si-based LIBs [190].

In summary, it must also be mentioned that, regardless of the method used, pre-lithiation needs to be regulated with a high exactitude. This is because over-lithiation would lead to an excess of lithium, resulting in Li metal plating on the top of the negative electrode surface (if the N:P ratio is not adjusted accordingly) and, subsequently, leading to safety-induced hazards during operation.

11.4.4 Electrolyte development

The performance of Si-based full-cells strongly depends on the electrolyte system, notably the salt anions, and additives. In this chapter, state-of-the-art carbonate-based electrolytes are predominantly considered, even though there is also immense research on alternative (co-)solvent concepts for Si-based anodes [191, 192]. In general, the electrolyte system will have a major impact on the SEI formation and characteristics, which in turn determines the amount of electrolyte consumption and active lithium losses, as discussed in Section 11.3.3. As mentioned above (Section 11.3.1), it is recommended that electrolyte studies are solely performed in LIB full-cells, to avoid the impact of a Li metal electrode (half-cell setup) [37].

11.4.4.1 Salt anions

Tuning the nature of salt anions is one of the approaches that can be utilized to boost the performance of Si-based LIB cells. In general, tailored electrolyte salts serve as designer precursors for the formation of high quality interphases, that is, SEI and/or cathode electrolyte interphase (CEI) [193]. Unlike to Gr, the mechanical and chemical stability of the SEI in Si-based full-cells necessities to endure the now-and-then large volume variation during the (de-)alloying process [18]. Ideally, salt anions, which up on reduction and/or oxidation lead to the formation of sufficiently high shear strength inorganic compounds with satisfactory ionic conductivity, are required to facilitate the practical deployment of Si-anode based LIB cells. However, though of supreme importance, efforts dealing with the exploration of new salt anions are still limited. Moreover, it has to be emphasized that salt anions that fit to the state-of-the-art Gr may not necessarily work for Si and derivative anodes, and new systematic studies are needed. For instance, though still the most widely employed salt for Si-based LIB cells, $LiPF_6$ is not a perfect salt, particularly considering its thermal instability and high sensitivity to traces of moisture, water and alcohols, leading to the formation of aggressive species such as HF, POF_3, and others [194–198]. For instance, HF could etch the silicon oxide layer in Si-based electrodes, thus, modifying the favorable and decisive interaction between the Si active particles and polymeric binders, accompanied by capacity decay upon long-term cycling. In addition, the decomposition products of HF lead to transition metal dissolution at the cathode and these ions can migrate to the (Si) anode and deteriorate the cell performance [199, 200]. Thus, searching for alternative salt anions as replacement for $LiPF_6$ needs to be considered. Among others, imides (e.g., lithium bis(fluorosulfonyl)imide (LiFSI)) [117], imidazole (lithium 2-trifluoromethyl-4,5-dicyanoimidazole (LiTDI))-based salts, and their advanced modified forms are considered as potential candidates owing to their higher tendency to form LiF on the Si surface [201]. However, issues such as instability and poor passivation against the Al current collector [202, 203], instability with high-voltage cathodes, and the cost-effectiveness associated with imide-based salts are remaining challenges that need an in-depth evaluation. Edström *et al.* evidenced that while $LiPF_6$ modifies the surface of Si particles, LiFSI-based electrolytes were found to keep the intended reactions between the binder and Si particles, attributed to the absence of fluorination of the Si particles' surface upon long-term cycling (Figure 11.17). This is again related to less sensitivity of LiFSI toward protonic impurities [117].

11.4.4.2 Electrolyte additives

The incorporation of a small dosage (usually $\leq 10\%$ by wt. or vol.) of foreign molecules, called additives, into the base electrolyte is hailed as one of the most effective, economical, and industrially scalable approaches to circumvent the above-mentioned predicaments, associated with the Si anode [3, 54, 90]. Owing to their limited quantities, these additives are not expected to substantially alter the electrolyte bulk characteristics (*e.g.*, ionic conductivity, viscosity, wettability) or the production cost or the battery cell production process as a whole. Once again, though acquiring the accumulated learned lessons from

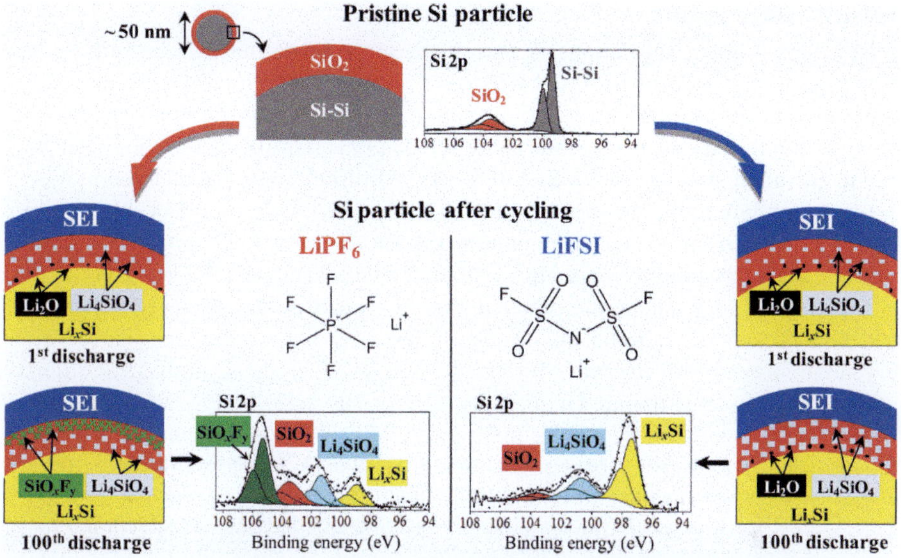

Figure 11.17 *Schematic comparison of the mechanisms occurring at the surface of a Si nanoparticle upon cycling of a Si || Li metal cell using either $LiPF_6$ or LiFSI salts. Reproduced from [117] with permission from the American Chemical Society, Copyright 2013.*

the Gr anode is crucial, additives that work best for Gr-based LIB cells may not necessarily conform to the requests for Si and/or Si-derivative anode-based LIB cells. Thus, a due attention is needed while designing molecular and/or ionic additives for Si-anode electrodes. On the anode compartment, the key function of additives could be summarized into two aspects: (i) formation of an effective protection layer (SEI) on the Si and/or derivative anode surface and thus dwindling the continuous (electro-)chemical-induced electrolyte decomposition reactions, and (ii) scavenging or transforming of the reactive and soluble species into more stable and relatively insoluble compounds. From molecular structure standpoint, molecules with electron deficient moieties (*e.g.*, –F, –C≡N, –O–C≡N, –N=C=O, –S–C≡N, –N=C=S, $-N^+{\equiv}C-$), reducible unsaturated multiple chemical bonds (*i.e.*, double and triple bonds), those leading to surface modification by forming cross-linked products (*e.g.*, silane, –S–O–) via chemical or physical means, ionic (salt-like, LiBOB, LiDFOB, LiFOB, etc.), and so forth are considered as effective filming additives [3, 90, 93]. To obviate the massive volume variation during (de-)alloying process, the mechanical properties of the SEI are crucial and, thus, entitled additives should serve as precursors for higher shear modules chemical species: LiF (55.1 GPa) > Li_2O (45.6 GPa) > Li_2CO_3 (28.9 GPa) > organic components [lower than 1 GPa, $(CH_2CH_2OCO_2Li)_2$ > $LiOCO_2C_2H_5$ > $LiOCO_2CH_3$ > $(-CH_2CH_2O-)_n$] [54].

As mentioned above (Section 11.3.3), fluoroethylene carbonate (FEC) is considered as the state-of-the-art electrolyte additive utilized with Si-based electrodes in combination with $LiPF_6$ and carbonate-based solvents [141]. FEC has proven to significantly improve

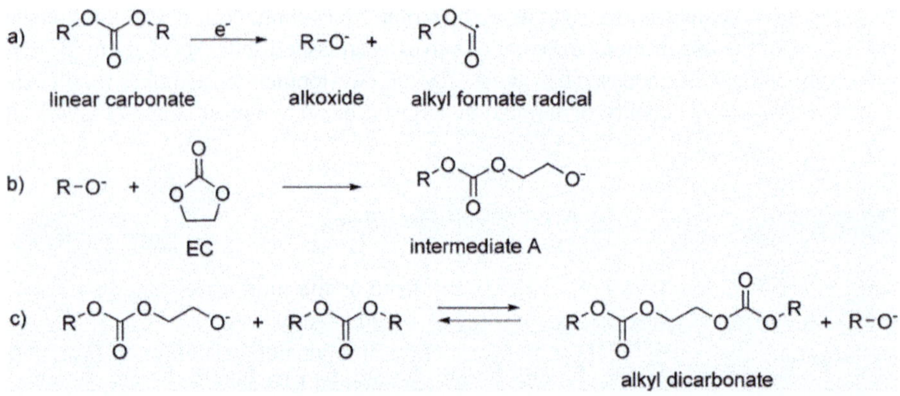

Figure 11.18 *Elementary steps for bis-carbonate degradation product generation in a linear carbonate and EC mixture-based electrolyte: (a) reduction of linear carbonates to reactive alkoxide species (RO⁻), (b) EC ring opening by alkoxide (RO⁻) leading to the formation of intermediate A, and (c) attack of intermediate A on the linear carbonate resulting in the formation of soluble alkyl dicarbonates [55].*

the capacity retention of Si-based LIB cells, which is a result of an effective SEI formation and reduction of active lithium losses [140, 141]. However, FEC is typically used in large amounts (*e.g.*, >10 wt.%), as it will be continuously consumed over charge/discharge cycling [141]. Furthermore, FEC is known to suffer from serious gassing issues and formation of thick interphases, especially at elevated temperatures and high cut-off voltages, which may result in impedance increase, as well as capacity fading and safety issues of Si-based full-cells [191, 204].

It is a well-established fact that the battery performance decay is also highly linked to the soluble compounds that are concurrently formed with the SEI. Initiated by alkoxides (RO^-Li^+), which in turn are generated from linear carbonates such as DMC or DEC, the intermediate soluble molecules may react continuously with the parent electrolyte solvents (*e.g.*, EC, DMC, DEC) as shown in Figure 11.18. Thus, halting or reducing the rate of formation of these reactive alkoxides is of supreme importance to mitigate the severe electrolyte degradation during the cycling process. Carbon dioxide (CO_2), incorporated, for example, in form of dry ice before sealing, as reported by Krause *et al.* [205] or *via* in situ generation, as proposed by Nölle *et al.* [93] is one example that transform the more soluble and reactive species into relatively less chemically reactive, and insoluble species. Thereby, Gasteiger *et al.* confirmed the positive impact of CO_2, which also evolves from reduction of VC and FEC, on the formation of Li_2CO_3 for an effective SEI on Gr anodes [206]. Furthermore, Nölle *et al.* presented a novel electrolyte additive with a synergistic functionality, *i.e.,* O-carboxyanhydride, which forms an effective SEI comprising poly(lactic)acid (PLA) and Li_2CO_3 by in situ releasing CO_2.

Existing realities attest that a blend of a variety of electrolyte additives is usually adopted in practical LIB cells. However, on the top of obtainable performance, the complexity, including issues such as compatibility, recycling, cost, and environmental impact, with the use of many different additives is an issue itself, especially in the case of highly reactive Si anodes. This entails that the development of additives that endow either multiple-functionality or tailored synergistic effect within a single-functionality compound are urgently needed [207].

11.5 Conclusion and future perspectives

Silicon (Si)-based materials are widely considered as the most appealing choice to be added to or even to replace state-of-the-art Gr-based negative electrode materials on the route toward high-energy LIB cells. Up to date, Si is already added in small amounts (≈3–8 wt.%) to Gr-based anodes within few commercial LIB cells. However, the incorporation of higher Si amounts, for example, ≥10–20 wt.%, is still hampered by the serious challenges in terms of cycle life. The outstandingly high interest in Si-based materials and cells is well reflected by massive research and patenting activities. Commercial LIB cells are still dominated by Gr-based electrodes and the penetration of Si toward such cells as well as toward higher Si contents progresses slowly, while Si/Gr blends seem to be the most promising material approach.

It has to be mentioned that very often superior performance is reported for novel Si material design approaches and concepts in the scientific battery community. However, these often "outstanding" results are typically achieved in laboratory cells and may not be directly translated into "practical" cells and commercial LIB cell formats. This is due to the fact that the conditions, under which such Si materials are evaluated, are usually not aligned with those for realistic "practical" applications (Figure 11.19). Further, the optimization of just a few of the key performance parameters seems to be relatively easy,

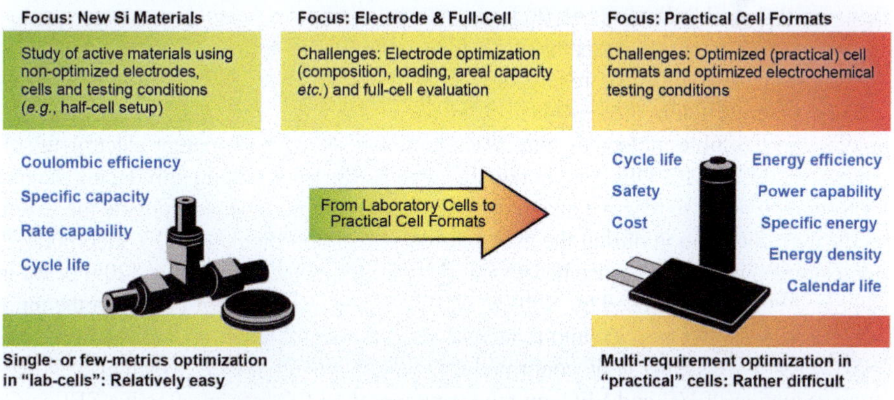

Figure 11.19 *Challenges for commercial application of Si-based materials in LIB cells. The improvement of single- or few-metrics in laboratory cells can be achieved relatively easy, while the improvement of multi-requirements in "practical" cells is rather difficult, as various requirements have to be fulfilled.*

especially when evaluated under non-optimized conditions. When it comes to practical application, there are many more parameters, which need to be optimized, such as high electrode mass loadings and high areal capacities, an optimized electrode composition with low contents of binder or conductive carbon, a reduced electrolyte amount per cell capacity, practically relevant testing conditions, gassing behavior, etc. As consequence, this multi-requirement optimization will directly have a (negative) impact on different performance parameters such as cycling stability, rate capability, or safety (Figure 11.19).

Therefore, future studies on Si-based materials should intensify the focus on practical studies in LIB full-cells and performance validation in practical LIB cell formats. In particular, there are still various approaches to further boost the performance of Si-based full-cells, such as electrode engineering by binder design (for high Si contents), commercially relevant pre-treatment approaches such as electrochemical pre-lithiation, electrolyte optimization (e.g., by design of novel electrolyte additives) as well as the verification of cycling conditions (e.g., limited depth-of-discharge) and the impact on cycling stability. Therefore, still immense research and development efforts are needed to achieve an improved performance of practical Si-based LIB cells with high Si contents.

References

[1] Winter M., Barnett B., Xu K. 'Before Li ion batteries'. *Chemical Reviews*. 2018;**118**(23):11433–56.

[2] Schmuch R., Wagner R., Hörpel G., Placke T., Winter M. 'Performance and cost of materials for lithium-based rechargeable automotive batteries'. *Nature Energy*. 2018;**3**(4):267–78.

[3] Eshetu G.G., Figgemeier E. 'Confronting the challenges of next-generation silicon anode-based lithium-ion batteries: role of designer electrolyte additives and polymeric binders'. *ChemSusChem*. 2019;**12**:2515–39.

[4] Andre D., Kim S.-J., Lamp P., *et al*. 'Future generations of cathode materials: an automotive industry perspective'. *Journal of Materials Chemistry A*. 2015;**3**(13):6709–32.

[5] Zeng X., Li M., Abd El-Hady D., *et al*. 'Commercialization of lithium battery technologies for electric vehicles'. *Advanced Energy Materials*. 2019;**9**(27):1900161.

[6] Wagner R., Preschitschek N., Passerini S., Leker J., Winter M. 'Current research trends and prospects among the various materials and designs used in lithium-based batteries'. *Journal of Applied Electrochemistry*. 2013;**43**(5):481–96.

[7] Arinicheva Y., Wolff M., Lobe S., *et al*. 'Ceramics for electrochemical storage' in Guillon O. (ed.). *Advanced Ceramics for Energy Conversion and Storage*. Elsevier; 2020. pp. 549–709.

[8] Freunberger S.A. 'True performance metrics in beyond-intercalation batteries'. *Nature Energy*. 2017;**2**(7):17091.

[9] Obrovac M.N., Chevrier V.L. 'Alloy negative electrodes for Li-ion batteries'. *Chemical Reviews*. 2014;**114**(23):11444–502.

[10] Obrovac M.N. 'Si-alloy negative electrodes for Li-ion batteries'. *Current Opinion in Electrochemistry*. 2018;**9**:8–17.

[11] Zuo X., Zhu J., Müller-Buschbaum P., Cheng Y.-J. 'Silicon based lithium-ion battery anodes: a chronicle perspective review'. *Nano Energy*. 2017;**31**(4):113–43.

[12] Jin Y., Zhu B., Lu Z., Liu N., Zhu J. 'Challenges and recent progress in the development of Si anodes for lithium-ion battery'. *Advanced Energy Materials*. 2017;**7**(23):1700715.

[13] Chae S., Choi S.-H., Kim N., Sung J., Cho J. 'Integration of graphite and silicon anodes for the commercialization of high-energy lithium-ion batteries'. *Angewandte Chemie International Edition*. 2020;**59**(1):110–35.

[14] Andre D., Hain H., Lamp P., Maglia F., Stiaszny B. 'Future high-energy density anode materials from an automotive application perspective'. *Journal of Materials Chemistry A*. 2017;**5**(33):17174–98.

[15] Ko M., Oh P., Chae S., Cho W., Cho J. 'Considering critical factors of Li-rich cathode and Si anode materials for practical Li-ion cell applications'. *Small*. 2015;**11**(33):4058–73.

[16] Li P., Zhao G., Zheng X., *et al*. 'Recent progress on silicon-based anode materials for practical lithium-ion battery applications'. *Energy Storage Materials*. 2018;**15**(3):422–46.

[17] Chae S., Ko M., Kim K., Ahn K., Cho J. 'Confronting issues of the practical implementation of Si anode in high-energy lithium-ion batteries'. *Joule*. 2017;**1**(1):47–60.

[18] Winter M. 'The solid electrolyte interphase – the most important and the least understood solid electrolyte in rechargeable Li batteries'. *Zeitschrift für Physikalische Chemie*. 2009;**223**(10-11):1395–406.

[19] Winter M., Besenhard J.O. 'Electrochemical lithiation of tin and tin-based intermetallics and composites'. *Electrochimica Acta*. 1999;**45**(1-2):31–50.

[20] Daniel C., Besenhard J.O. *Handbook of Battery Materials*. John Wiley & Sons; 2011.

[21] Winter M., Besenhard J.O., Albering J.H., Yang J., Wachtler M. 'Lithium storage alloys as anode materials for lithium ion batteries'. *Prog. Batteries Battery Mater*. 1998;**17**:208–14.

[22] Wachtler M., Besenhard J.O., Winter M. 'Tin and tin-based intermetallics as new anode materials for lithium-ion cells'. *Journal of Power Sources*. 2001;**94**(2):189–93.

[23] Trifonova A., Winter M., Besenhard J.O. 'Structural and electrochemical characterization of tin-containing graphite compounds used as anodes for Li-ion batteries'. *Journal of Power Sources*. 2007;**174**(2):800–4.

[24] Yang J., WinterM., BesenhardJ.O. 'Small particle size multiphase Li-alloy anodes for lithium-ion batteries'. *Solid State Ionics*. 1996;**90**(1-4):281–7.

[25] Yang J., WachtlerM., WinterM., BesenhardJ.O. 'Sub-microcrystalline Sn and Sn-SnSb powders as lithium storage materials for lithium-ion batteries'. *Electrochemical and Solid-State Letters*. 1999;**2**(4):161–3.

[26] Wagner M.R., Raimann P.R., Trifonova A., Möller K.-C., Besenhard J.O., Winter M. 'Electrolyte decomposition reactions on tin- and graphite-based anodes are different'. *Electrochemical and Solid-State Letters.* 2004;**7**(7):A201–5.

[27] Idota Y., KubotaT., MatsufujiA., MaekawaY., MiyasakaT. 'Tin-based amorphous oxide: a high-capacity lithium-ion-storage material'. *Science.* 1997;**276**(5317):1395–7.

[28] Sony [online]. Available from https://www.sony.net/SonyInfo/News/Press/200502/05-006E/ [Accessed 2 Jan 2020].

[29] Tarascon J.M. 'The Li-ion battery: 25 years of exciting and enriching experiences'. *Interface magazine.* 2016;**25**(3):79–83.

[30] Grey C.P., Tarascon J.M. 'Sustainability and in situ monitoring in battery development'. *Nature Materials.* 2016;**16**(1):45–56.

[31] Betz J., Bieker G., Meister P., Placke T., Winter M., Schmuch R. 'Theoretical versus practical energy: a plea for more transparency in the energy calculation of different rechargeable battery systems'. *Advanced Energy Materials.* 2019;**9**(6):1803170.

[32] Nishi Y. 'Lithium ion secondary batteries; past 10 years and the future'. *Journal of Power Sources.* 2001;**100**(1-2):101–6.

[33] Winter M., Besenhard J.O. 'Lithiated carbons' in Daniel C., Besenhard J.O. (eds.). *Handbook of Battery Materials.* Second Edition; 2011. pp. 433–78.

[34] Winter M., Appel W.K., Evers B., *et al.* 'Studies on the anode/electrolyte interface in lithium ion batteries'. *Monatshefte für Chemie/Chemical Monthly.* 2001;**132**(4):473–86.

[35] Ruttert M., Siozios V., Winter M., Placke T. 'Mechanochemical synthesis of Fe–Si-based anode materials for high-energy lithium ion full-cells'. *ACS Applied Energy Materials.* 2020;**3**(1):743–58.

[36] Chevrier V.L., Liu L., Le D.B., *et al.* 'Evaluating Si-based materials for Li-ion batteries in commercially relevant negative electrodes'. *Journal of the Electrochemical Society.* 2014;**161**(5):A783–91.

[37] Nölle R., Beltrop K., Holtstiege F., Kasnatscheew J., Placke T., Winter M. 'A reality check and tutorial on electrochemical characterization of battery cell materials: how to choose the appropriate cell setup'. *Materials Today.* 2020;**32**(4):131–46.

[38] Holtstiege F., Wilken A., Winter M., Placke T. 'Running out of lithium? A route to differentiate between capacity losses and active lithium losses in lithium-ion batteries'. *Physical Chemistry Chemical Physics.* 2017;**19**(38):25905–18.

[39] Liu N., Lu Z., Zhao J., *et al.* 'A pomegranate-inspired nanoscale design for large-volume-change lithium battery anodes'. *Nature Nanotechnology.* 2014;**9**(3):187–92.

[40] Liu N., Wu H., McDowell M.T., Yao Y., Wang C., Cui Y. 'A yolk-shell design for stabilized and scalable li-ion battery alloy anodes'. *Nano Letters.* 2012;**12**(6):3315–21.

[41] Meister P., Jia H., Li J., Klöpsch R., Winter M., Placke T. 'Best practice: performance and cost evaluation of lithium ion battery active materials with special emphasis on energy efficiency'. *Chemistry of Materials*. 2016;**28**(20):7203–17.

[42] Lin Z., Liu T., Ai X., Liang C. 'Aligning academia and industry for unified battery performance metrics'. *Nature Communications*. 2018;**9**(1):5262.

[43] Smith A.J., Burns J.C., Xiong D., Dahn J.R. 'Interpreting high precision Coulometry results on Li-ion cells'. *Journal of the Electrochemical Society*. 2011;**158**(10):A1136–42.

[44] Holtstiege F., Koç T., Hundehege T., Siozios V., Winter M., Placke T. 'Toward high power batteries: pre-lithiated carbon nanospheres as high rate anode material for lithium ion batteries'. *ACS Applied Energy Materials*. 2018;**1**(8):4321–31.

[45] Obrovac M.N., Christensen L., Le D.B., Dahn J.R. 'Alloy design for lithium-ion battery anodes'. *Journal of the Electrochemical Society*. 2007;**154**(9):A849–55.

[46] Lin D., Lu Z., Hsu P.-C., *et al*. 'A high tap density secondary silicon particle anode fabricated by scalable mechanical pressing for lithium-ion batteries'. *Energy & Environmental Science*. 2015;**8**(8):2371–6.

[47] Jantke D., Bernhard R., Hanelt E., Buhrmester T., Pfeiffer J., Haufe S. 'Silicon-dominant anodes based on microscale silicon particles under partial lithiation with high capacity and cycle stability'. *Journal of the Electrochemical Society*. 2019;**166**(16):A3881–5.

[48] Judez X., Eshetu G.G., Li C., Rodriguez-Martinez L.M., Zhang H., Armand M. 'Opportunities for rechargeable solid-state batteries based on Li-intercalation cathodes'. *Joule*. 2018;**2**(11):2208–24.

[49] Wu X., Song K., Zhang X., *et al*. 'Safety issues in lithium ion batteries: materials and cell design'. *Frontiers in Energy Research*. 2019;**7**(65).

[50] Eshetu G.G., Grugeon S., Laruelle S., *et al*. 'In-depth safety-focused analysis of solvents used in electrolytes for large scale lithium ion batteries'. *Physical Chemistry Chemical Physics*. 2013;**15**(23):9145–55.

[51] Profatilova I.A., Langer T., Badillo J.P., *et al*. 'Thermally induced reactions between lithiated nano-silicon electrode and electrolyte for lithium-ion batteries'. *Journal of the Electrochemical Society*. 2012;**159**(5):A657–63.

[52] Profatilova I.A., Stock C., Schmitz A., Passerini S., Winter M. 'Enhanced thermal stability of a lithiated nano-silicon electrode by fluoroethylene carbonate and vinylene carbonate'. *Journal of Power Sources*. 2013;**222**:140–9.

[53] Allcorn E., Nagasubramanian G., Fenton K.R. 'Materials safety study of practical nano-silicon + graphite anodes for lithium-ion batteries'. *Meeting Abstracts*. 2016;**MA2016-02**(6):887.

[54] Aupperle F., von Aspern N., Berghus D., *et al*. 'The role of electrolyte additives on the interfacial chemistry and thermal reactivity of Si-Anode based Li-ion battery'. *ACS Applied Energy Materials*. 2019;**2**(9):6513–27.

[55] Eshetu G.G., Grugeon S., Kim H., *et al.* 'Comprehensive insights into the reactivity of electrolytes based on sodium ions'. *ChemSusChem.* 2016;**9**(5):462–71.

[56] Pan K., Zou F., Canova M., Zhu Y., Kim J.-H. 'Systematic electrochemical characterizations of Si and SiO anodes for high-capacity Li-ion batteries'. *Journal of Power Sources.* 2019;**413**:20–8.

[57] Yao K.P.C., Okasinski J.S., Kalaga K., Almer J.D., Abraham D.P. 'Operando quantification of (de)lithiation behavior of silicon–graphite blended electrodes for lithium-ion batteries'. *Advanced Energy Materials.* 2019;**9**(8):1803380.

[58] Sick N., Bröring S., Figgemeier E. 'Start-ups as technology life cycle indicator for the early stage of application: an analysis of the battery value chain'. *Journal of Cleaner Production.* 2018;**201**(1):325–33.

[59] Sila Nanotechnologies [online]. Available from https://silanano.com/ [Accessed 2 Jan 2020].

[60] Enovix [online]. Available from https://enovix.com/ [Accessed 2 Jan 2020].

[61] Enevate [online]. Available from https://www.enevate.com/ [Accessed 2 Jan 2020].

[62] Amprius [online]. Available from https://www.amprius.com/ [Accessed 2 Jan 2020].

[63] Group14 [online]. Available from https://www.group14technologies.com/ [Accessed 2 Jan 2020].

[64] Jar L. [online]. Available from https://leyden-jar.com/ [Accessed 2 Jan 2020].

[65] Nexeon [online]. Available from https://www.nexeon.co.uk/ [Accessed 2 Jan 2020].

[66] EoCell [online]. Available from https://www.eocell.com/ [Accessed 2 Jan 2020].

[67] Ecellix Inc [online]. Available from https://ecellix.com/ [Accessed 14 Jan 2020].

[68] Nanotek Instruments [online]. Available from http://nanotekinstruments.com/ [Accessed 2 Jan 2020].

[69] StoreDot [online]. Available from https://www.store-dot.com/ [Accessed 2 Jan 2020].

[70] Sila Nanotechnologies will produce silicon anode batteries with 10-15% more capacity in wearable gadgets in 2019 and EV cars in 2023. [online]. Available from http://enwires.com/ [Accessed 2 Jan 2020].

[71] California Lithium Battery, Inc [online]. Available from https://clbattery.com/ [Accessed 2 Jan 2020].

[72] Zeptor [online]. Available from https://www.zeptoco.com/ [Accessed 2 Jan 2020].

[73] OneD Material [online]. Available from https://www.onedmaterial.com/ [Accessed 2 Jan 2020].

[74] E-magy [online]. Available from http://www.e-magy.com/ [Accessed 2 Jan 2020].

[75] Elkem [online]. Available from https://www.elkem.com/ [Accessed 2 Jan 2020].

[76] Cenate [online]. Available from https://www.cenate.com/ [Accessed 2 Jan 2020].

[77] Available from https://www.nextbigfuture.com/2018/08/sila-nanotechnologies-will-produce-silicon-anode-batteries-with-10-15-more-capacity-in-wearable-gadgets-in-2019-and-ev-cars-in-2023.html [Accessed 2 Jan 2020].

[78] Liu W.-R., Yen Y.-C., Wu H.-C., Winter M., Wu N.-L. 'Nano-porous SiO/carbon composite anode for lithium-ion batteries'. *Journal of Applied Electrochemistry*. 2009;**39**(9):1643–9.

[79] Bommier C., Chang W., Li J., *et al*. 'Operando acoustic monitoring of SEI formation and long-term cycling in NMC/SiGr composite pouch cells'. *Journal of the Electrochemical Society*. 2020;**167**(2):020517.

[80] Krueger S., Kloepsch R., Li J., Nowak S., Passerini S., Winter M. 'How do reactions at the anode/electrolyte interface determine the cathode performance in lithium-ion batteries?' *Journal of the Electrochemical Society*. 2013;**160**(4):A542–8.

[81] Kleiner K., Jakes P., Scharner S., Liebau V., Ehrenberg H. 'Changes of the balancing between anode and cathode due to fatigue in commercial lithium-ion cells'. *Journal of Power Sources*. 2016;**317**:25–34.

[82] Kasnatscheew J., Placke T., Streipert B., *et al*. 'A tutorial into practical capacity and mass balancing of lithium ion batteries'. *Journal of the Electrochemical Society*. 2017;**164**(12):A2479–86.

[83] Dose W.M., Piernas-Muñoz M.J., Maroni V.A., Trask S.E., Bloom I., Johnson C.S. 'Capacity fade in high energy silicon-graphite electrodes for lithium-ion batteries'. *Chemical Communications*. 2018;**54**(29):3586–9.

[84] Becking J., Gröbmeyer A., Kolek M., *et al*. 'Lithium-metal foil surface modification: an effective method to improve the cycling performance of lithium-metal batteries'. *Advanced Materials Interfaces*. 2017;**4**(16):1700166.

[85] Zhang M., Gui A.L., Sun W., *et al*. 'High capacity utilization of Li metal anodes by application of Celgard separator-reinforced ternary polymer electrolyte'. *Journal of the Electrochemical Society*. 2019;**166**(10):A2142–50.

[86] Liu J., Bao Z., Cui Y., *et al*. 'Pathways for practical high-energy long-cycling lithium metal batteries'. *Nature Energy*. 2019;**4**(3):180–6.

[87] Bieker G., Winter M., Bieker P. 'Electrochemical in situ investigations of SEI and dendrite formation on the lithium metal anode'. *Physical Chemistry Chemical Physics*. 2015;**17**(14):8670–9.

[88] Heine J., Hilbig P., Qi X., Niehoff P., Winter M., Bieker P. 'Fluoroethylene carbonate as electrolyte additive in tetraethylene glycol dimethyl ether based electrolytes for application in lithium ion and lithium metal batteries'. *Journal of the Electrochemical Society*. 2015;**162**(6):A1094–101.

[89] Rezqita A., Sauer M., Foelske A., Kronberger H., Trifonova A. 'The effect of electrolyte additives on electrochemical performance of silicon/mesoporous carbon (Si/MC) for anode materials for lithium-ion batteries'. *Electrochimica Acta*. 2017;**247**(7):600–9.

[90] Nölle R., Achazi A.J., Kaghazchi P., Winter M., Placke T. 'Pentafluorophenyl isocyanate as an effective electrolyte additive for improved performance of silicon-based lithium-ion full cells'. *ACS Applied Materials & Interfaces*. 2018;**10**(33):28187–98.

[91] Dalavi S., Guduru P., Lucht B.L. 'Performance enhancing electrolyte additives for lithium ion batteries with silicon anodes'. *Journal of the Electrochemical Society*. 2012;**159**(5):A642–6.

[92] Horowitz Y., Ben-Barak I., Schneier D., *et al*. 'Study of the formation of a solid electrolyte interphase (SEI) on a silicon nanowire anode in liquid disiloxane electrolyte with nitrile end groups for lithium-ion batteries'. *Batteries & Supercaps*. 2019;**2**(3):213–22.

[93] Nölle R., Schmiegel J.-P., Winter M., Placke T. 'Tailoring electrolyte additives with synergistic functional moieties for silicon negative electrode-based lithium ion batteries: a case study on lactic acid O-carboxyanhydride'. *Chemistry of Materials*. 2020;**32**(1):173–85.

[94] Qian Y., Schultz C., Niehoff P., *et al*. 'Investigations on the electrochemical decomposition of the electrolyte additive vinylene carbonate in Li metal half cells and lithium ion full cells'. *Journal of Power Sources*. 2016;**332**(10):60–71.

[95] Qian Y., Niehoff P., Börner M., *et al*. 'Influence of electrolyte additives on the cathode electrolyte interphase (CEI) formation on $LiNi_{1/3}Mn_{1/3}Co_{1/3}O_2$ in half cells with Li metal counter electrode'. *Journal of Power Sources*. 2016;**329**(7179):31–40.

[96] Vogl U.S., Lux S.F., Crumlin E.J., *et al*. 'The mechanism of SEI formation on a single crystal Si(100) electrode'. *Journal of the Electrochemical Society*. 2015;**162**(4):A603–7.

[97] Vogl U.S., Lux S.F., Das P., *et al*. 'The mechanism of SEI formation on single crystal Si(100), Si(110) and Si(111) electrodes'. *Journal of the Electrochemical Society*. 2015;**162**(12):A2281–8.

[98] Nie M., Chalasani D., Abraham D.P., Chen Y., Bose A., Lucht B.L. 'Lithium ion battery graphite solid electrolyte interphase revealed by microscopy and spectroscopy'. *The Journal of Physical Chemistry C*. 2013;**117**(3):1257–67.

[99] Vogl U.S., Das P.K., Weber A.Z., Winter M., Kostecki R., Lux S.F. 'Mechanism of interactions between CMC binder and Si single crystal facets'. *Langmuir*. 2014;**30**(34):10299–307.

[100] Wachtler M., Wagner M.R., Schmied M., Winter M., Besenhard J.O. 'The effect of the binder morphology on the cycling stability of Li–alloy composite electrodes'. *Journal of Electroanalytical Chemistry*. 2001;**510**(1–2):12–19.

[101] Dose W.M., Maroni V.A., Piernas-Muñoz M.J., Trask S.E., Bloom I., Johnson C.S. 'Assessment of Li-inventory in cycled Si-graphite anodes using $LiFePO_4$ as a diagnostic cathode'. *Journal of the Electrochemical Society*. 2018;**165**(10):A2389–96.

[102] Wetjen M., Pritzl D., Jung R., Solchenbach S., Ghadimi R., Gasteiger H.A. 'Differentiating the degradation phenomena in silicon-graphite

electrodes for lithium-ion batteries'. *Journal of the Electrochemical Society*. 2017;**164**(12):A2840–52.

[103] Wetjen M., Trunk M., Werner L., *et al.* 'Quantifying the distribution of electrolyte decomposition products in silicon-graphite electrodes by neutron depth profiling'. *Journal of the Electrochemical Society*. 2018;**165**(10):A2340–8.

[104] Waldmann T., Hogg B.-I., Wohlfahrt-Mehrens M. 'Li plating as unwanted side reaction in commercial Li-ion cells – a review'. *Journal of Power Sources*. 2018;**384**:107–24.

[105] Petzl M., Kasper M., Danzer M.A. 'Lithium plating in a commercial lithium-ion battery – a low-temperature aging study'. *Journal of Power Sources*. 2015;**275**:799–807.

[106] Buqa H., Blyth R.I.R., Golob P., *et al*. 'Negative electrodes in rechargeable lithium ion batteries — influence of graphite surface modification on the formation of the solid electrolyte interphase'. *Ionics*. 2000;**6**(3-4):172–9.

[107] Zhang W.-J. 'A review of the electrochemical performance of alloy anodes for lithium-ion batteries'. *Journal of Power Sources*. 2011;**196**(1):13–24.

[108] Klett M., Gilbert J.A., Pupek K.Z., Trask S.E., Abraham D.P., *et al.* 'Layered oxide, graphite and silicon-graphite electrodes for lithium-ion cells: effect of electrolyte composition and cycling windows'. *Journal of the Electrochemical Society*. 2017;**164**(1):A6095–102.

[109] Trask S.E., Pupek K.Z., Gilbert J.A., *et al*. 'Performance of full cells containing carbonate-based LiFSI electrolytes and silicon-graphite negative electrodes'. *Journal of the Electrochemical Society*. 2016;**163**(3):A345–50.

[110] Dupré N., Moreau P., De Vito E., *et al*. 'Multiprobe study of the solid electrolyte interphase on silicon-based electrodes in full-cell configuration'. *Chemistry of Materials*. 2016;**28**(8):2557–72.

[111] Delpuech N., Dupre N., Moreau P., *et al*. 'Mechanism of silicon electrode aging upon cycling in full lithium-ion batteries'. *ChemSusChem*. 2016;**9**(8):841–8.

[112] Chen Z., Zhang L., Wu X., *et al*. 'Effect of N/P ratios on the performance of $LiNi_{0.8}Co_{0.15}Al_{0.05}O_2||SiO_x$/graphite lithium-ion batteries'. *Journal of Power Sources*. 2019;**439**:227056.

[113] Louli A.J., Ellis L.D., Dahn J.R. 'Operando pressure measurements reveal solid electrolyte interphase growth to RANK Li-ion cell performance'. *Joule*. 2019;**3**(3):745–61.

[114] Reuter F., Baasner A., Pampel J., *et al*. 'Importance of capacity balancing on the electrochemical performance of $Li[Ni_{0.8}Co_{0.1}Mn_{0.1}]O_2$ (NCM811)/silicon full cells'. *Journal of the Electrochemical Society*. 2019;**166**(14):A3265–71.

[115] Shan J., Yang X., Yan C., Chen L., Zhao F., Ju Y. 'Promoting Si-graphite composite anodes with SWCNT additives for half and NCM811 full lithium ion batteries and assessment criteria from an industrial perspective'. *Frontiers in Energy*. 2019;**13**(4):626–35.

[116] Baasner A., Reuter F., Seidel M., *et al*. 'The role of balancing nanostructured silicon anodes and NMC cathodes in lithium-ion full-cells with

high volumetric energy density'. *Journal of the Electrochemical Society*. 2020;**167**(2):020516.

[117] Philippe B., Dedryvère R., Gorgoi M., Rensmo H., Gonbeau D., Edström K. 'Improved performances of nanosilicon electrodes using the salt LiFSI: a photoelectron spectroscopy study'. *Journal of the American Chemical Society*. 2013;**135**(26):9829–42.

[118] Smith A.J., Burns J.C., Dahn J.R. 'A high precision study of the Coulombic efficiency of Li-ion batteries'. *Electrochemical and Solid-State Letters*. 2010;**13**(12):A177–9.

[119] Smith A.J., Burns J.C., Zhao X., Xiong D., Dahn J.R. 'A high precision Coulometry study of the SEI growth in Li/graphite cells'. *Journal of The Electrochemical Society*. 2011;**158**(5):A447–52.

[120] Yim C.-H., Niketic S., Salem N., Naboka O., Abu-Lebdeh Y. 'Towards improving the practical energy density of Li-ion batteries: optimization and evaluation of silicon:graphite composites in full cells'. *Journal of the Electrochemical Society*. 2017;**164**(1):A6294–302.

[121] Kasnatscheew J., Röser S., Börner M., Winter M. 'Do increased Ni contents in $LiNi_xMn_yCo_zO_2$ (NMC) electrodes decrease structural and thermal stability of Li ion batteries? A thorough look by consideration of the Li^+ extraction ratio'. *ACS Applied Energy Materials*. 2019;**2**(11):7733–7.

[122] Reynier Y., Vincens C., Leys C., *et al*. 'Practical implementation of Li doped SiO in high energy density 21700 cell'. *Journal of Power Sources*. 2020;**450**(14):227699.

[123] Adelhelm P. 'Editorial: the energy challenge, batteries, and why simple math matters'. *Angewandte Chemie International Edition*. 2018;**57**(23):6710–1.

[124] Balducci A., Belanger D., Brousse T., Long J.W., Sugimoto W. 'Perspective—a guideline for reporting performance metrics with electrochemical capacitors: from electrode materials to full devices'. *Journal of the Electrochemical Society*. 2017;**164**(7):A1487–8.

[125] Mathis T.S., Kurra N., Wang X., Pinto D., Simon P., Gogotsi Y. 'Energy storage data reporting in perspective—guidelines for interpreting the performance of electrochemical energy storage systems'. *Advanced Energy Materials*. 2019;**9**(39):1902007.

[126] Walton K.S., Sholl D.S. 'Research challenges in avoiding "showstoppers" in developing materials for large-scale energy applications'. *Joule*. 2017;**1**(2):208–11.

[127] Chen S., Niu C., Lee H., *et al*. 'Critical parameters for evaluating coin cells and pouch cells of rechargeable Li-metal batteries'. *Joule*. 2019;**3**(4):1094–105.

[128] Baker D.R., Verbrugge M.W., Bower A.F. 'Swelling and elastic deformation of lithium-silicon electrode materials'. *Journal of the Electrochemical Society*. 2016;**163**(5):A624–31.

[129] Marinaro M., Yoon D.-hwan., Gabrielli G., *et al*. 'High performance 1.2 Ah Si-alloy/graphite|$LiNi_{0.5}Mn_{0.3}Co_{0.2}O_2$ prototype Li-ion battery'. *Journal of Power Sources*. 2017;**357**:188–97.

[130] Wetjen M., Solchenbach S., Pritzl D., Hou J., Tileli V., Gasteiger H.A. 'Morphological changes of silicon nanoparticles and the influence of cut-off potentials in silicon-graphite electrodes'. *Journal of the Electrochemical Society*. 2018;**165**(7):A1503–14.

[131] Wilkinson D.P., Blom H., Brandt K., Wainwright D. 'Effects of physical constraints on Li cyclability'. *Journal of Power Sources*. 1991;**36**(4):517–27.

[132] Wilkinson D.P., Wainwright D. 'In-situ study of electrode stack growth in rechargeable cells at constant pressure'. *Journal of Electroanalytical Chemistry*. 1993;**355**(1):193–203.

[133] Louli A.J., Li J., Trussler S., Fell C.R., Dahn J.R. 'Volume, pressure and thickness evolution of Li-ion pouch cells with silicon-composite negative electrodes'. *Journal of the Electrochemical Society*. 2017;**164**(12):A2689–96.

[134] Jeong G., Lee S.M., Choi N.S., Kim Y.-U., Lee C.K. 'Stabilizing dimensional changes in Si-based composite electrodes by controlling the electrode porosity: An *in situ* electrochemical dilatometric study'. *Electrochimica Acta*. 2011;**56**(14):5095–101.

[135] Gómez-Cámer J.L., Bünzli C., Hantel M.M., Poux T., Novák P. 'On the correlation between electrode expansion and cycling stability of graphite/Si electrodes for Li-ion batteries'. *Carbon*. 2016;**105**(9):42–51.

[136] Müller V., Scurtu R.-G., Richter K., *et al.* 'Effects of mechanical compression on the aging and the expansion behavior of Si/C-composite|NMC811 in different lithium-ion battery cell formats'. *Journal of the Electrochemical Society*. 2019;**166**(15):A3796–805.

[137] De Sutter L., Berckmans G., Marinaro M., *et al.* 'Comprehensive aging analysis of volumetric constrained lithium-ion pouch cells with high concentration silicon-alloy anodes'. *Energies*. 2018;**11**(11):2948.

[138] Willenberg L.K., Dechent P., Fuchs G., Sauer D.U., Figgemeier E. 'High-precision monitoring of volume change of commercial lithium-ion batteries by using strain gauges'. *Sustainability*. 2020;**12**(2):557.

[139] Lin Y.-M., Klavetter K.C., Abel P.R., *et al.* 'High performance silicon nanoparticle anode in fluoroethylene carbonate-based electrolyte for Li-ion batteries'. *Chemical Communications*. 2012;**48**(58):7268–70.

[140] Choi N.-S., Yew K.H., Lee K.Y., Sung M., Kim H., Kim S.-S. 'Effect of fluoroethylene carbonate additive on interfacial properties of silicon thin-film electrode'. *Journal of Power Sources*. 2006;**161**(2):1254–9.

[141] Jung R., Metzger M., Haering D., *et al.* 'Consumption of fluoroethylene carbonate (FEC) on Si-C composite electrodes for Li-ion batteries'. *Journal of the Electrochemical Society*. 2016;**163**(8):A1705–16.

[142] Petibon R., Chevrier V.L., Aiken C.P., *et al.* 'Studies of the capacity fade mechanisms of $LiCoO_2$/Si-alloy:graphite cells'. *Journal of the Electrochemical Society*. 2016;**163**(7):A1146–56.

[143] Holtstiege F., Bärmann P., Nölle R., Winter M., Placke T. 'Pre-lithiation strategies for rechargeable energy storage technologies: Concepts, promises and challenges'. *Batteries*. 2018;**4**(1):4–42.

[144] Liu Z., Yu Q., Zhao Y., *et al.* 'Silicon oxides: a promising family of anode materials for lithium-ion batteries'. *Chemical Society Reviews.* 2019;**48**(1):285–309.

[145] Chevrier V.L., Liu L., Wohl R., *et al.* 'Design and testing of prelithiated full cells with high silicon content'. *Journal of the Electrochemical Society.* 2018;**165**(5):A1129–36.

[146] Cui S., Chen S., Deng L. 'Si nanoparticles encapsulated in CNTs arrays with tubular sandwich structure for high performance Li ion battery'. *Ceramics International.* 2020;**46**(3):3242–9.

[147] Ding X., Wang Y. 'Bilayer-graphene-coated Si nanoparticles as advanced anodes for high-rate lithium-ion batteries'. *Electrochimica Acta.* 2020;**329**:134975.

[148] An W., Gao B., Mei S., *et al.* 'Scalable synthesis of ant-nest-like bulk porous silicon for high-performance lithium-ion battery anodes'. *Nature Communications.* 2019;**10**(1):1447.

[149] Chevrier V.L., Krause L.J., Jensen L.D., *et al.* 'Design of positive electrodes for Li-ion full cells with silicon'. *Journal of the Electrochemical Society.* 2018;**165**(13):A2968–77.

[150] Ma J., Sung J., Hong J., *et al.* 'Towards maximized volumetric capacity via pore-coordinated design for large-volume-change lithium-ion battery anodes'. *Nature Communications.* 2019;**10**(1):475.

[151] Nguyen D.-T., Kang J., Nam K.-M., Paik Y., Song S.-W. 'Understanding interfacial chemistry and stability for performance improvement and fade of high-energy Li-ion battery of $LiNi_{0.5}Co_{0.2}Mn_{0.3}O_2$//silicon-graphite'. *Journal of Power Sources.* 2016;**303**:150–8.

[152] Zhang P., Peng C., Liu X., *et al.* '3D lithiophilic "hairy" Si nanowire arrays @ carbon scaffold favor a flexible and stable lithium composite anode'. *ACS Applied Materials & Interfaces.* 2019;**11**(47):44325–32.

[153] Eom K., Joshi T., Bordes A., Do I., Fuller T.F. 'The design of a Li-ion full cell battery using a nano silicon and nano multi-layer graphene composite anode'. *Journal of Power Sources.* 2014;**249**:118–24.

[154] Wagner N.P., Asheim K., Vullum-Bruer F., Svensson A.M. 'Performance and failure analysis of full cell lithium ion battery with $LiNi_{0.8}Co_{0.15}Al_{0.05}O_2$ and silicon electrodes'. *Journal of Power Sources.* 2019;**437**:226884.

[155] Rodrigues M.-T.F., Gilbert J.A., Kalaga K., Abraham D.P. 'Insights on the cycling behavior of a highly-prelithiated silicon–graphite electrode in lithium-ion cells'. *Journal of Physics: Energy.* 2020;**2**(2):024002.

[156] Gao Y., Yi R., Li Y.C., *et al.* 'General method of manipulating formation, composition, and morphology of solid-electrolyte interphases for stable Li-alloy anodes'. *Journal of the American Chemical Society.* 2017;**139**(48):17359–67.

[157] Kierzek K., Machnikowski J. 'Factors influencing cycle-life of full Li-ion cell built from Si/C composite as anode and conventional cathodic material'. *Electrochimica Acta.* 2016;**192**:475–81.

[158] Klett M., Gilbert J.A., Trask S.E., *et al.* 'Electrode behavior re-visited: monitoring potential windows, capacity loss, and impedance changes in $Li_{1.03}(Ni_{0.5}Co_{0.2}Mn_{0.3})_{0.97}O_2$/silicon-graphite full cells'. *Journal of the Electrochemical Society*. 2016;**163**(6):A875–87.

[159] Gabrielli G., Marinaro M., Mancini M., Axmann P., Wohlfahrt-Mehrens M. 'A new approach for compensating the irreversible capacity loss of high-energy Si/C|$LiNi_{0.5}Mn_{1.5}O_4$ lithium-ion batteries'. *Journal of Power Sources*. 2017;**351**:35–44.

[160] Betz J., Nowak L., Winter M., Placke T., Schmuch R. 'An approach for pre-lithiation of $Li_{1+x}Ni_{0.5}Mn_{1.5}O_4$ cathodes mitigating active lithium loss'. *Journal of the Electrochemical Society*. 2019;**166**(15):A3531–8.

[161] Kwon T.-W., Choi J.W., Coskun A. 'The emerging era of supramolecular polymeric binders in silicon anodes'. *Chemical Society Reviews*. 2018;**47**(6):2145–64.

[162] Bresser D., Buchholz D., Moretti A., Varzi A., Passerini S. 'Alternative binders for sustainable electrochemical energy storage – the transition to aqueous electrode processing and bio-derived polymers'. *Energy & Environmental Science*. 2018;**11**(11):3096–127.

[163] Hochgatterer N.S., Schweiger M.R., Koller S., *et al.* 'Silicon/graphite composite electrodes for high-capacity anodes: influence of binder chemistry on cycling stability'. *Electrochemical and Solid-State Letters*. 2008;**11**(5):A76–80.

[164] Klamor S., Schröder M., Brunklaus G., *et al.* 'On the interaction of water-soluble binders and nano silicon particles: alternative binder towards increased cycling stability at elevated temperatures'. *Physical Chemistry Chemical Physics*. 2015;**17**(8):5632–41.

[165] Young B.T., Nguyen C.C., Lobach A., Heskett D.R., Woicik J.C., Lucht B.L. 'Role of binders in solid electrolyte interphase formation in lithium ion batteries studied with hard X-ray photoelectron spectroscopy'. *Journal of Materials Research*. 2018:1–10.

[166] Porcher W., Chazelle S., Boulineau A., *et al.* 'Understanding polyacrylic acid and lithium polyacrylate binder behavior in silicon based electrodes for Li-ion batteries'. *Journal of the Electrochemical Society*. 2017;**164**(14):A3633–40.

[167] Hays K.A., Ruther R.E., Kukay A.J., *et al.* 'What makes lithium substituted polyacrylic acid a better binder than polyacrylic acid for silicon-graphite composite anodes?' *Journal of Power Sources*. 2018;**384**:136–44.

[168] Nguyen C.C., Yoon T., Seo D.M., Guduru P., Lucht B.L. 'Systematic investigation of binders for silicon anodes: Interactions of binder with silicon particles and electrolytes and effects of binders on solid electrolyte interphase formation'. *ACS Applied Materials & Interfaces*. 2016;**8**(19):12211–20.

[169] Han Z.-J., Yamagiwa K., Yabuuchi N., *et al.* 'Electrochemical lithiation performance and characterization of silicon-graphite composites with lithium, sodium, potassium, and ammonium polyacrylate binders'. *Physical Chemistry Chemical Physics*. 2015;**17**(5):3783–95.

[170] Rehnlund D., Lindgren F., Böhme S., *et al.* 'Lithium trapping in alloy forming electrodes and current collectors for lithium based batteries'. *Energy & Environmental Science*. 2017;**10**(6):1350–7.

[171] Aravindan V., Lee Y.-S., Madhavi S. 'Best practices for mitigating irreversible capacity loss of negative electrodes in Li-Ion batteries'. *Advanced Energy Materials*. 2017;**7**(17):1602607–23.

[172] Sun Y., Lee H.-W., Seh Z.W., *et al.* 'High-capacity battery cathode prelithiation to offset initial lithium loss'. *Nature Energy*. 2016;**1**:15008.

[173] Solchenbach S., Wetjen M., Pritzl D., Schwenke K.U., Gasteiger H.A. 'Lithium oxalate as capacity and cycle-life enhancer in LNMO/graphite and LNMO/SiG full cells'. *Journal of the Electrochemical Society*. 2018;**165**(3):A512–24.

[174] Mancini M., Axmann P., Gabrielli G., Kinyanjui M., Kaiser U., Wohlfahrt-Mehrens M. 'A high-voltage and high-capacity $Li_{1+x}Ni_{0.5}Mn_{1.5}O_4$ cathode material: From synthesis to full lithium-ion cells'. *ChemSusChem*. 2016;**9**(14):1843–9.

[175] Tabuchi T., Yasuda H., Yamachi M. 'Li-doping process for Li_xSiO-negative active material synthesized by chemical method for lithium-ion cells'. *Journal of Power Sources*. 2005;**146**(1):507–9.

[176] Wang G., Li F., Liu D., *et al.* 'Chemical prelithiation of negative electrodes in ambient air for advanced lithium-ion batteries'. *ACS Applied Materials & Interfaces*. 2019;**11**(9):8699–703.

[177] Stefan I.C., Toulokhonova I., Roberts G. 'Prelithiation solutions for lithium-ion batteries'. *Amprius*. 2015:WO2015192051 A1.

[178] Zhao J., Lu Z., Liu N., Lee H.-W., McDowell M.T., Cui Y. 'Dry-air-stable lithium silicide-lithium oxide core-shell nanoparticles as high-capacity prelithiation reagents'. *Nature Communications*. 2014;**5**:5088–95.

[179] Burshtain D., Kedem N., Sella E., Aronov D. 'Polymer coatings and anode material pre-lithiation'. *Storedot Ltd*. 2017:US 2017 / 0294648 A1.

[180] Shellikeri A., Watson V.G., Adams D.L., *et al.* 'Pre-lithiation of carbon anodes using different lithium - sources'. *ECS Transactions*. 2017;**77**(11):293–303.

[181] Forney M.W., Ganter M.J., Staub J.W., Ridgley R.D., Landi B.J. 'Prelithiation of silicon–carbon nanotube anodes for lithium ion batteries by stabilized lithium metal powder (SLMP)'. *Nano Letters*. 2013;**13**(9):4158–63.

[182] Takezawa H., Ito S., Yoshizawa H., Abe T. 'Electrochemical properties of a SiO_x film anode pre-lithiated by evaporation of metallic Li in Li-ion batteries'. *Chemistry Letters*. 2017;**46**(9):1365–7.

[183] Tsuda T., Ando N., Mitsuhashi N., *et al.* 'Fabrication of porous graphite anodes with pico-second pulse laser and enhancement of pre-doping of Li^+ ions to laminated graphite anodes with micrometre-sized holes formed on the porous graphite anodes'. *ECS Transactions*. 2017;**77**(11):1897–903.

[184] Kim H.J., Choi S., Lee S.J., *et al.* 'Controlled prelithiation of silicon monoxide for high performance lithium-ion rechargeable full cells'. *Nano Letters*. 2016;**16**(1):282–8.

[185] Grant R.W., Sweetland M., Acharige A.M. 'Method for alkaliating anodes'. *Nanoscale Components*. 2017:US 9,598,789 B2.

[186] Grant R.W., Sweetland M., Acharige A.M. 'Methods for alkaliating roll anodes'. *Nanoscale Components*. 2018:US 10,128,487 B2.

[187] Ruttert M., Siozios V., Winter M., Placke T. 'Synthesis and comparative investigation of silicon transition metal silicide composite anodes for lithium ion batteries'. *Zeitschrift für Anorganische und Allgemeine Chemie*. 2019;**645**(3):248–56.

[188] Marinaro M., Weinberger M., Wohlfahrt-Mehrens M. 'Toward pre-lithiatied high areal capacity silicon anodes for lithium-ion batteries'. *Electrochimica Acta*. 2016;**206**:99–107.

[189] Domi Y., Usui H., Iwanari D., Sakaguchi H. 'Effect of mechanical prelithiation on electrochemical performance of silicon negative electrode for lithium-ion batteries'. *Journal of the Electrochemical Society*. 2017;**164**(7):A1651–4.

[190] Chang S., Moon J., Cho K., Cho M. 'Multiscale analysis of prelithiated silicon nanowire for Li-ion battery'. *Computational Materials Science*. 2015;**98**:99–104.

[191] Jia H., Zou L., Gao P., *et al*. 'High-performance silicon anodes enabled by nonflammable localized high-concentration electrolytes'. *Advanced Energy Materials*. 2019;**9**(31):1900784.

[192] Cekic-Laskovic I., von Aspern N., Imholt L., *et al*. 'Synergistic effect of blended components in nonaqueous electrolytes for lithium ion batteries'. *Topics in Current Chemistry*. 2017;**375**(2):37–100.

[193] Wagner R., Streipert B., Kraft V., *et al*. 'Counterintuitive role of magnesium salts as effective electrolyte additives for high voltage lithium-ion batteries'. *Advanced Materials Interfaces*. 2016;**3**(15):1600096.

[194] Aurbach D., Markovsky B., Shechter A., Ein-Eli Y., Cohen H. 'A comparative study of synthetic graphite and Li electrodes in electrolyte solutions based on ethylene carbonate-dimethyl carbonate mixtures'. *Journal of the Electrochemical Society*. 1996;**143**(12):3809–20.

[195] Eshetu G.G., Grugeon S., Gachot G., Mathiron D., Armand M., Laruelle S. 'LiFSI vs. $LiPF_6$ electrolytes in contact with lithiated graphite: comparing thermal stabilities and identification of specific SEI-reinforcing additives'. *Electrochimica Acta*. 2013;**102**(6):133–41.

[196] Heider U., Oesten R., Jungnitz M. 'Challenge in manufacturing electrolyte solutions for lithium and lithium ion batteries quality control and minimizing contamination level'. *Journal of Power Sources*. 1999;**81-82**:119–22.

[197] Nowak S., Winter M. 'Review—chemical analysis for a better understanding of aging and degradation mechanisms of non-aqueous electrolytes for lithium ion batteries: method development, application and lessons learned'. *Journal of the Electrochemical Society*. 2015;**162**(14):A2500–8.

[198] Terborg L., Nowak S., Passerini S., *et al*. 'Ion chromatographic determination of hydrolysis products of hexafluorophosphate salts in aqueous solution'. *Analytica Chimica Acta*. 2012;**714**:121–6.

[199] Evertz M., Horsthemke F., Kasnatscheew J., Börner M., Winter M., Nowak S. 'Unraveling transition metal dissolution of $Li_{1.04}Ni_{1/3}Co_{1/3}Mn_{1/3}O_2$ (NCM 111) in lithium ion full cells by using the total reflection X-ray fluorescence technique'. *Journal of Power Sources*. 2016;**329**(10):364–71.

[200] Gallus D.R., Schmitz R., Wagner R., *et al*. 'The influence of different conducting salts on the metal dissolution and capacity fading of NCM cathode material'. *Electrochimica Acta*. 2014;**134**:393–8.

[201] Lindgren F., Xu C., Niedzicki L., *et al*. 'SEI formation and interfacial stability of a Si electrode in a LiTDI-salt based electrolyte with FEC and VC additives for Li-ion batteries'. *ACS Applied Materials & Interfaces*. 2016;**8**(24):15758–66.

[202] Meister P., Qi X., Kloepsch R., *et al*. 'Anodic behavior of the aluminum current collector in imide-based electrolytes: Influence of solvent, operating temperature, and native oxide-layer thickness'. *ChemSusChem*. 2017;**10**(4):804–14.

[203] Krämer E., Passerini S., Winter M. 'Dependency of aluminum collector corrosion in lithium ion batteries on the electrolyte solvent'. *ECS Electrochemistry Letters*. 2012;**1**(5):C9–11.

[204] Schiele A., Breitung B., Hatsukade T., *et al*. 'The critical role of fluoroethylene carbonate in the gassing of silicon anodes for lithium-ion batteries'. *ACS Energy Letters*. 2017;**2**(10):2228–33.

[205] Krause L.J., Chevrier V.L., Jensen L.D., Brandt T. 'The effect of carbon dioxide on the cycle life and electrolyte stability of Li-ion full cells containing silicon alloy'. *Journal of the Electrochemical Society*. 2017;**164**(12):A2527–33.

[206] Schwenke K.U., Solchenbach S., Demeaux J., Lucht B.L., Gasteiger H.A. 'The impact of CO_2 evolved from VC and FEC during formation of graphite anodes in lithium-ion batteries'. *Journal of the Electrochemical Society*. 2019;**166**(10):A2035–47.

[207] Janssen P., Kasnatscheew J., Streipert B., *et al*. 'Fluorinated electrolyte compound as a bi-functional interphase additive for both, anodes and cathodes in lithium-ion batteries'. *Journal of the Electrochemical Society*. 2018;**165**(14):A3525–30.

Chapter 12

A silicon future

Kang Xu[1]

The birth of lithium-ion battery (LIB) in early 1990 marked a revolutionary milestone in the history of batteries. It not only set a record in energy density among all rechargeable batteries known at the time, more importantly, it was the first battery based entirely on intercalation chemistries [1]. It was this latter feature that ensures the unprecedented reversibility of the cell reactions, as reflected by the cycle-life in the order of over thousands of times.

Before LIB, rechargeable batteries, be it lead-acid, nickel-iron, nickel-cadmium, or nickel-metal hydride, were all constructed upon conversion-reaction chemistries, in which each discharge destroys the structures of the electrode materials and forms a new compound, while the reversing charge process tries to put everything back in their pristine place. Understandably, such bond-breaking and bond-reformation processes, accompanied by the simultaneous volume changes, encounter significant reversibility barriers, which is reflected in the universally poor rechargeability issues of these batteries (Figure 12.1a).

Intercalation chemistry circumvents the above reversibility barrier by minimizing drastic structural changes in electrodes. Leveraging the topotactic nature of "host-guest" chemistry, an electrochemical reaction occurring in intercalation compounds only involves ions entering or leaving the interstitials of the framework, without rearrangement of the basic lattice (Figure 12.1b). Excellent reversibility is hence expected from such reactions but at the expense of energy densities, because the intercalation lattice of the electrode does not participate in the reaction, therefore, presenting inert weight to the cell. Nevertheless, in most cases, this expense is worthwhile.

Whittingham demonstrated titanium sulfide (TiS_2) as the first intercalation electrode in 1976, whose layered structure can accommodate the entrance and exodus of Li^+ at high rates [2]. Soon after, Goodenough improved the voltage and stability of intercalation chemistry by switching from sulfides to transition metal oxides, with lithium cobalt oxide ($LiCoO_2$) as the most successful example of this class [3]. The lithiation potentials of both TiS_2 and $LiCoO_2$ reside at high values (3.0 and 4.0 V

[1]Battery Science Branch, Energy and Biotechnology, Division Sensor and Electron Devices Directorate, US Army Research Laboratory, 2800 Powder Mill Road, Adelphi, MD 20783, USA

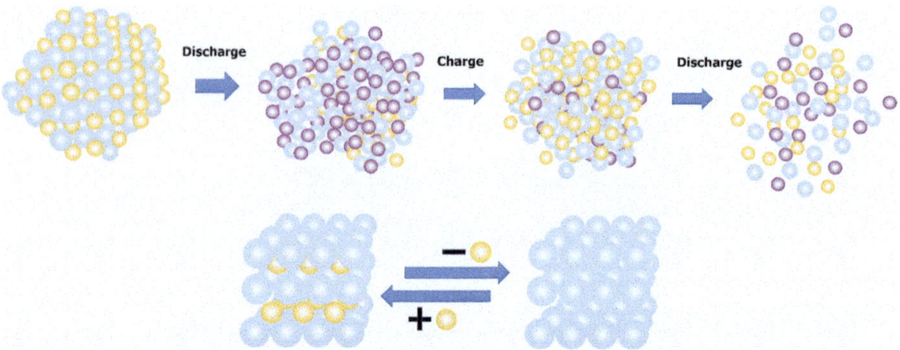

Figure 12.1 *Schematic illustration of electrochemical conversion-reactions (a) and electrochemical intercalation reactions (b). Conversion-reaction often promises much higher capacity and energy density than intercalation, but the difficulty of restoring the structure during repeated charge and discharge is a century-old challenge. On the other hand, the topotactic nature of the intercalation reactions requires the minimum structural change during each charge and discharge cycle, hence ensuring excellent reversibility.*

vs. Li, respectively), which qualify them as cathode rather than anode materials in a battery. Till today $LiCoO_2$ is still used in LIBs designed for electronics.

Despite the fact that a proper host for Li^+ at anode side did not exist at the time, Armand proposed an ingenious concept of a "dual-intercalation battery" structure in 1980, which predicted the basic architecture of modern LIB (Figure 12.2) [4]. It took another decade since then to finally identify the anode host of Li^+: the graphite or graphitic carbons, which were gradually synchronized to work with a nonaqueous electrolyte system to reversibly accommodate and release Li^+ at a very low potential (~0.15 V vs. Li) [1].

With the commercialization of the first generation LIB [5], the basic architecture for LIB has remained relatively static: a graphitic anode, a transition metal cathode, and a carbonate-based electrolyte (Figure 12.3). In the past three decades, LIB has significantly reshaped our lives by redefining how we communicate (consumer-portable electronics), transport (electrified vehicles), and inhabit (large-scale grid-storage). As a commercial product, it has experienced quite a few generations, each of which represented a new chemistry: $LiCoO_2$, lithium iron phosphate ($LiFePO_4$), nickel-manganese-cobalt (NMC) oxide ($LiNi_xMn_yCo_zO_2$), etc. However, the anode therein has always been graphite or graphitic derivatives, with only limited and not-so-successful attempts at new anode materials.

This cathode-centric bias can be witnessed in the research efforts on battery chemistry and materials. According to Web of Science, over 80 percent of the literature published in the first two decades after the commercialization of LIB (1990–2010) focused on developing new cathode chemistries. Only in the recent decade (2010–19) did this percentage decline to ~70 percent but mostly due to the increase

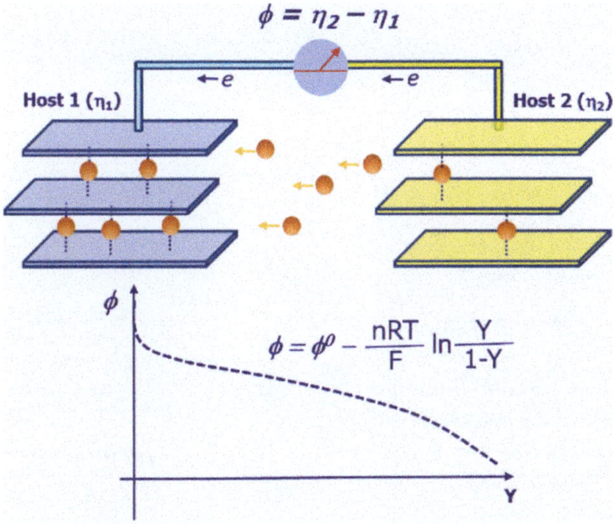

Figure 12.2 *The dual-intercalation battery concept proposed by Armand in 1980. At the time this was only a theoretical imagination, as people still did not know what material could serve as the ideal anode host for Li^+. The low operating potential of anode materials significantly hindered the efforts to identify such a host, because it will inevitably induce the formation of an interphase.*

in electrolyte materials research activities. Similar bias was also reflected in the distribution of large research funds granted by governments and foundations around the world.

Then what are the fundamental reasons behind such under-exploration of anode materials?

First of all, the urgency of replacing graphite with a new anode has not been pressing until recently.

For a very long time, the key to LIB energy density improvement resided on the cathode side instead of anode side, because the existing graphite anode was very perfect. Graphite provides a high gravimetric capacity of 372 mAh/g, and this capacity is delivered at a potential close to lithium metal Li^0 (−0.15 V vs. Li). The combination of those two characteristics make it the most ideal anode candidate only second to Li^0. In contrast, most cathode materials deliver a gravimetric capacity in a range between 100 and 170 mAh/g. In a battery, the capacities of cathode and anode are coupled in such a manner that the overall capacity of a full cell is determined by the relation:

$$C_{cell} = \frac{C_{cathode} \times C_{anode}}{C_{cathode} + C_{anode}} \tag{12.1}$$

If one finds an ideal electrode, be it cathode or anode, which has a capacity far above the other electrode, then any further improvement in the capacity of this ideal

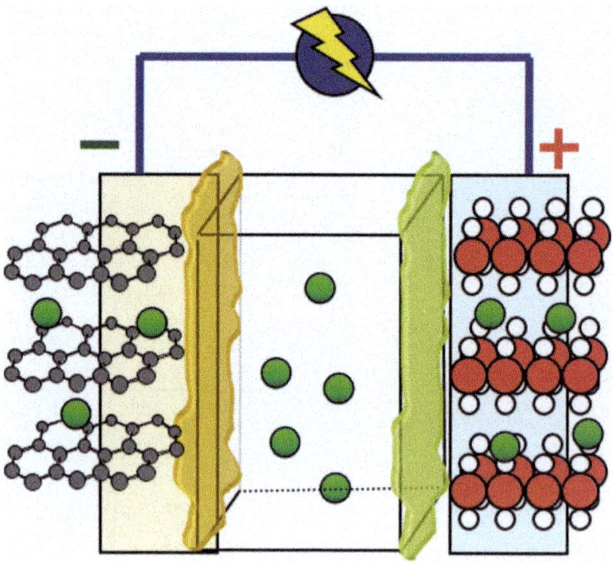

Figure 12.3 *The basic configuration of modern LIBs, which consists of a graphite anode, a transition metal oxide or iron phosphate cathode, nonaqueous electrolytes, and the interphases that protect electrolytes from the extreme potentials of anode and cathode. Both electrodes are of intercalation nature.*

electrode would be marginalized due to the constraint imposed by the other electrode. Such inter-dependence is best shown by Figure 12.4, which depicts how the capacity of a full cell changes with the varying capacity of an imaginary cathode material if this cathode is coupled with an anode of "mediocre" capacity (graphite) and an anode of high capacity (silicon). The theoretical capacities of graphite (372 mAh/g corresponding to LiC_6) and silicon (3 579 mAh/g corresponding to $Li_{15}Si_4$) were used in the calculation, and to keep the comparison simple, no consideration was given to the presence of supporting components such as electrolyte or cell packaging. Thus, the benefit of a high-capacity anode will only become apparent and meaningful when the cathode capacity becomes high enough, i.e., higher than 200 mAh/g. While various chemistries such as high-nickel NMC (811 and 622) or Li-rich manganese oxides have been able to approach or even surpass this number, only in recent years have those materials been adopted by the LIB industry with caution, apparently due to safety considerations. The breach of 200 mAh/g mark at the cathode side was still a fairly recent event.

Second, the low operation potential of anode materials constitutes an additional barrier to exploration of new materials, which is a fundamental challenge not faced by most cathode materials (Figure 12.5). In the old days (pre-1990s), such barrier actually prevented the emergence of graphitic anode for over four decades!

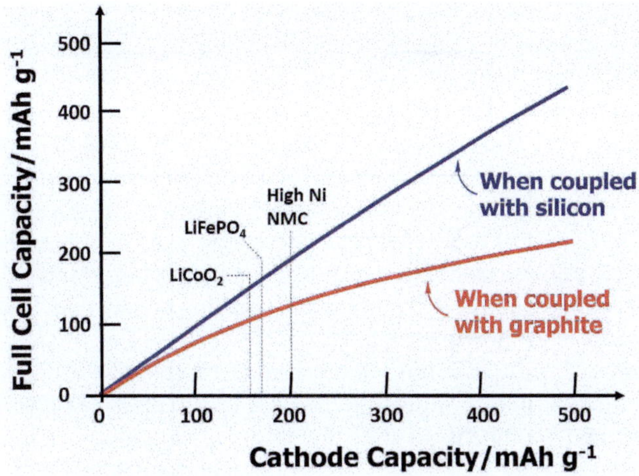

Figure 12.4 *The dependence of full cell capacity on the capacity of an imaginary cathode material when it is coupled with two different anode materials: graphite of mediocre capacity and silicon of high capacity.*

From energetic perspective, an ideal anode should accommodate Li^+ at a potential as low as possible. Because of this consideration, most anode materials are forced to work beyond the electrochemical stability window of electrolytes, where the formation of interphases is required. However, in early days, the knowledge about the existence of interphase and its relation to the electrolyte composition was unavailable. The most typical example is graphite. While the first lithiation of graphite into LiC_6 was successfully achieved in 1950s via chemical means [6], it took researchers nearly four decades to figure out how to replicate the same lithiation reaction via an electrochemical means in nonaqueous electrolytes [7]. The key

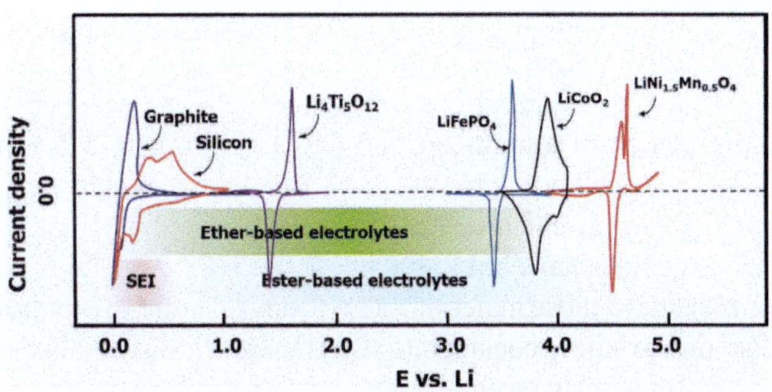

Figure 12.5 *Schematic illustration of the operating potential range for typical anode and cathode materials, and the electrochemical stability window that can be provided by the typical electrolytes based on ethers or carbonate esters*

know-how is a solid electrolyte interphase (SEI) formed by carbonate-based electrolytes at low potentials.

In contrast, most of these early cathode materials ($LiTiS_2$, $LiCoO_2$, and $LiFePO_4$) operate below the oxidation stability limits of carbonate electrolytes, until high-voltage, high-nickel cathode materials ($LiNi_{1.5}Mn_{0.5}O_2$, $LiCoPO_4$, and $LiNi_xMn_yCo_zO_2$ where x can be as high as 0.95) developed in recent years breached such limit, and then a cathode-electrolyte interphase became a necessity.

Now in 2020, most of these challenges no longer exist or at least become manageable. The new cathode materials with NMC-layer structure at high nickel content have reached capacities above 200 mAh/g, thus making the demand for a new anode to replace graphite ever more pressing.

The advances in fundamental understanding of electrolytes and interphases has also made it possible to tailor interphasial chemistries via electrolyte engineering or surface treatment so that the extreme potentials presented by the cathode materials above 4.5 V vs. Li can be accommodated. The most popular approaches included design of electrolyte additives that decompose to form interphases before bulk electrolyte components do, as well as various techniques that offer precise control of surface chemistry using atomic- or molecular-layer depositions.

Compared with the diversified chemistries discovered at the cathode side, the candidate materials as anode are much limited in numbers. The dominant position of graphite was never seriously threatened for a long time. Before the efforts of incorporating silicon into anode composites, the only commercialized alternative anode material was lithiated titanate $Li_4Ti_5O_{12}$, whose moderate capacity (150 mAh/g) and operating potential (1.50 V vs. Li) only found itself niche market in fast-charging applications.

From many aspects, silicon appears to be an ideal candidate to replace graphite. Its capacity is ten times as higher if full lithiation can be leveraged ($Li_{15}Si_4$), and its lithiation potential is as close to Li^0 as graphite (Figure 12.5). Its challenges, however, are also tremendous. In addition to the requirement for interphase protection due to its low potential, silicon experiences drastic volume changes during lithiation/delithiation. At its full lithaition state, a volume expansion of 300 percent is expected. Such expansion occurs in every charge step, destroys the formed SEI, and induces parasitic reactions, as summarized by the preceding chapters. In a sense, the lithiation of silicon is a conversion-reaction process (Figure 12.1a), because all Si–Si bonds in the pristine state would be broken at the fully lithiated state, where every Si atom is surrounded by Li^+. The reversibility has always been an issue with the conversion-reaction type electrodes.

Fortunately, we are now better armed with the knowledge of electrolytes and interphases, the technology of controlling and building materials at nanometric- or even molecular-scales, the advanced characterization techniques to look at materials at atomic length and under in situ and operando conditions, and the unprecedented computation power. Thanks to all these new knowledge and tools, the reversibility of silicon as anode host for Li^+ has been significantly improved in the past decade, which was started by Cui and coworkers in 2008 [8].

After a decade of exciting laboratory work, silicon gradually made into the commercial LIB. First as minority addition to the graphite-based anode composite, then as a major component, and finally as the sole active host material.

A silicon age dawns upon us.

… and the "Holy Grail," a rechargeable battery based on lithium metal anode, is on horizon.

References

[1] Winter M., Barnett B., Xu K. 'Before Li ion batteries'. *Chemical Reviews*. 2018;**118**(23):11433–56.

[2] Whittingham M.S. 'The role of ternary phases in cathode reactions'. *Journal of the Electrochemical Society*. 1976;**123**(3):315–20.

[3] Mizushima K., Jones P.C., Wiseman P.J., Goodenough J.B. 'Li_xCoO_2 (0 < x < -1): a new cathode material for batteries of high energy density'. *Materials research bulletin*. 1980;**15**:783–9.

[4] Armand M.B. 'Intercalation electrodes' in Murphy D. (ed.). *Materials for Advanced Batteries*. New York: Plenum Press; 1980. p. 145.

[5] Nagaura T.A. 'Lithium ion rechargeable battery'. Oral Presentation in the Third International Rechargeable Battery Seminar, Deerfield Beach, FL, March 5–7; 1990.

[6] Hérold A. 'Recherches sur les Composes D'insertion du Graphite'. *Bulletin de la Société Chimique de France*. 1955;**187**:999–1012.

[7] Fujimoto M., Yoshinaga N., Ueno K., Furukawa N., Nohma T., Takahashi M. 'Lithium secondary battery'. U.S. Patent 5,686,138, 1991 (Japanese Priority Date: November 12, 1991).

[8] Chan C.K., Peng H., Liu G., *et al*. 'High-performance lithium battery anodes using silicon nanowires'. *Nature Nanotechnology*. 2008;**3**(1):31–5.

Index